COST
AND OPTIMIZATION
ENGINEERING

COST
AND OPTIMIZATION
ENGINEERING

Second Edition

Edited by

Frederic C. Jelen
Fellow of the
American Association of Cost Engineers

James H. Black
Professor of Chemical Engineering
University of Alabama

Sponsored by the American Association of Cost Engineers

McGraw-Hill Book Company
New York St. Louis San Francisco Auckland Bogotá Hamburg
Johannesburg London Madrid Mexico Montreal New Delhi
Panama Paris São Paulo Singapore Sydney Tokyo Toronto

This book was set in Times Roman by York Graphic Services, Inc.
The editors were Julienne V. Brown and Susan Hazlett;
the production supervisor was John Mancia.
New drawings were done by Allyn-Mason, Inc.
The cover was designed by Scott Chelius.
R. R. Donnelley & Sons Company was printer and binder.

COST AND OPTIMIZATION ENGINEERING

1 2 3 4 5 6 7 8 9 0   DOCDOC   8 9 8 7 6 5 4 3 2

ISBN 0-07-032331-3

Library of Congress Cataloging in Publication Data
Main entry under title:

Cost and optimization engineering.

    Includes bibliographies and indexes.
    1. Costs, Industrial. 2. Cost control.
I. Jelen, Frederic C. (Frederic Charles), date
II. Black, James Hay, date   . III. American
Association of Cost Engineers.
TS167.C66   1983   658.1'5   82-15191
ISBN 0-07-032331-3   AACR2

ATIC
658.15
C837c2

Dedicated to the American Association of Cost Engineers, Incorporated, which has sponsored both editions of the book. Its effort in behalf of cost engineering has been an inspiration.

CONTENTS

Part 2 Optimization

Part 3 Cost Estimation and Control

18 Cost Engineering and Beyond

Appendixes

LIST OF CONTRIBUTORS

D. L. ANDERSON, B.S.
Deputy Project General Manager
Brown and Root, Inc.
Houston, Texas

H. JACK BEHRENS
Professional Engineer
Retired Cost Engineer,
    Northrop Corporation
Resides Marina Del Rey, California

JAMES H. BLACK, Ph.D.
Professor of Chemical Engineering
University of Alabama
University, Alabama

ARNOLD H. BOAS, Ph.D.
Manager Process Computer Systems
Foster Wheeler Energy Corp.
Livingston, New Jersey

J. R. BRAUWEILER, B.S.
Managing Consultant
Northwest Industries
Chicago, Illinois

W. B. HIRSCHMANN, M.Sc.
Valuation Consultant
Chicago, Illinois

WILLIAM A. JANDA, B.S.
Manager Cost and
    Scheduling Engineering
Flour Ocean Services, Inc.
Houston, Texas

FREDERIC C. JELEN, Ph.D.
Fellow of the
American Association
of Cost Engineers

O. P. KHARBANDA, D.Ch.E.
Managing Director
O. P. Kharbanda and Associates
Bombay, India

HARVEY C. NELSON, B.S.
Principal Engineering Specialist
    Corporate Engineering
Monsanto Company
St. Louis, Missouri

RAYMOND A. ROGERS, B.B.A.
Manager Corporate Productivity
Brown and Root, Inc.
Houston, Texas

O. JAMES VOGL, M.A.
Manager Value Engineering
    Corporate Office
Hughes Aircraft Company
Los Angeles, California

CARL L. YAWS, Ph.D.
Consultant
Beaumont, Texas

PREFACE

Cost engineering has expanded greatly in scope and attention since the first edition of this book appeared in 1970. The second edition brings the first edition up to date and attempts in one volume to present the principles of cost engineering for use as a textbook in colleges and universities, for continuing education, and for self-study.

The book covers principles and is not intended as a source of cost data. The allotment of space is based partly on the importance and timeliness of topics and partly on the requirement for an understandable presentation. Quantitative topics are illustrated by many examples worked out in detail. These examples should be fully assimilated if the principles are to be applied.

The book has been limited essentially to the size of the first edition, but has been rewritten completely.

Dr. James H. Black, who was unable to edit the first edition because of health, becomes a coeditor of the second edition. The editors are responsible for the scope and execution and must be blamed for any shortcomings. A diversified group of 13 contributors, including the editors, has assisted in making the wide coverage possible. The text has been carefully edited in an attempt to achieve a harmonious presentation.

If the reader considers the coverage in the book to be reasonably complete, the material up-to-date, the presentation clear and concise, and the format pleasant to use, our major ambitions will have been achieved.

Acknowledgment is made to the American Association of Cost Engineers for continuing to sponsor the book, and to sources as noted in the text who have given permission to use copyrighted material.

Frederic C. Jelen
James H. Black

MATHEMATICS OF COST COMPARISONS

ONE

INTRODUCTION

F. C. Jelen

1.1 Status of Cost Engineering

Cost is a major consideration in all human activities and gains in importance with increasing technology and the expanding complications of society. Cost is the glue that binds together a wide variety of components into a single structure. Tourists in a strange land may not know a single word in the native language, but they can bargain for their purchases by using simple gestures. Cost is elementary, basic, and understood universally.

As technology and society advance, it becomes necessary to estimate more closely to remain competitive. An estimate based on overdesign may be too high to win the award. If based on underdesign, it may well win the award and end in disaster.

Cost engineering is undergoing rapid growth. The American Association of Cost Engineers (AACE) was founded in 1956, and by the early 1980s it had 6000 members. They represented all states of the United States, 9 Canadian provinces, and over 65 other countries, and were organized into over 50 local sections and 7 foreign sections. Eleven other countries had cost engineering societies, and the list was growing rapidly. The first international conference on cost engineering was held in Montreal in 1971 and the second in Mexico City in 1972; since then they have been held semiannually on a worldwide basis.

Professional recognition of cost engineering has been established by the certification program of the AACE. A participant who passes a written examination will be recognized as a certified cost engineer (CCE), or if there is a conflict with

3

local governmental regulations then as a certified cost consultant (CCC). Over 600 individuals from over 25 countries are entitled to use the designation CCE or CCC.

1.2 Scope

In a broad sense, cost engineering is that area of engineering principles where engineering judgment and experience are utilized in the application of scientific principles and techniques to problems of cost estimating, cost control, profitability analysis, planning, and scheduling. In addition, cost engineering embraces the principles of several disciplines and branches into business and management, optimization, operations research, accounting, and economics, among others. Cost engineering covers a wide range of topics as listed in Table 1.1, which has been adapted from a study guide suggested by the AACE for applicants taking the examination for certified cost engineer. Cost engineering is increasing in scope and importance. The current expansion of information and communication techniques is particularly adaptable to cost engineering, and this has led to an explosive interest in the subject.

1.3 Cost Engineering in Education

Individual courses involving costs are offered at the college level in a few disciplines. Industrial engineers have featured costs in their engineering economy course since the inception of industrial engineering. Many civil engineering departments introduce costs in some courses. Chemical engineers include costs in design courses, and the profession has been very active in publications on costs and profitability. Individual courses that include costs are becoming more popular in several disciplines, and although no degree in cost engineering is currently available, a serious effort is under way to establish such degrees at both the undergraduate and graduate levels. Proposed curricula are shown in Tables 1.2 and 1.3.

An elementary first course based on this book should emphasize Chap. 2, on equivalence and cost comparisons, fully and the next chapter, on depreciation and taxes, but not quite so fully. Chapter 4 on continuous interest should be omitted. Chapter 5 on profitability should be included, but not all sections. Chapter 6 on inflation should be omitted. The first five sections of Chap. 7 on break-even analysis are well suited to a first course, but the remainder of Part 2 on optimization is best omitted. Chapter 14 on capital-cost estimation and Chap. 15 on operating cost estimation belong in a first course, as does Chap. 16 on cost control to some extent. An elementary first course would thus include the basic mathematics of engineering economy and the elements of cost estimation and control.

An intermediate course should include Chap. 3 on taxes fully and Chap. 4 on continuous interest. Chapter 6 on inflation, Chap. 8 on probability, and Chap. 9

Table 1.1 Component topics in cost engineering

Part I. Basic general knowledge
  A. Equivalence and cost comparisons
    1. Time value of money
    2. Discrete and continuous interest
    3. Detailed comparison of alternatives
    4. Replacement and displacement problems
    5. Life cycle costing
    6. Benefit-cost ratio
  B. Taxes
    1. Depreciation
    2. Depletion
  C. Inflation and cost escalation
  D. Learning curve and productivity
  E. Profitability
    1. Criteria for judging
    2. Risk analysis

Part II. Cost and cost control
  A. Construction costs
  B. Manufacturing and operating costs
  C. Capital costs
  D. Special topics in costs
    1. Labor
    2. Materials
    3. Utilities
    4. Transportation
    5. Costing plants and projects
    6. Costing products and services
    7. Waste disposal and avoidance of nuisances
  E. Cost control
    1. Planning
    2. Scheduling
    3. Controlling
    4. Network techniques

Part III. Associated cost topics
  A. Cost accounting
    1. Direct and indirect costs
    2. Standard costs
    3. Joint costs, transfer pricing
  B. Design considerations and cost
    1. Value engineering
    2. Economic balance
  C. Economics
    1. Market factors
    2. Plant site location
  D. Forecasting
  E. Productivity
  F. Cost indices
  G. Appraisals

Part IV. Management aspects
  A. Personnel administration, staffing, etc.
  B. Contractual, bidding, and legal considerations
  C. Budgeting
  D. Social, governmental, and ethical considerations
  E. Computers in estimation and control
  F. Management science and operations research
    1. Optimization techniques
    2. Inventory problems
    3. Queuing problems
    4. Linear, dynamic, and geometric programming
    5. Monte Carlo technique
    6. Uncertainty
    7. Break-even analysis

Source: Adapted from the AACE suggested study guide for the certified cost engineer examination.

on productivity should all be covered, though perhaps not fully. Chapter 17 on cost accounting, and Chap. 18 on cost engineering and beyond, belong in an intermediate course. The instructor can add some chapters on optimization, Chap. 10 being the most important.

An advanced course should, in addition, fully develop Chap. 6 on inflation and Chap. 10 on optimization, with many of the more difficult problems from the latter chapter being assigned. Chapters 11, 12, and 13 can be included according to the instructor's preference.

NOMENCLATURE

Nomenclature always presents a problem, particularly in books covering a wide range of topics. The nomenclature used is characterized by being highly delineated. In engineering problems use of an orderly system of nomenclature is good

Table 1.2 Undergraduate curriculum for cost engineering

First year	
Chemistry and Materials I	Chemistry and Materials
Engineering Graphics	II
Literature and	Computer Programming
Composition	History of the Modern
Calculus I	World
Physics I	Calculus II
Physical Education	Physics II
	Physical Education

Second year	
Calculus III	Differential Equations
Physics III	Statics and Dynamics
Economics	Literature
English Literature	Basic and Cost
Introduction to Cost	Accounting
Engineering	Industrial and
	Manufacturing
	Processes

Third year	
Cost Estimation Methods	Engineering Economics
I	Construction Methods
Probability and Statistics	Cost Estimation Methods
Work Organization and	II
Analysis	Cost Control and Analysis
Strength of Materials	Humanities Elective
Humanities	

Fourth year	
Finance	Project Management
Network Planning and	Computer Simulation
Scheduling	Systems Analysis
Management Science	Humanities Elective
Labor Relations	Law
Organizational Planning	

Source: C. Wolf, *Transactions of the American Association of Cost Engineers,* AACE, Inc., Morgantown, W. Va., 1978, pp. 338–341.

Table 1.3 Master's degree curriculum for cost engineering

Thesis and courses in the following:
1. Construction Management
2. Cost Estimating
3. Contract Law
4. Cost Analysis and Control

Four courses from the following:
1. Managerial Economics
2. Operations Research
3. Labor Relations
4. Organizational Planning
5. Financial Management
6. Computer Applications
7. Forecasting and Budgeting
8. Occupational Health and Safety
9. Advanced Cost Estimating
10. Econometrics

Source: C. Wolf, *Transactions of the American Association of Cost Engineers,* AACE, Inc., Morgantown, W. Va., 1978, pp. 338–341.

discipline, and it becomes a necessity in large problems such as those for computer programming. For example, in comparing two machines A and B on a present-value basis P, it is not good discipline to write

$$P_A \quad \text{and} \quad P_B$$

If each machine is compared on a 10-year basis, it is much more informative to write

$$P_{A10} \quad \text{and} \quad P_{B10}$$

Indeed, this kind of notation will avoid making a false comparison with say,

$$P_{A8} \quad \text{and} \quad P_{B10}$$

for the difference in duration is clearly evident.

Perhaps in the end nomenclature will imitate computer language, which manages with capitals and numbers alone but uses a high degree of delineation for distinction and detail. Hence, if the nomenclature and typography seem to be needlessly exact at times, it merely reflects the modern approach to engineering requirements for good characterization.

The nomenclature for a chapter is given immediately following the text of the chapter. A summary of nomenclature is given in Table 1.4.

Numerical tables in the Appendix are somewhat extensive, but the introduction of inexpensive, programmable pocket calculators has virtually eliminated the need for tables. The factors can easily be programmed on a magnetic card or cassette and recalled as needed. Interpolation is eliminated, and a wide choice of

Table 1.4 Nomenclature for text

a	Constant
a_n	Exponent in GP term
A	One of k kinds of events in the multinomial distribution, Chap. 8
A	Parameter in Eq. (10.25)
A	Total cost of all battery-limit investment, Eqs. (14.9) to (14.12), \$
A_{ij}	Coefficient in restraint equation, ith row, jth column, LP
A_0	$A/1000$, Eqs. (14.10) to (14.12), kilodollars
A_{t1}	Amount at time 1, \$
b	Constant in the equation for a straight line, Chap. 8
B	Book value, \$
B	Parameter in Eq. (10.25)
B	Constant in scale of relation, Chap. 13
B	Cost of all erected equipment, Eq. (14.9), \$
B/C	Benefit-cost ratio, dimensionless
B_i	Right side of the ith restraint equation, LP
C	Cost, Eqs. (14.1) and (14.6), \$
C	Incremental cost of alloys used for corrosion resistance, Eq. (14.9), \$
C_{bx}	Irregular cost at beginning of xth year, \$
C_d	Depreciable cost, \$
C_{ex}	Irregular cost at end of xth year, \$
C_f	Cost per period for a service facility which can service one item per period when working full time
C_F	Fixed cost, \$
C_h	Storage or holding cost per item per period, \$/(item)(period)
c_i	Constant in GP term
C_i	Initial or first cost, \$
C_i	Unit cost associated with basic variable in ith row, LP
c_j	Unit cost associated with x_j, LP
C_{nd}	Nondepreciable cost, \$
C_0	Cost of placing and receiving an order, or setup cost, \$, Chap. 11
C_{sal}	Salvage value, \$
C_t	Variable cost per period, \$/period
C_T	Total cost, \$
C_w	Cost for one item waiting one period, \$/(item)(period)
d	Inflation rate, decay of the \$, decimal per year, Chap. 6
D	Depreciation, \$
D	Production rate, items per period, Chap. 11
D	Distance from end of interval in Fibonacci search
D	Degree of difficulty, GP
D	Direct plant cost, Eq. (14.6), \$
D_f	Fractional depreciation, decimal
DB	Declining-balance method of depreciation
DCF	Discounted cash flow, \$
DCFRR	Discounted cash flow rate of return, decimal or percent per year
DP	Dynamic programming
e	Naperian constant 2.71828
e	Total heat-exchanger cost less incremental cost of alloy, Eqs. (14.10) and (14.11), \$
E	Delivered equipment cost, Eqs. (14.7) and (14.8), \$
E	Indirect cost factor, Eq. (14.9), dimensionless, normally assumed to be 1.4
E_N	Effort required to produce the Nth unit, e.g., work-hour per unit
E_T	Cumulative effort to produce units 1 through N, e.g., work-hour
E_T/N	Cumulative average unit, e.g., work-hour per unit
$E(X)$	Expected value of X
f	Ratio of cost indices, Eq. (14.6), dimensionless
f	Total cost of field-fabricated vessels, Eq. (14.10), \$

Table 1.4 (*Continued*)

F	A factor in this book
F	Future value by some writers, \$, Chap. 2
F	Objective, Chaps. 12, 13
F_a	Return from project A, LP
F_{DB}	Factor for declining-balance depreciation, decimal
F_{GP}	Factor to convert G to P, uniform-gradient-series present-value factor, years
F_i	Fibonacci number
F_I	Direct cost factor for instruments, Eq. (14.7), dimensionless
F_L	Cost factor for field labor, Eqs. (14.9) and (14.10), dimensionless
F_M	Cost factor for miscellaneous items, Eqs. (14.9) and (14.12), dimensionless
F_0	Indirect cost factor, Eq. (14.7), dimensionless
F_p	Cost factor for piping materials, Eqs. (14.9) and (14.11), dimensionless
F_{PK}	Factor to convert P to K, capitalized cost factor, dimensionless
F_{PR}	Factor to convert P to R, capital recovery factor, year^{-1}
$F_{P\bar{R},\bar{i},n}$	Factor to convert P to $\bar{R}$ with continuous compounding, year^{-1}
F_{PS}	Factor to convert P to S, compound interest factor, dimensionless
$F_{PS,\bar{i},n}$	Factor to convert P to S with continuous compounding, dimensionless
F_r	Fractional ratio of interval left to original interval, Chap. 13
F_{RP}	Factor to convert R to P, unacost present value factor, years
$F_{\bar{R}P,\bar{i},n}$	Factor to convert $\bar{R}$ to P with continuous discounting, years
F_{RS}	Factor to convert R to S, years
F_{SD}	Factor for sum-of-the-years-digits depreciation, decimal
F_{SDP}	Present value of \$1 for sum-of-the-years-digits depreciation, decimal
$F_{SDP,\bar{i},n}$	Factor to convert a unit total flow declining to zero at a constant rate over n years, starting with the reference point and with continuous discounting, decimal, dimensionless, approximates ψ_{SD}
F_{SLP}	Present value of \$1 for straight-line depreciation, decimal
F_{SP}	Factor to convert S to P, dimensionless
$F_{SP,\bar{i},n}$	Factor to convert S to P with continuous discounting, dimensionless
F_{SR}	Factor to convert S to R; sinking fund factor, year^{-1}
F_{UP}	Factor for units-of-production depreciation, decimal
FV	Future value
g	Constant in exponential-rate flow change, decimal
g	A posynomial, GP
G	Uniform-gradient amount, \$ per year
GP	Geometric programming
H	Number of objects in the population with desired property; used in hypergeometric distribution
i	Rate of return, decimal per year
i	The ith row
i_c	Interest rate, specifically emphasizing continuous compounding, decimal per year, Chap. 4
i_{eff}	Effective interest rate, decimal per year
i_p	Interest rate, specifically emphasizing periodic compounding, decimal per year, Chap. 4
I	Total battery-limits investment, Eq. (14.9), \$
I	Indirect costs, Eq. (14.9), \$
I	Instrument costs, Eq. (14.7), \$
j	The jth column
k	Number of kinds of events in the multinomial distribution
K	Capitalized cost, \$
K	Constant, Chap. 8
K	Constant in Eq. (9.10); theoretical effort required for first unit, e.g., work-hour, Chap. 9
L	Number of channels in service facility, dimensionless, Chap. 11

Table 1.4 (*Continued*)

L	Range of a variable, Chap. 13
L_D	Decimal learning ratio, dimensionless
L_P	Percentage learning ratio, dimensionless
LE	Lagrange expression
LP	Linear programming
m	A number, generally the mth year or mth term
M	Units of production, Chap. 3
M	Number of objects in the population, Chap. 8
M	Maximum inventory, number of items, Chap. 11
n	A number, usually the life of an article, years
n	The nth term or the last term, Chap. 13
n'	Useful life for tax purposes, years
N	Number of units of production, Chap. 7
N	Number of trials or data, Chap. 8
N	Cumulative number of units produced, dimensionless, Chap. 9
N	Cycles per month in Example 10.1; batches per year in Example 10.5
N	Number of periods per cycle, dimensionless, Chap. 11
N	Number of experiments or calculations allowed, Chap. 13
N	The Nth stage, DP
N	Number of variables, GP
N	Units of items or service per year, Chap. 18
N'	Number of units of production above 100% capacity, Chap. 7
N''	Number of units of production dumped, Chap. 7
NPV	Net present value, $
p	Number of periods per year, Chaps. 2, 4
p	Probability for success on a single trial in the binomial distribution, Chap. 8
p	Technological advancement rate, decimal per year, Chaps. 6, 9
p	Total pump plus driver cost less incremental cost of alloy, Eq. (14.11), $
P	Present value, $
P	Power to stirrer in Example 10.5, kilowatts
P_n	Present value for n years' duration, $
PV	Present value, $
PW	Present worth, $, not used
$P(X)$	Probability for event with value X
$P_c(X)$	Cumulative probability of X or less
$P_d(X)$	Probability density at X, probability per unit interval
Q	Total flow, $, Chap. 4
Q	Lot size, number of items, Chap. 11
Q	Quality coefficient, Chap. 12
Q	Capacity, Eq. (14.1), any convenient dimensions
Q_c	Cumulative meters drilled from start of cycle in Example 10.1
r	Rate of return, generally after taxes, decimal per year
r	Moles of D formed per mole of B in feed in Example 10.3; depends upon residence time
R	Uniform end-of-year annual amount, unacost, $ per year
R	Reorder level, number of items, Chap. 11
R	Regulated cost per unit of production, $ per unit, Chap. 7
R	The ratio of plant sizes, Eq. (14.6), dimensionless
$\bar{R}$	Uniform flow, $ per year
R_b	Uniform beginning-of-year annual amount, $ per year
R_D	Unaburden, a uniform end-of-year burden with inflation, increasing $ per year
R_e	Unacost, same as first R above
R_0	Initial flow rate, $ per year, Chap. 4
R_p	Yearly operating expense subject to improvement from technological advancement, $ per year, Chap. 9

Table 1.4 (*Continued*)

R_S	Sinking fund deposit, \$ per year
R_v	Revenue, end of year, in Example 10.2, \$ per year
R_x	Flow rate at time x, \$ per year, Chap. 4
RAI	Return on average investment, decimal or percent per year
ROI	Return on original investment, decimal or percent per year
s	Slope of straight line, Chap. 8
s	Slope constant, decimal, dimensionless, Chap. 9
S	Future value, \$
S	Net realizable sales value per unit sold, \$ per unit, Chap. 7
S	Capacity of reactor-mixer in Example 10.5, kilograms
S	Unit cost, \$ per unit, Chap. 18
S''	Net realizable sales value per unit dumped, \$ per unit, Chap. 7
S_L	Levelized cost, \$ per item
S_T	Total realizable sales value, \$, Chap. 7
SD	Sum-of-the-years-digits method of depreciation
SF	Sinking fund method of depreciation
SL	Straight-line method of depreciation
t	Tax rate, decimal
t	Total cost of tower shells less incremental cost of alloys, Eq. (14.12), \$
T	Time in years or periods
T	Residence time in reactor in Example 10.3, hours
T	Total time required for a batch in Example 10.5, hours
T	Number of terms in a posynomial, GP
T_a	Time between arrivals, periods, Chap. 8
T_d	Drilling time from start of cycle in Example 10.1, days
u	A term in a posynomial
u^n	$(1 + i)^n$, Chap. 2
U	The factor $1/(1 + i)$, Chap. 13
UP	Units-of-production method of depreciation
v	Dual function, GP
v^n	$(1 + i)^{-n}$, Chap. 2
V	Variable cost per unit of production, \$ per unit, Chap. 7
V	Volume of reactor in Example 10.3, cubic meters
V'	Variable cost per unit of production above 100% capacity, \$ per unit, Chap. 7
VA	Value analysis
VE	Value engineering
W	Number of objects withdrawn from population in hypergeometric distribution
W	Dependent variable, Chap. 13
X	Value of an event, Chap. 8
X	Total cost-capacity factor, Eqs. (14.1) and (14.6), dimensionless
X_a	Investment in project A, DP
X_i	Variable input, LP
X_N	Decision variable associated with Nth stage, DP
Y	Independent variable, generally
Y	The value of $(1 - 0.48F_{\text{SDP},r,25})F_{PR,r,25}$ in Example 10.2
Y_N	Output from Nth stage and input to $(N + 1)$ stage, DP
Z	Independent variable in standard normal distribution measured from mean value
Z_g	Gross profit before tax, \$, Chap. 7
Z_t	Net profit after tax, \$, Chap. 7
α	Weighting factor, GP
λ	Lagrange multiplier
λ	Average arrivals per period, Chap. 8
μ	Mean or average value, Chap. 8
σ	Standard deviation, Chap. 8
σ^2	Variance, Chap. 8

Table 1.4 (*Continued*)

ΣE_N	Same as E_T
$\Sigma E_N/N$	Same as E_T/N
ϕ	Cost factors, Eq. (14.8), dimensionless
ψ	Psi, present value of $1 of depreciation, decimal
ψ_{SD}	The ψ factor for sum-of-the-years-digits depreciation
ψ_{SL}	The ψ factor for straight-line depreciation

factors is available. For example, the present value of \$1 of future depreciation, in this book designated by ψ, appears in practice as $(1 - t\psi)$ where t is the decimal tax rate. The ψ factor is easily calculated and tabulated, as in the last two columns of Table A1.1, and is independent of the tax rate. It would not be practical to tabulate $(1 - t\psi)$ in the Appendix, since a change in the tax rate would negate the values; but a person using a programmable pocket calculator can always extract the factor $(1 - t\psi)$ and modify the program if the tax rate t should be changed.

EQUIVALENCE AND COST COMPARISONS

F. C. Jelen

TIME VALUE OF MONEY AND EQUIVALENCE

2.1 Compound Interest as an Operator

Investments are spread over a period of time, and time has a monetary value. An investor expects an investment, with due allowance for dividends, to increase with time in the form of appreciation. The appreciation of 1 year becomes an investment of the next and in turn should earn appreciation of its own. The increase in investment should be compounded with time, much like the increase of a deposit in a bank, but the time value of money in a practical problem applies without an actual banking transaction.

Suppose a present amount P earns at the rate i per year, expressed as a decimal. Also, let S be the future value at some later time. Then at the end of 1 year

$$S_1 = P + iP = P(1 + i)$$

$P(1 + i)$ becomes the starting value at the beginning of the second year, and at the end of the second year

$$S_2 = P(1 + i) + iP(1 + i) = [P(1 + i)](1 + i) = P(1 + i)^2$$

Repeating for another year gives

$$S_3 = P(1 + i)^2 + iP(1 + i)^2 = [P(1 + i)^2](1 + i) = P(1 + i)^3$$

The form of the relationship is obvious and for n years is

$$S = P(1 + i)^n \tag{2.1}$$

or
$$S = PF_{PS} \tag{2.2}$$

Future value = (present value)(compound-interest factor)

The relationships above are the *compound-interest law*. Factors in this book are designated by the letter F with a suitable subscript. The symbol F_{PS} is read "factor for converting P to S." It is more completely designated as $F_{PS,i,n}$, where i is the rate of return and n is years. Thus

$$F_{PS,i,n} = (1 + i)^n \tag{2.3}$$

$(1 + i)^n$ or F_{PS} is the *compound-interest factor,* values for which are given in Table A1.1. It should be noted that P is a single payment and that compounding is carried out periodically in steps.

In engineering economy, a year is usually taken as a unit period of time. Equations (2.1) and (2.2) are used with n in years and i the yearly rate of return expressed as a decimal, although i is usually stated as a percentage. If compounding occurs p times per year, the relationship is

$$S = P\left(1 + \frac{i}{p}\right)^{np} \tag{2.4}$$

where np is the number of periods and i/p is the rate of return per period, inasmuch as i always refers to the rate of return per year. The last equation can be written

$$S = P\left[\left(1 + \frac{i}{p}\right)^p\right]^n$$

where the expression in brackets is the value of 1 at the end of 1 year. The *increase* in value for 1 year

$$\left(1 + \frac{i}{p}\right)^p - 1 = \text{effective rate of return} \tag{2.5}$$

The *effective rate of return* can be used as if the compounding were done once a year instead of p times per year. For distinction i is referred to as the *nominal rate of return.*

The compound-interest law is used in engineering economy if the time is over 1 year. If the time is less than 1 year, *simple interest* can be used, with interest charged only on the original amount

$$S = P(1 + ni) \tag{2.6}$$

where n may be a fractional part of a year.

In this book, if interest is for 1 or more years plus a fractional part of a year, simple interest will be used for the fractional part and compound interest for the integer part unless otherwise indicated.

Example 2.1 Find the simple interest on $10,000 for 2 months at 10 percent per year.

SOLUTION

$$S = 10,000[(\tfrac{2}{12})(0.10)] = \$166.67$$

Example 2.2 A person borrows $1000 and repays $1050 at the end of 4 months. What interest rate was paid?

SOLUTION By Eq. (2.6):

$$S = 1000[1 + (\tfrac{4}{12}i)] = 1050$$

$$i = 0.15 \text{ or } 15 \text{ percent per year}$$

Example 2.3 How long will it take for money to double if compounded annually at 8 percent per year?

SOLUTION By Eq. (2.2):

$$S = PF_{PS,8\%,n} = 2P$$

$$F_{PS,8\%,n} = 2$$

From Table A1.1,

$$F_{PS,8\%,9} = 1.9990$$

That is, money will double in about 9 years at 8 percent per year.

Example 2.4 If $10,000 compounds to $14,120 in 4 years, what is the rate of return?

SOLUTION By Eq. (2.1):

$$14,120 = S = P(1 + i)^n = 10,000(1 + i)^4$$
$$(1 + i)^4 = 1.412$$
$$4 \log (1 + i) = \log 1.412 = 0.14983$$
$$\log (1 + i) = 0.03746$$
$$(1 + i) = 1.090$$
$$i = 0.09 \text{ or } 9 \text{ percent per year}$$

Example 2.5 Find the value of $10,000 compounded annually at 10 percent for 3 years and 4 months.

SOLUTION By Eq. (2.6) for 4 months:

$$S = 10,000[1 + (\tfrac{4}{12})(0.10)] = \$10,333$$

and, compounding by Eq. (2.2) for 3 years,

$$S = 10,333F_{PS,10\%,3} = 10,333(1.331) = \$13,753$$

Equation (2.1), the compound-interest law, can be written as an expression for P, present value, in terms of S, future value, and becomes

$$P = \frac{S}{(1 + i)^n} \tag{2.7}$$

or
$$P = S(1 + i)^{-n} \tag{2.8}$$

or
$$P = SF_{SP} \tag{2.9}$$

Present value = (future value)(present-value factor)

The factor $(1 + i)^{-n}$ or F_{SP}, which converts S to P, is the *single-payment present-value factor*. It will be referred to as the *present-value factor* and is tabulated in Table A1.1. The complete designation of F_{SP} is

$$F_{SP,i,n} = (1 + i)^{-n} \tag{2.10}$$

It will be noted that F_{SP} and F_{PS} are reciprocal relationships: i.e.,

$$F_{SP} = \frac{1}{F_{PS}} \tag{2.11}$$

Equations (2.8) and (2.9) show that a future amount is reduced when converted to a present amount. The procedure of reducing a future amount to a present amount is known as *discounting*. Discounting is merely an application of the principle that because of the time value of money, the future is not worth as much as the present when viewed from the present.

Equations (2.1) and (2.8) can be written as:

$S = P(1 + i)^n$ (*moving with the calendar*)

$P = S(1 + i)^{-n}$ (*moving against the calendar*)

and can be combined in a single relationship

$$A_{t2} = A_{t1}(1 + i)^n \tag{2.12}$$

Amount time 2 = (amount time 1)(compound-interest operator)

where n positive = with the calendar

n negative = against the calendar

The student should become familiar with Eq. (2.12), the *time-value conversion relationship*. The compound-interest factor $(1 + i)^n$ can be regarded as a mathematical operator which moves a unit value n years, with or against the calendar, depending on the sign of n.

In engineering economy, receipts and disbursements occur at different instants of time. They can be combined only after conversion to a single common instant. The time-value conversion relationship defines the mathematical operator that makes this possible. In solving problems, it is convenient to diagram expenditures and receipts as vertical lines positioned along a horizontal line representing time. Expenditures and receipts can point in opposite directions. The problem can be visualized in this manner and each amount moved by a routine technique to some common instant for consideration. The reference instant is called the *focal point* or *focal date*.

Example 2.6 A boy is celebrating his sixteenth birthday and will need $3000 on his seventeenth, eighteenth, nineteenth, and twentieth birthdays for his college education. His father agrees to lay aside a certain amount now and each year thereafter until the boy is 20 years old; moreover the contributions will form an arithmetical progression for all years, increasing by 20 percent after the first year. If money is worth 4 percent per year, what is the father's first contribution?

SOLUTION If x is the first contribution, succeeding contributions are $1.2x$, $1.4x$, etc. The diagram is:

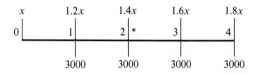

Take 2 years from now as the focal date. Use the time-value conversion relationship, Eq. (2.12), noting that contributions must equal withdrawals:

$$x(1.04^2) + 1.2x(1.04^1) + 1.4x(1.04^0) + 1.6x(1.04^{-1}) + 1.8x(1.04^{-2})$$
$$= 3000(1.04^1) + 3000(1.04^0) + 3000(1.04^{-1}) + 3000(1.4^{-2})$$

$$x = \$1699 \text{ first contribution}$$

2.2 Uniform Annual Amount, Unacost

Consider a uniform end-of-year annual amount R for a period of n years. The diagram is:

$$
\begin{array}{ccccccc}
R & R & R & & R & R \\
0 \quad 1 & 2 & 3 & \cdots & (n-1) & n
\end{array}
$$

where R will be referred to as *unacost*. The terms can be replaced by a single present value P by moving all terms to zero time using the time-value conversion relationship, Eq. (2.12)

$$P = R\left[\frac{1}{1+i} + \frac{1}{(1+i)^2} + \cdots + \frac{1}{(1+i)^n}\right] \tag{2.13}$$

The terms in the brackets can be condensed by the Σ (sigma) notation to

$$\sum_1^n \frac{1}{(1+i)^n}$$

and represent a geometrical series of n terms with first term $1/(1+i)$ and ratio $1/(1+i)$. For a geometric series

$$\text{Sum of geometric series} = \frac{1 - \text{ratio}^{\text{no. of terms}}}{1 - \text{ratio}} \text{ first term} \tag{2.14}$$

The previous summation becomes

$$\sum_1^n \frac{1}{(1+i)^n} = \frac{1 - [1/(1+i)]^n}{1 - [1/(1+i)]} \frac{1}{1+i} = \frac{(1+i)^n - 1}{i(1+i)^n} \tag{2.15}$$

so that Eq. (2.13) is

$$P = R\frac{(1 + i)^n - 1}{i(1 + i)^n} \qquad (2.16)$$

or
$$P = RF_{RP} \qquad (2.17)$$

Present value = (unacost)(unacost present-value factor)

F_{RP} must be called by its name, *unacost present-value factor*, to avoid confusion with F_{SP}, which is called simply the *present-value factor*. F_{RP}, which converts R to P, has for its complete designation

$$F_{RP,i,n} = \frac{(1 + i)^n - 1}{i(1 + i)^n} \qquad (2.18)$$

and is tabulated in Table A1.1. It is also known as the *equal-payment-series present-value factor* and as the *annuity present-value factor*.

Example 2.7 Find the equivalent value of the following series of receipts as of the end of the third year if money is worth 6 percent per year.

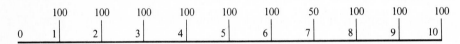

The series has an irregularity at the end of 7 years but with 50 added and subtracted at that time becomes:

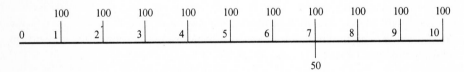

SOLUTION The present value of all the ascending lines is given by Eq. (2.17), and Eq. (2.9) gives the present value of the single descending line

$$P = 100F_{RP,6\%,10} - 50F_{SP,6\%,7}$$
$$P = 100(7.3601) - 50(0.66506) = 702.8 \qquad \text{(zero time)}$$

Move 3 years with the calendar

$$702.8F_{PS,6\%,3} = 702.8(1.1910) = \$837.00 \qquad \text{(end of third year)}$$

Equation (2.16) written as an expression for R in terms of P becomes

$$R = P\frac{i(1 + i)^n}{(1 + i)^n - 1} \qquad (2.19)$$

$$R = PF_{PR} \qquad (2.20)$$

Unacost = (present value)(capital-recovery factor)

These relationships are very important in engineering economy. The *capital-recovery factor* converts a single zero-time cost to an equivalent uniform end-of-year annual cost, *unacost*. The unacost repays the initial expenditure with interest, hence the name *capital-recovery factor*. The factor has the complete designation

$$F_{PR,i,n} = \frac{i(1 + i)^n}{(1 + i)^n - 1} \tag{2.21}$$

and is tabulated in Table A1.1. F_{PR} and F_{RP} are reciprocal relationships

$$F_{PR} = \frac{1}{F_{RP}} \tag{2.22}$$

Relationships between R and P have been established by Eqs. (2.16) and (2.17). Relationships between R and S, the future amount, are easily established. Noting that $P = S(1 + i)^{-n}$ from Eq. (2.8), and placing this expression for P in Eq. (2.16), gives

$$S = R \frac{(1 + i)^n - 1}{i} \tag{2.23}$$

or

$$S = RF_{RS} \tag{2.24}$$

Future amount = (unacost)(equal-payment-series future-value factor)

The factor

$$F_{RS,i,n} = \frac{(1 + i)^n - 1}{i} \tag{2.25}$$

called the *equal-payment-series future-value factor,* is not tabulated in the Appendix but can be obtained from the corresponding F_{RP} factor

$$F_{RS} = F_{RP}(1 + i)^n = F_{RP}F_{PS} \tag{2.26}$$

Also, Eq. (2.23) can be written as

$$R = S \frac{i}{(1 + i)^n - 1} \tag{2.27}$$

$$R = SF_{SR} \tag{2.28}$$

Unacost = (future amount)(sinking-fund factor)

The factor

$$F_{SR,i,n} = \frac{i}{(1 + i)^n - 1} \tag{2.29}$$

called the *sinking-fund factor,* is not tabulated in the Appendix but can be obtained from the corresponding F_{PR} factor

$$F_{SR} = (1 + i)^{-n}F_{PR} = F_{SP}F_{PR}$$

F_{SR} and F_{RS} are reciprocal relationships:

$$F_{SR} = \frac{1}{F_{RS}}$$

Instead of unacost, which is a uniform end-of-year annual amount, one might consider R_b, a uniform beginning-of-year annual amount. R and R_b are connected by the relationship

$$R = R_b(1 + i) \tag{2.30}$$

and substitution of this relationship for R in Eqs. (2.16), (2.19), (2.23), and (2.27) gives various relationships between R_b and P and between R_b and S.

So far, four amounts have been introduced:

P = present value, a single amount at zero time
S = future value, a single amount at the end of n years
R = unacost, a uniform end-of-year annual amount for n years
R_b = a uniform beginning-of-year annual amount for n years

The various relationships between these amounts are summarized in Table 2.1. The student should become familiar with the first four items of the table and understand fully the meaning of equivalence. The other relationships in the table are less important.

The symbol R implies an *equivalent* uniform end-of-year amount. For emphasis, it may be desirable to designate R as R_e where the subscript e designates end of year, in contrast to R_b, for an equivalent beginning-of-the-year amount.

2.3 Mathematics of Finance

There are numerous applications of the relationships of Table 2.1 in the mathematics of finance, sometimes called *business mathematics*. Terminology and nomenclature can be different from those in this book, but the principles are the same. *Present worth* and *future worth* are synonymous with and replaced respectively by *present value* and *future value* in this book. The factor $(1 + i)^n$ may be called the *accumulation factor* or *amount of 1* and is sometimes represented by u^n. The factor $(1 + i)^{-n}$ is sometimes called the *discount factor* or the *present value of 1* and may be represented by v^n. A series of equal payments at equal intervals is called an *annuity*. A periodic payment is called *rent*. An ordinary annuity has the payments at the end of the period, and an annuity due has payments at the beginning of the period. Amortization implies the substitution of a number of equal-period payments for a present obligation. A *sinking fund* implies a uniform end-of-year payment which will accumulate to some future value at the time of the last payment. A *deferred annuity* has the first payment delayed in date. Thus if an ordinary annuity has the first payment at the end of the sixth year, it would be deferred 5 years. A *forborne* annuity continues to earn interest after its termination when payments end. A *perpetuity* is an annuity that continues forever. A

Table 2.1 Summary of S, P, R, and R_b relationships

Item	Conversion	Algebraic relationship	Relationship by factor	Name of factor
1	P to S	$S = P[(1+i)^n]$	$S = PF_{PS,i,n}$	Compound-interest factor
2	S to P	$P = S[(1+i)^{-n}]$	$P = SF_{SP,i,n}$	Present-value factor
3	R to P	$P = R\dfrac{(1+i)^n - 1}{i(1+i)^n}$	$P = RF_{RP,i,n}$	Unacost present-value factor
4	P to R	$R = P\dfrac{i(1+i)^n}{(1+i)^n - 1}$	$R = PF_{PR,i,n}$	Capital-recovery factor
5	R to S	$S = R\dfrac{(1+i)^n - 1}{i}$	$S = RF_{RS,i,n}$ $S = RF_{RP,i,n}F_{PS,i,n}$	Equal-payment-series future-value factor
6	S to R	$R = S\dfrac{i}{(1+i)^n - 1}$	$R = SF_{SR,i,n}$ $R = SF_{SP,i,n}F_{PR,i,n}$	Sinking-fund factor
7	R_b to P	$P = R_b(1+i)\dfrac{(1+i)^n - 1}{i(1+i)^n}$	$P = R_b(1+i)F_{RP,i,n}$	
8	P to R_b	$R_b = \dfrac{P}{1+i}\dfrac{i(1+i)^n}{(1+i)^n - 1}$	$R_b = \dfrac{P}{1+i}F_{PR,i,n}$	
9	R_b to S	$S = R_b(1+i)\dfrac{(1+i)^n - 1}{i}$	$S = R_b(1+i)F_{RS,i,n}$ $S = R_b(1+i)F_{RP,i,n}F_{PS,i,n}$	
10	S to R_b	$R_b = \dfrac{S}{1+i}\dfrac{i}{(1+i)^n - 1}$	$R_b = S\dfrac{1}{1+i}F_{SR,i,n}$ $R_b = S\dfrac{1}{1+i}F_{SP,i,n}F_{PR,i,n}$	

balloon payment is an irregular payment practically always at the end of the term of the payments.

Interest may be compounded more than one period per year; thus for home mortgages the period is 1 month. Elaborate tables accurate to 10 decimals are available, as well as pocket editions. The principal tables are the amount of 1, the (future) amount of 1 per period, and the sinking fund deposit, which is the end-of-period payment that has a value of 1. Computations can be made from combinations of these tabulations.

Nomenclature has not been standardized, but some common usages with their equivalents as used in this book are: $s^n = F_{PS}$, $s_{n\rceil} = F_{RS}$, $1/s_{n\rceil} = F_{SR}$, and $a_{n\rceil} = F_{RP}$.

Some writers use terminology which employs principally capital letters to imitate computer programming. Thus PV and NPV are used for present value or net present value, FV for future value, CPIV for compound-value interest factor (our F_{PS}), PVIF for present-value interest factor (our F_{SP}), CVIF$_a$ for annuity interest factor (our F_{RS}), PVIF$_a$ for present-value interest factor of an annuity (our F_{RP}), PMT for payment, and BAL for balloon payment.

Coupon bonds have a *face value* or *denomination,* frequently $1000, to which are attached coupons which become redeemable periodically in the future. The *coupon rate* is an annual rate stated on the bond face. Thus a $1000 10-year bond paying 12 percent per year semiannually would have 20 coupons when issued, each worth $60 at 6-month intervals. The interest rate for the bond fixes the value of the coupons and has no other use. After a bond is issued, it will sell at a premium if the bond rate is greater than the going interest rate and, conversely, at a discount. The bond sells so that the *yield to maturity* equals the going interest rate. A buyer pays the present value at the current interest rate for the future coupons and future face value. Bond tables are available [R6], giving the purchase price for a $1000 bond for combinations of coupon rate, yield rate to maturity, and time to maturity.

Example 2.8 A person has $2000 for a down payment on a $10,000 lot. An annual end-of-year payment of $1174.11 for 12 years with money worth 12 percent per year will discharge the debt. However, the person elects to pay $1000 yearly and a balloon payment at the end. Find the balloon payment.

SOLUTION With a balloon payment X the $10,000 is now equivalent to the following time diagram for the payments:

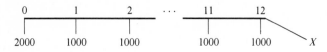

and using Table 2.1

$$10,000 = 2000 + 1000F_{RP, 10\%, 12} + XF_{SP, 10\%, 12}$$

$$= 2000 + 1000(6.8137) + X(0.31863)$$

$$X = \$3723.10 \quad \text{(in addition to the regular \$1000 payment)}$$

Example 2.9 A person plans to create a forborne annuity by depositing $1000 at the end of the year for 8 years and withdrawing the fund at the end of 14 years from now. If money is worth 10 percent per year, find the accumulated value at the end of the fourteenth year.

SOLUTION The time diagram for the payments is:

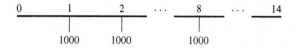

The present value, by Table 2.1, is

$$P = 1000F_{RP, 10\%, 8} = 1000(5.3349) = 5334.90$$

The future value at the end of 14 years is

$$S = 5334.90 \times F_{PS, 10\%, 14} = 5334.90(3.7975) = \$20,259$$

The problem can also be solved by considering the $1000 paid for 14 years less $1000 paid as annuity for the last 6 years. By Table 2.1, item 5,

$$S = 1000\left[\frac{(1.10)^{14} - 1}{0.10} - \frac{(1.10)^6 - 1}{0.10}\right] = 1000(27.975 - 7.7156) = \$20,259$$

Example 2.10 A $1000 face value bond bearing 4 percent per year interest payable semiannually, and due in 6 years, is to be bought so as to yield the purchaser a rate of 8 percent per year. What price should be paid for the bond?

SOLUTION There are 12 semiannual periods involved, with a 2 percent per period coupon rate. The 12 coupons are worth $20 each when due. Also, the purchaser receives the $1000 face value 12 periods hence. The diagram is:

The purchaser should pay the present value of these amounts at 8 percent per year, or 4 percent per period.

By Eqs. (2.17) and (2.9)

$$P = 20F_{RP, 4\%, 12} + 1000F_{SP, 4\%, 12}$$

$$= 20(9.3851) + 1000(0.62460) = \$812.30 \text{ purchase price}$$

The bond sells at a discount because the coupon rate is less than the yield rate. If the coupon rate is greater than the yield rate, the bond will sell at a premium.

Note that there are several rates associated with the example. The bond rate is 4 percent per year, the yield to maturity for the purchaser is 8 percent per year, which is the going rate for money, and the bond pays a current rate of 2(20/812.30)(100) or 4.92 percent.

COST COMPARISONS

2.4 Comparisons with Equal Duration

The preceding sections contain all the tools required to make a cost comparison between two machines or alternative investments having equal service lives or durations. The word machine is used in a general sense. Two machines generally have cost patterns differing on a time basis, but the costs can be reduced to some common instant and the comparison made on that basis. Any instant can be chosen for the comparison, but the present instant is usually preferred. Hence the comparison is made on the basis of present values for equal durations, which constitutes a fair comparison. The difference between the machines can have any degree of complexity, but the following example will suffice to illustrate the method.

In this and subsequent sections, a uniform expense is referred to as a *uniform end-of-year cost*. Actually, the expense may be spread over the year and is merely expressed as an equivalent end-of-year cost. With this understanding, the expense will exist for the year even if the machine is abandoned at the end of the year.

Example 2.11 Two machines, each with service life of 5 years, have the following cost comparison. If money is worth 10 percent per year, which machine is more economical?

	A	B
First cost, $	25,000	15,000
Uniform end-of-year maintenance, $ per year	2,000	4,000
Overhaul, end of third year, $		3,500
Salvage value, $	3,000	
Benefit from quality control as a uniform end-of-year amount, $ per year	500	

SOLUTION Construct the cost diagrams for each machine using descending lines for costs and ascending lines for receipts or benefits.

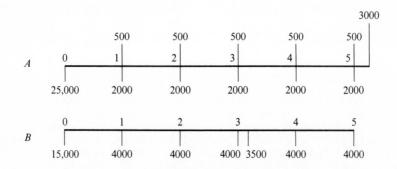

The present value of the costs for the diagrams on a 5-year basis is

$$P_{A5} = 25,000 + (2000 - 500)F_{RP,10\%,5} - 3000F_{SP,10\%,5}$$

$$= 25,000 + 1500(3.7908) - 3000(0.62092)$$

$$= \$28,823$$

$$P_{B5} = 15,000 + 4000F_{RP,10\%,5} + 3500F_{SP,10\%,3}$$

$$= 15,000 + 4000(3.7908) + 3500(0.75131)$$

$$= \$32,793$$

Machine A is more economical. The cost of machine A at $28,823 must be qualified by the information that the cost is the present value for 5 years ahead with money worth 10 percent per year.

2.5 Comparison with Unequal Durations—Common Denominator of Service Lives

If two machines have different service lives, a fair comparison can be made only on the basis of equal durations. It would not be equitable to compare the present value of the cost of 2 years of service for machine A with the present value of the cost of 3 years of service for machine B. One method of making the comparison is to compare them on the basis of a common denominator of their service lives, here 6 years. The procedure is merely a mathematical convenience without implying that either machine will be repeated.

Example 2.12 Two machines have the following cost comparison. If money is worth 10 percent per year, which machine is more economical?

	A	B
First cost, $	18,000	25,000
Uniform end-of-year maintenance cost, $ per year	4,000	3,000
Salvage value, $	500	1,500
Service life, years	2	3

SOLUTION The cost diagrams of each machine are first reduced to a single present value of the costs.

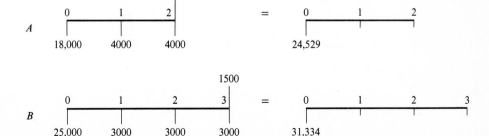

The simplified diagrams are now repeated to obtain 6 years' duration. Note that the present value of machine A is 24,529 as of its *time of installation*.

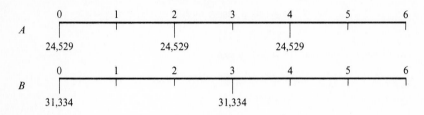

The present value of each of the above diagrams at 10 percent per year is

$$P_{A6} = \$61,555$$

$$P_{B6} = \$54,875$$

Machine B is more economical. The ratio of cost is

$$\frac{A}{B} = \frac{61,555}{54,875} = 1.1217$$

2.6 Cost Comparison by Unacost

The method of the preceding section is cumbersome because each problem requires a new duration for the common denominator of the service lives, and the calculation is tedious besides. It is better to use a common denominator for all service lives. One such service life is an equivalent end-of-year uniform annual cost which puts all systems on a 1-year basis.

Unacost fits the requirements exactly. Thus, converting any system lasting n years with a present value P_n to an equivalent 1-year cost as of the end of the year results in

$$R = P_n F_{PR,i,n} \tag{2.31}$$

which is a repeat of Eq. (2.20).

Most authors refer to the method simply as the *annual-cost method*. The term *annual cost* is, however, a general term and does not necessarily imply uniformity or the end of the year to all readers. The term *unacost* as used in this book definitely implies uniformity from year to year and the end of the year as part of the definition. Use of the term *unacost* avoids the circumlocutory expression *uniform end-of-year annual amount*.

Example 2.13 Solve Example 2.12 on the basis of unacost.

SOLUTION From the first solution:

$$P_{A2} = \$24,529$$

$$P_{B3} = \$31,334$$

By Eq. (2.31):

$$R = P_n F_{PR,i,n}$$

$$R_A = 24{,}529 F_{PR,10\%,2} = 24{,}529(0.57619) = \$14{,}133$$

$$R_B = 31{,}334 F_{PR,10\%,3} = 31{,}334(0.40211) = \$12{,}600$$

Machine B is more economical. The ratio of costs is

$$\frac{A}{B} = \frac{14{,}133}{12{,}600} = 1.1217$$

2.7 Cost Comparison by Capitalized Cost

Another common denominator for all service lives is infinity leading to the concept of capitalized cost. Capitalized cost is the present value on an infinite time basis. For a machine costing P_n and lasting n years, the present value replacing out to infinity is

$$K = P_n \sum_{x=0}^{\infty} \frac{1}{(1+i)^{xn}}$$

The Σ summation is an infinite series with first term 1 and ratio $1/(1+i)^n$. The sum is, by Eq. (2.14),

$$K = P_n \frac{1 - [1/(1+i)^n]^{\infty}}{1 - [1/(1+i)^n]} 1 = P_n \frac{(1+i)^n}{(1+i)^n - 1} = P_n F_{PK,i,n} \qquad (2.32)$$

where K is the capitalized cost and F_{PK} is the factor that converts a present value to capitalized cost.

$$\text{Capitalized cost} = (\text{present-value basis } n \text{ years' duration})(\text{capitalized-cost factor})$$

F_{PK}, which converts P to K, has the complete designation

$$F_{PK,i,n} = \frac{(1+i)^n}{(1+i)^n - 1} \qquad (2.33)$$

and is tabulated in Table A1.1.

The capital-recovery factor F_{PR} and the capitalized-cost factor are related very simply. By Eq. (2.21):

$$F_{PR,i,n} = \frac{i(1+i)^n}{(1+i)^n - 1}$$

Comparing this with Eq. (2.33) shows that

$$F_{PR,i,n} = i F_{PK,i,n} \qquad (2.34)$$

Capital-recovery factor $=$ (rate of return)(capitalized-cost factor)

Also, R and K are similarly related

$$R = iK \tag{2.35}$$

$$\text{Unacost} = (\text{rate of return})(\text{capitalized cost})$$

Example 2.14 Solve Example 2.12 on the basis of capitalized cost.

SOLUTION From the first solution:

$$P_{A2} = \$24,529$$

$$P_{B3} = \$31,334$$

By Eq. (2.32):

$$K = P_n F_{PK}$$

$$K_A = 24,529 F_{PK,\,10\%,\,2} = 24,529(5.7619) = \$141,330$$

$$K_B = 31,334 F_{PK,\,10\%,\,3} = 31,334(4.0211) = \$126,000$$

Machine B is more economical. The ratio of costs is

$$\frac{A}{B} = \frac{141,330}{126,000} = 1.1217$$

As a matter of information, it is possible to convert a present value P_{n1} of n_1 years' duration to an equivalent present value P_{n2} of n_2 years' duration. The unacosts must be equal to each other. Hence applying Eq. (2.20)

$$P_{n1} F_{PR,\,i,\,n1} = P_{n2} F_{PR,\,i,\,n2}$$

$$P_{n2} = P_{n1} \frac{F_{PR,\,i,\,n1}}{F_{PR,\,i,\,n2}} \tag{2.36}$$

Example 2.15 Solve Example 2.12 by the use of Eq. (2.36)

SOLUTION From the first solution:

$$P_A = 2\text{-year duration} = \$24,529$$

$$P_B = 3\text{-year duration} = \$34,334$$

Convert the present value of machine B to an equivalent present value for 2 years' duration using Eq. (2.36)

$$P_{B2} = P_{B3} \frac{F_{PR,\,10\%,\,3}}{F_{PR,\,10\%,\,2}} = 31,334 \frac{0.40211}{0.57619} = \$21,867$$

Machine B is more economical. The ratio of costs is

$$\frac{A}{B} = \frac{24,529}{21,867} = 1.1217$$

Example 2.14 has now been calculated in four ways, all equivalent. The results are summarized in Table 2.2.

Table 2.2 Comparison of solutions to Example 2.14

Method	Numerical value, $	Ratio of costs A/B
Present value on 6-year basis, common denominator of service lives; Example 2.12, first solution	$P_{A6} = 61,555$ $P_{B6} = 54,875$	1.1217
Unacost, equivalent uniform end-of-year annual cost; Example 2.13	$R_A = 14,133$ $R_B = 12,600$	1.1217
Capitalized cost, present worth for an infinite duration; Example 2.14	$K_A = 141,330$ $K_B = 126,000$	1.1217
Present value of machine B converted to 2-year duration; Example 2.15	$P_{A2} = 24,529$ $P_{B2} = 21,867$	1.1217

2.8 Examples of Cost Comparisons

It is desirable to develop expressions for unacost and capitalized cost for various types of costs or receipts. If the present value of the cost for a duration of n years is calculated, as from a time diagram, multiplication by the capital-recovery factor F_{PR} and the capitalized-cost factor F_{PK} will give the unacost and the capitalized cost. For example, consider a salvage value C_{sal}, a negative cost, due at the end of n years. The present value is

$$-C_{sal}\frac{1}{(1 + i)^n}$$

The equivalent unacost, by Eq. (2.19), is

$$R = -C_{sal}\frac{1}{(1 + i)^n}\frac{i(1 + i)^n}{(1 + i)^n - 1} = -C_{sal}\frac{i}{(1 + i)^n - 1}$$

The capitalized cost, by Eq. (2.33), is

$$K = -C_{sal}\frac{1}{(1 + i)^n}\frac{(1 + i)^n}{(1 + i)^n - 1} = -C_{sal}\frac{1}{(1 + i)^n - 1}$$

Table 2.3 is a summary of a number of relationships. Indeed, every type of cost is covered by the tabulation, inasmuch as all costs can be included as an irregular expense at some time and thus can be accommodated by the C_{ex} and C_{bx} items, which are for irregular costs.

Unacosts can be added together. Capitalized costs can also be added together.

Table 2.3 Summary of unacost and capitalized-cost relationships—no tax

Item no.	Item	Present value as a cost for n years duration	Unacost Algebraic expression	Unacost Factor expression	Capitalized cost Algebraic expression	Capitalized cost Factor expression
1	C_i Initial cost for n-year life	C_i	$C_i\dfrac{i(1+i)^n}{(1+i)^n - 1}$	$C_i F_{PR,i,n}$	Multiply corresponding algebraic expression for unacost by $\dfrac{1}{i}$	$C_i F_{PK,i,n}$
2	R Uniform end-of-year annual cost	$R\dfrac{(1+i)^n - 1}{i(1+i)^n}$	R	$R1$ (factor is unity)		$R\dfrac{1}{i}$
3	R_b Uniform beginning-of-year annual cost	$R_b(1+i)\dfrac{(1+i)^n - 1}{i(1+i)^n}$	$R_b(1+i)$	$R_b(1+i)$		$R_b F_{PK,i,1}$
4	C_{ez} Irregular cost at end of xth year for article lasting n years	$C_{ez}\dfrac{1}{(1+i)^x}$	$C_{ez}\dfrac{1}{(1+i)^x} \times \dfrac{i(1+i)^n}{(1+i)^n - 1}$	$C_{ez}\dfrac{1}{(1+i)^x}F_{PR,i,n}$		$C_{ez}\dfrac{1}{(1+i)^x}F_{PK,i,n}$
5	C_{bz} Irregular cost at beginning of xth year for article lasting n years	$C_{bz}\dfrac{1}{(1+i)^{x-1}}$	$C_{bz}\dfrac{1}{(1+i)^{x-1}} \times \dfrac{i(1+i)^n}{(1+i)^n - 1}$	$C_{bz}\dfrac{1}{(1+i)^{x-1}}F_{PR,i,n}$		$C_{bz}\dfrac{1}{(1+i)^{x-1}}F_{PK,i,n}$
6	C_{nd} Nondepreciable first cost, e.g., land or an article that lasts forever	$C_{nd}\dfrac{(1+i)^n - 1}{(1+i)^n}$	$C_{nd}i$	$C_{nd}i$		$C_{nd}1$ (factor is unity)
7	C_{sal} Salvage value at end of nth year, a receipt	$-C_{sal}\dfrac{1}{(1+i)^n}$	$-C_{sal}\dfrac{i}{(1+i)^n - 1}$	$-C_{sal}\dfrac{1}{(1+i)^n} \times F_{PR,i,n}$		$-C_{sal}(F_{PK,i,n} - 1)$ or $-C_{sal}\dfrac{1}{(1+i)^n}F_{PK,i,n}$

For a system

$$R_{SYS} = R_1 + R_2 + R_3 + \cdots$$

$$K_{SYS} = K_1 + K_2 + K_3 + \cdots$$

Example 2.16 A new machine costs $8000 and lasts 10 years with negligible salvage value. If money is worth 10 percent per year, how much can be spent now to repair an old machine to extend its life 3 years?

SOLUTION Unacost for a new machine by item 1, Table 2.3, is

$$R_{new} = C_i F_{PR, 10\%, 10} = 8000(0.16275) = \$1302.0$$

For a repair job having the same unacost,

$$R_{rep} = 1302.0 = C_i F_{PR, 10\%, 3} = C_i(0.40211)$$

$$C_i = \$3238$$

That is, up to $3238 can be spent for the repair job with money worth 10 percent per year.

Example 2.17 If, in the preceding example, the old machine can be repaired for $2500, what is the savings?

SOLUTION Unacost for a repair job by item 1, Table 2.3, is

$$R_{rep} = C_i F_{PR, 10\%, 3} = 2500(0.40211) = \$1005$$

$$R_{sav} = 1302 - 1005 = \$297$$

That is, $297/year is saved as of the end of each year for 3 years.

Example 2.18 A floor requires resurfacing at a cost of $5000 and will last 2 years. If money is worth 10 percent per year, how long must a new floor last if it costs $19,000?

SOLUTION The capitalized cost of resurfacing by item 1, Table 2.3, is

$$K = C_i F_{PK, 10\%, 2} = 5000(5.7619) = \$28,810$$

Assume a service life for the new floor and by repeated trials find the life for which the capitalized cost is $28,810:

$n = 11$: $\qquad K = C_i F_{PK, 10\%, 11} = 19,000(1.5396) = 29,252$

$n = 12$: $\qquad K = C_i F_{PK, 10\%, 12} = 19,000(1.4676) = 27,884$

By interpolation $n = 11.3$ for $K = 28,810$. That is, the new floor must last 11.3 years or longer.

Example 2.19 It will cost $5000 to line a stack and obtain a 2-year life. For $9400 a 4-year life can be obtained. At what rate of return are the two investments equal?

SOLUTION Try a 6 percent per year rate of return. Let R_2 and R_4 be unacost for 2- and 4-year linings. By item 1, Table 2.3,

$$R_2 = C_i F_{PR, 6\%, 2} = 5000(0.54544) = 2727.2$$

$$R_4 = C_i F_{PR, 6\%, 4} = 9400(0.28859) = 2712.7$$

For an 8 percent per year rate of return:

$$R_2 = C_i F_{PR,8\%,2} = 5000(0.56077) = 2803.9$$

$$R_4 = C_i F_{PR,8\%,4} = 9400(0.30192) = 2838.0$$

That is,

At 6 percent: $\qquad\qquad\qquad\qquad R_2 - R_4 = +14.5$

At 8 percent: $\qquad\qquad\qquad\qquad R_2 - R_4 = -34.1$

and by interpolation $R_2 - R_4 = 0$ at 6.6 percent per year rate of return. Below 6.6 percent the 4-year lining is more economical, and above 6.6 percent the 2-year lining is more economical. The example illustrates that a decision may be reversed depending upon the rate of return.

Example 2.20 A machine was purchased for $10,000 from a manufacturer who guaranteed the machine for 10 years and also agreed to include 6 percent per year for the time value of money. At the end of 8 years the machine becomes worthless. How much of a refund should the manufacturer pay as of that time?

FIRST SOLUTION The unacost acceptable to the purchaser was, by item 1, Table 2.3,

$$R = C_i F_{PR,6\%,10} = 10,000(0.13587) = \$1359$$

The equivalent time pattern is:

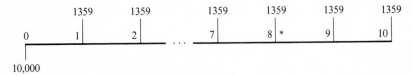

The purchaser gave $10,000 for $1359 worth of service as of the end of each year. At the end of 8 years he is owed $1359 1 year away and $1359 2 years away. The immediate value of these is

$$1359(1.06^{-1}) + 1359(1.06^{-2}) = \$2491$$

That is, $2491 would be a fair settlement.

SECOND SOLUTION The purchaser is owed the value of a 2-year machine based on $10,000 for a 10-year machine. For equal unacosts by item 1, Table 2.3,

$$C_{i,10} F_{PR,6\%,10} = C_{i,2} F_{PR,6\%,2}$$

$$10,000(0.13587) = C_{i,2}(0.54544)$$

$$C_{i,2} = \$2491 \qquad \text{(as before)}$$

THIRD SOLUTION The purchaser actually received an 8-year machine. The price for an 8-year machine by capitalized cost, item 1, Table 2.3, would be

$$C_{i,10} F_{PK,6\%,10} = C_{i,8} F_{PK,6\%,8}$$

$$10,000(2.2645) = C_{i,8}(2.6839)$$

$$C_{i,8} = \$8437$$

The purchaser's overpayment 8 years ago brought up to the present is $(10,000 - 8437)(1.06^8) = \$2491$, as before.

The three solutions illustrate the equivalence of various methods.

Example 2.21 Two machines have expenses as follows. If money is worth 10 percent per year, which is more economical and by how much?

	A	B
First cost, $	10,000	95,000
Uniform end-of-year annual maintenance cost, $ per year	3,000	1,000
Irregular cost, beginning of third year, $	4,000	
Irregular cost, beginning of fourth year, $	1,500	
Salvage value, $	0	20,000
Uniform end-of-year savings from quality control, $ per year	0	6,000
Life, years	4	10

SOLUTION In a more involved problem, the present value can be obtained first, after which conversion to a common-denominator basis can be made.

The present value of 4 years of machine A is

$$P_{A4} = 10,000 + 3000F_{RP, 10\%, 4} + 4000F_{SP, 10\%, 2} + 1500F_{SP, 10\%, 3}$$

$$= 10,000 + 3000(3.1699) + 4000(0.82645) + 1500(0.75131)$$

$$= 23,943$$

for which the unacost is, by item 1, Table 2.3,

$$R_A = 23,943F_{PR, 10\%, 4} = 23,943(0.31547) = \$7553$$

The present value of 10 years of machine B is

$$P_{B10} = 95,000 + (1000 - 6000)F_{RP, 10\%, 10} - 20,000F_{SP, 10\%, 10}$$

$$= 95,000 - 5000(6.1446) - 20,000(0.38554)$$

$$= 56,566$$

for which the unacost is

$$R_B = 56,566F_{PR, 10\%, 10} = 56,566(0.16275) = \$9206$$

Machine A is more economical. The unacost savings is

$$9206 - 7553 = \$1653$$

That is, machine A is more economical by $1653 per year as of the end of each of the 4 years that it is in service.

2.9 Finding the Best Alternative

Frequently a problem arises in such a way that several alternatives are possible, and the analyst must take precautions so as to arrive at the best answer, not just a better answer. The following problem illustrates the principle.

Example 2.22 A new tank costs $10,000 and lasts 5 years. An old tank can be repaired for $1500 to give 1 more year of service, then for $2000 to give 1 more year of service, then for $2500 to give another year of service, and so forth. If money is worth 10 percent per year, how long should the old tank be repaired?

SOLUTION At zero time, compare the cost of a new tank with the cost of repairing. Using capitalized cost and item 1, Table 2.3,

$$K_{new} = C_i F_{PK, 10\%, 5} = 10,000(2.6380) = \$26,380$$

$$K_{rep} = C_i F_{PK, 10\%, 1} = 1500(11.00) = \$16,500$$

It is more economical to repair. Repeat the calculation 1 year later.

$$K_{new} = \$26,380 \quad \text{(as above)}$$

$$K_{rep} = 2000 F_{PK, 10\%, 1} = 2000(11.00) = \$22,000$$

Again, it is more economical to repair. Repeat the calculation 1 year later, at the end of the second year.

$$K_{new} = \$26,380 \quad \text{(as before)}$$

$$K_{rep} = 2500 F_{PK, 10\%, 1} = 2500(11.00) = \$27,500$$

It will not pay to repair. Since the repair cost is rising each year, it follows that the tank should be repaired for 2 years and then replaced with a new tank.

The following solution is incorrect but sometimes offered. Let $K_{r1}, K_{r2}, \ldots$ be the capitalized cost of repairing for 1, 2, ... years. Taking the present values and converting to a capitalized cost by item 1, Table 2.3, gives

$$K_{r1} = 1500 F_{PK, 10\%, 1} = 1500(11.00) = 16,500$$

$$K_{r2} = \left(1500 + \frac{2000}{1.10}\right) F_{PK, 10\%, 2} = 3318.2(5.7619) = 19,119$$

$$K_{r3} = \left(1500 + \frac{2000}{1.10} + \frac{2500}{1.10^2}\right) F_{PK, 10\%, 3} = 5384.3(4.0211) = 21,650$$

Continuing,

Years repair	K, \$
1	16,500
2	19,119
3	21,650
4	24,096
5	26,457

The capitalized cost as found previously for a new tank is \$26,380, and the tabulation might be interpreted as a decision to repair for 4 years because only for 5 years' repairing does the capitalized cost of repairing exceed the capitalized cost of a new tank. The reasoning is incorrect, however. The capitalized cost of repairing has been taken by years as a group. The tabulation merely indicates that it is more economical to repair for 4 years than not to repair at all. The group of 4 years includes the savings of the first 2 years, which overcome the loss of the last 2 years.

The following alternative solution is correct. Compare the programs of buying a new tank every 5 years, buying a new tank every 6 years after repairing for 1 year, and so forth. Take present values and convert to capitalized costs.

$$K_5 = 10,000F_{PK,10\%,5} = 10,000(2.6380) = 26,380$$

$$K_6 = \left(10,000 + \frac{1500}{1.10^5}\right)F_{PK,10\%,6} = 10,931(2.2961) = 25,099$$

$$K_7 = \left(10,000 + \frac{1500}{1.10^5} + \frac{2000}{1.10^6}\right)F_{PK,10\%,7} = 12,060(2.0541) = 24,772$$

Continuing,

Total years	Years repair	K, \$
5	0	26,380
6	1	25,099
7	2	24,772
8	3	25,010
9	4	26,000
10	5	26,409

The minimum occurs for repairing for 2 years. Repairing for 3 or even 4 years is more economical than not repairing at all, but not the most economical.

The following variation also gives the correct analysis. Consider programs of repairing for various numbers of years and then purchasing a new tank to complete the cycle.

$$K_5 = 10,000F_{PK,10\%,5} = 10,000(2.6380) = 26,380$$

$$K_6 = \left(1500 + \frac{10,000}{1.10}\right)F_{PK,10\%,6} = 10,591(2.2961) = 24,318$$

$$K_7 = \left(1500 + \frac{2000}{1.10} + \frac{10,000}{1.10^2}\right)F_{PK,10\%,7} = 11,583(2.0541) = 23,793$$

Continuing,

Years repair	K, \$
0	26,380
1	24,318
2	23,793
3	24,174
4	25,122
5	26,427

Again, repairing for 2 years is most economical. Repairing for 3 or 4 years is more economical than not repairing at all, but not the most economical.

2.10 Considerations in Cost Comparison

In a practical problem all the pertinent costs must be included, and the analyst must be on guard to avoid omissions. If alternatives have costs equal in all respects, such costs may be omitted because only differences affect the choice. However, equality in all respects must be strictly true. Thus, suppose the comparison is between heat-exchanger tubes lasting 2 and 5 years. The installation cost for each may be the same, but it must be included. In one case the installation cost will be incurred every 2 years, in the other every 5 years. Conversely, if the cost for cleaning the tubes is the same for each on a yearly basis, this cost can be omitted in the comparison. Of course, inclusion of equal costs is never detrimental and may prevent their being overlooked in subsequent considerations.

It is not necessary that the machines or systems being compared give equal service. Benefits or penalties can include such considerations as extra production, better quality control, or even reduction of pollution. However, savings or losses associated with a machine should be included only if they are real. Downtime for repairs should not be considered a penalty for lost production if the machine can attain the overall production rate even with the downtime. Frequently the downtime of a machine can be a significant consideration.

Every practical comparison contains a number of intangibles. Two machines may have different capacities, although both can meet present demands. No immediate benefit can be given for the larger capacity, but it is an intangible benefit. In a comparison between air cooling and water cooling, air cooling has an intangible value because water is becoming more restricted and relatively more expensive with time. The intangibles can be an annoyance to the analyst. If they can be given a monetary value, they should be included to that extent. At least they can be enumerated so that they are there to be seen. The intangibles can easily swing the decision in a close comparison.

Sometimes such items as service life or future repair costs are not definitely known, but the comparison can be made on the basis of a range. For example, machine A can be calculated on a high, best estimate, and low basis. Also, results can be presented in a form that circumvents the uncertainty. In Example 2.18 the service life of the new floor is not known, but the calculation that the floor would have to last 11.3 years to be economical could facilitate making a decision.

2.11 The Two Time Factors

It has been emphasized that two time factors are involved in making a cost comparison. They are (1) the instant at which costs are incurred and (2) the duration over which costs extend. Unacost and capitalized cost both have built-in common denominators to accommodate any duration and convert it to a standard duration. If the durations for two machines are the same, both methods are still applicable. The methods are fundamentally present-value methods. Unacost uses a 1-year duration as an end-of-year cost, and capitalized cost uses an infinite duration with a zero-time cost. The methods do not imply respectively that an article will be used 1 year or replaced forever. The 1-year and infinite durations are chosen merely for mathematical convenience. Each method gives the correct comparison for any duration. The methods are readily converted from one to the other by the simple relationship (Eq. 2.35)

$$R = iK$$

In most problems the method used drops out. Management, however, may understand a unacost figure of, say, $10,000 per year better than $100,000 capitalized cost, although both are equivalent at a 10 percent per year rate of return.

The result of a comparison depends upon the time value of money as demonstrated by Example 2.19, where the decision is reversed above and below 6.6 percent. A discussion of what rate of return should be used is not presented here.

A subsequent chapter will discuss the analysis of whole projects or ventures for which all receipts and expenses are known, or at least assigned a definite value. In the present chapter the receipts associated with a particular machine or group of machines are usually not known. Thus, in a long assembly line, it is not possible to allocate the receipts of the entire operation to individual services or machines. This does not prevent the analyst, however, from using the methods of this chapter for comparing two competing machines. The rate of return used in the calculation is then merely the time value of money that the analyst places on investment for machines performing the indicated service. Generally it will not be the same as the rate of return for the venture.

2.12 Procedure for Cost Comparisons

When the time value of money is involved in the comparison of two or more machines or systems, the best procedure is to draw a time diagram of all money values incurred. Then convert each diagram to a present value, and the present value for each system to a unacost or capitalized cost, either of which will correct for different durations of the system.

Money put into the system can be displayed by lines below the time base of the time diagram, and money received by the system by lines above the time base. The length of the lines is not significant. In this chapter the emphasis is on cost and a cost is treated as a positive amount, with income being a negative amount. Later, in discussing profitability, income will be treated as positive, but this causes no confusion.

The fundamental aspects of the mathematics of cost comparisons have been presented in this chapter. Subsequent chapters will discuss additional considerations such as taxes.

The procedure stressed in this chapter emphasizes the conceptual nature and use of various factors, such as F_{PR}, rather than their algebraic form. The best engineering practice in solving a problem involves three distinct steps. Each step should be done independently so that there is no mental block caused by intermingling the steps. The first step is to draw a picture of the problem, here a time diagram. The next step is to set up the problem in a conceptual mathematical form, here the various amounts with the required F factors, to give an equation of value. The third step is to perform the calculations, here solving to obtain the actual numerical values.

NOMENCLATURE

A_{t1}	Amount at time 1, $
C_{bx}	Irregular cost at the beginning of the xth year, $
C_{ex}	Irregular cost at the end of the xth year, $
C_i	Initial or first cost, $
C_{sal}	Salvage value, $
F	Future value by some writers, $
F	Reserved for factors in this book
F_{PK}	Factor to convert P to K; capitalized cost factor, dimensionless
F_{PR}	Factor to convert P to R; capital recovery factor, year^{-1}
F_{PS}	Factor to convert P to S; compound interest factor, dimensionless
F_{RP}	Factor to convert R to P; unacost present-value factor, years
F_{RS}	Factor to convert R to S; equal-payment-series future-value factor, years
F_{SP}	Factor to convert S to P; single-payment present-value factor or present-value factor, dimensionless
F_{SR}	Factor to convert S to R; sinking fund factor, year^{-1}
FV	Future value
i	Rate of return, decimal per year
K	Capitalized cost, $
n	Number such as the life of an article, years
NPV	Net present value, $
p	Number of periods per year
P	Present value, $
PV	Present value, $
PW	Present worth, $; not used
R	Uniform end-of-year annual amount; unacost, $ per year
R_b	Uniform beginning-of-year annual amount, $ per year
R_e	Same as R
S	Single future value, $
u^n	$(1 + i)^n$
v^n	$(1 + i)^{-n}$

PROBLEMS

2.1 The mortgage on a house is $50,000 payable at 12 percent per year. If the end-of-month payments are $717.35, find how much of the first and second payments go to interest and how much to payment of the principal.

2.2 What is the present value of $1000 due in 6 months at 10 percent per year?

2.3 A person can discharge a debt by paying $1000 now or $1075 6 months from now. If money is worth 10 percent per year, what is the cash advantage in paying the debt now?

2.4 A person borrows $1000 from a credit union. At the end of each month, $100 is paid as a principal payment plus interest at 12 percent per year on the principal amount for the month. How much total interest is paid on the loan?

2.5 A person signs a note for $1000 due with interest in 8 months at 10 percent per year. Two months later the holder of the note sells it when the current interest rate is 12 percent. What does the note sell for?

2.6 A person owes $1000 due in 4 months and $500 due in 1 year. What single payment 6 months from now will discharge the debt if money is worth 8 percent per year?

2.7 Find the effective rate of return if money at 8 percent per year is compounded
    (*a*) quarterly (*b*) monthly (*c*) daily (*d*) hourly (*e*) continuously.

2.8 Due now is $5000, and $10,000 is due 6 months from now. What amount will discharge the debt 3 years and 9 months from now if money is worth 12 percent per year?

2.9 What will $15,000 amount to in 4 years at 8 percent per year compounded
   (*a*) annually   (*b*) quarterly?

2.10 What sum will yield $1850 in 5 years at
   (*a*) 6 percent per year   (*b*) 5 percent per year?

2.11 In 1626 Peter Minuit bought Manhattan Island from the Indians for $24 worth of trinkets. What would be the value in 1985 with interest compounded at 6 percent per year?

2.12 What is the rate of return if $10,000 in 7 years amounts to
   (*a*) $17,138   (*b*) $16,400?

2.13 If money is worth 10 percent per year, find the present value of a $10,000 receipt due in 4 years, plus a $15,000 receipt due in 6 years, plus an $8000 debt due in 9 years, plus a $12,000 receipt due in 10 years.

2.14 In Prob. 2.13 find a single debt due in 7 years which would make the receipts and debts equivalent.

2.15 What is the present value of a series of equal end-of-year payments of $1000 for 8 years if money is worth 10 percent per year?

2.16 A series of end-of-year payments forms a geometric series with first payment $1000, second payment $1210, etc. Find the present value of the payments for 6 years if money is worth 10 percent per year.

2.17 What is the present value of a series of equal beginning-of-year payments of $1000 for 8 years if money is worth 10 percent per year?

2.18 If $10,000 is owed, what equal yearly payments for 8 years will discharge the debt if money is worth 6 percent per year and payments are made
   (a) at the end of the year
   (*b*) at the beginning of the year?

2.19 A note has a maturity value of $10,000 due in 4 years and 6 months. What is the present value if money is worth 12 percent per year?

2.20 A debtor signs a note for $10,000 due 5 years from now, with interest to accumulate at 10 percent per year. The holder of the note sells it 2 years and 6 months from now when money is worth 8 percent per year. What is the note sold for?

2.21 If $30,000 is owed and equal payments of $5500 at the end of each year for 8 years will discharge the debt, what is the rate of return?

2.22 If $20,000 is owed with money worth 6 percent per year, and equal end-of-year payments of $3000 are made, how long will it take to discharge the debt?

2.23 A lot can be bought for $10,000 cash now, or for $4000 cash now plus $7000 cash at the end of 4 years. If money is worth 10 percent per year, which plan is better and by how much?

2.24 The population of a city increased from 24,852 to 32,173 in 10 years. Find the average annual rate of increase during the period.

2.25 A person bought a house for $50,000 with various closing fees of $800. Three years later the house was sold for $72,000 with $1200 in various fees and a real estate broker's fee of 6 percent of the selling price. What average annual rate of return was realized on the investment?

2.26 The sales of a business are $100,000 per year and are increasing at a rate of 20 percent per year. How long will it take for sales to be $1,000,000 per year?

2.27 A certain organism can double its population every 6 hours. Assuming no deaths, how long will it take 100 of the species to increase to 1 million?

2.28 A house sells for $50,000 and 3 years later for $75,000. Assume that it increases in value at a constant rate. Find
   (*a*) what it will sell for 4 years after it sells for $75,000
   (*b*) when it will sell for $250,000.

2.29 A person dies and leaves an estate of $100,000 which is invested and earns 8 percent per year. Three children aged 8, 10, and 13 survive the person and at age 18 each child will receive an equal amount from the estate. What is the amount?

2.30 A building sells for $100,000, for which $15,000 is paid in cash. A note for $15,000 due in 1 year, a note for $20,000 due in 2 years, and two equal notes payable in 3 and 4 years are given. Money is worth 10 percent per year. What is the maturity value of the notes due in 3 and 4 years?

2.31 A person owes $10,000 due 1 year from now, owes $30,000 due 5 years from now, will receive $20,000 3 years from now, and will receive $26,500 8 years from now. At what rate of return will the person be even?

2.32 A person pays an ordinary annuity of $1000 per year for 10 years and makes no further payments. If money is worth 10 percent per year, what is the value in the fund at the end of 15 years?

2.33 Repeat Prob. 2.32 for an annuity due.

2.34 A person deposits $1000 at the end of the year for 20 years and then withdraws a uniform amount at the end of each year for the next 5 years. If money is worth 8 percent per year, what is the amount of each withdrawal?

2.35 A person deposits $1000 at the end of each year for 10 years. If money is worth 8 percent per year, find the interest earned by the fund for the sixth year.

2.36 The money payment on a house is $300 payable on the first of the month, with money worth 12 percent per year compounded monthly. A person misses the February, March, April, May, and June payments. What single payment can be made on 1 July to become up to date?

2.37 Show that the present value of an ordinary annuity can be written

$$P = \frac{1 - v^n}{i} \qquad \text{where } v = (1 + i)^{-1}$$

2.38 Find the equivalent cash cost for an automobile that can be purchased for $1000 down and $300 a month for 36 months if money is worth 18 percent per year compounded monthly.

2.39 A person sells a lot for $100,000 but for tax purposes takes 29 percent down and the balance in 14 equal payments starting 1 year from now and continuing every 3 months, with money worth 10 percent per year compounded quarterly. Find the amount of the quarterly payments.

2.40 A person wishes to accumulate exactly $10,000 by making end-of-year payments of $1000 with money worth 8 percent. How long a period is required? If the last payment is to be at the end of a year and is to make the fund exactly $10,000, what is the amount of the last payment?

2.41 A dealer sells an automobile for $9500 less a $1000 allowance for a used automobile. Payments of $276.20 per month are to be made for 36 months starting 1 month from today. What is the monthly rate of interest charged, the nominal rate, and the effective rate?

2.42 A person inherits $100,000 but elects to take 10 yearly payments starting 5 years from now. If money is worth 8 percent per year, find the yearly payments.

2.43 A debt of $12,000 is to be amortized, with equal end-of-year payments of $2000 and money worth 10 percent per year. Payments will be made only at the end of the year. Find the amount of the last payment.

2.44 Lenders sometimes charge points on a loan to avoid exceeding a legal limit on interest rates. Thus for five points the lender would loan only $95 for each $100 mortgage. The borrower borrowing $1000 would receive only $950 but would have to make payments just as if he or she had received $1000. A borrower makes a $100,000 loan payable at the end of each year for 5 years with an interest rate of 12 percent per year, but has been charged five points. What is the true interest rate?

2.45 A bond has a face value of $1000, pays 6 percent per year semiannually, and becomes due in 4.5 years. Find the purchase price if it is bought to yield
    (a) 8 percent   (b) 4 percent

2.46 A bond has a face value of $1000, pays 8 percent per year semiannually, becomes due in 3 years, and is purchased for $1100. What is the yield rate to the purchaser?

2.47 A bond with a face value of $1000 pays interest semiannually and has 3 years to go to maturity. A purchaser will net 12 percent per year by buying the bond for $901.65. Find the annual rate appearing on the face of the bond.

2.48 A bond with a face value of $1000 pays 8 percent per year semiannually and has 4 years to maturity. Although it should have eight coupons attached, it has only six, the next two due having been removed. A purchaser wishes to earn 10 percent on the money. What should be paid for the bond?

2.49 A bond with a face value of $1000 pays 8 percent per year semiannually. It is purchased for $922.78, so that the yield to maturity is 10 percent. Find the time to maturity for the bond.

2.50 A machine costs $15,000 and will have a $3000 salvage value at the end of its life in 10 years. A sinking fund is established at 8 percent per year to purchase a replacement.
    (*a*) What is the sinking fund payment?
    (*b*) What is the value of the sinking fund at the end of 7 years?
    (*c*) How much does the sinking fund increase during the eighth year?

2.51 If $10,000 is owed and it is agreed to pay $1740 annually at the end of the year for 8 years to discharge the debt, but the fourth payment is missed, what equal end-of-year payments thereafter will discharge the debt on time?

2.52 If $5000 is spent now to avoid $15,000 repairs 2 years from now, what rate of return is earned by repairing now?

2.53 An operation is performed manually at a cost of $10,000 per year. A machine costing $60,000 and lasting 12 years will perform the operation and eliminate the labor cost.
    (*a*) What is the rate of return earned by the investment?
    (*b*) If in addition the machine is worth $5000 per year because of added production, what is the rate of return on the investment?

2.54 A company can produce 1 million units per year on a product which infringes a patent. The company can pay a royalty of $0.08 per unit or invest $200,000 in equipment and incur a cost of $0.05 per unit to avoid the patent. If money is worth 12 percent per year, which course should the company take if only 10 years ahead are to be considered? What are the intangibles?

2.55 Three machines are available for an operation. Machine A costs $100,000 and lasts 5 years. Machine B costs $150,000 and machine C costs $200,000. If money is worth 20 percent per year, how long must machines B and C last to be economically equal to machine A?

2.56 Two machines have the following costs:

	A	B
First cost, $	10,000	25,000
Salvage value, $	1,100	1,500
Uniform end-of year expense, $ per year	3,000	2,000
Irregular expense, end first year, $	1,000	
Irregular expense, end third year, $		2,500
Benefit from quality control, as of end of each year, $ per year		600
Life of machine, years	2	5

Compare the machines on the following basis:
    (*a*) Present value for some common denominator of the service lives with money worth 8 percent per year.
    (*b*) Present value, basis 4 years, with money worth 8 percent per year.
    (*c*) Unacost with money worth 8 percent per year.
    (*d*) Capitalized cost with money worth 8 percent per year.

(*e*) The service life of machine B that would be economically equal to machine A with money worth 8 percent per year.

(*f*) The rate of return at which the two are equal.

2.57 Expressions for irregular costs C_{bx} and C_{ex} are given in Table 2.3. Derive the expressions for (*a*) C_{bx} and (*b*) C_{ex}.

2.58 Derive the expressions for a nondepreciable first cost such as land given in Table 2.3.

2.59 A new tank costs $50,000 and lasts 5 years. An old tank can be patched now for $7000 to give 1 more year of life, then for $9500 to give another year of life, then for $12,000 to give 1 more year of life, etc. How long should the old tank be patched if money is worth 10 percent per year?

2.60 A liquid stream in an oil refinery requires servicing at a cost of $0.04 per barrel in a reactor. Operation is at 10,000 barrels per day for 350 days per year. A new catalyst is available at a cost of $180,000 which would reduce the servicing cost by one-half. If money is worth 8 percent per year, how long will it take for an investment in the new catalyst to pay off?

2.61 A pipeline was built 3 years ago to last 6 years. It develops leaks according to the relationship

$$\log N = 0.07T - 2.42 \qquad T > 30$$

where N is the total number of leaks from installation and T is the time in months from installation. It costs $500 to repair a leak. If money is worth 8 percent per year, how much can be spent now for a cathodic system that will reduce leaks by 75 percent?

2.62 Plan A is to build an addition having a capacity of 600 Mg/yr at a cost of 3 M$ and 3 years from now to add to the addition at a cost of 2.4 M$ to double production; M is mega-.

Plan B is to build a full-scale plant now for 4.5 M$ with a capacity of 1200 Mg/yr.

The product sells for $3/kg. For plan A the variable operating cost is $0.50/kg produced for all the years, and for plan B it is $0.40/kg produced for all the years. Sales demand for the years is expected to be:

Year	Mg
1	600
2	800
3	1,000
4	1,200
5	1,200

If money is worth 10 percent per year, which plan is the more economical?

2.63 Which of the following machines is more economical, and by how much per year if money is worth 10 percent per year?

	A	B
First cost, $	12,000	20,000
Annual expense, end year 1	3,000	—
2	3,000	—
3	3,000	4,000
4		—
5		—
Life, years	3	5

2.64 A machine costs $100,000 and lasts 5 years. Past records show:

Year	Repairs as of beginning of year, k$	Salvage value, end of year, k$
1	0	50
2	22	20
3	28	8
4	38	3
5	65	2

If money is worth 10 percent per year, how long should the machine be kept?

2.65 An operation can be performed manually at an end-of-year cost of $72,000/year. How much can be spent for a machine which lasts 10 years if the operating cost is $18,000 at the end of the year and an overhaul costing $24,000 is required at the beginning of the fifth year? Money is worth 10 percent per year.

2.66 A machine is bought which will yield a 10 percent per year return based on the following end-of-year savings:

Year	Savings, $
1	10,000
2	15,000
3	17,000
4	12,000
5	10,000

The savings for the first 3 years are obtained as predicted, but the machine has to be abandoned with a $5000 value at the end of the third year. What rate of return was actually realized?

2.67 It is required to produce 1 million gears per year for a small automobile part. The study will look ahead 12 years, with money worth 14 percent per year.

Plan A is to build a production unit in the United States at a cost of 7 M$, with expenses of 2 M$ annually. There will be a 0.5 M$ investment at the end of the sixth year for better pollution control. M is mega-.

Plan B is to build a production unit in Asia for 4 M$ having a yearly expense of 1.2 M$, but with a duty of 30 percent of the selling price of that in the United States. At what selling price are the two plans equal?

2.68 A company makes 2 million parts per year of a product. The company pays $0.10 per part as a royalty or can circumvent the royalty by investing 300 k$ and incurring a cost of $0.05 per part. Money is worth 10 percent per year, and 8 years of operation is expected. How much is saved or lost per year by paying the royalty?

2.69 A machine costs $10,000 and lasts 12 years. If money is worth 10 percent per year, what is the loss on abandonment at the end of 8 years?

2.70 Two types of pipe are available for carrying water with costs as follows:

	A	B
First cost, $	100,000	200,000
Annual end-of-year cost, $ per year	25,000	20,000
Salvage value, $	0	0
Life, years	15	20

Type A pipe must be repaired at times as part of the annual cost, but the water can become contaminated at these times. Contamination with type B pipe can be neglected. If money is worth 6 percent per year, how much benefit must be given to type B pipe to make it economically equal to type A pipe?

2.71 A company manufactures 1 million units per year which sell at $2 less than a competitor's superior product. How much can be spent on research as of the end of this year, assuming a 40 percent chance for success in attaining the extra $2 per unit? Money is worth 10 percent per year, and a period of 8 years beyond this year is to be considered.

2.72 A machine which lasts 10 years can be bought for $10,000. Operating expenses for the machine form an arithmetic progression and are $2000 for the first year as an end-of-year expense, $4000 for the next year, and so forth. At what age should the machine be exchanged for a new one if money is worth 10 percent per year? Ignore the salvage value.

2.73 A company predicts and then actually realizes receipts from an operation as follows:

End of year	Predicted, k$	Actual, k$
1	1,000	1,200
2	1,300	1,400
3	1,500	1,000
4	1,700	300
5	1,200	100
6	800	0
7	400	0
8	100	0

A unit is set up at a cost that will yield 20 percent per year on the basis of the 8-year prediction.
   (*a*) What was the actual rate of return?
   (*b*) What was the loss on abandonment at the end of the fifth year based on the original 20 percent per year rate of return?

2.74 A company owns its office building, which can be sold for $10,000,000 now. Total operating expenses for the building are $600,000 per year, and every 5 years from now a modernization outlay for the building will cost $1,000,000. The company expects to use the building for 20 years, at which time it should sell for $8,000,000. The company can earn 10 percent on money invested in its business. A real estate firm offers to buy the building and lease it back to the company. What is the maximum end-of-year rent that the company could afford to pay?

REFERENCES

1. Barish, N. N., and S. Kaplan: *Economic Analysis for Engineering and Managerial Decision Making*, 2d ed., McGraw-Hill Book Company, New York, 1978.
2. Canada, J. R.: *Intermediate Economic Analysis for Management and Engineering*, Prentice-Hall, Inc., Englewood Cliffs, N.J., 1971.
3. Cissell, R., and H. Cissell: *Mathematics of Finance*, 4th ed., Houghton Mifflin Company, Boston, 1973.
4. DeGarmo, E. P., J. R. Canada, and W. G., Sullivan: *Engineering Economy*, 6th ed., The Macmillan Company, New York, 1979.
5. Estes, J. C.: *Handbook of Interest and Annuity Tables*, McGraw-Hill Book Company, New York, 1976.
6. Financial Publishing Company: *Comprehensive Bond Value Tables*, 4th desk ed., Publication No. 61, Boston, 1958.

7. Grant, E. L., W. G. Ireson, and R. S. Leavenworth: *Principles of Engineering Economy,* 6th ed., The Ronald Press Company, New York, 1976.

8. Gushee, C. H., ed.: *Financial Compound Interest and Annuity Tables,* 5th ed., Publication No. 376, Financial Publishing Company, Boston, 1970.

9. Jelen, F. C., ed.: *Project and Cost Engineers' Handbook,* American Association of Cost Engineers, Morgantown, W.Va., 1979.

10. Kasner, E.: *Essentials of Engineering Economics,* McGraw-Hill Book Company, New York, 1979.

11. Minrath, W. R.: *Handbook of Business Mathematics,* 2d ed., D. Van Nostrand Company, Princeton, N.J., 1967.

12. Newnan, D. G.: *Engineering Economics Analysis,* rev. ed., Engineering Press, San Jose, Calif. 1980.

13. Ostwald, P. F.: *Cost Estimating for Engineering and Management,* Prentice-Hall, Inc., Englewood Cliffs, N.J., 1974.

14. Riggs, J. L.: *Engineering Economics,* McGraw-Hill Book Company, New York, 1977.

15. Smith, G. W.: *Engineering Economy,* 2d ed., Iowa State University Press, Ames, Iowa, 1973.

16. Taylor, G. A.: *Managerial and Engineering Economy,* 3d ed., D. Van Nostrand Company, Inc., New York, 1980.

17. Thorndike, D., and D. S. Benton: *Thorndike Encyclopedia of Banking and Financial Tables,* rev. ed., Warren, Gorham, & Lamont, Inc., Boston, 1981.

18. Woods, D. R.: *Financial Decision Making in the Process Industry,* Prentice-Hall, Inc., Englewood Cliffs, N.J., 1975.

THREE †

DEPRECIATION AND TAXES: EQUIVALENCE AFTER TAXES

F. C. Jelen and C. L. Yaws

DEPRECIATION AND TAXES

3.1 Nature of Depreciation

In any economic endeavor expenditures are incurred that have various durations. For accounting and tax purposes, a sharp demarcation exists between expenditures that are paid for and used up in a single accounting period and expenditures for which the duration exceeds one accounting period. A year is accepted as the period for both accounting and tax purposes; hence there is a sharp distinction between expenditures consumed within 1 year and expenditures which outlast 1 year. Investment in a building will last many years. It would be incorrect to charge the cost of the building entirely to operations for the first year and make no charge for the cost of the building for subsequent years. A means must be found for spreading over the years expenditures that outlast 1 year. An expenditure that decreases in value with time must be apportioned over its life. The term used to describe this loss in value is "depreciation."

Depreciation arises from two causes: wear and tear and obsolescence. A machine may wear out and eventually be unable to perform its function. A machine may show little wear and tear but become economically inefficient due to technological advancement. The latter is called *obsolescence* and represents a loss in value, just as real as wear and tear. Wear and tear and obsolescence are both recognized by tax rules as depreciation.

†See end-of-chapter note on page 70.

Depreciation is an expense. Indeed, the accountant's charge for depreciation is depreciation expense. Depreciation is a fundamental aspect of accounting and cost engineering which must be considered for manufacturing costs, pricing, taxes, and cost comparisons. A company may use one depreciation method for tax purposes and another method for accounting purposes.

Popular presentations of depreciation can lead the student to narrow interpretations. The popular conception of the existence of a depreciation fund to which payments are made so that the original investment is restored when an article is worn out is not in accord with practice. In this book depreciation is used in its mathematical sense. Depreciation arises when an expenditure outlasts 1 year, and a depreciation schedule is a means of allocating the expenditure as an expense over the years.

A depreciation schedule has two mathematical requirements: the life of the depreciable equipment and the mathematical procedure used in allocating the depreciation expense for each year of life of the equipment.

For tax purposes the allowable life of equipment is given by the CLADR System (Class and Depreciation Range System). Table 3.1 is a brief selection of classes taken from *Tax Information on Depreciation,* published by the IRS [2]. The useful life of property can also be determined by the Class Life System on the basis of particular operating conditions and experience. The analyst may choose any service life for internal company use.

A schedule for allocating depreciation over the years for tax purposes must meet government rules. Generally the government will accept any consistent method provided the total depreciation at the end of each year does not exceed during the first two-thirds of the useful life the total depreciation that would result if twice straight-line declining-balance depreciation were used.

The use of accelerated depreciation predates the tax situation. An asset, such as a machine, represents two types of cost, the initial depreciable cost and the operating cost. Operating costs tend to be low in the early years and high in the later years; hence the practice arose of using high depreciation in the early years and low depreciation in the later years to keep the sum of depreciation and operating costs more nearly constant. Accelerated depreciation, too, is more in line with the resale value of machinery at intermediate years.

3.2 Depreciation Terms

Suppose an asset has an initial cost C_i, a life of n years, and a terminal, or salvage, value C_{sal}. Let $D_{f1}, D_{f2}, \ldots$ be the fractional depreciation for each year. The total depreciation, or the *depreciable first cost C_d,* is

$$C_d = C_i - C_{sal} \tag{3.1}$$

The depreciation for any arbitrary year, the mth year, is

$$D_m = D_{f,m} C_d \tag{3.2}$$

Table 3.1 Asset depreciation ranges

	Lower limit, yr	Guideline period, yr	Upper limit, yr
Assets used in all businesses			
Office furniture and equipment	8	10	12
Information system, including computers	5	6	7
Heavy general-purpose trucks	5	6	7
Railroad cars and locomotives	12	15	18
Vessels, barges, tugs	14.5	18	21.5
Depreciable assets used in the following activities			
Agriculture	8	10	12
Mining	8	10	12
Exploration for and production of petroleum	11	14	17
Petroleum refining	13	16	19
Manufacture of wood products and furniture	8	10	12
Manufacture of pulp and paper	10.5	13	15.5
Manufacture of chemical and allied products	9	11	13
Manufacture of primary ferrous metals	14.5	18	21.5
Manufacture of machinery	8	10	12
Manufacture of electrical equipment	9.5	12	14.5
Manufacture of motor vehicles	9.5	12	14.5
Manufacture of aerospace products	8	10	12
Ship- and boat-building machinery and equipment	9.5	12	14.5
Railroad machinery and equipment	11	14	17
Motor transport—freight	6.5	8	9.5
Water transport	16	20	24
Air transport	9.5	12	14.5
Electric utility hydraulic production plant	40	50	60
Electric utility nuclear production plant	16	20	24
Electric utility steam production plant	22.5	28	33.5
Wholesale and retail service assets	2	2.5	3
Personal and professional services	8	10	12

Source: Tax Information on Depreciation, U.S. Internal Revenue Service, Publication No. 534, published yearly.

The *book value* of the asset at the end of m years B_m is the original value less the accumulated depreciation expense to that time

$$B_m = C_i - C_d \sum_{m=1}^{m=m} D_{f,m} \tag{3.3}$$

Book value = initial cost − accumulated depreciation

Subtracting and adding C_d to Eq. (3.3) gives

$$B_m = C_i - C_d + C_d - C_d \sum_{1}^{m} D_{f,m} = C_{\text{sal}} + C_d \left(1 - \sum_{1}^{m} D_{f,m}\right) \tag{3.4}$$

Book value = salvage value + future depreciation

The book value may bear no relation to the resale value.

When the time value of money is a factor, it becomes necessary to consider the present value of the yearly depreciation.

3.3 Straight-line Depreciation (SL)

Allocation of the depreciable cost is uniform for all years in this method

$$D_{f1} = D_{f2} = \cdots = D_{f,n}$$

or
$$D_f = \frac{1}{n} \tag{3.5}$$

Depreciation for the mth year is

$$D_m = \frac{C_d}{n} \tag{3.6}$$

Accumulated depreciation for m years is

$$\sum D_m = C_d \frac{m}{n} \tag{3.7}$$

Book value at the end of m years is

$$B_m = C_i - C_d \frac{m}{n} \tag{3.8}$$

also
$$B_m = C_d \left(1 - \frac{m}{n}\right) + C_{\text{sal}} \tag{3.9}$$

Depreciation remaining for future years is

$$C_d \sum_{m}^{n} D_{f,m} = C_d \sum_{m=m}^{m=n} \frac{1}{n} = C_d (n - m) \frac{1}{n} = C_d \left(1 - \frac{m}{n}\right) \tag{3.10}$$

The present value of $1 of depreciation is

$$F_{\text{SLP},i,n} = \frac{1}{n}\left[\frac{1}{1+i} + \frac{1}{(1+i)^2} + \cdots + \frac{1}{(1+i)^n}\right]$$

where the symbol $F_{\text{SLP},i,n}$ converts straight-line depreciation to a present value. The terms within the brackets are a geometrical series and are summed by Eq. (2.15)

$$F_{\text{SLP},i,n} = \frac{1}{n}\frac{(1+i)^n - 1}{i(1+i)^n} = \frac{1}{n}F_{RP,i,n} \tag{3.11}$$

$F_{\text{SLP},i,n}$ is the straight-line present-value factor and is tabulated in Table A1.1. Table 3.2 is a summary of relationships for various depreciation methods.

3.4 Sinking-fund Depreciation (SF)

This method gives a yearly depreciation schedule which increases systematically with time and hence is opposite to accelerated depreciation. Suppose an annual deposit R_S is made to a sinking fund to restore the depreciable value at the end of n years. Then by item 6, Table 2.1,

$$R_S = C_d\frac{1}{(1+i)^n - 1} = C_d F_{SP,i,n}F_{PR,i,n} \tag{3.12}$$

The depreciation for any year is the sinking-fund increase for that year, which is the deposit for the year plus interest earned by the fund for the year.

The relationships for sinking-fund depreciation are given in Table 3.2. The method is employed by some governmental agencies and public utilities but is not widely used in industrial operations. It has at times been referred to as *theoretical depreciation* because if the sinking fund bears the same interest rate as that used in an economy study, the book value at any time will be the theoretical value of an asset having the same remaining life and salvage value. This aspect should not be overemphasized, but it is of interest.

Example 3.1 An asset costs $20,000, lasts 10 years, and has $1000 salvage value. If money is worth 8 percent per year, find the book value at the end of 7 years using sinking-fund depreciation. Compare this to the equivalent cost of an asset having a 3-year life and $1000 salvage value with money worth 8 percent per year.

SOLUTION The sinking-fund payment, by Eq. (3.12), is

$$R_S = (20,000 - 1000)F_{SP,8\%,10}F_{PR,8\%,10} = 19,000(0.46319)(0.14903) = \$1311.6$$

The amount in the sinking fund at the end of 7 years by item 5, Table 2.1, is

$$S = RF_{RP,8\%,7}F_{PS,8\%,7} = 1311.6(5.2064)(1.7138) = \$11,703$$

The book value at the end of 7 years, by Eq. (3.3), is

$$B_7 = 20,000 - 11,703 = \$8297$$

Table 3.2 Relationships for depreciation, *n*-year life*

Item no.	Method	Annual depreciation for mth year	Accumulated depreciation for m years	Present value of m years of depreciation
1	Straight line (SL)	$\dfrac{C_i - C_{sal}}{n}$	$(C_i - C_{sal})\dfrac{m}{n}$	$\dfrac{C_i - C_{sal}}{n}F_{RP,i,m}$
2	Sinking fund (SF)	$(C_i - C_{sal})F_{SP,i,n}F_{PR,i,n}F_{PS,i,m-1}$	$(C_i - C_{sal})\dfrac{F_{PS,i,m}-1}{F_{PS,i,n}-1}$	$(C_i - C_{sal})\dfrac{m}{1+i}F_{SP,i,n}F_{PR,i,n}$
3	Declining balance (DB)	$C_i(1 - F_{DB})^{m-1}F_{DB}$	$C_i[1 - (1 - F_{DB})^m]$	$C_i F_{DB}\dfrac{1-\left(\dfrac{1 - F_{DB}}{1 + i}\right)^m}{i + F_{DB}}$
4	Sum-of-the-years-digits (SD)	$(C_i - C_{sal})\dfrac{n - m + 1}{0.5n(n + 1)}$	$(C_i - C_{sal})\dfrac{m}{n(n + 1)}(2n + 1 - m)$	$\dfrac{2(C_i - C_{sal})}{n(n + 1)i}\times[n - F_{RP,i,m} - (n - m)F_{SP,i,m}]$
5	Units of production (UP)	$(C_i - C_{sal})\dfrac{M_m}{M}$	$(C_i - C_{sal})\dfrac{\sum\limits_{1}^{m}M_m}{M}$	$\dfrac{C_i - C_{sal}}{M}\sum\limits_{1}^{m}\dfrac{M_m}{(1 + i)^m}$

*Book value at end of mth year $= C_i$ less accumulated depreciation for m years. Depreciation remaining for $n - m$ future years $=$ book value at end of mth year less C_{sal}. For declining-balance depreciation, terminal value is not related to salvage value, but total write-off should be to salvage value. At the end of useful life, $m = n$.

51

This is what a 3-year machine with $1000 salvage value is worth. Use unacost items 1 and 7, Table 2.3.

$$R_{10} = 20{,}000 F_{PR,8\%,10} - 1000(1/1.08)^{10} F_{PR,8\%,10}$$
$$= 20{,}000(0.14903) - 1000(0.46319)(0.14903) = \$2911.5$$

$$R_3 = 8297 F_{PR,8\%,3} - 1000(1/1.08)^3 F_{PR,8\%,3}$$
$$= 8297(0.38803) - 1000(0.79383)(0.38803) = \$2911.5$$

3.5 Declining-balance Depreciation (DB)

In this accelerated depreciation method the yearly depreciation is a fixed percentage or fixed fraction of the book value at the beginning of the year. Let F_{DB} be the fixed fraction; then,

$$D_1 = C_i F_{DB}$$

$$B_1 = C_i - C_i F_{DB} = C_i(1 - F_{DB})$$

$$D_2 = C_i(1 - F_{DB})F_{DB}$$

$$B_2 = B_1 - D_2 = C_i(1 - F_{DB}) - C_i(1 - F_{DB})F_{DB} = C_i(1 - F_{DB})^2$$

For the mth year, the relationships are:

$$D_m = C_i(1 - F_{DB})^{m-1}F_{DB} \tag{3.13}$$

$$B_m = C_i(1 - F_{DB})^m \tag{3.14}$$

The fixed fraction F_{DB} is limited to $2/n$ where n is the useful life in years. Any value less than $2/n$ may be used and in a few special cases is mandated; thus 1.5 times the straight-line rate is the maximum that can be taken on used tangible personal property. For emphasis in terminology, the declining-balance method with $F_{DB} = 2/n$ can be referred to as *twice the straight-line rate* or *double-declining-balance depreciation*.

3.6 Declining-balance Depreciation Reducing to Salvage Value

The declining-balance method will not in general reduce the final book value to the salvage value, nor is this requirement necessary, as explained below. If one wishes to choose a declining-balance fraction so that the final book value and salvage are equal, then by Eq. (3.14) for $m = n$,

$$B_m = C_{sal} = C_i(1 - F_{DB})^n$$

whence
$$F_{DB} = 1 - \sqrt[n]{\frac{C_{sal}}{C_i}} \tag{3.15}$$

Eq. (3.15) is of historical interest but has little current significance.

3.7 Sum-of-the-years-digits Depreciation (SD)

This is an accelerated depreciation method having fractional depreciation for which the numerator is the years of life remaining as of the beginning of the year and the denominator is the sum of the years digits for the total life. For example, for a 10-year life $1 + 2 + 3 \cdots + 10 = 55$, the fractional depreciation for the first year is $10/55$, for the second year $9/55$, and so forth. For the mth year in a total life of n years, the factor is

$$F_{SD} = \frac{n - m + 1}{0.5n(n + 1)} \qquad (3.16)$$

The present value of $1 of depreciation is

$$F_{SDP,i,n} = \sum_{m=1}^{m=n} \frac{n - m + 1}{0.5n(n + 1)} \frac{1}{(1 + i)^m} = \frac{n - F_{RP,i,n}}{0.5n(n + 1)i} \qquad (3.17)$$

$F_{SDP,i,n}$ converts sum-of-the-years-digits depreciation to a present value and is the sum-of-the-years-digits present-value factor. It is tabulated in Table A1.1.

The relationships for sum-of-the-years-digits depreciation are given in Table 3.2.

3.8 Units-of-production Depreciation (UP)

The methods discussed previously allocate depreciation on a time basis without regard to use. The units-of-production method allocates depreciation on a use basis without regard to time. If M is the total units produced and M_m that for the mth year, the fractional depreciation for the mth year is

$$F_{UP,m} = \frac{M_m}{M}$$

The relationships for units-of-production depreciation are given in Table 3.2.

3.9 Taxes and Depreciation

The federal profits tax is based on the income after all costs have been deducted. A depreciation cost, accordingly, is of considerable interest to the Treasury Department because it affects the profit tax. Thus the modified life of equipment has been established, as shown in Table 3.1. The regulations are subject to revision at any time, and the most recent information available should be obtained from the Treasury Department in making final computations.

A summary of corporation income taxes is shown in Table 3.3. In determining the influence of depreciation upon the profits tax, it should be recognized that all depreciation costs reduce the tax base and are a tax savings or credit in any year

Table 3.3 Federal profits tax summary (1980)

Tax on first $25,000 profit	17%
Tax on next $25,000 profit	20%
Tax on next $25,000 profit	30%
Tax on next $25,000 profit	40%
Tax on profit above $100,000	46%

amounting to tD, where t is the decimal tax rate and D is the depreciation in dollars for the year.

The tax situation is quite involved. The investment credit is particularly important, as a taxpayer can take 10 percent of the value of investments when acquired as a reduction in the income tax and, moreover, still depreciate fully. The investment must be in tangible items such as machinery and equipment used in production, and the maximum 10 percent is allowed only for investments lasting 7 years or longer, but no credit is given for a useful life of less than 3 years. For a useful life of 3 to 4.99 years, one-third of the investment qualifies, and for a useful life of 5 to 6.99 years, two-thirds qualifies. A special windfall profits tax on the oil industry has a significant effect and cannot be ignored. The overall tax situation is quite involved and is a complicated mixture of revenue acquisition and politics. There are a large number of special considerations such as work incentive and job credits, preference items, and so forth which need not be discussed here.

The declining-balance and sum-of-the-years-digits methods are limited to acquisitions with a life of 3 years or more. Under existing law straight-line, sinking-fund, declining-balance, and sum-of-the-years-digits methods of depreciation are all acceptable.

A switch to straight-line depreciation can be made generally if acceptable to the government but in some cases at the company's option. That is particularly true for the declining-balance method so as to get the terminal value equal to the salvage value. On the other hand, the book value must never fall below the salvage value, as this would mean taking excessive depreciation.

It is permissible to depreciate individual properties by combining two or more assets into an account as well as by treating each asset as a separate account. These multiple-assets accounts are treated as a group for depreciation purposes referred to as *group depreciation*. Group depreciation poses some problems for economic evaluation. These problems can be resolved by noting that for an item in a group, the life for tax purposes is that of the group, but the life for the economic comparison can be something else. Group depreciation is used widely for such items as pumps, motors, and the like, which would be an unnecessary burden if depreciated on an individual basis. Assets may be combined in numerous ways, e.g., same useful life, type of asset, function of asset, dates of acquisition, cost, location.

The normal tax was given in Table 3.3. The capital gains tax results from the sale or exchange of a capital asset held over 1 year, and for a large corporation the tax rate is about 28 percent. However, a gain on disposal of a capital asset, even though held over 1 year, is taxed at the regular rate.

In this book it will generally be assumed that the company as a whole is operating at a profit, so that a loss sustained by a single project will reduce the tax base and generate a tax offset for which it should be credited. For a large corporation, the tax bracket will be assumed to be above $100,000, so that a 46 percent tax rate prevails. State income taxes should be added if applicable.

Local property taxes and other taxes which are not levied on the basis of profit are just an expense and do not enter the tax considerations of this chapter.

3.10 Comparison of Methods

Table 3.4 shows the yearly depreciation by five depreciation schedules for an asset costing $1100 with a salvage value of $100. In column (4) the declining-balance factor is 0.38096, as calculated from Eq. (3.15), to make the final book value equal to the salvage value. In column (5), with the declining-balance factor at the maximum, here $2/5 = 0.4$, the depreciation for the fifth year is limited to $42.6 to prevent the book value from falling below the salvage value.

Note that the declining-balance and the sum-of-the-years-digits methods have accelerated depreciation at the beginning and the highest present values. Conversely, sinking-fund depreciation has depreciation per year highest at the end and the lowest present value.

As shown in Table 3.4, when using the declining-balance method, care must be taken not to drop the book value below the salvage value. On the other hand, a switch may be required to reach the salvage value. A switch to straight-line depreciation can be made at any time without prior approval. The switch point for best policy depends on the ratio of salvage value to initial cost, and the useful life.

3.11 Special Considerations

The importance of accelerated depreciation in contributing to profitability lies in the greater present value of larger depreciation allotments in the early years of an economic venture. In practice many companies use one depreciation method for tax purposes and another for internal use to maintain a better perspective for true costs and pricing. Sometimes a liability account is set up to be drawn against later when delayed taxes can become a burden.

The student should understand depreciation in its true perspective and recognize its fundamental nature, although it is intermingled with taxes and even profitability. Because depreciation is an expense, the manner of allocation over the years has an effect on the profit for individual years. Cash flow, which is essentially the sum of depreciation plus profit, is discussed in a following chapter.

A depreciable asset may not be in service for the time set up for the depreciation program. If a machine becomes inoperable before it has been completely depreciated, there are provisions in the accounting procedure to allow for the

Table 3.4 Comparison of depreciation methods*

(1) End-of-year	(2) Straight line	(3) Sinking fund 10% per year	(4) Declining balance $F_{DB} = 0.38096$	(5) Declining balance $F_{DB} = 0.4$	(6) Sum-of-the-years-digits
			Yearly depreciation, $		
1	200.0	163.8	419.1	440.0	333.3
2	200.0	180.2	259.4	264.0	266.7
3	200.0	198.2	160.6	158.4	200.0
4	200.0	218.0	99.4	95.0	133.3
5	200.0	239.8	61.5	42.6†	66.7
Total	1000	1000	1000	1000	1000
Present value of 5 years of depreciation at 10% per year					
0	758.2	744.5	822.2	828.6	806.1

*Original cost $1100, salvage value $100, life 5 years.
†Brought down to $100 salvage value.

unrecovered capital, such as loss on disposal. Conversely, if equipment is useful for longer periods than anticipated, it is not possible to charge additional depreciation for tax purposes. For pricing, however, some kind of cost for wear and tear might very well be included.

If on retirement or abandonment a machine has depreciation available for future years, immediate disposal may give a present value advantage which can be considered.

Example 3.2 A machine has $330,000 book value, $30,000 salvage value, 3 years of life remaining, and is being depreciated by the straight-line method. If abandoned, it can be taken as an immediate loss on disposal at a 46 percent tax rate. Assume that the machine can be sold for $30,000 at any time. If money is worth 10 percent per year, find the present value of the benefit by abandonment now.

SOLUTION Depreciation is $100,000 per year. This reduces the tax base. Recovery of the salvage value does not affect the tax base. If the machine is retained, the present value after taxes is

$$P_{ret} = 100,000 \times 0.46 \left(\frac{1}{1.10} + \frac{1}{1.10^2} + \frac{1}{1.10^3} \right) + \frac{30,000}{1.10^3} = 114,395 + 22,539 = \$136,940$$

The present book value is $330,000. Of this, $30,000 recovered now does not affect the tax base; but loss of $330,000 minus $30,000 = $300,000 does affect the tax base. The present value after taxes is

$$P_{aban} = 300,000 \times 0.46 + 30,000 = \$168,400$$

The present value advantage by abandoning now is

$$168,400 - 136,940 = \$31,460$$

If the machine were part of a group, then it would not be possible to take the loss on disposal now, and the present value advantage would be simply

$$30,000 - \frac{30,000}{1.10^3} = \$7460$$

3.12 Depletion

When exhaustible resources are sold, part of the sales realization is a return of capital, and the income tax should adjust for that. Also, it is desirable to have an incentive to encourage exploration for new resources as existing resources are used up.

Depletion is a term used to describe the write-off of certain exhaustible natural resources such as minerals, oil and gas, and timber. Depletion applies to production units withdrawn from the property, whereas depreciation is limited to the original cost less the estimated salvage value.

There are two methods for computing depletion: cost depletion and percentage depletion.

In the cost method the value of a depletion unit, say, a ton of ore, is arrived at by calculating the total value to be depleted divided by the tons of ore to be depleted. Deduction for a tax year is the depletion unit multiplied by the number of units sold within the year. Cost depletion can be applied to all types of depleted property and is often called *valuation depletion*.

In the percentage method, the depletion allowance for the year is a specified percentage of the "gross income from the property" but must not exceed 50 percent of the taxable income figured without the depletion allowance.

The cost method can always be used, but the percentage method has certain limitations; thus timber is exempt. The deduction should be computed both ways, if applicable, and the larger deduction taken.

Percentage depletion varies from 5 percent to a maximum of 22 percent. Oil and gas wells have recently lost the percentage depletion allowance, except for certain small producers subject to limitations.

It should be noted that under cost depletion, when the total cost and accumulated depletion are equal, no further cost depletion is allowed. However, percentage depletion is not limited to original cost less salvage, as is ordinarily true with depreciation of assets.

Accounting for depletion can be quite complicated because shifting expenses and production from one year to another can result in a tax advantage. Also, there are complications arising from multiple properties, additions to capital, reevaluation of resources, election to expense or capitalize, production limits for oil and gas wells, and even the disqualification of a small producer who controls a retail outlet properly defined.

Example 3.3 A mining property with an estimated 1 megaton (Mt) of ore originally cost $5,000,000. In one year 100 kilotons (kt) of ore is sold for $16/t with expenses of $1,000,000. The percentage depletion allowance is 22 percent, and the tax rate is 46 percent. Which is more advantageous, cost depletion or percentage depletion?

SOLUTION The cost per ton of ore for cost depletion is

$$\frac{5,000,000}{1,000,000} = \$5/t \quad \text{(adjusted basis)}$$

	Cost depletion, $	Percentage depletion, $
1. Gross income	1,600,000	1,600,000
2. Expenses, excluding depletion	1,000,000	1,000,000
3. Taxable income before depletion	600,000	600,000
4a. Cost depletion at $5/t	500,000	
4b. Percentage depletion at 22%		300,000*
5. Actual taxable income (3)–(4)	100,000	300,000
6. Tax at 46%	46,000	138,000
7. Cash flow (3)–(6)	554,000	462,000

*22% of $1,600,000 is $352,000, but (4b) cannot exceed 50% of (3).

Cost depletion is better than percentage depletion.

Example 3.4 A mining property using percentage depletion at 22 percent and a tax rate of 46 percent has gross income and expenses before depletion for 2 consecutive years as follows, with all amounts in kilodollars:

	Year 1	Year 2
Gross income	1,000	1,000
Expenses, excluding depletion	400	700

Is there any advantage in shifting 50 k$ of expenses to year 1, all other things remaining the same?

SOLUTION

	Plan A		Plan B	
	Year 1	Year 2	Year 1	Year 2
1. Gross income	1,000	1,000	1,000	1,000
2. Expenses, excluding depletion	400	700	450	650
3. Taxable income before depletion	600	300	550	350
4. Percentage depletion	220	150*	220	175*
5. Actual taxable income	380	150	330	175
6. Tax at 46%	174.8	69	151.8	80.5
7. Cash flow (3)–(6)	425.2	231	398.2	269.5
8. Cash flow for 2 years	656.2		667.7	

*Limited to 50% of (3).

Plan B is better. One forgoes $425.2 - 398.2 = 27$ k$ to gain $269.5 - 231 = 38.5$ k$ 1 year later, which is a 42.6 percent rate of return.

Example 3.5 A mining property is using cost depletion. It was bought for 10 M$ with 1 Mt of estimated ore. After 400,000 t of ore is sold, the mine is then reevaluated and estimated to have 1.2 Mt of ore remaining. The next year 50,000 tons of ore is sold. Find the cost depletion for that year.

SOLUTION The original adjusted basis was

$$\frac{10,000,000}{1,000,000} = \$10/t$$

With 400,000 t of ore sold,

$$400,000 \times 10 = \$4,000,000 \text{ cost depletion}$$

On the new basis,

$$10,000,000 - 4,000,000 = \$6,000,000 \quad \text{(value for 1.2 Mt)}$$

$$\frac{6,000,000}{1,200,000} = \$5/t \quad \text{(new adjusted basis)}$$

$$50,000 \times 5 = \$250,000 \quad \text{(cost depletion for the year)}$$

COST COMPARISON AFTER TAXES

3.13 Present Value after Taxes

In Chap. 2 cost comparisons were made on a no-tax basis. In practice taxes enter, and the remainder of this chapter will develop the comparisons of alternatives on an after-tax basis.

One effect introduced by the corporation tax is a reduction of the rate of return. If i is the rate of return before taxes, t the tax rate, and r the rate of return after taxes, all expressed as decimals, then

$$r = i(1 - t) \tag{3.18}$$

If an investment compounds at a rate i before taxes, it will compound at a rate r after taxes.

Income taxes are actually paid quarterly, but in this chapter they will be treated as being paid once a year at the end of the year. Assets bought during a year can be averaged as of the middle of the year or, alternatively, given full year depreciation if purchased in the first half of the year. In this chapter assets will be regarded as purchased at the first of the year.

Consider now an article that lasts n years with initial cost C_i, salvage value C_{sal}, and depreciable cost C_d, where

$$C_d = C_i - C_{sal} \tag{3.19}$$

There is no tax consideration at the time the article is purchased. At the end of the first year depreciation amounting to $D_{f1}C_d$, where D_{f1} is the fractional depreciation, becomes available as an expense recognized for tax purposes. The taxable base is reduced $D_{f1}C_d$ and a savings or reduction in taxes amounting to $D_{f1}C_d t$ is realized. The time-cost diagram is:

The present value is:

$$P = C_d - C_d t \left[\frac{D_{f1}}{1 + r} + \frac{D_{f2}}{(1 + r)^2} + \cdots + \frac{D_{fn}}{(1 + r)^n} \right] \tag{3.20}$$

The sum of the terms within the brackets is the present value of $1 of depreciation and depends upon the depreciation method selected. Whatever the method used, the equation can be written

$$P = C_d - C_d t \psi = C_d(1 - t \psi) \tag{3.21}$$

where
$$\psi = \sum_{1}^{n} \frac{D_{fm}}{(1 + r)^m} \tag{3.22}$$

The ψ factor depends upon r, n, and the method chosen for depreciation. It represents the present value of \$1 of depreciation and can be calculated for any value of r and n for any depreciation method. In particular, for sum-of-the-years-digits and straight-line depreciation,

$$\psi_{\text{SD},r,n} = F_{\text{SDP},r,n}$$

$$\psi_{\text{SL},r,n} = F_{\text{SLP},r,n}$$

F_{SDP} converts sum-of-the-years-digits depreciation to a present value, and F_{SLP} converts straight-line depreciation to a present value. Both factors are tabulated in Table A1.1.

Sum-of-the-years-digits depreciation will be used for the remainder of this section, with the understanding that any other method can be used. Equation (3.21) becomes

$$P = C_d(1 - tF_{\text{SDP},r,n}) \tag{3.23}$$

It should be noted that for all depreciation methods

$$\psi_{1r} = \frac{1}{1 + r} = F_{\text{SDP},r,1} \tag{3.24}$$

$$\psi_{\alpha_r} = 0 = F_{\text{SDP},r,\alpha} \tag{3.25}$$

Equation (3.23) can be converted to a unacost or capitalized cost by Eq. (2.20) or (2.32), respectively, giving

$$R = C_d F_{PR,r,n}(1 - tF_{\text{SDP},r,n'}) \tag{3.26}$$

$$K = C_d F_{PK,r,n}(1 - tF_{\text{SDP},r,n'}) \tag{3.27}$$

where n' is used to designate that the life for tax purposes can be different from n, the life for an economic study.

It is now possible to duplicate Table 2.3 on an after-tax basis utilizing the method just described for first determining present value. Table 3.5 has been prepared in this manner.

For item 2 in the table, R is a uniform end-of-year cost occurring simultaneously with the tax instant. The cost after taxes is $R(1 - t)$ as of the end of each year for n years. The present value, by Eq. (2.16), with r substituted for i, is

$$P = R(1 - t)\frac{(1 + r)^n - 1}{r(1 + r)^n} \tag{3.28}$$

For item 3, R_b, a uniform beginning-of-year cost, the tax credit $R_b t$ occurs at the end of the year. Combining the two as of the end of the year gives

$$R_b(1 + r) - R_b t = R_b\left(1 - \frac{t}{1 + r}\right)(1 + r) \tag{3.29}$$

Table 3.5 Summary of unacost and capitalized-cost relationships with tax

Item no.	Item	Present value as a cost for n-year duration	Unacost — Algebraic expression
1	C_d Fully depreciable part of an initial cost for n-year life, n' years for tax purposes; $C_d = C_i - C_{sal}$	$C_d(1 - t\psi_{rn'})$	$C_d(1 - t\psi_{rn'})$ $\times \dfrac{r(1 + r)^n}{(1 + r)^n - 1}$
2	R Uniform end-of-year annual cost, depreciated fully at instant incurred	$R(1 - t)\dfrac{(1 + r)^n - 1}{r(1 + r)^n}$	$R(1 - t)$
3	R_b Uniform beginning-of-year annual cost, fully depreciated at end of year	$R_b\left(1 - \dfrac{t}{1 + r}\right)(1 + r)$ $\times \dfrac{(1 + r)^n - 1}{r(1 + r)^n}$	$R_b\left(1 - \dfrac{t}{1 + r}\right)(1 + r)$
4	C_{ex} Irregular cost at end of xth year for article lasting n years, fully depreciated at instant incurred	$C_{ex}\dfrac{1 - t}{(1 + r)^x}$ or $\dfrac{C_{ex}}{(1 + r)^x}$ $\times [1 - t(1 + r)\psi_{r1}]$	$C_{ex}\dfrac{1 - t}{(1 + r)^x}$ $\times \dfrac{r(1 + r)^n}{(1 + r)^n - 1}$
5	C_{bx} Irregular cost at beginning of xth year for article lasting n years, fully depreciated at end of year x	$C_{bx}\left(1 - \dfrac{t}{1 + r}\right)$ $\times \dfrac{1}{(1 + r)^{x-1}}$ or $\dfrac{C_{bx}}{(1 + r)^{x-1}}(1 - t\psi_{r1})$	$C_{bx}\left(1 - \dfrac{t}{1 + r}\right)$ $\times \dfrac{1}{(1 + r)^{x-1}}$ $\times \dfrac{r(1 + r)^n}{(1 + r)^n - 1}$
6	C_{nd} Nondepreciable first cost such as land or an article that lasts forever	$C_{nd}\dfrac{(1 + r)^n - 1}{(1 + r)^n}$	$C_{nd}\, r$
7	C_{sal} Salvage value at end of nth year; treated as a nondepreciable first cost, an expense	$C_{sal}\dfrac{(1 + r)^n - 1}{(1 + r)^n}$	$C_{sal}\, r$

Table 3.5 (*Continued*)

Unacost	Capitalized cost	
Factor expression for sum-of the-years-digits depreciation	Algebraic expression	Factor expression for sum-of-the-years-digits depreciation
$C_d(1 - tF_{\text{SDP},r,n'})F_{PR,r,n}$	Multiply corresponding algebraic expression for unacost by $\dfrac{1}{r}$	$C_d(1 - tF_{\text{SDP},r,n'})F_{PK,r,n}$
$R(1 - t)$		$R(1 - t)\dfrac{1}{r}$
$R_b(1 - tF_{\text{SDP},r,1})(1 + r)$		$R_b(1 - tF_{\text{SDP},r,1})F_{PK,r,1}$
$C_{ex}\dfrac{1}{(1 + r)^x}$ $\times\,[1 - t(1 + r)F_{\text{SDP},r,1}]F_{PR,r,n}$		$C_{ex}\dfrac{1}{(1 + r)^x}$ $\times\,[1 - t(1 + r)F_{\text{SDP},r,1}]F_{PK,r,n}$
$C_{bx}\dfrac{1 - tF_{\text{SDP},r,1}}{(1 + r)^{x-1}}F_{PR,r,n}$		$C_{bx}\dfrac{1 - tF_{\text{SDP},r,1}}{(1 + r)^{x-1}}F_{PK,r,n}$
$C_{\text{nd}}\,r$		C_{nd}
$C_{\text{sal}}\,r$		C_{sal}

which can be converted to a present value by Eq. (2.16) with r substituted for i.

For item 4, C_{ex}, an irregular cost occurring at the end of the xth year for an article lasting n years, the cost occurs simultaneously with the tax instant. The present value is

$$P = C_{ex}(1 - t)\frac{1}{(1 + r)^x} \tag{3.30}$$

For item 5, C_{bx}, an irregular cost occurring at the beginning of the xth year, the tax credit $C_{bx}t$ when brought forward to the beginning of the year becomes $C_{bx}t/(1 + r)$. The cost as of the beginning of the year is then reduced to a present value at zero time, giving

$$P = \left(C_{bx} - \frac{C_{bx}t}{1 + r}\right)\frac{1}{(1 + r)^{x-1}} = C_{bx}\left(1 - \frac{t}{1 + r}\right)\frac{1}{(1 + r)^{x-1}} \tag{3.31}$$

If C_{ex} and C_{bx} cannot be written off in 1 year for tax purposes, substitute $F_{\text{SDP},r,n'}$ for $F_{\text{SDP},r,1}$ in items 3 and 4 of Table 3.5.

For item 6, C_{nd}, a nondepreciable expense such as land or working capital, an amount C_{nd} is invested now and fully recovered at the end of n years with no tax consideration. The present value is

$$P = C_{nd} - \frac{C_{nd}}{(1 + r)^n} = C_{nd}\frac{(1 + r)^n - 1}{(1 + r)^n} \tag{3.32}$$

Item 7 in Table 3.5 for salvage value C_{sal} is correct in showing the item as positive, an expense. A sum is invested, and although it is fully recovered, the expense arises from the loss of interest.

3.14 Procedure for Cost Comparison after Taxes

Costs before taxes must be corrected to the cost after taxes. Some costs will be the same before and after taxes. For example, $10,000 invested in inventory is a swap of money for goods which are later recovered. When invested and recovered, the $10,000 is not subject to tax and is $10,000 before and after tax when invested and when recovered. If a cost occurs at the same instant as the tax payment, the cost after taxes is merely $(1 - t)$ times the cost before taxes. At times, the cost and the tax consideration have different timing. A first depreciable cost is the same before and after taxes at the instant it occurs, as it is merely a swap of money for assets, but each year the depreciation creates a tax benefit that must be recognized. The best procedure is to use a time diagram, include all costs and their timing, include all tax and tax benefits and their timing, and reduce the entire system to a present value after taxes. Finally, convert to an unacost or a capitalized cost to compare with other systems having different service lives. Although all comparisons should be made on an after-tax basis, any value reported back to management should be on a before-tax basis. In any case, the before- and after-tax amounts should be well labeled.

Example 3.6 Two machines have the following cost comparison. If money is worth 10 percent per year after a 46 percent tax, and sum-of-the-years depreciation is used, which machine is more economical and by how much per year?

	A	B
First cost, $	18,000	24,000
Uniform end-of-year expense, $/yr	1,000	0
Salvage value, $	500	0
Service life, years	2	3
Life for tax purposes, years	5	5

SOLUTION Refer to Table 3.5 and find the present value of machine A by items 1, 2, and 7 in order.

$$C_d(1 - tF_{\text{SDP}, 10\%, 5}) = (18,000 - 500)[1 - 0.46(0.80614)] = \quad 11,011$$

$$R(1 - t)F_{RP, 10\%, 2} = 1000(1 - 0.46)(1.7355) = \quad 937$$

$$C_{\text{sal}}[1 - (1.10)^2] = 500(1 - 0.82645) = \quad\underline{\quad 87}$$

$$P = \$12,035$$

The present value of machine B by item 1, Table 3.5, is

$$C_d(1 - tF_{\text{SDP}, 10\%, 5}) = 24,000[1 - 0.46(0.80614)] = \$15,000$$

Convert to unacosts by factor F_{PR}.

$$R_A = 13,247F_{PR, 10\%, 2} = 12,035 \times 0.57619 = \$6934 \text{ after tax}$$

$$R_B = 15,000F_{PR, 10\%, 3} = 15,100 \times 0.40211 = \$6072 \text{ after tax}$$

Machine B is cheaper by $(6934 - 6072) = \$864/\text{year}$ on an after-tax basis. The report to management should be that machine B saves

$$\frac{864}{1 - 0.46} = \$1600/\text{year before tax}$$

Example 3.7 A project requires purchase of a machine that costs $100,000, lasts 10 years, and will be written off in 10 years using sum-of-the-years-digits depreciation. The tax rate is 46 percent. The machine qualifies for an immediate 10 percent investment credit. The operating cost is $10,000 per year at end of year. Inventory worth $10,000 is purchased at zero time and will be recovered at the end of the project. At the end of 8 years the project will be abandoned and the machine sold for $15,000, but the machine is part of a group and will continue to be depreciated. If money is worth 10 percent per year after taxes, find the present value of the project.

SOLUTION Depreciation continues for 10 years. The present value of the first cost by Table 3.5, item 1, is

$$C_d(1 - tF_{\text{SDP}, 10\%, 10}) = 100,000[1 - 0.46(0.70099)] = \$67,754$$

The present value of the investment credit benefit is

$$-100,000(0.10) = \$-10,000$$

The present value of the yearly cost by Table 3.5, item 2, is

$$10,000(1 - 0.46)F_{RP, 10\%, 8} = 10,000(0.54)(5.3349) = \$28,808$$

The inventory goes in and out, with no tax consideration. The present value is

$$10,000[1 - (1.10)^{-8}] = \$5335$$

Book value at the end of 8 years with 2 years remaining is

$$\frac{2 + 1}{55}100,000 = \$5445$$

The machine is sold for $15,000, which is $9545 over book value. However, the $5455 is part of the group depreciation and will be recovered in the next 2 years. The entire $15,000 becomes income subject to tax, and the present value is

$$-15,000(1 - 0.46)(1.10)^{-8} = \$-3778$$

The net present value as a cost after taxes is

$$67,754 - 10,000 + 28,808 - 3778 = \$82,784$$

which to the company as a whole would cost a present value of

$$\frac{82,784}{1 - 0.46} = \$153,300 \qquad \text{(before taxes)}$$

If the machine were not part of a group, the $15,000 received at the end of the eighth year would be $9545 subject to tax and $5455 tax free. However, in that case the present value of the first cost after taxes would have to be revised so as not to include the tax benefit from depreciation originally allowed for years 9 and 10.

NOMENCLATURE

B	Book value, \$
C_{bx}	Irregular cost at beginning of xth year, \$
C_d	Depreciable cost, \$
C_{ex}	Irregular cost at end of xth year, \$
C_i	Initial or first cost, \$
C_{nd}	Nondepreciable cost, \$
C_{sal}	Salvage value, \$
D	Depreciation, \$
DB	Declining-balance method of depreciation
D_f	Fractional depreciation, decimal
F_{DB}	Factor for declining-balance depreciation, decimal
F_{PK}	Factor to convert P to K; capitalized-cost factor, no units
F_{PR}	Factor to convert P to R; capital-recovery factor, year^{-1}
F_{PS}	Factor to convert P to S; compound-interest factor, no units
F_{RP}	Factor to convert R to P; unacost present-value factor, years
F_{SD}	Factor for sum-of-the-years-digits depreciation, decimal
F_{SDP}	Present value of $1 for sum-of-the-years-digits depreciation, decimal
F_{SLP}	Present value of $1 for straight-line depreciation, decimal
F_{SP}	Factor to convert S to P; present-value factor, decimal
F_{UP}	Factor for units-of-production depreciation, decimal
i	Rate of return, generally before taxes, decimal per year
K	Capitalized cost, \$
m	A number, generally the mth year
M	Units of production
n	A number, generally the service life of an item, years

n'	Service life for tax purposes, years
P	Present value, \$
r	Rate of return, generally after taxes, decimal per year
R	Uniform end-of-year annual amount; unacost, \$ per year
R_b	Uniform beginning-of-year annual amount, \$ per year
R_s	Sinking-fund deposit, \$ per year
S	Future value, \$
SD	Sum-of-the-years-digits method of depreciation
SF	Sinking-fund method of depreciation
SL	Straight-line method of depreciation
t	Tax rate, decimal
UP	Units-of-production method of depreciation
ψ	Psi, present value of \$1 of depreciation, decimal

PROBLEMS

3.1 A furniture manufacturer buys special new manufacturing equipment for 1.8 M\$ that has an estimated salvage value of 0.3 M\$.† Find

 (a) the useful life for tax purposes from tables

 (b) the book value at the end of 4 years for twice the straight-line rate declining-balance depreciation.

3.2 A box manufacturer uses cutting dies that cost \$45,000 with a salvage value of \$5000. The dies are expected to produce 32 million boxes before being replaced. If production for 4 consecutive years is 2, 8, 16, and 6 million boxes, compare the annual depreciation by twice the straight-line rate declining-balance and units-of-production depreciation.

3.3 Compare the present value for the depreciation using straight-line, twice the straight-line rate declining-balance with switch at the end of the fourth year, and sum-of-the-years-digits depreciation for an asset valued at \$14,000 with \$2000 salvage value after 6 years with money worth 12 percent per year.

3.4 An asset costs \$14,000 and has a useful life of 10 years and salvage value of \$2000. Find the declining-balance depreciation factor which will give the same book value at the end of 5 years as does straight-line depreciation.

3.5 An asset costs \$20,000 with a salvage value of \$2000 and has a useful life of 8 years. Find a declining-balance factor which will reduce the book value to \$4000 at the end of 5 years.

3.6 A variation for the calculation of the sum-of-the-years-digits method called the *remaining life plan* uses a fraction with the numerator being the useful life in years remaining as of the beginning of the year and the denominator being the sum of the numbers representing the remaining useful life. This fraction is applied to the unrecovered cost. An asset costs \$12,000 with \$2000 salvage value and lasts 5 years. Show that the depreciation for each year by the remaining life plan is the same as that given by the ordinary calculation for the sum-of-the-years-digits method.

3.7 For assets acquired during a year, it is permissible to figure depreciation for each 12-month period and then allocate each amount in the proportion of the service year that occurs in the tax year. For example, if an article were bought on 1 December and lasts 5 years, the sum-of-the-years-digits depreciation fractions would be 5/15 for 1/12 of the first year. For the second year it would be 5/15 for 11/12 of the year plus 1/12 of 4/15. An asset costs \$10,000, is bought on 1 April, and lasts 5 years with no salvage value. Prepare a table showing the depreciation for each of the 6 calendar years that will occur for tax purposes.

3.8 A machine costs \$100,000 with an estimated \$10,000 salvage value. It has an estimated operating life of 15,000 h. During the third year it is used 2500 h. Find the depreciation by the UP method.

† M is mega.

3.9 A company can purchase a machine for $75,000 with $10,000 salvage value and a useful life of 4 years. Annual operating costs are $7000/year, and the machine will generate $40,500/year additional revenue. The income tax rate is 46 percent. Find the net cash flows resulting from the purchase of the machine for each of the 4 years using
  (a) straight-line depreciation
  (b) sum-of-the-years-digits depreciation

3.10 A company has an asset worth $12,000 with a salvage of $2000 at the end of 5 years. It will use sum-of-the-years-digits depreciation for tax purposes and straight-line depreciation for internal use to stockholders. The accountants set up an account called Reserve for Future Taxes from Depreciation, the purpose of which is to set aside revenue from the underpaid (though legal) tax in the early years which can be used as a bookkeeping fund to offset the increase tax in later years. Such a reserve will help to make the apparent profit for each year more uniform despite the variations from straight-line depreciation. Find the transfers in and out of the reserve fund for each of the 5 years if the tax rate is 46 percent.

3.11 An accountant imitating sum-of-the-years-digits depreciation suggests a sum-of-the-squared years-digits method. The first year fractional depreciation would be $n^2/\Sigma n^2$, the second year $(n - 1)^2/\Sigma n^2$, and so forth. Set up the fractions for a useful life of 5 years and comment on the method.

3.12 The gross income for a mineral property is 1.2 M$ for 1 year, and the income tax is 46 percent. What is the annual cash flow if the annual expenses for the operation, excluding depletion, are
  (a) 0.8 M$ (b) 0.6 M$?
The depletion allowance is 22 percent. (M is mega-.)

3.13 A mining property cost 12 M$ for an estimated 6 Mt of ore. In one year 250 kt of ore is sold for $20/t with expenses of $10/t. If the depletion allowance is 22 percent, which is more advantageous, cost depletion or percentage depletion? The income tax rate is 46 percent. (M is mega-.)

3.14 A mining property has a gross income of 6 M$ and an adjusted basis of $3/t for cost depletion. It sells 300 kt and, using cost depletion at an income tax rate of 46 percent, a cash flow of $2.574 M$ is realized. Find the cash flow if a percentage depletion at 22 percent were used. (M is mega-.)

3.15 Refer to Example 3.4. Find the minimum amount of expenses that should be shifted from year 2 to year 1 to maximize the cash flow for the 2 years.

3.16 A mining company sells ore at $10/t with expenses of $2/t. The income tax rate is 46 percent. Currently the company uses percentage depletion at 22 percent. The company next year will be limited to 15 percent depletion, which will cost the company 322 k$ in cash flow compared to this year for the same production. Find how much the company must charge per ton of ore next year to suffer no reduction in cash flow for the same production.

3.17 An individual is in a 49 percent income tax bracket and for capital gains held over 1 year pays the income tax on 40 percent of the capital gains. A $1000 bond paying semiannually at a rate of 12 percent per year has 3 years to run and is available for $900. What rate of return after taxes does the purchaser realize from the bond?

3.18 In Prob. 3.17, the purchaser wishes to earn 8 percent per year on the money after taxes. What should he pay for the bond?

3.19 A municipal bond is available for which interest is exempt from the federal income tax, but 40 percent of capital gains are subject to the tax. The bond has $1000 face value, interest payable semiannually at 6 percent per year, with 4 years to go. An individual in the 49 percent tax bracket wishes to obtain an 8 percent per year rate of return after taxes on the bond. What should she pay for the bond?

3.20 Repeat Example 2.16. Use sum-of-the-years-digits depreciation, with money worth 10 percent per year after a 46 percent tax and a 10-year life for tax purposes for the new machine. Repair expense, for tax purposes, can be written off in 1 year, like maintenance.

3.21 In Prob. 3.20, what is the savings if the old machine can be repaired for $2500?

3.22 Repeat Example 2.18. Use sum-of-the-years-digits depreciation with money worth 10 percent per year after a 46 percent tax, a resurfacing job written off in 1 year for tax purposes, and a new floor written off in the years of its life.

3.23 Repeat Example 2.19. Use sum-of-the-years-digits depreciation with a 46 percent tax rate. A relining job can be written off in 1 year for tax purposes, a 4-year stack in 4 years.

3.24 Repeat Example 2.21. Money is worth 10 percent per year after a 46 percent tax using straight-line depreciation. Machine A will be written off in 8 years for tax purposes, machine B in 10 years. The irregular expenses can be written off in 1 year. Savings from quality control are treated as ordinary income.

3.25 Repeat Prob. 2.53 using straight-line depreciation and a 12-year life for tax purposes for the machine. The tax rate is 46 percent.

3.26 Repeat Prob. 2.54. Royalty payments can be treated as a regular expense. The equipment will be capitalized with a 12-year life for tax purposes using straight-line depreciation, with money worth 12 percent after a 46 percent tax.

3.27 Repeat Prob. 2.55. The tax rate is 46 percent, all machines will be written off in 10 years using straight-line depreciation, and there is a 20 percent rate of return after taxes.

3.28 Repeat Prob. 2.59. The new tank can be written off in 10 years for tax purposes. Use straight-line depreciation and money worth 10 percent per year after a 46 percent tax.

3.29 Repeat Prob. 2.60. Money is worth 8 percent after a 46 percent tax. The catalyst can be written off in 5 years using sum-of-the-years-digits depreciation.

3.30 Repeat Prob. 2.61. Money is worth 8 percent per year after a 46 percent tax. Repair costs for leaks can be written off in the year incurred. The cathodic protection system can be written off in 4 years using sum-of-the-years-digits depreciation.

3.31 Repeat Prob. 2.62. Money is worth 10 percent per year after a 46 percent tax. Use straight-line depreciation. All plans, including additions, are written off in 10 years for tax purposes.

3.32 Repeat Prob. 2.63. Money is worth 10 percent per year after a 46 percent tax. Use straight-line depreciation and write off both machines in 6 years for tax purposes.

3.33 Repeat Prob. 2.64. Money is worth 10 percent per year after a 46 percent tax. The machine can be written off in 10 years for tax purposes using sum-of-the-years-digits depreciation. Treat the overhaul as a maintenance expense for the year.

3.34 A machine is bought that is written off in 5 years using straight-line depreciation and with no salvage value. The machine will return 10 percent after a 46 percent tax rate based on the following end-of-year savings:

Year	Savings, $
1	10,000
2	15,000
3	17,000
4	12,000
5	10,000

The savings for the first 3 years are realized, but the machine must be abandoned at the end of 3 years and sold for $7000. What rate of return was actually realized

  (a) if the new machine can be taken as a loss on disposal at the end of 3 years

  (b) if the machine is part of a group and must continue to be depreciated for 2 additional years?

3.35 A machine costs $10,000 and lasts 12 years, with a useful life of 10 years for tax purposes. Money is worth 10 percent per year after a 46 percent tax using sum-of-the-years-digits depreciation. What is the loss on abandonment at the end of 8 years if the machine is junked for no value and is

  (a) depreciated as a single-asset account

  (b) depreciated as a multiple-asset account?

3.36 Repeat Prob. 2.70. Money is worth 8 percent per year after a 46 percent tax. Use straight-line depreciation and a useful life of 15 years for tax purposes for both kinds of pipe.

3.37 Repeat Prob. 2.71. Money is worth 10 percent per year after a 46 percent tax. Use straight-line depreciation for costs that are capitalized. How much can be spent on the research at a 40 percent chance for success if
     (*a*) the research can be written off in 1 year for tax purposes
     (*b*) the research must be capitalized for a 6-year life for tax purposes?

3.38 Repeat Prob. 2.72. Money is worth 10 percent per year after a 46 percent tax. The machine can be written off in 10 years for tax purposes using sum-of-the-years-digits depreciation.

3.39 Repeat Prob. 2.74. Money is worth 10 percent per year after a 46 percent tax. The building has already been written off for tax purposes.

3.40 A machine cost $252,000 5 years ago and was estimated to have $25,000 salvage value tax purposes. The life was estimated at 7 years, and sum-of-the-years-digits depreciation has been used. At the present time, it will be scrapped and will bring $15,000 on disposal. Money is worth 10 percent after a 46 percent tax rate. What is the value in being allowed to write off the machine in full now compared to holding it on the books for 2 additional years?

REFERENCES†

1. *Prentice-Hall Federal Tax Course,* Prentice-Hall, Inc., Englewood Cliffs, N.J., published yearly.
2. U.S. Internal Revenue Service: *Tax Information on Depreciation,* U.S. Department of the Treasury Publication No. 534, published yearly.

   † See also references for Chap. 2.

     *   *   *

The foregoing discussion applies to assets acquired prior to January 1, 1981. The Economic Recovery Tax Act of 1981 changed drastically the depreciation rules such that assets put into service after that date must be depreciated under the Accelerated Cost Recovery System (ACRS).

The new rules are more beneficial to business than those previously allowed because capital equipment can be depreciated over a shorter period, a greater percentage of original cost can be recovered in the earlier years, and parts of the ACRS rules are simpler than previous rules. A complete discussion of the new law is contained in IRS Publication 534 (U.S. Government Printing Office, Washington, DC).

The production schedule for this book precluded a detailed discussion of these changes. Future reprintings, however, will reflect the new law.

FOUR

CONTINUOUS INTEREST AND DISCOUNTING

W. B. Hirschmann and
J. R. Brauweiler

4.1 Logic for Continuous Interest

Interest can be compounded periodically, e.g., annually, semiannually, or even daily, or it can be compounded continuously. Annual discounting is appropriate for handling mortgages, bonds, and similar financial transactions, which require payments or receipts at discrete times. In most businesses, however, transactions occur throughout the year. These circumstances suggest a continuous flow of money, for which continuous compounding and discounting are more realistic to some than are periodic compounding and discounting. This chapter illustrates how continuous interest can cope with a variety of cash flows.

4.2 Continuous Interest as an Operator

Table 4.1 shows the effect of increasing the number of compounding periods in 1 year, as developed in Sec. 2.1. Note that there is little difference between the factors for monthly and continuous compounding, so that continuous factors could be used with little error to determine mortgage payments, for example.

The future value S of a present amount P with periodic compounding after n years with a nominal decimal interest rate i compounded p times per year is given by Eq. (2.4):

$$S = P\left(1 + \frac{i}{p}\right)^{np} \tag{4.1}$$

Table 4.1 Comparison of compounding factors

Period	Relationship	For $i = 0.06$	Factor for $i = 0.06$
Annually	$(1 + i)^1$	1.06^1	1.06000
Semiannually	$\left(1 + \dfrac{i}{2}\right)^2$	1.03^2	1.06090
Quarterly	$\left(1 + \dfrac{i}{4}\right)^4$	1.015^4	1.0613635
Monthly	$\left(1 + \dfrac{i}{12}\right)^{12}$	1.005^{12}	1.0616778
Daily	$\left(1 + \dfrac{i}{365}\right)^{365}$	1.00016^{365}	1.0618305
Continuously	e^i	$e^{0.06}$	1.0618365

As p approaches infinity for continuous compounding,

$$\left(1 + \frac{i}{p}\right)^p = \left[\left(1 + \frac{i}{p}\right)^{p/i}\right]^i = e^i \tag{4.2}$$

since the limit of $(1 + 1/x)^x = e$ where e is the naperian constant 2.71828. From Eqs. (4.1) and (4.2)

$$P = Se^{-in} \tag{4.3}$$

$$S = Pe^{in} \tag{4.4}$$

Thus the factor e^{in} is an operator that moves \$1 n years *with* the calendar at a nominal decimal rate i per year. Similarly, the factor e^{-in} is an operator that moves \$1 n years *against* the calendar at a nominal decimal rate i per year.

Generally, there is no confusion between periodic and continuous interest. However, a bar over a letter will be used, when necessary, to emphasize that continuous interest or continuous flow is intended. Thus, in keeping with the terminology of Chap. 2,

$$F_{PS,\bar{i},n} = e^{in} \tag{4.5}$$

$$F_{SP,\bar{i},n} = e^{-in} \tag{4.6}$$

The factor $F_{PS,\bar{i},n}$ converts (compounds) \$1 to a future value n years from now at nominal decimal rate i per year and is tabulated in Table A2.1. The factor $F_{SP,\bar{i},n}$ is used for discounting and is tabulated in Table A2.2. Continuous compounding gives compact tables since in appears as a product and not separately, as in periodic interest tables of Appendix 1.

Note that for values of i less than 0.01 $e^i = 1 + i$, and that plus other properties of exponentials permits ready extension of Tables A2.1 and A2.2. For example,

$$e^{10.613} = e^5 e^5 e^{0.61} e^{0.003} = 148.41(148.41)(1.8404)(1.0030) = 40{,}666$$

In this chapter i is the decimal annual rate. The discounting or compounding interval determines the effective annual rate. The effective interest rate is the interest rate which, if used for periodic once-a-year compounding, gives the same end-of-year amount as a more frequent compounding rate. Thus, $1 at decimal yearly effective rate i will amount to

$$1 + i_{\text{eff}} \tag{4.7}$$

at the end of the year. With a decimal continuous rate i_c the value of $1 at the end of one year, by Eq. (4.5), is

$$e^{i_c}$$

For the two to be equal (equivalence),

$$e^{i_c} = (1 + i_{\text{eff}})$$

$$i_{\text{eff}} = e^{i_c} - 1 \tag{4.8}$$

$$i_c = \ln (1 + i_{\text{eff}}) \tag{4.9}$$

where ln refers to natural logarithm. Equations (4.8) and (4.9) permit conversion from continuous to *equivalent* effective rates, and the opposite. Some equivalent values are:

```
 5 percent annual =  4.9 percent continuous
10 percent annual =  9.5 percent continuous
20 percent annual = 18.2 percent continuous
30 percent annual = 26.2 percent continuous
40 percent annual = 33.7 percent continuous
```

Figure 4.1 shows a comparison of equivalent annual and continuous rates. Note that for rates below 15 percent there is very little difference between annual and continuous compounding, but for rates above 30 percent the difference becomes significant.

4.3 Uniform Flow

In the previous section, compounding and discounting were performed on a single amount. In this section the operations will be performed on a continuous flow. Suppose that an amount flows at the rate $\bar{R}$ per year for n years. Consider a small interval of time dX starting X years from now, as in Fig. 4.2. The flow during this interval is given by rate multiplied by time and is $\bar{R}\,dX$. The present value for this small element of time, from Eq. (4.3), is

$$P_{\text{elem}} = \bar{R}\,dX\,e^{-iX}$$

and for all the elements

$$P = \bar{R} \int_0^n e^{-iX}\,dX = \bar{R}\left[\frac{e^{-iX}}{-i}\right]_0^n = \bar{R}\,\frac{1 - e^{-in}}{i} \tag{4.10}$$

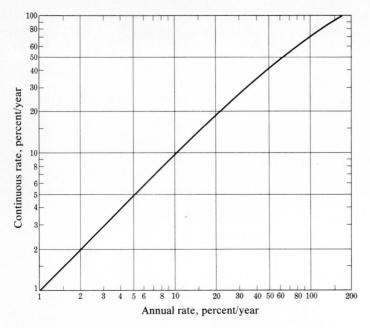

Figure 4.1 Equivalent annual and continuous rates.

If the relationship above is multiplied and divided by n, it becomes

$$P = n\bar{R}\left[\frac{1 - e^{-in}}{in}\right] \tag{4.11}$$

The value $n\bar{R}$ is the *total flow* for the period. The factor within the brackets now appears as a function of in only and can be tabulated compactly. In the terminology of this book

$$F_{\overline{RP},\bar{i},n} = \frac{1 - e^{-in}}{in} \tag{4.12}$$

and

$$F_{\overline{PR},\bar{i},n} = \frac{in}{1 - e^{-in}} \tag{4.13}$$

The factor $F_{\overline{RP},\bar{i},n}$, which converts $n\bar{R}$ to P, is tabulated in Table A2.3, as the evaluation of $(1 - e^{-x})/x$, where $x = in$.

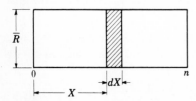

Figure 4.2 Discounting a uniform flow.

Example 4.1 A mine is expected to yield a cash income after taxes of $26,000/year continuously for the next 15 years. If the minimum acceptable rate of return on investment is 12 percent per year after taxes, find the maximum amount that can be economically justified for buying the mine.

SOLUTION By Eq. (4.11) and Table A2.3:

$$P = (n\bar{R})F_{\bar{R}P, 0.12, 15} = 15(26,000)(0.4637) = \$180,843$$

Example 4.2 If $1/day is invested as received at 8 percent per year interest, what will the sum be in 15 years?

SOLUTION First, find the present value of the uniform flow of $365/year by Eq. (4.11) and Table A2.3:

$$P = 15(365)F_{\bar{R}P, 0.08, 15} = 15(365)(0.5823) = \$3188$$

Next, convert to a future value by Eq. (4.4) and Table A2.1:

$$S = 3188e^{(0.08)(15)} = 3188(3.3201) = \$10,584$$

Example 4.3 Parents plan to save enough money to send their child to college. How much must they invest monthly in 5 percent per year continuous-interest tax-free bonds to accumulate the $20,000 they believe will be needed 17 years hence?

SOLUTION Present value of the $20,000 needed is, by Eq. (4.3) and Table A2.2,

$$P = 20,000e^{-(0.05)(17)} = 20,000(0.4274) = \$8548$$

which, in turn, can be converted to a uniform flow by Eq. (4.11) and Table A2.3:

$$8548 = 17\bar{R}F_{\bar{R}P, 0.05, 17} = 17\bar{R}(0.6736)$$

$$\bar{R} = \$746.47/\text{year}$$

$$\frac{746.47}{12} = \$62.21/\text{month}$$

4.4 Flow Changing at an Exponential Rate

In practice, the cash flow will not be uniform from year to year. If the flow changes at an exponential rate, increasing or decreasing, the use of continuous compounding of interest leads to compact relationships for the present value P.

If an initial flow of R_0 dollars per year increases continuously at a rate g per year, expressed as a decimal, the rate of flow at any time X by analogy with Eq. (4.4) is

$$R_x = R_0 e^{gx}$$

Consider a small time interval dX starting X years from now, as in Fig. 4.3. The flow for this interval is $R_0 e^{gX} dX$. The present value for this small element of flow from Eq. (4.3) is

$$P_{\text{elem}} = R_0 e^{gX} dX e^{-iX} = R_0 e^{(g-i)X} dX$$

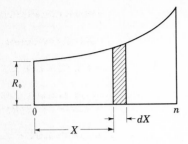

Figure 4.3 Discounting a flow changing at an exponential rate.

and for all the elements

$$P = R_0 \int_0^n e^{(g-i)X} \, dX = R_0 \frac{e^{(g-i)n} - 1}{g - i}$$

$$= nR_0 \frac{1 - e^{-(i-g)n}}{(i - g)n} = (nR_0)F_{\overline{RP}, \overline{i-g}, n} \tag{4.14}$$

Thus, the present value is easily calculated from a knowledge of the initial flow rate and from the factors tabulated in Table A2.3, with $(i - g)n$ replacing in as the argument.

When the rate of increase of g is greater than the interest rate i, $(i - g)$ becomes negative and Table A2.3 cannot be used directly, but Eq. (4.14) still applies.

Example 4.4 Repeat Example 4.1, but with a forecast that inflation will raise prices 3 percent per year continuously.

SOLUTION By Eq. (4.14) and Table A2.3

$$P = nR_0 F_{\overline{RP}, 0.12-0.03, 15} = 15(26,000)(0.5487) = \$213,993$$

Example 4.5 Repeat Example 4.1 with the condition that the mine will become gradually depleted so that its net income declines at the rate of 5 percent per year.

SOLUTION Here $i = 0.12$ and $g = -0.05$. Thus $i - g = 0.12 - (-0.05) = 0.17$. By Eq. (4.14)

$$P = 15(26,000)F_{\overline{RP}, 0.17, 15} = 15(26,000)(0.3615) = \$140,985$$

Example 4.6 Repeat Example 4.1 subject to both an inflation rate of 3 percent per year and a depletion rate of 5 percent per year.

SOLUTION Here $i - g = 0.12 - 0.03 - (-0.05) = 0.14$, and Eq. (4.14) becomes:

$$P = 15(26,000)F_{\overline{RP}, 0.14, 15} = 15(26,000)(0.4179) = \$162,981$$

4.5 Flow Declining in a Straight Line to Zero

Consider Fig. 4.4, in which an initial flow, R_0 dollars per year, declines to zero by a straight-line relationship in n years. At time X the flow is R_x, and by similar

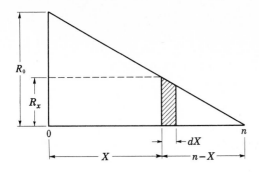

Figure 4.4 Discounting a flow declining in a straight line to zero.

triangles

$$\frac{R_x}{n-X} = \frac{R_0}{n}$$

or

$$R_x = R_0\left(1 - \frac{X}{n}\right) \tag{4.15}$$

In a small time interval dX starting X years from now, the flow for the interval is

$$R_x\,dX = R_0\left(1 - \frac{X}{n}\right)dX$$

and the present value for this small element of flow is, from Eq. (4.3),

$$P_{\text{elem}} = R_0\left(1 - \frac{X}{n}\right)dX\,e^{-iX}$$

For all the elements

$$P = R_0\int_0^n\left(1 - \frac{X}{n}\right)e^{-iX}\,dX = R_0\int_0^n e^{-iX}\,dX - \frac{R_0}{n}\int_0^n Xe^{-iX}\,dX \tag{4.16}$$

The first integral on the right has already been evaluated and is

$$R_0\frac{1 - e^{-in}}{i}$$

Tables of integrals show

$$\int Xe^{-aX}\,dX = -\frac{e^{-aX}}{a^2}(aX + 1)$$

so that the second integral on the right of Eq. (4.16) is

$$-\frac{R_0}{n}\left[-\frac{e^{-iX}}{i^2}(iX + 1)\right]_0^n = \frac{R_0}{i}\left(e^{-in} + \frac{e^{-in}}{in} - \frac{1}{in}\right)$$

The combined integrals on the right of Eq. (4.16) become

$$P = \frac{R_0}{i}\left(1 - e^{-in} + e^{-in} + \frac{e^{-in}}{in} - \frac{1}{in}\right) = \frac{R_0}{i}\left(1 - \frac{1 - e^{-in}}{in}\right)$$

The latter can be written

$$P = \frac{nR_0}{2}\frac{2}{in}\left(1 - \frac{1 - e^{-in}}{in}\right) \tag{4.17}$$

The total flow Q is the area of Fig. 4.4 and is $nR_0/2$. Finally, Eq. (4.17) becomes

$$P = Q\left[\frac{2}{in}\left(1 - \frac{1 - e^{-in}}{in}\right)\right] \tag{4.18}$$

A table of discount factors for such a flow is the evaluation of the bracketed terms on the right; i.e.,

$$\frac{2}{x}\left(1 - \frac{1 - e^{-x}}{x}\right) \qquad \text{with } x = in$$

and is tabulated in Table A2.4. This type of flow approximates sum-of-the-years-digits (SD) depreciation and in symbols is

$$F_{\text{SDP},\bar{i},n} = \frac{2}{in}\left(1 - \frac{1 - e^{-in}}{in}\right) \tag{4.19}$$

Table A2.4 is commonly referred to as the *years-digits table*.

Example 4.7 A machine costs $150,000 and can be depreciated over 20 years by sum-of-the-years-digits depreciation. Find the present value of the depreciation, before taxes, if the discount rate is 16 percent per year.

SOLUTION By Eqs. (4.18) and (4.19) and Table A2.4:

$$P = QF_{\text{SDP},\bar{i},n} = 150,000F_{\text{SDP},0.16,20} = 150,000(0.4376)$$
$$= \$65,640$$

4.6 Equivalent Uniform Flow

In Sec. 4.3 a uniform flow was converted to a present value. The inverse of that procedure, the conversion of a present value to a uniform flow, will be considered in this section. Solving Eq. (4.11) for $\bar{R}$ gives

$$\bar{R} = \frac{P}{n}\frac{1}{(1 - e^{-in})/in} \tag{4.20}$$

which, by Eq. (4.12), becomes

$$\bar{R} = \frac{P}{n}\frac{1}{F_{\overline{RP},\bar{i},n}} \tag{4.21}$$

Equations (4.20) and (4.21) are important. They permit transformation of a present value P having n years' duration to a uniform flow. $\bar{R}$ will be referred to as a *uniform flow*. $\bar{R}$ can be called the *continuous capital-recovery amount*. It can also be called *unaflow* and is analogous to unacost in periodic compounding.

Equivalent annual flow is important since it permits a comparison when service lives are different. All articles or systems are reduced to a cost for one year expressed as an equivalent annual flow.

Example 4.8 A firm has the option of getting a patent license by making either a single payment of $50,000 or royalty payments of $5000/year for the 17-year life of the patent. If the payments can be expensed in either case, and if the firm earns 15 percent per year after taxes, which is the more attractive choice?

SOLUTION Uniform flow for royalty payments is $5000/year, as given. Uniform flow for purchase of a patent, by Eq. (4.21) and Table A2.3, is

$$\bar{R} = \frac{50,000}{17} \frac{1}{F_{\bar{R}P, 0.15, 17}} = \frac{50,000}{17} \frac{1}{0.3615} = \$8136$$

The annual royalties of $5000 are thus cheaper for this firm. The ratio of costs, purchase to lease, is $8136/5000 = 1.6272$. The calculation assumes that in both cases the tax depreciation will be taken at a uniform rate and will cancel out as affecting both alternatives equally.

Example 4.9 A $15,000 mortgage is to be repaid over 20 years at 6 percent per year interest. Find the monthly payments.

SOLUTION By Eq. 4.21 and Table A2.3, unaflow is

$$\bar{R} = \frac{15,000}{20} \frac{1}{F_{\bar{R}P, 0.06, 20}} = \frac{15,000}{20} \frac{1}{0.5823} = \$1288$$

That is, the flow must be $1288 per year, or

$$\frac{\$1288}{12} = \$107.33/\text{month}$$

4.7 Capitalized Cost

Capitalized cost, like uniform annual flow, can be used to compare articles or systems having different service lives. It reduces all service lives to a common denominator, i.e., present value on the basis, for mathematical purposes, of service forever.

Consider an article that has an initial cost C and lasts n years. The present value of supplying service forever is

$$P_\infty = Ce^{-i0} + Ce^{-in} + Ce^{-2in} + Ce^{-3in} + \cdots$$

which is an infinite geometrical series with the first term C and the ratio e^{-in}. The sum is given by Eq. (2.14), and letting $P_\infty = K$,

$$K = \left[\frac{1 - (e^{-in})^\infty}{1 - e^{-in}}\right] C = \frac{1}{1 - e^{-in}} C \tag{4.22}$$

The bracketed term on the right converts a present value of n years' duration to a capitalized cost; i.e.,

$$K = P_n \frac{1}{1 - e^{-in}} \tag{4.23}$$

or
$$K = P_n \frac{e^{in}}{e^{in} - 1} \tag{4.24}$$

where the symbol P_n emphasizes that P is a present value representing n years' duration.

Equations (4.23) and (4.24) are important because they are the basis for using the capitalized-cost concept with continuous interest. The reader is referred to Chap. 2 for a more complete discussion of capitalized cost.

A relationship between capitalized cost K and uniform flow $\bar{R}$ is easily derived. The present value of a uniform flow $\bar{R}$ for n years is, by Eq. (4.11),

$$P_n = n\bar{R} \frac{1 - e^{in}}{in}$$

and the capitalized cost of this present value becomes, by Eq. (4.23),

$$K = n\bar{R} \frac{1 - e^{-in}}{in} \frac{1}{1 - e^{-in}} = \frac{\bar{R}}{i}$$

That is,

$$\bar{R} = iK \tag{4.25}$$

Equation (4.25) for continuous interest and unaflow is analogous to the corresponding relationship $R = iK$, Eq. (2.35), for periodic interest.

Example 4.10 Repeat Example 4.8 on the basis of capitalized cost.

SOLUTION Capitalized cost of the royalty payments, by Eq. (4.25), is

$$K = \frac{\bar{R}}{i} = \frac{5000}{0.15} = 33,333$$

Capitalized cost of purchase is given by Eq. (4.24), which, using Table A2.1, becomes

$$K = 50,000 \frac{e^{(0.15)(17)}}{e^{(0.15)(17)} - 1} = 50,000 \frac{12.807}{12.807 - 1} = 54,235$$

It is cheaper to pay the royalties. The ratio of costs, purchase to lease, is $54,235/33,333 = 1.6271$. This checks the calculation by uniform flow in Example 4.8.

Example 4.11 In a given exposure, a paint job lasts 4 years and costs $0.20/ft². A supplier offers a new coating which is claimed to last 20 years but costs $0.60/ft². Is it economically attractive to change to the coating which lasts five times as long and costs only three times as much, if money is worth 10 percent? Neglect taxes.

SOLUTION Capitalized costs can be calculated from Eq. (4.24) and Table A2.1 and are, for the 4- and 20-year jobs, respectively,

$$K_4 = \$0.20 \frac{e^{(0.10)(4)}}{e^{(0.10)(4)} - 1} = 0.20 \frac{1.4918}{0.4918} = 0.6067$$

$$K_{20} = \$0.60 \frac{e^{(0.10)(20)}}{e^{(0.10)(20)} - 1} = 0.60 \frac{7.3891}{6.3891} = 0.6939$$

The 4-year coating is more economical. The saving as an annual flow per year per square foot can be obtained from Eq. (4.25) and is

$$\bar{R} = i(K_{20} - K_4) = 0.10(0.6939 - 0.6067) = 0.00872$$

That is, use of the 4-year coating saves $0.00872/year $\cdot$ ft^2 in comparison with the 20-year coating.

4.8 Income Tax

The reader is referred to Chap. 3 for a detailed discussion of the effect of income tax using periodic interest. This section is concerned with the inclusion of income tax using continuous interest. A before-tax amount will always transform to an after-tax amount using the expression

$$(\text{After-tax amount}) = (\text{before-tax amount})(1 - t\psi) \qquad (4.26)$$

where t is the decimal tax rate and ψ is the present value of $1 of depreciation. The value for ψ depends upon the discounting rate and how it is taken, the depreciation method, the useful life of the article for tax purposes, and the timing of tax payments. If the tax timing and expense or receipt are simultaneous, the $\psi = 1$; otherwise ψ will be less than unity.

In practice, income tax is usually paid on a quarterly basis, which agrees neither with yearly periodic discounting and a single end-of-year tax payment nor with continuous discounting and continuous tax payment. However, in practice, there is little error in using approximations. In particular, since comparisons are always being made, simplifying assumptions tend to balance out. For this section, it is necessary only to summarize the ψ factor for some circumstances using continuous interest.

For a uniform yearly flow $\bar{R}$, assuming a continuous tax applied simultaneously with the continuous flow,

$$\psi = 1$$

For straight-line depreciation and continuous tax, the tax benefit is a uniform flow and the ψ factor is related to Eq. (4.11):

$$\psi = F_{\text{SLP},\bar{i},n} = F_{\overline{RP},\bar{i},n} = \frac{1 - e^{-in}}{in} \qquad (4.27)$$

which is tabulated in Table A2.3.

For an approximation to sum-of-the-years-digits depreciation and continuous tax, the ψ factor was given by Eq. (4.19) and is tabulated in Table A2.4.

The ψ factor can always be calculated from its definition—the present value of $1 of future depreciation. For example, calculate the ψ factor for sum-of-the-years-digits depreciation using continuous interest but a periodic tax at the end of the year for a useful life of n years. Sum-of-the-years-digits is $n(n + 1)/2$, and the present value is

$$P = \frac{2}{n(n + 1)}[ne^{-i} + (n - 1)e^{-2i} + (n - 2)e^{-3i} + \cdots + e^{-in}] \qquad (4.28)$$

The summation can be done by special techniques, giving

$$\psi = \frac{2}{n(n+1)(e^i - 1)}\left(n - \frac{1 - e^{-in}}{e^i - 1}\right) \tag{4.29}$$

Equation (4.28) is not exactly the same as Eq. (4.19). They both use continuous interest, but Eq. (4.19) uses an approximation to sum-of-the-years-digits depreciation, and they differ in the timing of the tax.

Example 4.12 A $1000 investment has an expected life of 20 years and is to be depreciated over a 15-year life using sum-of-the-years-digits depreciation with money worth 10 percent per year after a 46 percent tax. Find (a) the present value of the capital charge after taxes and (b) the equivalent annual flow.

SOLUTION
(a) By Eqs. (4.26) and (4.19) and Table A2.4:

$$P = \$1000(1 - 0.46F_{\text{SDP},0.10,15}) = \$1000[1 - 0.46(0.6438)] = \$704.31 \tag{4.30}$$

(b) The after-tax present value given above can be converted to an after-tax uniform flow by Eq. (4.11) and Table A2.3:

$$P = \$704.31 = n\bar{R}\left(\frac{1 - e^{-in}}{in}\right) = 20\bar{R}F_{P\bar{R},0.10,20}$$

$$\$704.31 = 20\bar{R}(0.4323)$$

$$\bar{R} = \$81.46 \quad \text{(after tax)} \tag{4.31}$$

$\bar{R}$ in Eq. (4.31) is figured on an after-tax basis. The ψ factor is 1 for a uniform flow; hence $\bar{R}$ before tax can be calculated from

$$\bar{R}(1 - 0.46) = \$81.46$$

$$\bar{R} = \$150.85/\text{year} \quad \text{uniform annual flow before tax}$$

4.9 Equivalence

The purpose of this chapter, using continuous interest, and that of previous chapters, using periodic interest, has been to demonstrate how to compare two or more systems differing in receipts and expenses, as well as their timing, the evaluation of present value after taxes, and the conversion of the present values to some common denominator of economic life, such as equivalent uniform flow or capitalized cost, to correct for different durations. Comparisons can also be made on the basis of rate of return, discounted cash flow, as developed in the following chapter.

The method to be used for comparing alternatives or ventures, and the choice between periodic and continuous interest, are left to the analyst. Although an economic specialist may prefer one method, the client is often a manager who is a generalist by necessity. Results must be presented in terms familiar to such a client and must be simple enough to be grasped on the run. Experience shows that if a solution is presented in unfamiliar or seemingly unrealistic terms, it will not be understood; if not understood, it will not be believed; and if not believed, it will not be accepted.

In practice, real problems are complex, not so much in computation as in defining what the cash flows will be. Often 95 percent of the total time in solving a problem is required for determining costs and incomes, the applicable tax and other government regulations, and projecting sales, costs, and so on.

A summary of the various relationships using continuous interest is given in Table 4.2. The ψ factor, the present value of $1 of future depreciation, is given in Table 4.3.

One article [5] attempts to coordinate periodic and continuous interest. It shows that both methods of compounding will give equivalent results for a comparison between alternatives if the three following aspects are recognized:

1. Continuous and periodic interest, when equivalent, are not alike numerically. The equivalency is given by Eq. (4.8).
2. A flow of funds may be continuous or periodic, and the nature of the flow must be considered independently of the method used to compound the rate of return.
3. The timing of the tax situation must be consistent in both methods.

A simple example will show what is meant by equivalency and how relationships between continuous and periodic interest can be made compatible.

Example 4.13 Derive the relationship for ψ with continuous interest as given by Table 4.3, item 3, from the known relationship with yearly compounding given by Eq. (3.11).

SOLUTION Equation (3.11) can be written

$$\psi = \frac{1}{n} \frac{(1 + i_{\text{eff}})^n - 1}{i_{\text{eff}}(1 + i_{\text{eff}})^n}$$

Substituting from Eq. (4.8) with i used for i continuous gives

$$\psi = \frac{1}{n} \frac{e^{in} - 1}{(e^i - 1)e^{in}} = \frac{1}{n} \frac{1 - e^{-in}}{e^i - 1}$$

as given in the table. Here the ψ factor is also given by a geometric series which can be summed by Eq. (2.14):

$$\psi = \frac{1}{n} \sum_{n=1}^{n=n} e^{-in} = \frac{1}{n} \frac{1 - e^{-in}}{e^i - 1}$$

Table 4.2 Summary of relationships for continuous interest

Item no.	Item	Description	Algebraic relationship	Factor relationship
1	P to S	Moves a fixed sum P to another instant of time n years with the calendar	$S = Pe^{in}$	$S = PF_{PS,\bar{i},n}$ (Table A2.1)
2	S to P	Moves a fixed sum S to another instant of time n years against the calendar	$P = Se^{-in}$	$P = SF_{SP,\bar{i},n}$ (Table A2.2)
3	$\bar{R}$ to P	Converts a uniform flow $\bar{R}$ for n years to present value at the start of the flow	$P = n\bar{R}\dfrac{1 - e^{-in}}{in}$	$P = n\bar{R}F_{RP,\bar{i},n}$ (Table A2.3)
4	$\bar{R}$ for 1 year to P	Present value of 1 year of uniform flow starting X years hence	$P = \bar{R}e^{-ix}\dfrac{1 - e^i}{i}$	$P = \bar{R}F_{SP,\bar{i},X}F_{RP,\bar{i},1}$
5	P of flow changing at an exponential rate for n years	Present value of $$R_z = R_0 e^{\pm gz}$$ for n years	$P = n\bar{R}_0\dfrac{1 - e^{-(i \mp g)n}}{i \mp g}$	$P = nR_0 F_{RP,(\bar{i} \mp g),n}$
6	P of flow declining in a straight line to zero	Flow goes from $\bar{R}_0$ at zero time to zero in n years; total flow Q is $nR_0/2$	$P = Q\left[\dfrac{2}{in}\left(1 - \dfrac{1 - e^{-in}}{in}\right)\right]$	$P = QF_{SDP,\bar{i},n}$ (Table A2.4)
7	P to $\bar{R}$	Converts a present value to a uniform flow of n years	$\bar{R} = \dfrac{P}{n}\dfrac{1}{(1 - e^{-in})/in}$	$R = \dfrac{P}{n}\dfrac{1}{F_{RP,\bar{i},n}}$
8	$\bar{R}$ to S	Converts a uniform flow for n years to a future amount n years hence	$S = n\bar{R}e^{in}\dfrac{1 - e^{-in}}{in}$	$S = n\bar{R}F_{PS,\bar{i},n}F_{RP,\bar{i},n}$
9	S to $\bar{R}$	Converts a future sum S, n years from now, to a uniform flow; sinking-fund payment	$\bar{R} = \dfrac{Se^{-in}}{n}\dfrac{1}{(1 - e^{-in})/in}$	$R = \dfrac{S}{n}F_{PS,\bar{i},n}\dfrac{1}{F_{RP,\bar{i},n}}$
10	P to K	Converts a present value representing n years to a capitalized cost	$K = P\dfrac{e^{in}}{e^{in} - 1}$	
11	K to $\bar{R}$	Converts a capitalized cost to a uniform flow	$\bar{R} = iK$	

Table 4.3 The ψ factor, the present value of \$1 of depreciation

Item no.	Type of depreciation	Tax timing	$\psi_{i,n}$ with continuous interest
1	Instantaneous	Instantaneous	1
2	Uniform	Continuous	$\dfrac{1 - e^{-in}}{in}$ (Table A2.3)
3	Uniform	Periodic, end of year	$\dfrac{1 - e^{-in}}{n(e^i - 1)}$
4	Flow declining in a straight line to zero; simulated sum-of-the-years-digits	Continuous	$\dfrac{2}{in}\left(1 - \dfrac{1 - e^{-in}}{in}\right)$ (Table A2.4)
5	Sum-of-the-years-digits	Periodic, end of year	$\dfrac{2}{n(n + 1)(e^i - 1)}\left(n - \dfrac{1 - e^{-in}}{e^i - 1}\right)$
6	Declining balance without switch	Periodic, end of year	$\dfrac{F_{DB}}{e^i - (1 - F_{DB})}\left[1 - \left(\dfrac{1 - F_{DB}}{e^i}\right)^n\right]$

NOMENCLATURE

C_d	Depreciable first cost, \$
e	Naperian constant 2.71828 $\cdots$
$F_{P\bar{R},\bar{i},n}$	Factor to convert P to $\bar{R}$ with continuous compounding; reciprocal of $F_{\bar{R}P,\bar{i},n}$, year^{-1}
$F_{PS,\bar{i},n}$	Factor to convert P to S with continuous compounding, e^{in}, Table A2.1, dimensionless
$F_{\bar{R}P,\bar{i},n}$	Factor to convert $\bar{R}$ to P with continuous discounting; Table A2.3, years
$F_{SP,\bar{i},n}$	Factor to convert S to P with continuous discounting, e^{-in}, Table A2.2, decimal, dimensionless
$F_{SDP,\bar{i},n}$	Factor to convert a unit total flow declining to zero at a constant rate over n years starting with the reference point and with continuous discounting, decimal, dimensionless, approximates ψ_{SD}
g	Constant in exponential-rate flow change, decimal
i	Interest rate, decimal per year; generally nominal rate
i_c	Interest rate, specifically emphasizing continuous compounding, decimal per year
i_{eff}	Effective interest rate, decimal per year
i_p	Interest rate, specifically emphasizing periodic compounding, decimal per year
K	Capitalized cost
n	Time, years
n'	Time for tax depreciation, years
p	Periods per year
P	Present value, \$
P_n	Present value for n years' duration, \$
Q	Total flow, \$
r	Nominal rate of return after taxes, decimal per year
$\bar{R}$	Uniform flow, \$/year
R_0	Initial flow rate, \$/year
R_x	Flow rate at time x, \$/year
S	Future value, \$
SD	Sum-of-the-years-digits depreciation
SL	Straight-line depreciation
t	Income tax rate, decimal
T	Time, years

ψ Factor representing present value of \$1 of future depreciation

ψ_{SD} The ψ factor for sum-of-the-years-digits depreciation

ψ_{SL} The ψ factor for straight-line depreciation

PROBLEMS

4.1 Develop a relationship for discounting a flow increasing in a straight line from zero at zero time to $\bar{R}_n$ at time n.

4.2 Develop a relationship for discounting a flow increasing in a straight line from $\bar{R}_1$ at zero time to $\bar{R}_2$ at time n.

4.3 Develop a relationship for discounting a series of periodic cash flows of k payments, Y each, at intervals of n years, the first one beginning n years hence.

4.4 A firm has a contributory savings plan whereby each employee can set aside 5 percent of his or her gross salary. The firm will match this amount, invest the sums in its capital stock, and reinvest all dividends in capital stock. If an employee's salary is consistently \$12,000 per year, how much will the employee accumulate after 20 years if the company's net earnings average 8 percent per year and the stock consistently sells at book value?

4.5 What is the average rate of growth of the employee's \$600 per year portion of the contribution?

4.6 Suppose the employee finds an alternative proposition which promises to double the money every 5 years. Is it better to participate in the savings plan or forgo the company's contribution and invest the employee's contribution in the alternative?

4.7 If the parents in Example 4.3 continue their monthly savings during the 4 years their child attends college, i.e., for 21 instead of 17 years, how much must their monthly savings be to permit \$3000 per year to be withdrawn uniformly over the 4 years from the seventeenth to the twenty-first birthday?

4.8 A new machine costs \$8000 and lasts 10 years, using sum-of-the-years-digits depreciation and a 10-year life for tax purposes. If money is worth 10 percent per year after a 46 percent tax, how much can be spent to repair an old machine to extend its life 3 years? The repair job can be written off at once for tax purposes. Compare with Prob. 3.20.

4.9 Repeat Example 2.21 using continuous discounting. Money is worth 10 percent per year after a 46 percent tax rate. Use straight-line depreciation. Machine A will be written off in 8 years for tax purposes, machine B in 10 years. Maintenance costs and savings from quality control are uniform flows in the years in which they occur. The salvage value is anticipated and cannot be depreciated for tax purposes.

	A	B
First cost, \$	10,000	95,000
Maintenance, \$ per year	3,000	1,000
Extra maintenance, year 3, \$	4,000	
Extra maintenance, year 4, \$	1,500	
Savings from quality control, \$ per year		6,000
Salvage value		20,000
Life, years	4	10

4.10 A company completed a plant 10 years ago. It was expected to be serviceable for 25 years, but technical advances and accumulated know-how suggest that obsolesence may have progressed faster than expected, so that it may be profitable to replace it now. Assume, for simplicity, that (1) a new plant would have the same capacity as the old and would produce the same array of products with the same initial revenue for both, so that the advantage of the new plant is reflected only in its lower operating costs; (2) these savings in operating costs are $180,000 per year; (3) depreciation on the old plant is $35,000 per year on a straight-line basis, and present salvage value is zero; (4) the tax rate is 46 percent per year; and (5) any loss on disposal can be taken immediately for tax purposes. If the investment required for the new plant is $1,000,000 and both its economic life and its life for tax purposes are 15 years with sum-of-the-years-digits depreciation, what is the rate of return to be earned by investment in a new plant?

4.11 Establish the relationship for the ψ factor using continuous interest and a periodic end-of-year tax given by Table 4.3, item 5.

4.12 An item will be depreciated over a 10-year life with money worth 15 percent per year after a 46 percent tax. Use continuous discounting. Compare the ψ factors obtained from Table 4.3, item 4, for simulated sum-of-the-years-digits depreciation and item 5 for periodic end-of-year depreciation. Also, compare $(1 - t\psi)$ with $t = 0.46$ since that is the expression in which ψ will be used.

4.13 Let R_e be an equivalent end-of-year amount and R_c a continuous uniform flow during the year, compounded with continuous interest at nominal rate i. For equivalency, show that

$$R_e = \frac{R_c}{i}(e^i - 1)$$

4.14 Some analysts use a continuous uniform flow for 1 year as if it were a single total flow at midyear, and use simple interest for $\frac{1}{2}$ year to obtain a single end-of-year amount. Let the flow be $1000/year and the periodic interest rate 20 percent per year.

(a) Find the single amount at the end of the year.

(b) The analyst now recalculates the amount using the continuous interest rate equivalent to the periodic rate of 20 percent but continuous flow of the $1000. What single amount is found as of the end of that year?

4.15 A manufacturer currently makes a profit M of $100,000/year on an item. Money is worth 8 percent per year, and interest is to be compounded continuously. Twenty years of operation is anticipated, and there is no income tax.

(a) Find the present value of the profit.

(b) Because of increased productivity, the profit per year increases continuously and can be expressed as

$$M = 100,000(2 - e^{-0.1T})$$

where T is time in years. Find the present value of the profit.

(c) Productivity increases as in (b), but competition reduces the profit continuously by 3 percent per year. Find the present value of the profit.

REFERENCES

1. Hirschmann, W. B.: "Profit from the Learning Curve," *Harvard Business Review*, January–February 1964, pp. 125–139.
2. Hirschmann, W. B., and J. R. Brauweiler: "Investment Analysis: Coping with Change," *Harvard Business Review*, May–June 1965, pp. 62–72.
3. Hirschmann, W. B., and J. R. Brauweiler: "Continuous Discounting for Realistic Investment Analysis," *Chemical Engineering*, July 19, 1965, pp. 210–214.
4. Hirschmann, W. B., and J. R. Brauweiler: "Realistic Investment Analysis, II," *Chemical Engineering*, August 16, 1965, pp. 132–136.
5. Jelen, F. C., and C. L. Yaws: "Unify Interest Compounding," *Hydrocarbon Processing*, vol. 60, April 1981, pp. 223–226.

FIVE

PROFITABILITY

F. C. Jelen

5.1 Nature of Profitability

In a broad sense, profitability is a measure of the total income for a project compared to the total outlay. Profitability can be applied to a short period of time, say, year by year, or to the entire life of a project. It can be applied to a single cost center, an entire project, or an organization as a whole. Profitability always applies to a monetary measure, although intangible benefits and costs must not be ignored. Inherently, a measure of profitability should be adjusted for size, such as the profit per dollar invested, rather than just the profit.

In this chapter criteria for profitability will be developed, followed by a discussion of profitability and some associated topics.

5.2 Criteria for Profitability

The major classifications for profitability are given in Table 5.1, and there are several variations for some of the entries. The criterion in group I is expressed as time, such as time to recover the investment. Group II uses rate of return as a criterion but does not recognize the timing of cash flows. The discounted-cash-flow rate of return in group IV also uses rate of return as a criterion but does recognize the timing of cash flows. Group III, based on present value, recognizes the timing of cash flows but expresses the profitability as a premium (or deficiency) net present value compared to that at some referenced rate of return.

Table 5.1 Criteria for profitability

I. Payout time
   A. Without interest
   B. With interest
   C. Equivalent maximum investment period
   D. Positive equivalent recovery period

II. Return on investment
   A. Return on original investment
   B. Return on average investment

III. Present value
   A. Net present value
   B. Excess present-value index

IV. Discounted-cash-flow rate of return

5.3 A Simple Project

The classifications in Table 5.1 will now be applied to a simple project, A, having cash flows, as in Table 5.2. The cash flow here is the sum of the profit after taxes plus depreciation and represents an in-pocket amount after taxes. Amounts are in kilodollars. Money going into the project is taken as negative and money coming back from the project as positive, following the customary designations in profitability studies.

    Project A involves swapping 1000 k$ invested at zero time for the after-tax amounts given by years in the last column, starting with 475 k$ at the end of year 1. The numbers in Table 5.2 alone determine the profitability.

5.4 Payout Time without Interest

Payout time is also known as *payback time*. It is the time required to reduce the investment to zero.

    Example 5.1 Find the payout time without interest for the data in Table 5.2.
        The tabulation on top of page 91 shows that the investment will be reduced to zero between 2 and 3 years, or approximately 2.4 years by interpolation.

Table 5.2 Cash flow for project A

Time, end year	After-tax profit, k$	Depreciation, k$	Cash flow, k$
0	−1,000	0	−1,000
1	275	200	475
2	200	200	400
3	130	200	330
4	70	200	270
5	0	200	200

Time, end year	Cumulative cash flow, k$
0	− 1,000
1	− 525
2	− 125
3	+ 205
4	+ 475
5	+ 675

The values are plotted in Fig. 5.1.

Payout time without interest does not consider working capital, e.g., inventory, since it is assumed that working capital can be recovered at any time and does not affect the time required to get back the fixed investment.

5.5 Payout Time with Interest

The method allows for a return on investment and is subject to variations. The following variation applies an interest charge only on the remaining fixed investment. A variation also includes an interest charge on working capital but recognizes that working capital itself is fully retrievable at any time.

In practice, payout time with interest is seldom used, and the term *payout time* should mean "without interest" unless interest is specifically mentioned.

Example 5.2 Using the data in Table 5.2, find the payout time with interest on the remaining investment at 10 percent per year.

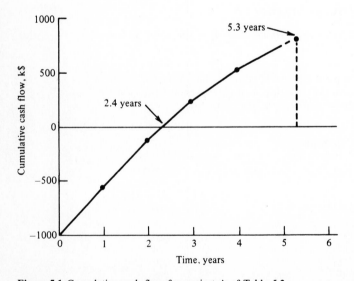

Figure 5.1 Cumulative cash flow for project A of Table 5.2.

SOLUTION The following tabulation shows that the payout time is between 2 and 3 years or, more exactly, 2.95 years by interpolation.

(1) End year	(2) Investment for year	(3) Charge on investment for year, 10% of (2)	(4) Cash flow	(5) Cash flow after investment charge, (4) − (3)	(6) Cumulative net cash flow
0			−1,000		−1,000
1	1,000	100	475	375	−625
2	625	62.5	400	337.5	−287.5
3	287.5	28.75	330	301.25	+13.75

The investment for each year is the investment for the previous year less the cash flow after the investment charge for the previous year.

5.6 Equivalent Maximum Investment Period

Payout time without interest for project A was found to be 2.4 years by the tabulation in Sec. 5.4 or from Fig. 5.1. The value 2.4 years does not disclose how rapidly or slowly the curve approaches zero at this time, i.e., how "fat" the area is between the curve and the zero axis from time 0 to time 2.4 years. The average ordinate between time 0 and 2.4 years provides a measure of how bloated the curve is. The equivalent maximum investment period has been suggested as a criterion for disclosing how rapidly the investment is reduced.

Example 5.3 Using the data in Table 5.2, find the equivalent maximum investment period.

SOLUTION The average investment for the first year is

$$\frac{1000 + 525}{2} = 762.5$$

and for 0 to 2.4 years the cumulative value for the product of investment multiplied by time is

$$\frac{1000 + 525}{2}(1) = 762.5$$

$$\frac{525 + 125}{2}(1) = 325.0$$

$$\frac{125 + 0}{2}(0.4) = \underline{25.0}$$

$$1112.5 \text{ k\$/year}$$

The initial investment of 1000 k\$ has been invested on the average:

$$\frac{1112.5}{1000} = 1.11 \text{ years}$$

The average time of 1.11 years for the initial (maximum) investment supplements the 2.4 years required to get back the initial investment.

5.7 Positive Equivalent Recovery Period

The methods used to calculate payout time discussed up to this point do not include the period after the payout time. One variant that does is known as the *positive equivalent recovery period.*

Refer to Fig. 5.1. The data extend only to 5 years but are extrapolated slightly beyond that point. The area between the curve and the zero axis up to 2.4 years is easily calculated, as in the previous section. By trial, it is found that at 5.3 years the same area exists above the zero axis and the curve from 2.4 to 5.3 years. The time 5.3 years supplements the payout time of 2.4 years. At 5.3 years the positive area of the curve between 2.4 and 5.3 years is equal to the negative area of the curve between 0 and 2.4 years.

5.8 Advantages and Disadvantages of Payout Time

The advantages of payout time are that it is a simple, easily understood method and is commonly used in the business world. The method is particularly adaptable to simple problems, such as replacing an old, inefficient pump that wastes energy with a new, energy-efficient pump. A payout time of 9 months in this case would be an adequate and convincing criterion for purchasing a new pump.

Perhaps the chief disadvantage of payout time is that, as ordinarily used, it gives no credit for the period beyond the payout time. The previous section demonstrated a variant which included the period beyond the payout time, but such variants are not easily understood and are seldom used.

When payout time is used as a criterion on large projects, it can encourage shoddy investment, favor skimpy design, lead to high start-up costs, and promote a policy-inhibiting perspective. In the end, it could lead investors to abandon their investment and become merchandisers for the products of others, since no investment assures the minimum payout time—zero.

5.9 Return on Original Investment (ROI)

Return on original investment, also called the *DuPont* or *engineer's method,* is the percentage relationship of the average annual profit (sometimes the average annual cash flow) to the original investment, including nondepreciable items such as working capital. Thus it is given by

$$\frac{\text{Average yearly profit during earning life}}{\text{Original fixed investment} + \text{working capital}}(100) \tag{5.1}$$

Example 5.4 Using the data in Table 5.2, find the return on original investment.

SOLUTION The average profit is

$$\frac{275 + 200 + 130 + 70 + 0}{5} = 135 \text{ k\$/year}$$

By Eq. (5.1), the return on original investment is

$$\frac{135}{1000 + 0}(100) = 13.5 \text{ percent per year}$$

The time value of money is not considered, since only the average profit is used, not its timing. The profits from years 1 through 5 could be reversed, and the return on original investment would be the same.

5.10 Return on Average Investment (RAI)

The return on average investment is similar to the return on original investment except that the divisor in Eq. (5.1) is the average outstanding investment.

Example 5.5 Using the data in Table 5.2, find the return on average investment.

SOLUTION Average investment is found from the following tabulation:

Year	Investment, $
1	1,000
2	1,000 − 200 = 800
3	800 − 200 = 600
4	600 − 200 = 400
5	400 − 200 = 200
	Av = 600

$$\frac{135}{600}(100) = 22.5 \text{ percent per year}$$

The return on average investment at 22.5 percent is higher, as it must be, than the return on original investment at 13.5 percent. The two are independent criteria and must be used separately. Comparing the ROI of one system with the RAI of another system would be unfair. The two must not be intermixed in a comparison.

5.11 Net Present Value (NPV)

The return on original investment and the return on average investment do not reflect the time value of money. Comparing projects on the basis of present value does include the time value of money, since all cash flows are referred to zero time before an evaluation is made. The original concept was known as *venture worth* but is now generally referred to as *net present value (NPV)*. Reduction to present value, discounting, is done at a fixed rate, usually the minimum acceptable rate of return on capital. The net present value is then the present value of all inflows less the present value of all outflows. Accordingly, the net present value is a single amount referred to zero time and represents a premium if positive, or a deficiency if negative, at some chosen fixed rate of return.

Example 5.6 Using the data in Table 5.2, find the net present value at 10 percent per year and at 25 percent per year.

SOLUTION At 10 percent per year, the net present value is

$$\frac{-1000}{1.10^0} + \frac{475}{1.10^1} + \frac{400}{1.10^2} + \frac{330}{1.10^3} + \frac{270}{1.10^4} + \frac{200}{1.10^5} = +318.92$$

At 25 percent per year, the net present value is

$$\frac{-1000}{1.25^0} + \frac{475}{1.25^1} + \frac{400}{1.25^2} + \frac{330}{1.25^3} + \frac{270}{1.25^4} + \frac{200}{1.25^5} = -18.91$$

The project has a premium present value of 318.92 k\$ at 10 percent per year and a deficiency in present value of 18.91 k\$ at 25 percent per year. The negative present value at 25 percent per year does not mean that the project is a loser, but only that it does not pay off at 25 percent per year.

Net present value is not corrected for size. Thus two of the projects would have a net present value of $2 \times 318.92 = 638.8$ k\$ at 10 percent per year, but fundamentally the profitability has not changed. This objection can be overcome by using an excess present value index defined as

$$\frac{\text{Present value receipts}}{\text{Present value outlays}}$$

which in this case, at 10 percent per year, would be

$$\frac{1318.92}{1000} = 1.319$$

Net present value does not correct for a difference in the service lives of projects. In the present example, at 10 percent per year the net present value for 5 years ahead is 318.92 k\$. If the life of the project is to be 10 years, obtained by repeating the project at the end of 5 years, the net present value is changed numerically to

$$318.92 + \frac{318.92}{(1.10)^5} = 516.94 \text{ k\$}$$

If projects have different durations, the net present value should be converted to an annual cost, such as unacost, or a capitalized cost, both of which have built-in factors for different service lives.

Example 5.7 Compare project A, having an initial investment of 1000 k\$, a 5-year life, and a net present value of 318.92 k\$, with project W, having an initial investment of 1500 k\$, a 7-year life, and a net present value of 675 k\$. Both present values are at 10 percent per year. Correct for the difference in size by referring to an initial investment of 1 k\$, and correct for the difference in duration by converting to a premium unacost.

SOLUTION For project A:

$$\frac{318.92}{1000} F_{PR,\,10\%,\,5} = 0.3192(0.26380) = 0.0842$$

For project W:

$$\frac{675}{1500} F_{PR,\,10\%,\,7} = 0.4500(0.20541) = 0.0924$$

Thus project A has a premium of 0.0842 k\$/year per 1 k\$ of original investment against 0.0924 for project W. Project W is more profitable than project A at 10 percent per year. However, the comparison does depend upon the rate of return used, and may even reverse at another rate of return.

One method for correcting for the duration of a project is to evaluate all projects at some fixed duration. Thus one company compares all projects on a 12-year life. The procedure is acceptable for projects lasting 12 years or longer, since the time beyond 12 years contributes little to the present value at practical rates of return.

Net present value, or any variation thereof, requires only a direct calculation. No trial and error procedure is required since the rate of return is chosen before the calculation is begun. Originally, the avoidance of a trial and error calculation was an advantage. Now, however, with the introduction of computers and the ubiquitous programmable calculator, trial and error calculations are easily made, and some of the advantage of net present value has been lost.

5.12 Discounted Cash Flow Rate of Return (DCFRR)

The discounted cash flow rate of return is a sophisticated criterion for profitability that corrects for differences in the duration and size of projects, includes the time value of money and the timing of all cash flows, and recognizes that investment changes with time. It is also known as the *interest-rate-of-return method, profitability index (PI), internal rate of return,* and *investor's method.* It includes all cash flows over the entire life of the project and adjusts them to one point fixed in time, usually the original investment or start-up time, using the compound-interest procedure; but the calculated rate of return is independent of the point in time chosen for discounting. A trial and error calculation is required to determine the compound interest rate at which the sum of the time-adjusted cash outflows equals the sum of the time-adjusted inflows. The present value of all cash flows is zero at the discounted cash flow rate of return:

$$0 = \sum_{n=1}^{n=n} \frac{\text{Cash flow for year } n}{(1 + r)^n} \tag{5.2}$$

Example 5.8 Using the data in Table 5.2, find the discounted cash flow rate of return.

SOLUTION The present value of the cash flows must be zero. Hence,

$$P = 0 = \frac{-1000}{(1 + r)^0} + \frac{475}{(1 + r)^1} + \frac{400}{(1 + r)^2} + \frac{330}{(1 + r)^3} + \frac{270}{(1 + r)^4} + \frac{200}{(1 + r)^5}$$

$$P = \begin{cases} +75.16 & \text{for } r = 0.20 \\ -18.91 & \text{for } r = 0.25 \end{cases}$$

By interpolation, $P = 0$ for $r = 0.239$

The discounted cash flow rate of return is 23.9 percent.

The discounted cash flow rate of return on a project is a constant annual rate of return on that part of the investment in the project outstanding each year over its life. The investment base declines over time as returns accumulate to pay interest and retire principal. Therefore, the dollar amount returned each year varies with time, although the rate of return remains constant. In other words, the rate of return is a constant fraction of the nonconstant investment.

In some special cases, there may be no rate of return at which Eq. (5.2) reduces to zero; or there may be multiple roots, and thus more than one rate of return. These cases are not discussed here, but see Probs. 5.20 and 5.21.

5.13 Cash Flow

Cash flow is the clue to profitability studies and will be discussed in some detail. Cash flow is simply the movement and timing of cash with respect to a project. Movement can be further defined as in- or out-of-pocket movement. To set up the cash flow for a project, it is necessary to establish all the in- and out-of-pocket amounts and the time at which they occur. Local taxes are easily included. A more complicated task is the evaluation of the corporate income tax which must be included, since cash flow is always calculated on an after-tax basis. The income tax is best calculated separately and is included in the cash flow as a single-line item.

Cash flow itself is not a measure of profitability and may merely reflect a high turnover for wear and tear, i.e., depreciation. However, cash flow is used to calculate profitability, as in the discounted cash flow rate of return method, which is a measure of profitability.

Project D, now to be discussed, should be studied in detail since it embodies virtually all the principles involved in cash flow calculations.

A plant will cost 10.4 M$, with payment in equal amounts at 9, 6, 3, and 0 months before start-up taken as zero time. The project will last 5 years, with 0.4 M$ salvage value. The sum-of-the-years-digits method of depreciation will be used for tax purposes for a 5-year life. Land at a cost of 1 M$ will be paid for 9 months prior to start-up and is expected to be sold at the end of 5 years for 1.7 M$. Although there will be an internal profit of 0.7 M$ when the land is sold which normally would become subject to the prevailing 48 percent income tax, some allowance for improvements can be anticipated, and the effective tax rate can be taken as 26 percent of the 0.7 M$ gain.

Production, sales, and regular expenses will vary, and the following values can be used as end-of-year amounts:

Year	Activity as % of full production	Sales, M$	Regular expenses, M$
1	33.3	5	3
2	66.6	10	5
3	100	15	7
4	100	15	7
5	100	15	7

A special expense is classified separately and refers to some rental equipment. It amounts to 0.5 M\$/year and must be treated in that manner for tax purposes. However, 3 full years' expense, totaling 1.5 M\$, must be paid at start-up time, and 0.5 M\$ at the end of the fourth and fifth years.

Working capital, consisting chiefly of inventory and cash, will be proportional to sales and will be 1 M\$ for full production. Thus 0.33 M\$ must be invested at the beginning of the first and second years and 0.34 M\$ at the beginning of the third year. Working capital will be recovered in full at the end of the project. Movement of working capital into and out of the project is not subject to income tax. It is merely a bookkeeping exchange of cash and inventory, with no effect on profit.

A start-up cost of 0.7 M\$ taken at zero time will be capitalized along with the plant cost.

Part of the project is subject to depletion and can be taken as 25 percent of the sales, with corresponding expenses of 25 percent of the regular expense. The depletion allowance is 22 percent of the annual gross applicable income, limited to 50 percent of the taxable income computed without the deduction for depletion.

Periodic interest compounded on a yearly basis with simple interest for less than 1 year will be used. Income will be taken as positive and outgo as negative. The cash flows for the project and the discounted cash flow rate of return are required.

The calculation for the cash flows is given in Table 5.3. The first seven items are easily calculated from the statements of the project, and they represent the items in the cash flow before the corporate income tax. The income tax is calculated separately in Table 5.4 and becomes item 8 in Table 5.3. It is advantageous to calculate the income tax separately because the income tax can be complicated, and there is no need to destroy the simplicity of Table 5.3 with the details of the tax situation.

The income tax calculation given in Table 5.4 includes provision for the regular tax rate of 48 percent, special attention to the sale of the property, and a depletion allowance of 22 percent for part of the project. Its entire gist is to reduce the calculation to a tax for each year to the last line in Table 5.4.

Depreciation is an expense for tax purposes, although it is not an out-of-pocket outlay at the time it is taken. Similarly, the depletion allowance reduces the tax base, with no actual expenditure being made for the period. Working capital is not part of the tax situation, since it consists merely of funds frozen in the project and ultimately regained. The salvage value itself is not depreciable, but it does reduce the depreciable cost of the plant. Land which will be sold at a profit at the termination of the project is given special attention for the tax calculation.

The timing for tax purposes may be different from the timing of the actual expenditures. The special rental expense requires an outlay of 1.5 M\$ at zero time, but it is a prepaid expense. For tax purposes, it is regarded as a 0.5 M\$ yearly expense.

The tax situation can result in an income from savings in taxes for the project

Table 5.3 Calculation of cash flow

Item	M$ at years from start-up									
	$-\frac{3}{4}$	$-\frac{1}{2}$	$-\frac{1}{4}$	0	1	2	3	4	5
1. Plant	−2.60	−2.60	−2.60	−2.60	—	—	—	—	0.40
2. Land	−1.0	—	—	—	—	—	—	—	1.70
3. Sales	—	—	—	—	5.00	10.00	15.00	15.00	15.00
4. Regular expenses	—	—	—	—	−3.00	−5.00	−7.00	−7.00	−7.00
5. Special rental expense	—	—	—	−1.50	—	—	—	−0.50	−0.50
6. Working capital	—	—	—	−0.33	−0.33	−0.34	—	—	1.00
7. Start-up expense-capitalized	—	—	—	−0.7	—	—	—	—	—
8. Income tax from Table 5.4	—	—	—	—	1.11	−0.53	−2.17	−2.51	−3.04
9. Cash flow—sum of above items	−3.60	−2.60	−2.60	−5.13	2.78	4.13	5.83	4.99	7.56

Table 5.4 Calculation of income tax by years

Item	M$ at end of year				
	1	2	3	4	5
1. Sales, not subject to depletion	3.75	7.50	11.25	11.25	11.25
2. Regular expenses, not subject to depletion	−2.25	−3.75	−5.25	−5.25	−5.25
3. Special rental expense, not subject to depletion	−0.50	−0.50	−0.50	−0.50	−0.50
4. Depreciable value for project 10.4 − 0.4 + 0.7 = 10.7	—	—	—	—	—
5. Depreciation factor for year, decimal 1 + 2 + 3 + 4 + 5 = 15	5/15	4/15	3/15	2/15	1/15
6. Depreciation for year, (4) × (5)	−3.57	−2.85	−2.14	−1.43	−0.71
7. Partial base for income tax, (1) + (2) + (3) + (6)	−2.57	0.40	3.36	4.07	4.79
8. Partial income tax at 48% of (7)	1.23	−0.19	−1.61	−1.95	−2.30
9. Sales, subject to depletion	1.25	2.50	3.75	3.75	3.75
10. Expenses, subject to depletion	−0.75	−1.25	−1.75	−1.75	−1.75
11. Calculated taxable income re depletion, (9) + (10)	0.50	1.25	2.00	2.00	2.00
12. Depletion allowance, 22% of (9), but not over 50% of (11)	0.25	0.55	0.83	0.83	0.83
13. Partial base for income tax, (11) − (12)	0.25	0.70	1.17	1.17	1.17
14. Partial income tax at 48% of (13)	−0.12	−0.34	−0.56	−0.56	−0.56
15. Income from sale of property	—	—	—	—	0.7
16. Tax on (15) at 26% effective rate	—	—	—	—	−0.18
17. Income tax, (8) + (14) + (16)	1.11	−0.53	−2.17	−2.51	−3.04

for some of the years. During the first year the company as a whole is making money, but the project itself is losing money. The loss by the project reduces the tax which otherwise would be paid by the company. Thus, the project gains an income tax credit of 1.11 M$.

If the project has an investment tax credit, it can easily be included on the last line of Table 5.4 at the time the tax credit is taken, say at zero time or the end of year 1.

The discounted cash flow rate of return is that rate of return which makes the present value of all cash flows equal to zero at zero time. The cash flows are given on the last line of Table 5.3. Using simple interest for less than 1 year and compound interest for more than 1 year gives

$$0 = -3.60(1 + 0.75r) - 2.60(1 + 0.50r) - 2.60(1 + 0.25r)$$

$$-5.13 - \frac{2.78}{(1 + r)^1} + \frac{4.13}{(1 + r)^2} + \frac{5.83}{(1 + r)^3} + \frac{4.99}{(1 + r)^4} + \frac{7.56}{(1 + r)^5}$$

By trial and error $r = 0.180$, and the discounted cash flow rate of return for the project is 18.0 percent.

A simplified expression for cash flow is sometimes given as

$$\text{Cash flow} = \text{profit after taxes} + \text{depreciation} \tag{5.3}$$

where depreciation includes the depletion allowance. The expression is not a definition of cash flow, however, nor is it always exact. Consider Tables 5.3 and 5.4 and calculate the cash flow at the end of year 2 using Eq. (5.3). The calculation is:

Sales	10.00 M$
Regular expenses	−5.00
Special rental expense	−0.50
Depreciation expense	−2.85
Allowance for depletion	−0.55
Profit before income tax	1.10
Less income tax at 48%	0.53
Net profit after taxes	0.57
Add depreciation	2.85
Add depletion allowance	0.55
Calculated cash flow	3.97 M$

which does not agree with the correct value, 4.13 M$, given in Table 5.3. The calculation above includes 0.50 M$ for the special rental expense, and it certainly must be considered in the tax and profit, but the expense was actually prepaid and is not a cash outflow at the end of year 2. Hence, 0.5 M$ should be added to the cash flow as calculated above. Again, at the end of year 2 there was an outlay of 0.34 M$ for increased working capital. The outlay is merely money frozen in the project and does not enter the tax or profit consideration, but it is an outlay and should be subtracted from the cash flow calculated above. If the two corrections are made, the cash flow at the end of year 2 becomes

$$3.97 + 0.50 - 0.34 = 4.13 \text{ M\$}$$

which agrees with the value given in Table 5.3.

A more compact format for a simple project follows.

Example 5.9 A venture requires 1000 k\$ now, with an investment tax credit of 20 k\$ at the end of the first year and no salvage value. Receipts and expenses, excluding depreciation, are:

End year	Receipts, k\$	Expenses, k\$
1	1,600	1,000
2	1,700	1,000
3	1,140	800

Working capital of 200 k\$ is invested at zero time. Sum-of-the-years-digits depreciation will be used for tax purposes for a useful life of 3 years. The income tax rate is 46 percent. Find the discounted cash flow rate of return.

The cash flow is calculated in Table 5.5.

SOLUTION For the present value of the cash flow to be zero,

$$P = 0 = -1200 + \frac{574}{(1 + r)} + \frac{531.18}{(1 + r)^2} + \frac{460.42}{(1 + r)^3}$$

$$P = \begin{cases} +23.005 & \text{for } 0.14 \\ -15.447 & \text{for } 0.16 \end{cases}$$

By interpolation,

$$P = 0 \qquad r = 0.152$$

The discounted cash flow rate of return is 15.2 percent.

Table 5.5 Cash flow for Example 5.9 in k\$

	Time 0	End year 1	End year 2	End year 3
1. Receipts		1,600	1,700	1,140
2. Expenses, excluding depreciation		−1,000	−1,000	−800
3. Receipts for tax purposes (1) + (2)		600	700	340
4. Depreciation				
(3/6)(1,000)		500		
(2/6)(1,000)			333	
(1/6)(1,000)				167
5. Taxable income (3) − (4)		100	367	173
6. Tax at 46%		46	168.82	79.58
7. Fixed capital	−1,000			0
8. Working capital	−200			200
9. Investment tax credit		20		
10. Cash flow (3) − (6) + (7) + (8) + (9)	−1,200	574	531.18	460.42

Table 5.6 Effect of economic yardsticks on project ranking†

Time, end year	Project A, declining income		Project B, steady income		Project C, increasing income	
	Cash flow	Profit after taxes	Cash flow	Profit after taxes	Cash flow	Profit after taxes
—	−1,000		−1,000		−1,000	
1	475	275	355	155	200	0
2	400	200	355	155	300	100
3	330	130	355	155	400	200
4	270	70	355	155	450	250
5	200	0	355	155	490	290
		Av = 135		Av = 155		Av = 168
		Rank		Rank		Rank
Return on original investment	13.5	3	15.5	2	16.8	1
Payout time, years	2.38	1	2.82	2	3.22	3
Discounted cash flow	23.9%	1	22.8%	2	20.8%	3
Present value at 10%	319	3	346	1	342	2

†Depreciation in all cases is $200/year.

5.14 Three Projects Compared

Project A, as given in Table 5.2, features a declining income. Table 5.6 repeats project A together with project B, which has a steady income, and project C, which has an increasing income. All amounts are in kilodollars. All projects have an initial investment of − 1000 k$ and employ straight-line depreciation. Table 5.6 summarizes the calculations for income taken as discrete amounts at the end of the year, and interest is compounded periodically at the end of the year. Profitability calculated by four common methods—return on original investment, payout time without interest, discounted cash flow rate of return, and net present value—is compared for the three projects. The rank varies with the criterion used for profitability. It will be shown in Sec. 5.16 that a single criterion is not adequate as an absolute measure for judging profitability.

5.15 Continuous Interest

Periodic interest has been used so far in this chapter to avoid confusion for the reader who understands no other method. The use of continuous interest in profitability studies is becoming more common, and the reader is referred to Chap. 4 for a discussion of its development and use. Recalculation of projects A, B, and C of Table 5.6 on the basis of continuous interest gives the following result when the rate expressed is the nominal rate:

	A	Rank	B	Rank	C	Rank
Discounted cash flow	21.5%	1	20.5%	2	18.9%	3
Present value at 10%	304	3	328	1	322	2

The rank for both discounted cash flow and present worth at 10 percent with continuous compounding is the same in this case as the rank with discrete compounding (cf. Table 5.6). Generally, discrete and continuous compounding give the same order of rank.

To facilitate the computation with continuous interest, the more commonly used discount or present-value factors are tabulated in Appendix 2. Table A2.1 covers cash effects occurring at a point in time before time zero, the present. Table A2.2 covers cash effects occurring at a point in time after time zero. Table A2.3 covers a cash flow occurring uniformly over time. The table lists factors which discount $1 cash flow to its present value at the start of the flow. Table A2.4 covers a cash flow declining to zero at a uniform rate discounted to a present value at the start of the flow. Tax credit generated by sum-of-the-years-digits depreciation is an approximation of such a cash flow. Factors for most other special growth and decline patterns can be developed from these.

Example 5.10 An investment gives the following data:

Initial investment $122,000 spent uniformly from $-\frac{1}{2}$ to 0 years, all capitalized
Annual savings $200,000 uniform
10-year project life
49 percent income tax
7 percent investment credit
20-year sum-of-the-years-digits depreciation

Find the discounted cash flow rate of return, using kilodollars.

SOLUTION Try 60 percent for the first rate. The investment, -122, is treated as a uniform cash flow at 60 percent for $\frac{1}{2}$ year, or $in = 0.30$. The present value at the beginning of its time, $\frac{1}{2}$ year ago, is taken from Table A2.3, and the factor is 0.8639. This must now be moved with the calendar $\frac{1}{2}$ year at 60 percent, to get to zero time, the factor for which (by Table A2.1, with $in = 0.30$) is 1.350. The combined factor is $0.8639(1.350) = 1.17$

The investment credit amounts to $0.07(122) = 9$ and is taken as a savings in tax reduction at zero time.

Tax credit from depreciation is $0.49(122) = 60$, which for sum-of-the-years-digits depreciation declines to zero at a uniform rate and requires Table A2.4, discount factor, here 0.1528 for $in = 0.60(20) = 12$.

Savings are treated as a uniform cash flow amounting to $200(10) = 2000$ before taxes or $2000(0.51) = 1020$ after taxes. The discount factor is obtained from Table A2.3, and for $in = 0.60(10) = 6$ is 0.1663.

A tabulation of the calculations for 60, 70, and 80 percent follows.

	Cash flow after taxes	60% rate		70% rate		80% rate	
		Factor	Present value	Factor	Present value	Factor	Present value
Investment	−122	1.17	−143	1.20	−146	1.23	−150
Investment credit	9	1.00	9	1.00	9	1.00	9
Depreciation	60	0.153	9	0.133	8	0.117	7
Savings	1,020	0.166	169	0.143	146	0.125	127
Net			+44		+17		−7

The rate of return by graphical interpolation is 77 percent.

5.16 Inadequacy of the Single Criterion

The question arises, which criterion for judging profitability is best? Before answering this question, it is well to recognize that even a single criterion is not unique. Consider three ventures, A, B, and C, having end-of-year cash flows as follows, with all amounts in kilodollars:

	Zero time	Year 1	Year 2	Year 3
A	−1,000	750	390	180
B	−1,000	350	470	660
C	−1,000	533	467	400

Each of these ventures has a discounted cash flow rate of return of 20 percent:

$$\text{Venture A: } -1000 + \frac{750}{(1.20)^1} + \frac{390}{(1.20)^2} + \frac{180}{(1.20)^3} = 0$$

$$\text{Venture B: } -1000 + \frac{350}{(1.20)^1} + \frac{470}{(1.20)^2} + \frac{660}{(1.20)^3} = 0$$

$$\text{Venture C: } -1000 + \frac{533}{(1.20)^1} + \frac{467}{(1.20)^2} + \frac{400}{(1.20)^3} = 0$$

Ventures A, B, and C require a 1000 k$ investment, last 3 years, and give a calculated 20 percent per year discounted cash flow rate of return. However, they differ in the pattern of the cash flows. Obviously, the single criterion, 20 percent per year, has washed out the individuality of the ventures.

Table 5.7 gives a breakdown of the three ventures on a year-by-year basis. The difference between the three ventures becomes apparent. All have 1000 k$

Table 5.7 Details of ventures A, B, and C at a 20 percent per year rate of return in k$

	A	B	C
1. Investment at zero time	1,000	1,000	1,000
2. Less income end year 1	−750	−350	−533
3. Plus 20% of item 1	200	200	200
4. Investment beginning year 2	450	850	667
5. Less income end year 2	−390	−470	−467
6. Plus 20% of item 4	90	170	133
7. Investment beginning year 3	150	550	333
8. Less income end year 3	−180	−660	−400
9. Plus 20% of item 7	30	110	67
10. Investment end year 3	0	0	0

invested in the first year, but the investments remaining for years 2 and 3 are different, as shown by items 4 and 7 in the tabulation. Figure 5.2 shows the investment as a function of time. Venture A can be classified under "accelerated reduction of investment" since the investment is reduced most in the early years. This implies, too, the need for accelerated reinvestment if the initial investment is to be maintained. Venture B can be classified under "retarded reduction of invest-ment" since the investment is reduced most in the later years. This implies, too, a delay in generating funds for reinvestment. Venture C is intermediate and reduces the investment uniformly from year to year. Each venture has its advantages and disadvantages, and the only purpose here is to show that the difference between the ventures shows up in an analysis on a year-to-year basis.

A single criterion is not unique, and there are an infinite number of cash flow patterns that will produce the same numerical value for the criterion. A single criterion is like an average, and it wipes out the individuality of a venture.

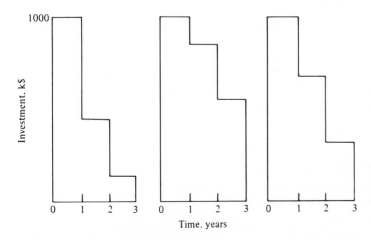

Figure 5.2 Investment as a function of time.

**Table 5.8 Time, cost, and revenue
elements of a typical project**

1. Costs incurred before construction
2. Fixed capital
3. Start-up costs
4. Working capital
5. Subsequent investment
6. Sales price and volume trends
7. Annual expenses
8. Tax effects
9. Salvage value

The most popular single criteria are discounted cash flow rate of return, pay-out time, return on original investment, and net present value. In practice, an analyst will not use any single criterion but will consider, in addition, the individuality of the project, such as by breaking it down to a year-by-year analysis; review what the project means to the company in the future; analyze how uncertainty affects the project's profitability; and consider a host of other issues. There is no single best criterion since no one criterion alone can reveal all the details.

5.17 Projects for Profitability Studies

A checklist of the major cost and revenue elements of a typical project is shown in Table 5.8, with the items arranged roughly by time of occurrence. Table 5.9 lists the common project types subject to profitability studies. The mathematical treatment is the same for all. The comparisons can vary—e.g., abandonment or neglect versus shutting down—but the principles do not vary. The difficult task is to arrive at numerical values for all the pertinent cost items involved and not to overlook any applicable costs or benefits.

Table 5.10 lists some underlying characteristics on which the project types depend. Sunk costs are costs incurred in the past and generally do not enter a

**Table 5.9 Project types for profitability
studies**

1. Abandonment
2. Process modification and cost
   reduction
3. Replacement and displacement
4. Expansion
5. New process, product, or market
6. Purchased know-how vs. internal development
7. Joint venture
8. Lease vs. buy
9. Make vs. buy

**Table 5.10 Some important
characteristics of projects**

1. Sunk costs
2. Differing lives
3. Cost sharing
4. Cost savings
5. Quality improvement
6. New business
7. Special financing
8. Risk and uncertainties

profitability analysis since profitability studies are concerned with the future. However, the tax situation associated with them may influence the future and should not be omitted. For example, if a piece of equipment is abandoned, without any value the original cost is a sunk cost and does not enter a replacement analysis. However, the depreciation on its book value and its timing cannot be ignored since present and future tax offsets may be involved. For differing lives, criteria such as unacost, capitalized cost, and discounted cash flow rate of return make fair comparisons, but comparisons based on straight present value can be misleading unless the difference in service lives has been compensated for. Cost-sharing projects can introduce difficult decisions about how such costs are to be allocated.

Example 5.11 The laboratory must decide whether to buy or lease a highly specialized instrument which is undergoing considerable technological advancement.

A new instrument can be bought for $50,000. It has a technological life of 4 years and can be disposed of for $2000 to be taken as regular income. It is fully depreciable in 4 years using sum-of-the-years-digits depreciation for tax purposes. The maintenance cost, at the end of each year, is expected to be $6000 per year. If the instrument is owned, it can be modified and used in connection with other instruments, and this fact is given a value of $8000/year, as of the end of the year.

A rented instrument costs $15,000/year, payable at the beginning of the year with free servicing. Rental is on an annual basis, renewable each year at the lessee's option. Renting gives the advantage of deferring buying a year at a time in a period of high technological advancement, and this is judged to be worth $20,000 on a present basis for a 4-year period.

If money is worth 10 percent per year after a 52 percent tax, should the instrument be bought or rented?

SOLUTION Unacost, the equivalent uniform end-of-year cost, on a purchased-instrument basis (Table 3.5, items 1, 4, and 2 in order, and Table A1.1) is

$$C_d(1 - tF_{\text{SDP},10\%,4})F_{PR,10\%,4} = 50,000[1 - 0.52(0.83013)]0.31547 = \$8964$$

The disposal income, negative cost, is:

$$-C_{ex}\frac{1-t}{(1.10)^4}F_{PR,10\%,4} = -2000\frac{1-0.52}{1.4641}0.31547 = -207$$

$$R(1 - t) = (6000 - 8000)(1 - 0.52) = \underline{-960}$$

$$R = \$7797$$

Unacost for a rented instrument (by Table 3.5, items 3 and 1 in order, and Table A1.1) is

$$R_b\left(1 - \frac{t}{1 + r}\right)(1 + r) = 15{,}000\left(1 - \frac{0.52}{1.10}\right)1.10 = \$8{,}700$$

Delay value treated as a first cost lasting 4 years with an instant tax benefit is

$$-20{,}000(1 - t)F_{PR, 10\%, 4} = -20{,}000(1 - 0.52)(0.31547) = \underline{-3028}$$
$$R = \$5672$$

In this example, it is more economical to rent. Note that the credit given to the rented article for its value in deferring a purchase until later is the determining factor in the analysis. It is difficult to assign this value. When renting on a longer-term basis without annual renewal options, the financing effects of the rental contract must be recognized as equivalent to financing with debt. In most of these cases, purchase is more economical than renting.

5.18 Classification System for Economic Appraising

Several economic evaluations are normally required during the development of a project. They arise when a decision for a different order of magnitude of capital and effort is required, when significant new data become available, or when new alternatives are conceived.

The AACE has prepared a classification system dividing these studies into four classes, depending on the amount of information available [3]: (1) final, (2) intermediate, (3) preliminary, and (4) screening.

5.19 Benefit-Cost Analysis

A benefit-cost analysis is a special type of decision making and is closely related to decisions on profitability. The method is most applicable to public works projects, but it can arise in private enterprise for such projects as building a baseball field at a plant for the use of company personnel. The main feature of benefit-cost analysis is that the benefits are highly subjective and usually involve social, political, and special-interest ramifications.

Table 5.11 summarizes a benefit-cost procedure, Table 5.12 is a checklist for project costs, and Table 5.13 is a checklist for project benefits and disbenefits. The

Table 5.11 Summary of the benefit-cost analysis procedure

---

1. Define the set of feasible, mutually exclusive public-sector alternatives to be compared
2. Define the objectives to be used
3. Develop the cost-savings and benefit-disbenefit profiles for each alternative
4. Specify the interest rate to be used
5. Specify the measure of merit to be used
6. Compare the alternatives using the measure of merit
7. Perform supplementary analyses
8. Select the preferred alternative

Table 5.12 Checklist for project costs and savings

I. Project costs (positive costs or disbursements)
   A. Capital outlay
      1. Land, rights-of-way, easements, severance damages, or water-rights acquisition
      2. Equipment and buildings
      3. Design
      4. Filing fees
      5. Construction
      6. Relocation costs
      7. Engineering
      8. Working capital
   B. Annual expenses
      1. Operating costs
      2. Taxes forgone
      3. Maintenance costs
      4. Replacement costs
      5. Depreciation or amortization costs
      6. Administrative costs

II. Project savings (negative costs or receipts)
   A. Capital savings
      1. Residual or salvage values
      2. Paid-up rents, royalties, or license fees
   B. Annual savings
      1. Sale of any products or publications
      2. Sale of any services
      3. Toll revenues
      4. Fines collected
      5. Increased taxes resulting from property appreciation and increased business activity
      6. Charges to users, such as running royalties, rentals, or license fees

principles involved in a benefit-cost analysis are elementary to a cost engineer, but analysts involved in this work must be highly specialized. The intimate contact with the public's interest required by this work demands that the analyst be responsive to public opinion and expert in evaluating intangible benefits. Benefit-cost analysis usually involves the public interest, a matter subject to considerable subjective evaluation. The analyst must be an expert in evaluating intangible benefits and be adept and fair in judging the value of the public interest.

Example 5.12 A canal can be built from city D to city E at a cost of 10 M\$, and the annual cost of maintenance and operation will be 0.2 M\$. Annual favorable consequences to the general public are estimated at 3.62 M\$, offset by adverse consequences of 0.5 M\$ to a small segment of the population. The canal is expected to last 10 years, and the current interest rate is 6 percent per year. The annual cost of capital recovery will be 1 M\$.
(*a*) Calculate the B/C ratio if the disbenefits are taken as negative benefits.
(*b*) Change the disbenefits to costs and recalculate the B/C ratio.
(*c*) Find the excess of benefits over costs.

Table 5.13 Checklist for project benefits and disbenefits

I. Benefits
  A. Primary or direct benefits
     1. Reduced damage to health
     2. Reduced damage to property
     3. Reduced damage to aesthetic qualities
     4. Reduced time requirement for performance
     5. Reduced operating cost of equipment
     6. Reduced insurance costs
     7. Increased convenience
     8. Increased safety
     9. Increased direct employment (increased income)
    10. Increased land or other property values
    11. Increased satisfaction or enjoyment
    12. Value of increased production
    13. Preservation of areas of unique natural beauty and scenic, historical, or scientific interest
  B. Secondary or indirect benefits
     1. Increased availability of health care facilities
     2. Increased availability of property for intended use
     3. Increased recreational use
     4. Increased indirect (service) employment
     5. Increased income from broader new uses (e.g., low-flow augmentation for water quality control might achieve navigational uses)
     6. Increased income from reduced interruptions in secondary activities (e.g., flood control might reduce wage losses caused by flooded factories)
     7. Increased income from forward production linkages (e.g., increased cotton production resulting from flood control projects would lead to increased employment to process the cotton into products)
     8. Increased income from backward production linkages (e.g., increased cotton production resulting from flood control projects would lead to increased employment in farm machinery, fertilizer, and pesticide industries)

II. Disbenefits
  A. Primary or direct disbenefits
     1. Any private investment to produce or utilize the project output (e.g., new machinery to utilize water cleaned up in a water pollution project)
     2. Any operating costs associated with private investment to produce or utilize the project output (e.g., the operating costs of schools required by population increases resulting from the project)
     3. Increased costs from project construction (e.g., increased cost of transportation around a reservoir being built for flood control)
     4. Any damage to health, property, or aesthetic qualities
     5. Any increase in the time requirements or operating costs of equipment
     6. Any decrease in property values, satisfaction, or enjoyment
     7. Any loss in production
     8. Any destruction of areas of natural beauty and scenic, historical, or scientific interest

Table 5.13 (*Continued*)

B. Secondary or indirect disbenefits
1. Increased costs for, or reduction in, public services
2. Increased damage or hazards created by the project
3. Decreases in land values or natural resources
4. Decrease in business activity
5. Any decrease in availability of health care facilities, or availability of the property for its intended use or for recreational use
6. Decrease in employment
7. Decreased income as a result of interruptions in secondary activities because of project operations
8. Loss in income and employment from the loss in forward or backward production linkages

---

SOLUTION Part (*a*), converting to present values, is

B: $\qquad (3.62 - 0.5)F_{RP,6\%,10} = 3.12(7.3601) = 22.964$

C: $\qquad 10.0 + (1.0 + 0.2)F_{RP,6\%,10} = 10.0 + 1.2(7.3601) = 18.832$

$$B/C = 22.964/18.832 = 1.22$$

Part (*b*) is

B: $\qquad 3.620F_{RP,6\%,10} = 3.62(7.3601) = 26.644$

C: $\qquad 10.0 + (1.0 + 0.2 + 0.5)F_{RP,6\%,10} = 10.0 + 1.7(7.3601) = 22.512$

$$B/C = 26.644/22.512 = 1.18$$

Part (*c*) is

$$B - C = 22.964 - 18.832 = 4.132 \text{ M\$} \qquad \text{present value}$$

$$B - C = 26.644 - 22.512 = 4.132 \text{ M\$} \qquad \text{present value}$$

5.20 Life Cycle Cost

Life cycle cost is not fundamentally an index of profitability. It is a concept used for making decisions between alternatives and is a term used principally by the government for expressing the total cost of an article or system. A life cycle cost indicates that the first cost alone is not enough to evaluate fully an article or system, but that all costs incurred over the life cycle must be considered. The concept is nothing new in cost engineering, but an analyst should be aware of its growing importance in contacts with the government. General Administration Bulletin No. FPMR E-153 suggests that all federal agencies abandon the traditional practice of procuring supplies and services on the basis of lowest competitive bids and substitute the criterion of lowest life cycle cost.

Life cycle costs can be compared, after allowing for a difference in service lives, as absolute measures of cost. The life cycle cost emphasizes two principles. First, all costs throughout the life cycle must be included. Second, no costs should be excluded because they might be regarded as common costs. Thus, life cycle costs can be judged in their entirety, yet with inclusion of all details.

The engineer has an inherent disposition to reduce a problem to its basic costs. The final decision will be based on the pertinent costs, ignoring common costs to some extent. Life cycle costing will lead to the same decision, and the extra detail may have advantages. The analysis is less restricted and is not tempered by considerations of what an article is being compared with; hence it offers the best opportunity for avoiding errors or omissions. A checklist for life cycle costing is given in Table 5.14.

5.21 Cost per Unit or Cost Effectiveness

Sometimes it is not possible to measure return, service, or improvement in terms of dollars. Typical examples are gasoline octane number improvement, effectiveness of a weapon, and effectiveness of an antipollution device. It is possible, however, to measure unit costs, or cost effectiveness. This method is used widely in pollution and defense projects when the outputs or improvements cannot be measured in dollars. The results are measured in terms such as dollars per unit of output or per unit of improvement.

> **Example 5.13** If \$51 per year is spent on control measures, the deposition of sulfuric acid droplets can be reduced from 94 to 78 kg/100 m². Find the cost effectiveness in dollars per percentage decrease in the weight of acid deposition per 100 m² per year.

Table 5.14 Checklist for life cycle costing

Purchase or manufacturing cost
Transportation cost
Installation cost
Direct costs
Indirect costs
Maintenance
Inventory for materials
Inventory for parts
Periodic overhauls
Supervision
Service reliability
Benefits and penalties for quality
Salvage value
Distribution expense
General and administrative expense
Conformity with trends
Safety and ecological considerations
Governmental rules and regulations
Uncertainty and risk

SOLUTION

$$\frac{94 - 78}{94}100 = 17 \text{ percent decrease}$$

$$\text{Cost effectiveness} = \frac{\text{annual cost control for } 100 \text{ m}^2}{\text{percentage decrease in deposition}}$$

$$= \frac{51}{17} = \$3/\text{year for } 100 \text{ m}^2 \text{ for a 1 percent reduction}$$

5.22 Sensitivity Analysis

In evaluating a project, a large number of estimates may be made. It becomes necessary to recognize how sensitive the final estimate is to the individual component estimates. Knowing this, attention can be concentrated on the components having the greatest effect.

For example, the total cost of cooling for a unit in a plant includes a component cost for inhibitors for the cooling water. The cost of the inhibitors can vary twofold depending upon operations. However, this cost is only 1 percent of the total cost at most. Clearly, the total cost is insensitive to the inhibitor cost. Sensitivity analysis is becoming an important tool in cost analysis; thus it is an integral part of linear programming.

Sensitivity analysis is not the same as uncertainty analysis. In the above case, the cost for the inhibitors can vary twofold, a high degree of uncertainty, but the total cost, as noted, is insensitive to the inhibitor cost. Sensitivity analysis is discussed quantitatively in Chap. 8.

5.23 Uncertainty and Risk Analysis

Modern times are dominated by rapid change and instability, creating an atmosphere of uncertainty and risk. Frequently, the terms *uncertainty* and *risk* are used as equivalents, but the purist views them differently. Risk is a measure of reaction to uncertainty. For example, a new office copier is introduced at one-third the present price, but whether it is built well enough to be worth buying is uncertain. A large company using many copiers can buy one at little risk. An individual who uses only one copier would take a much greater risk. The uncertainty about the copier is the same in each case, but the risk to the buyers is different.

A profitability study always involves some uncertainty since not all the components that go into the study are exactly known. Most are estimates. The trend now is to demand attention to and inclusion of uncertainty. Suppose that the cost for a project is reported as 10 M$. That value is the expected value only; it must now be supplemented with a measure of the uncertainty. What is the probability that the cost will be 12 M$ or more, or between 11 M$ and 13 M$? Management will even want to know how sensitive the uncertainty of the project is to the uncertainty of the component elements. Methods for attacking these problems on a quantitative basis are discussed in Chap. 8.

5.24 Pitfalls in Profitability Analysis

Profitability is a concern of high-level management but requires conscious attention at all levels of an organization if it is to be attained. A list of pitfalls in profitability analysis is given in Table 5.15, and comments on some of the items follow.

Profit and profitability are different concepts. Accounting principles determine the measure of profit. *Profit* is fundamentally a short-term evaluation—e.g., an income statement for 1 year—and is therefore amenable to distortion. It is possible for a company to show a profit for 1 year when in truth it is losing money

Table 5.15 List of pitfalls in profitability analysis

Profitability is a long-term concept
A single criterion is inadequate
Circumvent the criterion
Estimation of cash flow
    Sales volume
    Sales price
    Raw materials costs
    Raw materials availability
Operating costs
Investment
Depreciation
Inflation
Replacement frequency and cash flow
Evolutionary factor
    Position at end of project
    Availability and commitment of funds
Quality control
Reliability of production
Allocation problems
    Utility costs
    Transfer pricing
    Residual and scrap value
    Overhead
Influence of government
    Subsidies
    Control of wages and materials
    Safety, ecology, and pollution
    Regulating power
    Social factors
Inventory pricing
Uncertainty
Productivity rate gain
    Diffusion
    Fringe benefits
    Personnel factors
Reviews and post mortems

over the long run. The recognition of time as a fundamental constituent of profitability is paramount, and *profitability* is a long-term concept.

Profitability can be evaluated only *after* time has elapsed. Profitabilities for the future are only estimates. The automobile industry showed a profit for the late 1970s, but it failed to tool up for the oncoming demand for small, economical automobiles. The year 1980 was a disaster for profits and revealed in retrospect that the profits of the late 1970s were false because the investment necessary for change and modernization had been disregarded. Again, nothing should be given an ultimate judgment on a short-term basis, and even a short-term venture should be evaluated on the basis of its long-term effect since it will influence the future beyond its termination.

A single criterion is inadequate for a full evaluation of profitability, as discussed in Sec. 5.16.

A common error in comparing profitability estimates is the failure to use a common denominator of service lives, for example, using net present value when there is a difference in the duration of projects.

Once a fixed value for a criterion is established, it becomes a record to be challenged and beaten. Demanding a short payout time will encourage cheap investment, which actually may be more costly in the end. Criteria and their usage must be flexible.

The most important estimates in the evaluation of profitability are the predictions of cash flows, with sales volume and sales price the most critical factors. The possibility of disastrous change in raw materials costs and availability is becoming commonplace in a world that seems to be running short of everything.

Operating costs are more critical than investment since the former are repetitive and the latter is made only once. The depreciation schedule used in a profitability study that considers the time value of money will affect the calculated profitability. The effect of inflation is discussed in Chap. 6.

Some restraint should be used when comparing cash flows. Although not a definition, cash flow is principally the sum of profit plus depreciation. A high cash flow may merely signify a high depreciation expense, that is, a high cost for wear and tear of equipment caused by frequent replacement.

All activities become part of an evolutionary process. A decision made today will influence the physical plant and marketing position of a company after the economic life of the equipment has expired. The company's position at the end of the project transcends the merely physical aspect of the project, such as its salvage value.

Future improvement in quality control must be anticipated since a competitor may offer a superior product. Reliability of production may be a factor, involving attention to inventories from raw materials to finished goods and investment in spare parts and spare equipment.

Allocation of overhead costs can have an important influence on apparent costs and profitability of individual products and services. The problem is one of the most acute in cost engineering and is discussed in Chap. 17.

The influence of government is omnipresent. Plants have closed because they were unable to meet antipollution requirements. Although laws cannot be made retroactive, plants will have to meet future requirements. Whether government interferes as a guardian of antipollution, social justice, or military preparedness, a cost exists whatever its origin. Costs tend to increase with time. It is necessary to recognize the increasing influence of government and to make judgments that are in tune with anticipated governmental action. It is better to swim with the political tide than be drowned by it.

Inventory pricing can have an effect upon short-term profits, as discussed in Chap. 17. Uncertainty and risk analysis are developed in Chap. 8 since an application of probabilities and distributions is involved.

Productivity commands considerable attention and is discussed in Chap. 9, particularly the dissipation of productivity by a factor called *diffusion*.

Whether a current expense is charged as such or capitalized and written off as a future expense is important to apparent current profitability. Capitalizing expenses will increase current profits, but at the expense of the profitability picture in the future.

During periods of inflation, there is a temptation to capitalize expenses. A fixed dollars amount, written off in part or wholly as a future expense, will appear to be less at a future time, when the value of the dollar is less.

Reviews and postmortems can be instructive but are seldom made, as there is a widespread unwillingness to search for causes of mistakes.

Profitability must emphasize the long term versus the short term. Emphasis on the short term leads to a policy of "milking the franchise" and patching. Maintenance, research, modernization, expansion, and necessary costs are reduced or eliminated. Short-term profitability may appear enhanced, but real profitability may be destroyed.

A company must have a policy for profitability. Single criteria are useful tools, but success is more an attitude than a devotion to mathematical criteria. It is the uncertainties and intangibles that will dominate eventual profitability.

The essential ingredients for profitability for a manufacturing company are as follows:

1. Profitability should be judged on a long-term basis.
2. Operations must be as efficient as possible, recognizing that technology is always in flux. Modernize or pass into history.
3. The diffusion effect of peripheral activities should be held to a minimum.

Real growth is measured by the proportion of necessary personnel employed with respect to the total available population. A conflict exists here. Operating with the highest technical efficiency possible while avoiding diffusion effects will tend to reduce personnel. A progressive organization faces a constant demand for growth in markets and activities just to stay even.

The only certainty in a profitability study is that profitability evolves in unexpected ways. Decisions should try to be in tune with the times and anticipate the future.

NOMENCLATURE

B/C	Benefit-cost ratio, dimensionless
C_d	Depreciable first cost, $
C_{ex}	Irregular cost at end of xth year, $
DCF	Discounted cash flow, $
DCFRR	Discounted cash flow rate of return, decimal or percent per year
F_{PR}	Factor to convert P to R, capital recovery factor, year^{-1}
F_{RP}	Factor to convert R to P, years
F_{SDP}	Present value of $1 for sum-of-the-years-digits depreciation, decimal
i	Rate of return, decimal per year
n	Number, such as life of an article, years
NPV	Net present value, $
P	Present value, $
r	Rate of return, generally after taxes, decimal per year
R	Uniform end-of-year annual amount, unacost, $ per year
R_b	Uniform beginning-of-year annual amount, $ per year
RAI	Return on average investment, decimal or percent per year
ROI	Return on original investment, decimal or percent per year
t	Tax rate, decimal

PROBLEMS

5.1 A venture has cash flows after taxes, excluding working capital and depreciation, as follows:

Time, yrs	Amount, M$
0	−5
1	1.2
2	1.3
3	1.4
4	1.5
5	1.1
6	0.9
7	0.5

Working capital is 1.5 M$ invested at zero time. Depreciation is constant at 1 M$ for 5 years. Find the payout time without interest.

5.2 In Prob. 5.1, find the payout time with interest at 10 percent per year (*a*) excluding working capital, (*b*) including working capital.

5.3 In Prob. 5.1, find the net present value at 10 percent per year on a 5-year life and the discounted cash flow rate of return on a 5-year life.

5.4 Calculate the results for projects A, B, and C in Table 5.6, assuming that incomes occur uniformly throughout the year, rather than at the end of the year, using continuous compounding of interest.

5.5 It is necessary to produce 1.1 million units per year of an automobile part, looking 12 years ahead and with money worth 14 percent per year compounded periodically and with no tax.

Plan A is to build a plant in the United States at a cost of 7 M$, with end-of-year expenses at 2 M$/year. It is expected that 0.5 M$ will have to be invested for pollution control, but this can be delayed until the end of the fifth year.

Plan B is to build a plant outside the United States for 4 M$, with end-of-year expenses at 1.2 M$/year, but there will be a duty charge of 30 percent of the selling price in the United States.

Find the selling price in the United States at which the two plans are equal.

5.6 Two plans are available for constructing a plant. Income will be the same for both plans for all the years and cancels out in the analysis. There is no tax consideration, and a period 8 years ahead is the basis for comparison, with money worth 12 percent per year using periodic interest.

Plan A is to build a half-size plant now at a cost of 1 M$ with an operating cost of 200 k$ as of the end of the year for 4 years. Expansion costing 0.9 M$ will occur at the end of year 4. End-of-year operating costs will then become 350 k$/year.

Plan B is to build a full-scale plant now at a cost of 1.5 M$, with an end-of-year operating cost of 250 k$.

Which plan should be chosen?

5.7 An investment of $20,000 will be made 2 years from now in a project lasting 4 years. It will permit before-tax savings of $10,000 per year for 4 years. Sum-of-the-years-digits depreciation will be used for an 8-year life for tax purposes. Salvage value at the end of year 4 will be the undepreciated balance. Money is worth 12 percent per year after a 46 percent tax rate. Use periodic interest and find the present value as of today.

5.8 A company can choose between two plans. Plan A is to build a plant now, having a capacity of 600 Mg/year at a cost of 1 M$, and 3 years from now add to the plant at a cost of 800 k$ to double the production capacity.

Plan B is to build a full-scale plant now for 1.5 M$ having a capacity of 1.2 Mkg/year.

The product sells for $1/kg. For plan A the operating cost is $0.5/kg, and for plan B it is $0.4/kg, in both cases for all the years. Sales demand for the years is expected to be

Year	Mg
1	600
2	800
3	1,000
4	1,200
5	1,200

If money is worth 10 percent per year and there is no tax consideration, which plan is more economical based on present value using periodic interest?

5.9 The investment for a new project is:

Fixed depreciable capital at zero time, k$	250
Start-up costs at zero time, capitalized, k$	175
Working capital at zero time, k$	130
Investment credit for tax purposes at zero time, k$	17.5
Project life, years	20
Tax rate, %	46
End-of-year operating cost, k$	813
End-of-year insurance, etc., as % of total investment, including start-up	2.5
End-of-year sales, k$	1,000
Depreciation method	sum-of-the-years-digits

Use periodic interest and find the discounted cash flow rate of return.

5.10 A venture requires an investment of 10 M$, 15 percent of which is nondepreciable land. Net annual end-of-year income is 2.2 M$, excluding depreciation. The plant will last 10 years, with straight-line depreciation for tax purposes. The tax rate is 46 percent. Find the discounted cash flow rate of return using periodic interest.

5.11 A project requires a depreciable investment of 1.5 M$ having no salvage value and 0.5 M$ for nondepreciable land. The depreciation will be over 4 years, with a tax at 46 percent. Yearly depreciation is:

End year	Fractional depreciations
1	0.4
2	0.3
3	0.2
4	0.1

Sales realization and expenses, before depreciation, will be:

End year	Sales, M$	Expenses, M$
1	0.3	0.2
2	0.75	0.25
3	0.9	0.3
4	0.9	0.3
5	0.9	0.3

Find the discounted cash flow rate of return of the project using periodic interest.

5.12 A new facility costs $100,000, $40,000 of which is for land and $60,000 for construction, and has a 20-year economic life. Sales volume is estimated at $150,000/year in the third and subsequent years. The annual cash return after taxes, including depreciation, is $9000/year at full production. Working capital is $15,000 at full production. In the first and second years, volume is estimated at one-third and two-thirds of capacity, respectively. Assume that the working capital and annual cash return are proportional to volume. The salvage value is negligible. Find the present value at 0, 4, and 6 percent and the discounted cash flow rate of return using periodic interest.

5.13 A project requires an immediate investment of 1 M$, 22 percent of which is for working capital. The profit-and-loss statement for a year will show a 100 k$ net profit after taxes for each of the 10 years of the project life. The economic life and tax life are the same. Use straight-line depreciation and a 46 percent tax.

    (*a*) Find the discounted cash flow rate of return.
    (*b*) Find the payout time without interest.

5.14 The investment for a new project is:

Fixed capital	
Depreciable portion, k$	250
Expensed proportion, k$	175
	425
Working capital, k$	130
	555

The depreciable and expensed portions are spent uniformly from -1 to 0 years, and working capital is spent at zero time. A 7 percent investment credit is allowed on the depreciable portion taken at zero time. The project life is 20 years and the annual before-tax margin is $400,000, excluding depreciation. The tax rate is 46 percent. Use periodic interest and 20 year sum-of-the-years-digits depreciation. Find the discounted cash flow rate of return for the project.

5.15 A project will have a construction cost of 4 M$ occurring uniformly 1 year prior to start-up. Another 1 M$ for construction must be paid at start-up. Start-up costs occurring uniformly during the first year at 0.5 M$ will be capitalized. Salvage at the end of year 4 will be 0.2 M$. The life for economic and tax purposes will be 4 years using straight-line depreciation and a 46 percent tax. Other data in Megadollars are as follows:

	Time 0	End year 1	End year 2	End year 3	End year 4
Sales		3.0	3.5	3.8	2.8
Expenses before depreciation		0.6	0.8	0.7	0.5
Value of inventory	0.8	1.0	0.9	0.8	0

Find the discounted cash flow rate of return using periodic interest.

5.16 Project A requires a capital investment at zero time of 1 M$ which will be written off in 5 years for tax purposes using twice the straight-line declining-balance depreciation method for the first 3 years with a switch to straight-line depreciation for the last 2 years. The project lasts 5 years, and the tax rate is 46 percent. Income taken at the end of the year, before taxes and depreciation, will be:

Year	k$
1	600
2	700
3	700
4	700
5	600

Project B, which competes with project A, shows a 20 percent per year discounted cash flow rate of return with periodic interest. Is project A more acceptable than project B?

5.17 A venture will have the following financial considerations in kilodollars:

	Time 0	End year 1	End year 2	End year 3	End year 4
Sales		1,500	1,800	1,700	1,400
Expenses before depreciation		300	400	350	250
Value of inventory	400	500	450	425	0

The investment at the start is 2000 k$, having 100 k$ salvage value to be written off in 4 years using sum-of-the-years-digits depreciation. The tax rate is 46 percent. Find the discounted cash flow rate of return using periodic interest.

5.18 The Jones Company is considering the replacement of 5-year-old machines with some improved units which recently became available. Management is concerned with the high operating cost of the existing machines, but expects to replace all of them in 5 years when a new system will be installed. From the information below, determine whether or not the equipment should be replaced now. Money is worth 9 percent per year after a 46 percent tax. Use continuous interest and straight-line depreciation.

	Old machines	New machines
Number of machines	6	5
Cost of each machine, $	4,200	7,800
Depreciated value at present each (tax books), $	2,100	
Trade-in value on new machines, $	1,500	
Market value of new machine in 5 years, $		3,000
Variable costs of operation, annual, $	24,000	17,000

5.19 Show that the rate of return at which the equivalent uniform annual cost (unacost) or capitalized cost is zero is the discounted cash flow rate of return.

5.20 When an investment occurs in the future rather than at zero time, the discounted cash flow rate of return may have more than one value, or there may be no value.

A project requires the expenditure of 15 M$ 5 years from now and will have an annual cash flow of 1 M$/year starting 1 year from now and continuing through year 20.

Find two discounted cash flow rates of return for the project.

5.21 Repeat Prob. 5.20, changing the expenditure 5 years from now to 5 M$, and show that there is no finite rate of return for the project.

5.22 An airport costing $750,000 is to be built. The annual cost of capital recovery for the investment in land, buildings, and runways will be $75,000, and the annual maintenance cost will be $100,000. Annual favorable consequences to the general public will be $625,000, and annual adverse consequences to a small segment of the public will be $50,000. The prevailing interest rate is 8 percent, and the project life is 10 years.

(a) Consider the adverse consequences as a cost and find the benefit-cost ratio.
(b) Find the benefit-cost ratio if the adverse consequences are considered a reduction in benefits.
(c) Find the excess of benefits over costs.

5.23 Total environmental damage from iron and steel operations is 528 M$/year. By spending 687 M$/year on control measures, the damage could be reduced by 496 M$/year. Find the cost effectiveness as measured by the dollars of control cost per dollar of damage reduction.

5.24 It costs 35 M$ to install a limestone-scrubbing unit on a 500-MW power plant to meet environmental standards for sodium dioxide removal from stack gas. It would cost 60 M$ to install a sodium carbonate scrubbing unit on a 750-MW power plant to perform the same service. Find the cost effectiveness for each alternative in dollars per kilowatt and state which system is more cost effective based on the investment cost.

REFERENCES

1. Abrams, H. J.: "Economics of Investment Criteria, Part I," *Chemical Engineer* (London), September 1970, pp. CE252–259.
2. Agarwal, J. C., and I. V. Klumpar: "Profitability, Sensitivity, and Risk Analysis for Project Engineers," *Chemical Engineering*, vol. 29, September 1975, pp. 66–72.
3. American Association of Cost Engineers: "Classification System for Economic Appraisals," *Cost Engineers Notebook*, sec. D-3, 1963, pp. 1–3.

4. Comber, N. H., and A. K. Biswas: *Evaluation of Environmental Intangibles,* Genera Press, Bronxville, N.Y., 1973.
5. Federal Register: "Cost-Effectiveness Analysis," *Appendix A,* vol. 38, no. 174, September 10, 1973, pp. 24639–24640.
6. Holland, F. A., F. A. Watson, and J. K. Wilkinson: "Probability Techniques for Estimates of Profitability," *Chemical Engineering,* January 7, 1974, pp. 105–108.
7. Jelen, F. C.: "Analyze Ventures This Way," *Petroleum Refiner* (now *Hydrocarbon Processing*), March 1964, pp. 125–128.
8. Jelen, F. C.: "Pitfalls in Profitability Analysis," *Hydrocarbon Processing,* January 1976, pp. 111–115.
9. Jelen, F. C., and C. L. Yaws: "How Energy Affects Life Cycle Costs," *Hydrocarbon Processing,* July 1977, pp. 89–91.
10. Jelen, F. C., and C. L. Yaws: "Project Cash Flow Description and Interpretation," *Hydrocarbon Processing,* March 1978, pp. 77–81.
11. Levin, R. I., and C. A. Kirkpatrick: *Quantitative Approaches to Management,* 3d ed., McGraw-Hill Book Company, New York, 1975.
12. Mishan, E. J.: *Cost-Benefit Analysis,* Frederick A. Praeger, Inc., New York, 1976.
13. Newnan, D. G.: *Engineering Economic Analysis,* Engineering Press, Inc., San Jose, Calif. 1980.
14. Park, W. R.: *Cost Engineering Analysis,* John Wiley & Sons, Inc., New York, 1973.
15. Phung, D. L.: "Cost Analysis Methodologies," *Cost Engineering,* May–June 1980, pp. 139–143.
16. Reul, R. I.: "Algorithms vs. Concepts," *American Association of Cost Engineers Bulletin,* vol. 7, June 1965, pp. 46–49.
17. Riggs, J. L.: *Engineering Economics,* McGraw-Hill Book Company, New York, 1977.
18. Strickland, T. H., and E. C. Grady: "Capital Budgeting in Project Evaluation: Part I," *Hydrocarbon Processing,* March 1981, pp. 179–204.
19. Thorne, H. C.: "Post Installation Appraisals: What American Oil Looks For," *Petroleum Refiner* (now *Hydrocarbon Processing*), March 1965, pp. 109–111.
20. White, J. A., M. H. Agee, and K. E. Case: *Principles of Engineering Economic Analysis,* John Wiley & Sons, Inc., New York, 1977.
21. Zinn, C. D., and W. G. Lesso: "Analyze Risks with This Method," *Hydrocarbon Processing,* December 1979, pp. 74–79.

SIX

INFLATION

F. C. Jelen

6.1 Cost Comparisons under Inflation

Inflation is a factor in cost evaluation and cost comparisons. This chapter is not concerned with the causes of inflation, the social, economic, and political consequences, or the problem of mitigating inflation. Its purpose is to develop the mathematical relationships so that inflation can be included along with rate of return, taxes, and productivity gain.

When inflation is a factor, costs change each year as inflation increases or decreases. Although a varying inflation rate can be accommodated, the rate is, at best, a prediction. It will be assumed that the inflation rate is constant and that all costs and revenues are equally affected. A cost of A now will become $A(1 + d)$ 1 year from now, $A(1 + d)^2$ 2 years from now, and so forth, where d is the decay or inflation rate per year expressed as a decimal. The relationship is

$$A_{t2} = A_{t1}(1 + d)^n \tag{6.1}$$

where A_{t1} is a cost now and A_{t2} is the cost n years from now.

6.2 Unaburden

When there is no inflation unacost, a uniform end-of-year cost is a satisfactory criterion for comparing different systems. With inflation, a satisfactory end-of-year criterion is not a uniform cost in dollars but becomes a uniform burden or cost, called *unaburden,* that increases each year by the factor $(1 + d)$. Thus the burden will be R_D in dollars at the end of the first year and $R_D(1 + d)^{n-1}$ in dollars at the end of the nth year. The tax offsets from depreciation are not affected by inflation since depreciation is based on the original cost.

Suppose an article now has a depreciable first cost C_d and lasts n years. The present value of the first cost and its tax credits is given in Eq. (3.21) and is

$$P = C_d(1 - t\psi) \tag{6.2}$$

The problem is to find a unaburden R_D which on an after-tax basis is equivalent to the value given in Eq. (6.2). The present value of the end-of-year costs, where R_D is the actual cost at the end of the first year, is

$$P = \sum_1^n R_D \frac{(1 + d)^{n-1}}{(1 + r)^n} \tag{6.3}$$

Equation (6.3) is a geometric series with the ratio $(1 + d)/(1 + r)$, first term $R_D/(1 + r)$, and having n terms. It can be evaluated by Eq. (2.14) and is

$$P = R_D \frac{(1 + r)^n - (1 + d)^n}{(1 + r)^n} \frac{1}{(r - d)} \tag{6.4}$$

For equal present values, Eqs. (6.2) and (6.4) are equal, leading to

$$R_D = C_d(1 - t\psi_{r,n'}) \frac{(1 + r)^n}{(1 + r)^n - (1 + d)^n}(r - d) \tag{6.5}$$

Equation (6.5) is the relationship that converts a depreciable first cost C_d into a unaburden R_D. Note that the life for tax purposes n' can be different from the economic life n. In Eq. (6.5) the unaburden is calculated on an after-tax basis, but it is possible to refer to a before-tax unaburden as well.

To emphasize, unaburden R_D is the actual value at the end of the first year and increases by the factor $(1 + d)$ each year. The equivalent value at zero time is $R_D/(1 + d)$. No confusion should result between R_D and $R_D/(1 + d)$. Either would be satisfactory as a definition, but unaburden is defined as the actual cost at the end of the first year.

Unaburden is an important concept. It permits rate of return, tax rate, and inflation rate to be included independently. It allows articles and systems with different durations to be compared. Unaburden corrects for different durations, as can be shown by reworking Eqs. 6.2 and 6.3 based on a duration of $2n$ years. Equation 6.2 becomes

$$P = C_d(1 - t\psi) + \frac{C_d(1 + d)^n}{(1 + r)^n}(1 - t\psi) \tag{6.6}$$

and Eq. (6.3) becomes

$$P = \sum_1^{2n} R_D \frac{(1 + d)^{n-1}}{(1 + r)^n} \tag{6.7}$$

Setting the two equal and executing the algebraic operations gives Eq. (6.5) as before.

6.3 Unaburden for Various Costs

Using the previous procedure, relationships for other types of costs can be derived. A summary is given in Table 6.1. In using the table, all costs are entered as their present cost. The relationships will correct for the expected change in cost with inflation when the costs are actually incurred. For item 7, t is the tax rate for the increase in salvage value arising from inflation over the expected salvage value without inflation. Note that t is usually the same rate as the tax rate for regular income.

Normally, the rate of return r will be greater than the inflation rate d. Table 6.1 also gives the correct relationships if the inflation rate is greater than the rate of return.

A special case arises when the inflation rate is equal to the rate of return, that is, when $r = d$. Some of the relationships in Table 6.1 assume the form 0/0, but evaluation by the calculus leads to the following for the items in the table taken consecutively where $r = d$:

$$R_D = \frac{C_d}{n}(1 - t\psi_{r,n'})(1 + r) \tag{6.8}$$

$$R_D = R(1 - t)(1 + r) \tag{6.9}$$

Table 6.1 Summary of unaburden relationships

Item no.	Item		Unaburden
1	C_d	Depreciable first cost*	$C_d(1 - t\psi_{rn'})\dfrac{(1 + r)^n}{(1 + r)^n - (1 + d)^n}(r - d)$
2	R	Uniform end-of-year burden*	$R(1 - t)(1 + d)$
3	R_b	Uniform beginning-of-year burden*	$R_b(1 - t\psi_{r1})(1 + r)$
4	C_{ex}	Irregular cost at end of xth year*	$C_{ex}(1 - t)(1 + d)^x\dfrac{(1 + r)^{n-x}}{(1 + r)^n - (1 + d)^n}(r - d)$
5	C_{bx}	Irregular cost at beginning of xth year*	$C_{bx}(1 - {}'t\psi_{r1})(1 + d)^{x-1}\dfrac{(1 + r)^{n-x+1}}{(1 + r)^n - (1 + d)^n}(r - d)$
6	C_{nd}	Nondepreciable first cost*	$C_{nd}(r - d)$
7	C_{sal}	Salvage value*	$C_{sal}\left[1 + t\dfrac{(1 + d)^n - 1}{(1 + r)^n - (1 + d)^n}\right](r - d)$

*All costs are calculated in terms of the present cost.

$$R_D = R_b(1 - t\psi_{r1})(1 + r) \tag{6.10}$$

$$R_D = \frac{C_{ex}}{n}(1 - t)(1 + r) \tag{6.11}$$

$$R_D = \frac{C_{bx}}{n}(1 - t\psi_{r1})(1 + r) \tag{6.12}$$

$$R_D = C_{nd}(0) = 0 \tag{6.13}$$

$$R_D = C_{sal}\frac{t}{n}\frac{(1 + r)^n - 1}{(1 + r)^n}(1 + r) \tag{6.14}$$

Example 6.1 A machine lasts 2 years with negligible salvage value, can be written off in 1 year for tax purposes, and can be purchased now for $20,000. How much can be spent now for a new machine which lasts 10 years with negligible salvage value and can be written off in 10 years for tax purposes using sum-of-the-years-digits depreciation, when money is worth 10 percent per year after a 46 percent tax, if the inflation rate is (a) 4 percent per year, (b) 10 percent per year, and (c) 12 percent per year?

SOLUTION

(a) For 4 percent per year inflation, use item 1, Table 6.1. Unaburden for the 2-year machine with $\psi = F_{SDP, 10\%, 1} = 0.90909$, is

$$R_D = 20,000[1 - 0.46(0.90909)]\frac{(1.10)^2}{(1.10)^2 - (1.04)^2}(0.10 - 0.04)$$

$$= \$6,579$$

For the same unaburden for the 10-year machine, with $\psi = F_{SDP, 10\%, 10} = 0.70099$,

$$6579 = C_d[1 - 0.46(0.70099)]\frac{(1.10)^{10}}{(1.10)^{10} - (1.04)^{10}}(0.10 - 0.04)$$

$$C_d = \$69,480$$

(b) For 10 percent per year inflation, use Eq. (6.8). For the 2-year machine

$$R_D = \frac{20,000}{2}[1 - 0.46(0.90909)](1.10) = \$6400$$

For the same unaburden for the 10-year machine

$$6400 = \frac{C_d}{10}[1 - 0.46(0.70099)](1.10)$$

$$C_d = \$85,870$$

(c) For a 12 percent per year inflation rate, use item 1, Table 6.1.

$$R_D = 20,000[1 - 0.46(0.90909)]\frac{(1.10)^2}{(1.10)^2 - (1.12)^2}(0.10 - 0.12)$$

$$= \$6342$$

For the same unaburden for the 10-year machine

$$6342 = C_d[1 - 0.46(0.70099)]\frac{(1.10)^{10}}{(1.10)^{10} - (1.12)^{10}}(0.10 - 0.12)$$

$$C_d = \$92,400$$

Example 6.2 A new tank costs $10,000, lasts 3 years, and belongs to a group depreciated in 5 years with sum-of-the-years-digits depreciation. If money is worth 10 percent per year after a 46 percent tax and the inflation rate is 4 percent per year, how much can be spent for repairing the tank at the end of each year as an equal burden so as to obtain a 5-year life?

SOLUTION Unaburden for a new tank lasting 3 years, by item 1, Table 6.1, with $\psi = F_{SDP, 10\%, 5} = 0.80614$, is

$$R_D = 10,000[1 - 0.46(0.80614)]\frac{(1.10)^3}{(1.10)^3 - (1.04)^3}(0.10 - 0.04)$$

$$= \$2437.5$$

Unaburden for a new tank, plus repairs, is given by items 1 and 2, Table 6.1, and for equal economy is

$$2437.5 = 10,000[1 - 0.46(0.80614)]\frac{(1.10)^5}{(1.10)^5 - (1.04)^5}(0.10 - 0.04) + R(1 - 0.46)(1.04)$$

$$R = \$1592 \quad \text{present dollars per year before taxes}$$

That is, $1592(1.04) is the actual cost at the end of the first year, $1592(1.04)^2 is the actual cost at the end of the second year, and so forth.

6.4 Capitalized Cost

Capitalized cost can be used when inflation is a factor in comparing systems, and it will make a correct adjustment for different service lives. To convert a unaburden R_D to a capitalized cost K, evaluate Eq. (6.3) out to infinity.

$$K = \sum_1^\infty R_D \frac{(1 + d)^{n-1}}{(1 + r)^n} \tag{6.15}$$

which can be summed similarly to Eq. (6.3) and, if $r > d$, becomes

$$K = \frac{R_D}{r - d} \quad \text{or} \quad R_D = (r - d)K \tag{6.16}$$

Thus, for $r > d$, Table 6.1 can be used to obtain the capitalized cost by dividing each unaburden by $(r - d)$.

If the inflation rate is greater than the rate of return, then a concept called the *extended capitalized cost* is obtained from Table 6.1 by dividing the unaburden by $(r - d)$, as above, and then taking the negative of that expression. The extended capitalized cost remains positive and is just as suitable for the circumstance $d > r$ as is the regular capitalized cost for the circumstance $r > d$. For details, see the original articles [1].

For the special case in which the rate of return and the inflation rate are equal, Eqs. (6.8) through (6.14) based on unaburden must be used.

6.5 Inclusion of Productivity

If the productivity gain is equal to p each year, expressed as a decimal, costs will reduce by a factor $(1 - p)$ each year due to productivity alone. With both productivity and inflation, costs will change by the factor $(1 - p)(1 + d)$ each year. The intriguing problem of including all four rates—rate of return, tax rate, inflation rate, and productivity gain rate—has been evaluated using capitalized cost as the criterion for equivalence.

The reader is referred to the original articles [5] for details of the method, which includes all kinds of costs. The capitalized cost for a depreciable first cost C_d is given by

$$K = C_d(1 - t\psi_{r,n'})\frac{(1 + r)^n}{(1 + r)^n - (1 - p)^n(1 + d)^n} \tag{6.17}$$

Example 6.3 An article that lasts 1 year can be purchased for $10,000 and written off in 1 year. How much can be paid for an article that lasts 5 years if it can be written off in 3 years for tax purposes using straight-line depreciation when the rate of return is 16 percent per year after a 46 percent tax and the productivity gain rate is 3 percent per year with an inflation rate of 10 percent per year?

SOLUTION Capitalized cost for the 1-year article, by Eq. (6.17), with $\psi = F_{SLP, 16\%, 1} = 0.86207$, is

$$K = 10,000[1 - 0.46(0.86207)]\frac{1.16}{1.16 - [(0.97)(1.10)]} = \$75,269$$

For the same capitalized cost for the 5-year article, with $\psi = F_{SLP, 16\%, 3} = 0.74863$,

$$75,269 = C_d[1 - 0.46(0.74863)]\frac{(1.16)^5}{(1.16)^5 - [(0.97)^5(1.10)^5]}$$

$$C_d = \$39,210$$

That is, $39,210 could be paid for the 5-year article.

The above solution is correct because the same productivity gain rate applies to the 1-year and 5-year articles. Now and in the future, it will be possible to pay 3.921 times as much for the 5-year article as for the 1-year article. However, if the productivity gain rates for the articles are different, the ratio of justifiable costs will change with time. The reader should refer to the original reference.

It is interesting to rework the example for different combinations of the productivity gain rate p and the inflation rate d, with the following results:

Productivity rate, p	Inflation rate, d	Justifiable first cost for 5-year article
0	0	$34,959
0.03	0	33,218
0	0.10	41,500
0.03	0.10	39,210

The tabulation shows that productivity and inflation have opposing effects. Inflation adds to what can be paid for the longer-lasting article by buying into the future at today's cost. Productivity gain penalizes the longer-lasting article because the purchaser will be delayed in entering the market to obtain the reduced cost resulting from technological advancement.

At the present time, the inflation rate is much greater than the productivity gain rate; hence inflation dominates the comparison. Also, the inflation rate as deduced from cost indices is the *net* inflation rate after the productivity gain rate. Under the present circumstances, it is adequate to use the net inflation rate as a combination rate and omit the productivity gain rate, except for a few cases of new technology in which the productivity gain rate might be significant.

6.6 Allowance for Inflation

Inflation has another serious consequence: It causes companies to reinvest funds on an after-tax basis just to stay even. Thus depreciation is based on the original investment and not on the current cost. The depreciation expense for accounting purposes is less than the true current cost. Consequently, the tax base is higher than it should be, and taxes are overpaid. The true profits are less than those indicated. The inadequate allowance for depreciation is not the sole culprit. Inventory increases in value even if it remains the same physically, and reinvestment of money is required just to stay even.

The concept of an *allowance for inflation* has been proposed as an accounting term to recognize the hidden costs created by inflation [3]. The hidden cost of inflation is particularly insidious since it must be met on an after-tax basis, and at the present 46 percent tax rate a profit of almost $2 before taxes is required to meet a $1 inflation cost. The hidden cost of inflation becomes a double burden and strikes hardest at those industries having high long-term investments—the very industries that need the most help in fighting inflation. Reference 3 shows how to calculate the allowance for inflation on a yearly basis; furthermore, it emphasizes that the current use of net profit after taxes is really a profit after taxes but *before* inflation. The true profit is the profit after taxes and after inflation.

In summary, inflation is now an important factor, since it affects costs directly and the compound rate becomes high in a few years. The hidden costs created by inflation are equally serious. Inflation and technological advancement have opposing effects, but inflation now predominates, with technology having only a small mitigating effect.

Deflation can be calculated using the mathematical equations of this chapter by giving d, the inflation rate, a negative value. However, deflation may be of historical interest only.

In engineering economy and cost comparisons future inflation is assumed and therefore must be predicted. There is no excuse for omitting inflation merely because it cannot be predicted precisely. Indeed, many contracts now contain various escalation clauses in order to remain realistic in terms of the changing dollar.

Inflation, in the last analysis, is not a matter of changing prices but rather of changing currency, and is one of the most important current topics in cost engineering.

NOMENCLATURE

A_{t1}	Amount at time 1, $
C_{bx}	Irregular cost at beginning of year x, $
C_d	Depreciable first cost, $
C_{ex}	Irregular cost at end of year x, $
C_{nd}	Nondepreciable cost, $
C_{sal}	Salvage value, $
d	Inflation rate, decay of $, decimal per year
F_{SDP}	Present value of $1 of future sum-of-the-years-digits depreciation, decimal
F_{SLP}	Present value of $1 of future straight-line depreciation, decimal
K	Capitalized cost, $
n	A number, such as number of years for economic analysis
n'	Number of years of life for tax purposes
p	Technological advancement rate, decimal per year
P	Present value, $
r	Rate of return after taxes, decimal per year
R	Uniform end-of-year annual amount without inflation, $/year
R_D	Unaburden, a uniform end-of-year burden with inflation, $/year
t	Tax rate, decimal
ψ	Present value of $1 of future depreciation, decimal

PROBLEMS

6.1 An article costs $20,000 and lasts 1 year. How much can be spent for an article that lasts 3 years if money is worth 20 percent per year after a 46 percent tax, with a 10 percent per year inflation rate and a 3-year life with straight-line depreciation?

6.2 A machine costs $100,000, lasts 2 years, and is written off in 3 years with straight-line depreciation for tax purposes. Using unaburden, find how much can be spent for a machine that lasts 5 years and is written off in 3 years using straight-line depreciation for tax purposes if the rate of return is 16 percent per year after a 46 percent tax and the inflation rate is 10 percent per year.

6.3 A factory floor can be resurfaced for $10,000 and lasts 1 year. Money is worth 16 percent per year after a 46 percent tax, straight-line depreciation will be used, and the inflation rate is 10 percent per year. How long must a special resurfacing job last if it costs $40,000 and can be written off in 3 years for tax purposes?

6.4 Use the data of Prob. 2.63. Money is worth 16 percent per year after a 46 percent tax, and the inflation rate is 10 percent per year. All costs are given in terms of the present dollar. Use a 3-year life for each machine for tax purposes and straight-line depreciation. Which machine is cheaper, and by how much in before-tax dollars?

6.5 Show on a time diagram how Eqs. (6.6) and (6.7) are related. Evaluate the present values and show the algebraic calculations leading to Eq. (6.5), thereby demonstrating that unaburden R_D is independent of duration.

6.6 Derive the relationship for item 2, Table 6.1.

6.7 Derive the relationship for item 3, Table 6.1.

6.8 Derive the relationship for item 4, Table 6.1.

6.9 Derive the relationship for item 5, Table 6.1.

6.10 Derive the relationship for item 6, Table 6.1.

6.11 Derive the relationship for item 7, Table 6.1.

6.12 Derive Eq. (6.8) from item 1, Table 6.1, for the special case $r = d$.

6.13 Maintenance repairs for a machine cost $10,000 at the beginning of the year and are written off in full at the end of the year, and the repairs have a 1-year life. If a new machine costs $37,620, lasts 5 years, and is written off in 5 years using straight-line depreciation, and if money is worth 16 percent per year after a 46 percent tax, at what inflation rate would the 5-year machine be justified?

6.14 Sometimes a shortcut period is used in which the desired rate of return is added to the inflation rate and the problem worked out using the combined rate. The method thus reduces to a single rate which can be evaluated by referring to Chap. 3. Rework Prob. 6.1 on this basis using a single rate of return of 30 percent (20 + 10 = 30 percent) and find how much can be spent for the 3-year article. Do you agree that the rate of return and the inflation rate can be combined?

6.15 Rework Prob. 6.14 on the basis of a 30 percent rate of return, 20 percent over the inflation rate, but including the 10 percent inflation rate using the method given in this chapter. Find how much can be spent for the 3-year article.

6.16 Repeat Prob. 6.2 on the basis of capitalized cost. Note that the method used, whether based on unaburden or capitalized cost, disappears.

6.17 In using Eq. (6.17) for a system, the rate of return r, the tax rate t, and the inflation rate d should be given the same value throughout the system. Does this require that the productivity gain rate p be constant throughout the system?

REFERENCES

1. Jelen, F. C.: "Consider Inflation in Comparative Cost Analyses," *Chemical Engineering,* May 1956, pp. 164–169; June 1956, pp. 247–252.
2. Jelen, F. C.: "Remember All Three in Cost Analyses," *Chemical Engineering,* January 27, 1958, pp. 123–128.
3. Jelen, F. C.: "Include Inflation in Cost Accounting," *Petroleum Refiner* (now *Hydrocarbon Processing*), July 1959, pp. 113–116.
4. Jelen, F. C., and M. S. Cole.: "Can New Equipment Cost Less?," *Hydrocarbon Processing,* July 1971, pp. 97–100.
5. Jelen, F. C., and M. S. Cole.: "Methods for Economic Analysis," *Hydrocarbon Processing,* Part 1, July 1974, pp. 133–139; Part 2, September 1974, pp. 227–233; Part 3, October 1974, pp. 161–163.

OPTIMIZATION

SEVEN

BREAK-EVEN AND MINIMUM-COST ANALYSIS

F. C. Jelen

7.1 Break-even Analysis

Break-even analysis is a method for presenting costs and profits in a form designed to aid interpretation and analysis. The presentation can be graphical or tabular. Typically, this information presents profits as a function of a percentage of plant utilization, reveals plants where utilization operations are showing no profit and are just breaking even, and determines the point at which it will pay to suspend operations.

As a simple example, consider a plant having *fixed costs* C_F in dollars per year which do not change with the production rate, such as depreciation and local taxes. Other costs will be variable and will change with production, such as materials and labor. Suppose each unit of production has a cost of V \$ per unit where V is the *variable cost per unit of production*. Then, if a linear relationship exists, the variable cost itself will be NV where N is the number of units per production. Note that V is the cost per unit of production and may be constant and that NV is the actual variable cost. If C_T is the total annual cost, then

$$C_T = NV + C_F \qquad (7.1)$$

In Fig. 7.1 C_T is a straight line because V is assumed to be independent of N, and C_F is constant. Other situations are considered below in which the variable cost per unit of production V depends upon N, and the variable cost and total cost become curved.

Fixed costs include expenditures which are independent of production rate, such as (1) indirect plant costs including depreciation, (2) management expense, and (3) selling expense.

The variable cost is made up of items such as materials and labor. In general,

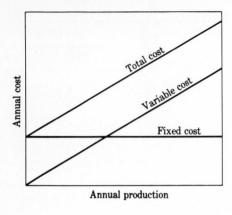

Annual production

Figure 7.1 Diagram of relations between costs and production.

the variable cost per unit of production is essentially constant. However, some variable costs per unit of production depend upon factors such as power, which is sold on a declining-rate scale, or labor costs, which may increase per unit of production at high production rates because of inefficient labor or overtime.

7.2 Economic-Production Charts

The economic-production chart is a geometric chart which utilizes cost and price data and reduces them to a simple visual model. The economic-production, or break-even, chart was proposed by Rautenstrausch [6] and can be utilized in either graphical or analytical form to evaluate an economic operation. The charts make visible the interrelationship of fixed costs, variable costs, sales revenue, and profits. The cost engineer can evaluate the overall economy of the operation from the charts and determine the relative effects of changes in the components.

The mathematical relationships take a very simple form if the following assumptions are made for the purpose of this section:

1. V is constant; hence the variable costs NV are linearly dependent on production.
2. Fixed costs are independent of production.
3. There are no financial costs.
4. There is no income other than that from operations.
5. All units are sold at the same price per unit.
6. All units produced are sold.

If S is the net sales realization *per unit* and Z_g the gross profit, then using the expression for C_T obtained from Eq. (7.1),

$$Z_g = NS - C_T = NS - (NV + C_F) \qquad (7.2)$$

If Z_t is net profit, i.e., profit after taxes, and the decimal tax rate is t,

$$Z_t = Z_g(1 - t) \qquad (7.3)$$

The various relationships all plot as straight lines against N, units of production, and are shown in Fig. 7.2. The point at which the sales line crosses the total-cost line is called the *break-even point* and is the point below which operation results in a loss and above which it yields a profit.

The break-even point can be found mathematically by setting $Z_g = 0$ in Eq. (7.2)

$$N \text{ break-even} = \frac{C_F}{S - V} \qquad (7.4)$$

This point is not a function of the tax rate since no profit is obtained at this production level.

The quantitative effects of various selling prices can be visualized from the chart by sketching a separate net-sales line for each different unit sales price, and

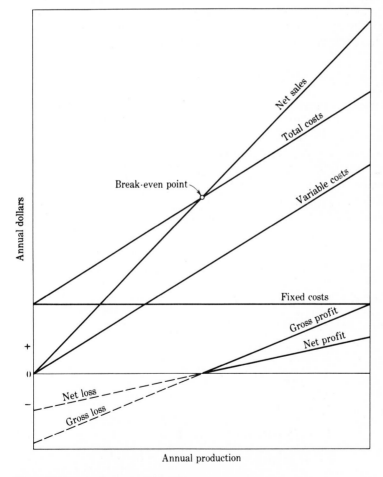

Figure 7.2 Economic-production chart.

the procedure can be used for other variables. The charts can be utilized for any period of time and for any combination of products provided the pertinent costs are prorated.

7.3 Economic-Production Chart above 100 Percent Capacity

Usually a project operates at maximum efficiency when producing at 100 percent capacity. However, even if the variable cost per unit V increases at operations above 100 percent of capacity, the total profit probably will continue to increase. These points are illustrated in Fig. 7.3. If the two parts of the plot are considered as two steps, all the previous relations apply for each step, the proper terms being included for operations above 100 percent of capacity, here denoted by primes. For example, gross profit can be written

$$Z_g = (N + N')S - (NV + N'V' + C_F) \tag{7.5}$$

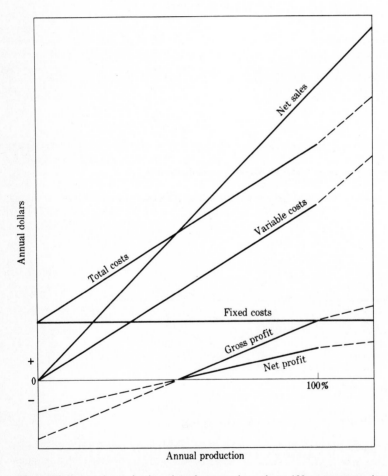

Figure 7.3 Economic-production chart for operations above 100 percent capacity.

where N' and V' are the number of production units above 100 percent capacity and the associated variable cost per unit and N and V are the corresponding values at 100 percent capacity.

Example 7.1 A project can produce 12,000 units per year at 100 percent of capacity. Fixed costs are $10,000/year. The variable cost per unit is $3 up to 100 percent of capacity and $3.30 above 100 percent of capacity. All units are sold at a net selling price of $5. The tax rate is 52 percent. Find the break-even point and the profit per year after taxes at 80 and 120 percent of production capacity.

SOLUTION The break-even point is given by Eq. (7.3):

$$N = \frac{C_F}{S - V} = \frac{10,000}{5 - 3} = 5000 \text{ units per year}$$

The gross profit at 80 percent of capacity is given by Eq. (7.2):

$$Z_g = 0.80(12,000)(5) - [0.80(12,000)(3) + 10,000] = \$9200/\text{year}$$

which after a 52 percent tax is

$$Z_t = 9200(1 - 0.52) = \$4416/\text{year}$$

The gross profit at 120 percent of capacity is given by Eq. (7.5):

$$Z_g = 1.20(12,000)(5) - [12,000(3) + 0.2(12,000)(3.30) + 10,000] = \$18,080/\text{year}$$

which after a 52 percent tax is

$$Z_t = 18,080(1 - 0.52) = \$8678/\text{year}$$

Example 7.2 The *average* net sales realization for a unit of product in dollars for a production of M kilounits per year is given by $2000 - 40M$. Find the net sales realization at a production rate of 2500/year and the incremental sales income at that production rate.

SOLUTION Let $N = 1000M$ where N represents one unit. Then the average net sales is:

$$2000 - 40M = 2000 - 40\frac{N}{1000} = 2000 - 0.04N$$

For $N = 2500$,

$$\frac{\Sigma NV}{N} = 2000 - 0.04N$$

$$\Sigma NV = 2000N - 0.04N^2 = 2000(2500) - 0.04(2500)^2 = \$4,750,000 \text{ net sales realization}$$

The incremental sales realization is:

$$\frac{d}{dN}(2000N - 0.04N^2) = 2000 - 0.08N = 2000 - 0.08(2500) = \$1800$$

That is, each additional unit N brings in $1800.

7.4 Economic-Production Chart for Dumping

The principles discussed in connection with Fig. 7.3 apply to the practice of *dumping,* which occurs when a manufacturer sells N units of production at one sales price, S per unit, but because demand is not sufficient to take all the production, he or she sells N'' units at a lower price, S'' per unit. The manufacturer obtains greater sales and keeps the plant operating at higher output. The diagram is similar to Fig. 7.3 and shows a break in the sales line, gross profit, and net profit at N. The gross profit is given by

$$Z_g = NS + N'S' - [(N + N'')V + C_F] \tag{7.6}$$

Example 7.3 For the data given in Example 7.1, the manufacturer can sell only 80 percent of rated capacity at $5 per unit. How much should the manufacturer charge for additional units if production is brought up to 100 percent of capacity and profits after taxes are increased by an additional $1000?

SOLUTION The net profit after taxes at 80 percent of production was $4416. The gross profit at 100 percent of production will be

$$\frac{4416 + 1000}{1 - 0.52} = 11,283$$

The manufacturer will sell $0.80(12,000) = 9600$ units at $5 and $0.20(12,000) = 2400$ units at S''. By Eq. (7.6),

$$11,283 = 9600(5) + 2400S'' - [(9600 + 2400)(3) + 10,000]S'' = \$3.87 \text{ per unit}$$

7.5 Nonlinear Economic-Production Charts

In practice, it is necessary to consider the analysis of economic-production charts in which some entries vary with production in a nonlinear manner. Costs which change disproportionately with increased production are classified as *regulated costs,* here denoted by NR. Included in this category are labor, maintenance, and utilities. The following example illustrates a general case.

Example 7.4 A plant operation for 1 year has fixed costs C_F, variable costs NV, regulated costs NR, and sales realization NS, as follows, where N is the operating rate in percent:

$$C_F = 20,000$$

$$NV = 500N$$

$$\begin{cases} NR = 15,000 + 30N^{1.5} & N > 0 \\ NR = 0 & N = 0 \end{cases}$$

$$NS = 2000N - 7.5N^2$$

A tabulation and a plot of various items are shown in Table 7.1 and Fig. 7.4.
   At the break-even point there is no profit.

$$0 = Z_g = NS - (NV + NR + C_F)$$

$$= 2000N - 7.5N^2 - (500N + 15,000 + 30N^{1.5} + 20,000)$$

$$= 1500N - 30N^{1.5} - 7.5N^2 - 35,000$$

There is a lower break-even point at $N = 32.1$, as read from the tabulation, the graph, or by a trial and error solution of the last equation. It is the lowest production rate that will not result in a loss.

Table.7.1 Tabulation for Example 7.4, dollar basis, 1 year

(1) Line no.	(2) N percent operation	(3) C_F fixed cost	(4) NV variable cost	(5) NR regulated cost	(6) Total cost, (3) + (4) + (5)	(7) NS sales realization	(8) Profit before taxes, (7) − (6)
1	0	20,000	0	0	20,000	0	−20,000
2	0.1	20,000	50	15,001	35,051	200	−34,851
3	5	20,000	2,500	15,335	37,835	9,813	−28,022
4	10	20,000	5,000	15,949	40,949	19,250	−21,699
5	11.4	20,000	5,700	16,155	41,855	21,825	−20,030
6	20	20,000	10,000	17,683	47,683	37,000	−10,683
7	30	20,000	15,000	19,930	54,930	53,250	−1,680
8	32.1	20,000	16,050	20,456	56,506	56,472	0
9	40	20,000	20,000	22,589	62,589	68,000	5,411
10	50	20,000	25,000	25,606	70,606	81,250	10,644
11	60	20,000	30,000	28,943	78,943	93,000	14,057
12	70	20,000	35,000	32,570	87,570	103,250	15,680
13	74.2	20,000	37,100	34,175	91,275	107,108	15,833
14	80	20,000	40,000	36,466	96,466	112,000	15,534
15	90	20,000	45,000	40,614	105,614	119,250	13,636
16	100	20,000	50,000	45,000	115,000	125,000	10,000
17	110	20,000	55,000	49,610	124,610	129,250	4,640
18	116.8	20,000	58,400	52,869	131,269	131,283	0
19	120	20,000	60,000	54,436	134,436	132,000	−2,436
20	130	20,000	65,000	59,467	144,467	133,250	−11,217
21	138.5	20,000	69,250	63,899	153,149	133,133	−20,016
22	140	20,000	70,000	64,695	154,695	133,000	−21,695
23	150	20,000	75,000	70,114	165,114	131,250	−33,864

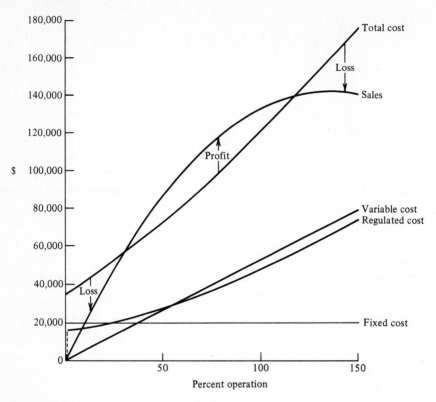

Figure 7.4 Chart for Example 7.4.

There is also an upper break-even point at $N = 116.8$ known as a *profit limit point*. A production level *above* this point will result in a loss.

The maximum profit is $15,833 and occurs at $N = 74.2$. It can be obtained from the tabulation, the figure, by maximizing the right side of the last equation, or by differentiating the last equation, setting it equal to zero, and solving for N by trial and error.

Another concept of interest is the *shutdown point*. At times it may be more economical to shut down completely and absorb the fixed costs rather than operate. In this plant operation, there is a shutdown point at $N = 11.4$. Operation below this point will result in a loss larger than the $20,000 fixed cost at shutdown.

There is a second shutdown point at $N = 138.5$. It is an upper shutdown point; i.e., it would be more economical to shut down than to operate *above* that point.

There are other calculations of interest. The maximum average unit profit is

$$\frac{Z_g}{N} = \frac{1500N - 30N^{1.5} - 7.5N^2 - 35,000}{N}$$

$$= 1500 - 30N^{0.5} - 7.5N - \frac{35,000}{N}$$

Differentiating and setting equal to zero gives

$$0 = -15N^{-0.5} - 7.5 + 35,000N^{-2}$$

for which $N = 61.0$ and $Z_g/N = \$234.42$.

The minimum average unit cost is

$$\frac{C_F + NV + NR}{N} = \frac{20,000 + 500N + 15,000 + 30N^{1.5}}{N}$$

$$= \frac{35,000}{N} + 500 + 30N^{0.5}$$

This has a minimum value of $1096.90 at $N = 176$, which obviously is outside the range of interest in this problem.

The minimum average variable cost is

$$\frac{NV}{N} = \frac{500N}{N} = \$500$$

and is a constant for all values of N.

The minimum average regulated cost is

$$\frac{NR}{N} = \frac{15,000 + 30N^{1.5}}{N} = \frac{15,000}{N} + 30N^{0.5}$$

which has a minimum value of 450 at $N = 100$.

The maxima, minima, and averages can mean different things and should not be confused. Thus maximum profit does not correspond to maximum average profit.

However, there is one similarity that should be noted. Unit incremental net sales revenue and unit incremental total costs are equal at maximum profit, and the equality is not accidental. Thus,

$$\frac{\Delta NS}{\Delta N} = \frac{d}{dN}(2000N - 7.5N^2) = 2000 - 15N$$

and is equal to 887 for $N = 74.2$ at maximum profit. Also,

$$\frac{\Delta(C_F + NV + RN)}{\Delta N} = \frac{d}{dN}(20,000 + 500N + 15,000 + 30N^{1.5})$$

$$= 500 + 45N^{0.5}$$

and is also equal to 887 for $N = 74.2$

The incremental expressions are also known as *marginal expressions*. Thus the marginal cost is the additional cost that results from producing one more unit at a certain production level.

The unit incremental, or marginal, cost is an important criterion in analyzing production charts and can be used to allocate production between competing production facilities.

7.6 One-Product Multiple-Machine Example

Up to this point, the minimum-cost analyses were concerned only with the level at which a facility making one product should be operated. Letting the term *machine* signify a process or plant or any single producing unit, consider now two or more machines being used to make the same product with different cost relationships. The problem is to select the level at which each machine should operate so that overall production costs will be minimal. The following example† has been modified from one originally presented by Schweyer [7] and is used here with the permission of the publisher, Van Nostrand Reinhold Company, New York.

†With permission from Van Nostrand Reinhold Company, New York.

Example 7.5 Total annual costs for three machines, A, B, and C, producing N_A, N_B, and N_C annual units, are

$$C_{TA} = N_A V_A + C_{FA} = 0.06 N_A^{1.1} + 1200 \tag{7.7}$$

$$C_{TB} = N_B V_B + C_{FB} = 0.02 N_B^{1.3} + 1000 \tag{7.8}$$

$$C_{TC} = N_C V_C + C_{FC} = 0.002 N_C^{1.6} + 600 \tag{7.9}$$

If 400 annual production units are made, how should production be allotted to the three machines?

SOLUTION The total cost for all three machines is obtained by adding Eqs. (7.7), (7.8), and (7.9).

$$C_T = 0.06 N_A^{1.1} + 0.02 N_B^{1.3} + 0.002 N_C^{1.6} + 2800 \tag{7.10}$$

If $N = 400$ is the total production from all three plants, then

$$400 = N_A + N_B + N_C \tag{7.11}$$

The problem now is to minimize Eq. (7.10) subject to the restraint, Eq. (7.11). Using the method of Lagrange multipliers as developed in Chap. 10, the Lagrange expression for the problem at hand is

$$LE = 0.06 N_A^{1.1} + 0.02 N_B^{1.3} + 0.002 N_C^{1.6} + 2800 + \lambda(N_A + N_B + N_C - 400)$$

The partial derivatives are set equal to zero for a minimum.

$$\frac{\partial LE}{\partial N_A} = 0.066 N_A^{0.1} + \lambda = 0 \tag{7.12}$$

$$\frac{\partial LE}{\partial N_B} = 0.026 N_B^{0.3} + \lambda = 0 \tag{7.13}$$

$$\frac{\partial LE}{\partial N_C} = 0.0032 N_C^{0.6} + \lambda = 0 \tag{7.14}$$

$$\frac{\partial LE}{\partial \lambda} = N_A + N_B + N_C - 400 = 0 \tag{7.15}$$

These equations show that

$$0.066 N_A^{0.1} = 0.026 N_B^{0.3} = 0.0032 N_C^{0.6} \tag{7.16}$$

Comparing with the partial derivatives of Eqs. (7.7) to (7.9) gives

$$\frac{\partial C_{TA}}{\partial N_A} = \frac{\partial C_{TB}}{\partial N_B} = \frac{\partial C_{TC}}{\partial N_C} \tag{7.17}$$

These results indicate that minimum total costs in a one-product multiple-machine program are achieved when the incremental, or marginal, costs are the same for all machines. Figure 7.5 illustrates the variation of the incremental costs with production. Optimum operation is obtained by drawing a horizontal line (equal incremental costs) and reading the value of the abscissa at the intersection of the horizontal line and each of the incremental-cost curves. The horizontal line is adjusted in position until the sum of the abscissas corresponds to the total production desired.

An analytic solution can be made by trial and error by assuming, say, N_C and calculating N_A and N_B from Eq. (7.16) and repeating until here $N_A + N_B + N_C = 400$. This leads to $N_A = 39$, $N_B = 75$, and $N_C = 285$, which checks with the indicated graphical solution.

The treatment given in this section will give the correct answer if the incremental curves shown in Fig. 7.5 are regular; i.e., among other things they are continuous and single-valued. In practice, the curves do resemble those of Fig. 7.5 and the method is generally applicable.

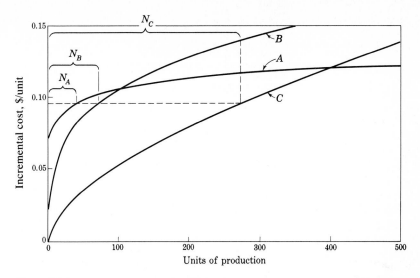

Figure 7.5 Incremental costs for Example 7.5.

7.7 Multiple-Products One-Machine Example

The analysis up to this point implies that production involves one product. A more elaborate problem occurs when there are two or more activities, i.e., a mixture of products, which compete for production time on a machine. When there are multiple activities and all the relationships vary in a linear, or first-order, manner, linear programming as developed in Chap. 12 may be used. For cases in which one or more relationships are nonlinear, an analysis can be made using the principles of economic-production charts. The following† is modified from one given by Schweyer [7], and is used here with the permission of the publisher, Van Nostrand Reinhold Company, New York.

Example 7.6 A machine can make product A at 1000 per hour or product B at 2000 per hour. The available total production time is 2400 h/year. Annual fixed costs are $80,000. Other data are:

Product	Average variable cost per unit, $ per unit	Average net sales per unit, $ per unit
A	$6N_A \times 10^{-8} + 0.2$	0.50
B	0.15	0.20

where N_A is the yearly production units of product A. Find the allocation of products which gives the maximum gross profit per year.

†With permission from Van Nostrand Reinhold Company, New York.

SOLUTION The total annual cost is

$$C_T = NV + C_F = 6N_A^2 \times 10^{-8} + 0.2N_A + 0.15N_B + 80{,}000 \tag{7.18}$$

where N_B is the annual production of product B. The total revenue S_T is

$$S_T = 0.5N_A + 0.2N_B \tag{7.19}$$

and the gross profit Z_g is the difference between Eqs. (7.19) and (7.18):

$$Z_g = 0.3N_A + 0.05N_B - 6N_A^2 \times 10^{-8} - 80{,}000 \tag{7.20}$$

The profit Z_g is to be maximized subject to the following restraint equation, which limits total hours for products A and B to 2400 per year:

$$\frac{N_A}{1000} + \frac{N_B}{2000} = 2{,}400 \tag{7.21}$$

or

$$2N_A + N_B - 4{,}800{,}000 = 0 \tag{7.22}$$

The Lagrange expression is

$$LE = 0.3N_A + 0.05N_B - 6N_A^2 \times 10^{-8} - 80{,}000 + \lambda(2N_A + N_B - 4{,}800{,}000) \tag{7.23}$$

The partial derivatives are set equal to zero for a maximum.

$$\frac{\partial LE}{\partial N_A} = 0.3 - 12 \times 10^{-8}N_A + 2\lambda = 0 \tag{7.24}$$

$$\frac{\partial LE}{\partial N_B} = 0.05 + \lambda = 0 \tag{7.25}$$

$$\frac{\partial LE}{\partial \lambda} = 2N_A + N_B - 4{,}800{,}000 = 0 \tag{7.26}$$

whence $N_A = 1{,}667{,}000$ units per year and $N_B = 1{,}467{,}000$ units per year.

The example can also be solved using the incremental approach. Equations (7.18) for total costs and (7.19) for total revenue as written involve N_A and N_B, but N_B can be eliminated by Eq. (7.22), giving equations for C_T and S_T as a function of N_A only. The derivatives of these equations must be equal at the maximum profit, as shown in the previous section. Setting

$$\frac{dC_T}{dN_A} = \frac{dS_T}{dN_A}$$

will lead to the value of N_A.

The example can be solved analytically by trial and error by assuming, say, N_A, obtaining N_B from Eq. (7.22), and calculating and tabulating the total profit Z_g from Eq. (7.20). Repeating for various values of N_A will establish the maximum profit.

7.8 Critique of the Use of Break-Even and Minimum-Cost Analysis

Break-even analysis is a valuable tool. In general, it is well illustrated by Example 7.4, which was analyzed by a tabulation, a chart, and supplemental calculations and analyses. By showing variations in the curves as the parameters are modified, a single chart can display a range of circumstances, and multiple charts permit wide variations for study. The technique is easily understood and is a good means for presentation.

Utilization of the functional equation for maximum profit is a good introduction to optimization. Other criteria for profitability have been developed in Chap.

5. One of the most serious limitations of the simple economic evaluations presented in this chapter is lack of consideration for the time value of money.

Another simplifying assumption involves the projection of sales prices at different production levels. In practice, probabilities must be utilized to predict the future values of the variables. A general comment on mathematical models seems appropriate at this point.

Mathematical models are an abstraction of a real system. In reality, the systems are much more complex than any mathematical model which could be built. Therefore, it is the task of the analyst to select the significant aspects of the real system and ignore those aspects which make a negligible contribution at the level of accuracy desired.

The analyst must always compromise between constructing the most accurate model and keeping it simple enough to have a manageable solution. The computer has made possible the development of very complex models the solution of which would not have been feasible by analytical or other manual means. Thus, it has increased the degree of accuracy with which real systems can now be simulated.

NOMENCLATURE

C_F	Fixed cost, $
C_T	Total cost, $
LE	Lagrange expression
N	Number of units of production
N'	Number of units of production above 100 percent of capacity
N''	Number of units of production dumped
R	Regulated cost per unit of production, $ per unit
S	Net realizable sales value per unit sold, $ per unit
S''	Net realizable sales value per unit dumped, $ per unit
S_T	Total realizable sales value, $
t	Tax rate, decimal
V	Variable cost per unit of production, $ per unit
V'	Variable cost per unit of production above 100 percent of capacity, $ per unit
Z_g	Gross profit before tax, $
Z_t	Net profit after tax, $
λ	Lagrange multiplier

PROBLEMS

7.1 Waste water can be processed for pollution control by two methods, A and B. Pertinent costs are:

	A	B
Labor, $/year	70,000	95,000
Depreciation, $/year	50,000	28,000
Chemicals, $/m³ treated	0.15	0.10

How much water must be treated per year for the two methods to be equal?

7.2 The average variable cost for a unit of product in dollars for a production of N million units per year is given by

$$1.75 + 0.40N - 0.1N^2$$

Find the variable cost at a production rate of 2.5 million units of production per year. What is the incremental variable cost per unit of production at that production rate?

7.3 A boiler plant now recovers 5000 kg of fly ash per day at a collection cost of $3/t and a disposal cost by dumping in the ocean at $2/t. A plant costing $175,000 with a life of 20 years must be built to recover an additional 6000 kg of fly ash at the same collection cost per metric ton. The plant would process the old and new ash collections. If the plant were not built, the new ash would have to be dumped in the ocean. Find the required selling price per metric ton of fly ash if all the fly ash processed in the new plant could be sold and the processing cost is $1.80/t.

7.4 A plant operating at 80 percent of capacity has fixed costs of $55,000/year, direct costs of $120,000/year, and sales of $200,000/year. What is the break-even point?

7.5 An office intercom system will cost $1000 + 150N$ dollars for N connections. The system will last 8 years, with a maintenance cost of $20/year for each connection. A telephone company will charge $4/month for each intercom outlet by using existing telephone connections. How many connections are necessary to justify the intercom system?

7.6 A company has fixed costs of $300,000/year. At 75 percent of capacity, the variable cost is $800,000 and the sales realization is $1,500,000. On a single break-even chart, show the effect of increasing and decreasing the selling price per unit by 10 percent. What is the break-even point for (*a*) the selling price as is, (*b*) the selling price increased by 10 percent, (*c*) the selling price decreased by 10 percent and (*d*) extra profit when the selling price is decreased by 10 percent, permitting capacity production as opposed to the present rate at the present selling price?

7.7 A company now sells 1000 units per week at a profit of $20,000/week, with fixed costs at $10,000/week and variable costs at $5000/week. If the fixed costs increase by 15 percent, by what fraction must the selling price be raised to maintain the existing profit if output is the same?

7.8 A company with fixed costs of $200,000 and a break-even point of 60 percent for net sales of $400,000 will show what profit before taxes at 100 percent of capacity? If the profits tax is 46 percent, what is the total capitalized value, or market value, if prospective purchasers desire an 18 percent return after taxes based on full-capacity operations? Does the depreciation charge affect the capitalized value?

7.9 The profit at 65 percent of capacity is 50 percent greater than the profit at 50 percent of capacity. Fixed costs are $100,000, and net sales are 300 percent greater than the variable cost at both operating rates. Find the profit at 100 percent of capacity.

7.10 A meat-waste processing plant is designed for operations of 20,000 kg/day at 100 percent of capacity when the average variable costs are $0.48/kg. At 75 percent of capacity, the average variable costs are $0.56/kg. The fixed costs are $1040 per operating day.

(*a*) Compute the unit costs for operations at both 75 and 100 percent of capacity.

(*b*) What is the incremental cost in dollars per day per kilogram in going from 75 percent to 100 percent of capacity?

(*c*) What is the average unit incremental cost in going from the lower to the higher capacity?

7.11 If the ratio of direct costs to sales dollars is 0.4 and the fixed costs are $250,000 annually for a product selling at $0.50/kg, what is the unit cost of the product when annual net sales are (*a*) $1,250,000 and (*b*) at a maximum capacity of 4,000,000 kg?

7.12 For net sales of a company amounting to $600,000 annually, when the fixed costs are $350,000 and the direct costs are 35 percent of the net sales dollars:

(*a*) What is the gross profit?

(*b*) What is the break-even point in terms of sales dollars?

(*c*) What sales are required for a profit of $80,000?

7.13 Under normal conditions, an operation with a capacity of 10,000 units annually has a selling price of $100, which represents $15 in profit, $30 in fixed costs, and $55 in variable costs. Demand drops to

2000 units annually. Fixed costs can be reduced 10 percent below normal under these conditions and 25 percent below normal if the plant shuts down.

(a) Should the plant remain open or shut down?

(b) How low can the selling price go before the plant should shut down when demand is 2000 units annually?

7.14 A manufacturer can produce a part for $5 per part, with fixed annual costs of $15,000. He can buy the part for $8 for the first 10,000 and for $4 for all additional parts used in a year. At what production rate would it be cheaper to buy the part?

7.15 A hardware store pays $10 for a tool and plans to run an advertising campaign costing $500 for the allocated space. The number of sales expected at various prices is:

Expected sales	Selling price
100	$20
200	$16
500	$15
1000	$12

Find the selling price that will give the largest profit.

7.16 In Prob. 7.15, assume that the number to be sold at each selling price is proportional to the square root of the amount spent on advertising. Find the selling price and advertising cost that will maximize the profit, and find the maximum profit.

7.17 An operation rated at 10,000 units per year has fixed costs of $8000/year, variable costs per unit at $5 for production up to 10,000 units per year, variable costs per unit at $6 for production above 10,000 units per year, and a selling price of $9 for each unit of production. Find the break-even point, and the profit after a 46 percent tax at capacity production and at 50 percent above capacity production.

7.18 In the previous problem, assume that all units above 10,000 per year must be sold at $4 to fulfill a prior legal contract. At what production rate will the operation cease to make a profit and show a profit limit point?

7.19 A study of a plant operation having annual fixed costs of $100,000/year has variable costs and net sales expressed as

$$NV = N^3 \times 10^{-7} - 0.0012N^2 + 21.6N$$

$$NS = 76N - 0.0036N^2$$

where N represents tons and costs are for 1 year.

(a) Find the break-even point and the profit limit point.

(b) Find the maximum annual profit and the production rate at which it occurs.

(c) Find the average profit and the incremental profit at a production rate of 9000 ton/year.

7.20 Economic analysis of a plant operation has shown the following empirical relationships, where V is a variable cost per unit and S is net sales realization per unit, with N being units of production:

$$\text{Average sales price} = A - BN^2 \quad \$/\text{kg}$$

$$\text{Average variable cost} = CN^3 + DN^2 \quad \$/\text{kg}$$

If C_F is the total constant annual fixed costs and A, B, C, and D above are constants, develop equations to evaluate N for the following in terms of A, B, C, D, and C_F:

(a) The break-even point.

(b) The production rate for maximum profit.

(c) The maximum profit per kilogram.

(d) The minimum unit cost if N differs from the value in (c).

(e) The marginal cost, or incremental cost.

7.21 The marginal or incremental revenue for a part can be expressed as

$$100 - 0.02N$$

where N is the number of parts produced per period. Total costs, including fixed costs, can be expressed as

$$2 \times 10^{-4}N^2 + 10,000$$

Find (a) the break-even point, (b) the maximum profit and the production rate at which it occurs, and (c) the minimum average cost and the production rate at which it occurs.

7.22 The average revenue for a quantity Q per kilogram is given by

$$100 - 0.01Q$$

and the total cost for Q per period is given by

$$50Q + 30,000$$

Determine (a) the marginal revenue, (b) the marginal cost, and (c) the point of maximum profit and the maximum profit per period.

7.23 Machine A has a capacity of 20,000 units per period, is operating at 30 percent of capacity, and has total costs per period expressed as

$$C_T = 300N + 4N^2 + 12,000$$

where N is the percent of operating capacity.

Machine B has a capacity of 10,000 units per period, is operating at 55 percent of capacity, and has total costs per period expressed as

$$C_T = 75N + 2N^2 + 8000$$

where N is the percent of operating capacity.

(a) If one more unit per period is to be produced, to which machine should it be assigned, i.e., which will have the smaller added cost?

(b) How should the current production be allotted between the machines for minimum cost?

(c) In (b), what will be the incremental cost for each machine?

(d) Find a total production such that the machines share the production load equally and have equal incremental costs. This is equivalent to finding a point at which the machines share the load equally and also have a minimum cost.

7.24 A machine is available for 8 h/day to make products A and B. The machine can make 800 units of product A exclusively, or 2400 units of product B exclusively, or a mixture of both. Product A sells for $11 per unit and product B for $7 per unit. The machine has a fixed cost of $100/8-h day. The variable costs per 8-h day for N_A and N_B units are

$$3N_A + 0.002N_A^2$$

$$5N_B + 50$$

Find the maximum profit possible per day if all units can be sold, and show that the incremental costs for A and B are equal at the optimum mix when an equivalent comparison is used.

7.25 A one-product, two-mix operation is being considered with a limit of $100,000 per time period. Labor A costs $2000 per unit of labor per period, and feedstock B costs $1000/ton. The amount of labor and material used determines the production Q, which in tons per period is given by

$$Q = 5AB$$

(a) What amounts should be spent for labor and materials, and what is the optimum production?

(b) Show how the Lagrange multiplier solves the problem.

REFERENCES

1. Brighton, E. F., and J. L. Pappas: *Managerial Economics,* 2d ed., The Dryden Press, Hinsdale, Ill., 1976.
2. Kasner, E.: *Essentials of Engineering Economics,* McGraw-Hill Book Company, New York, 1979.
3. Levin, R. I., and C. A. Kirkpatrick: *Quantitative Approaches to Management,* 4th ed., McGraw-Hill Book Company, New York, 1978.
4. Newnan, D. G.: *Engineering Ecomomic Analysis,* rev. ed., Engineering Press, Inc., San Jose, Calif., 1980.
5. Park, W. R.: *Cost Engineering Analysis,* John Wiley & Sons, Inc., New York, 1973.
6. Rautenstrausch, W.: *Economics of Business Enterprise,* John Wiley & Sons, Inc., New York, 1939.
7. Schweyer, H. E.: *Analytic Models for Managerial and Engineering Economics,* Van Nostrand Reinhold Company, New York, 1964.
8. Taylor, G. A.: *Managerial and Engineering Economy,* 3d ed., D. Van Nostrand Company, Inc., New York, 1980.

EIGHT

PROBABILITY, SIMULATION, UNCERTAINTY

F. C. Jelen

DISCRETE FREQUENCY DISTRIBUTIONS

8.1 Distribution Functions

Many practical problems involve uncertainty and are probabilistic as opposed to deterministic. It is no longer sufficient to report that a new venture is expected to have a capital investment of 10 M$. It may be necessary to give some measure indicating that it may cost 12 M$ or more. The components going into the estimate have uncertainty and a range of values. The final estimate has an expected value, but also has uncertainty and a range of possible values. The analysis of such problems depends on the distribution that makes up the average or expected value and is part of the mathematics of probability and statistics. The purpose of this chapter is to review those aspects of the mathematics which are essential for an understanding of the application to cost engineering.

Consider the case of throwing dice and noting the total shown. Table 8.1 and Fig. 8.1 give data computed for one, two, and three dice, and the probabilities $P(X)$ for the occurrence of various sums.

The three distributions shown are examples of a *discrete distribution*, for only integer values can appear; i.e., the sums are not a continuous variable. The probabilities in Table 8.1 are easily calculated. Thus in the case of two dice there are 36 equally likely ways the dice can turn up, and the sum 6 can be obtained in five ways: 1—5, 2—4, 3—3, 4—2, and 5—1. The probability for the sum 6 is $5/36 = 0.1389$.

The distributions shown are symmetrical, but in practice distributions may be nonsymmetrical, or skewed.

Table 8.1 Frequency distribution with dice

Sum of dice X	One die — Number of ways for appearance	One die — Probability P(X) of X	Two dice — Number of ways for appearance	Two dice — Probability P(X) of X	Two dice — Cumulative probability $\Sigma P(X)$ of X for Sec. 8.3	Three dice — Number of ways for appearance	Three dice — Probability P(X) of X
1	1	0.1667					
2	1	0.1667	1	0.0278	0.0278		
3	1	0.1667	2	0.0556	0.0833	1	0.0046
4	1	0.1667	3	0.0833	0.1667	3	0.0139
5	1	0.1667	4	0.1111	0.2778	6	0.0278
6	1	0.1667	5	0.1389	0.4167	10	0.0463
7	6		6	0.1667	0.5833	15	0.0694
8			5	0.1389	0.7222	21	0.0972
9			4	0.1111	0.8333	25	0.1157
10			3	0.0833	0.9167	27	0.1250
11			2	0.0556	0.9722	27	0.1250
12			1	0.0278	1.0000	25	0.1157
13			36			21	0.0972
14						15	0.0694
15						10	0.0463
16						6	0.0278
17						3	0.0139
18						1	0.0046
						216	

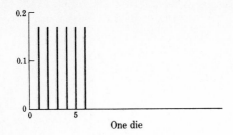

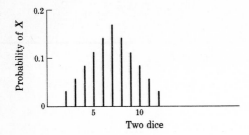

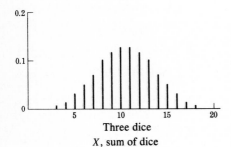

Three dice

X, sum of dice

Figure 8.1 Distributions of sums with dice.

8.2 Properties of Discrete-Distribution Functions

An inspection of Fig. 8.1 shows differences, and it is desirable to develop parameters which show these differences. Only two are considered here, the mean and the variance.

The *mean* is the arithmetic mean of the distribution or a number of samples taken from it. If X_i is the value of the ith in a total of N equally likely events, the mean μ is

$$\mu = \frac{\sum\limits_{i=1}^{i=N} X_i}{N} \tag{8.1}$$

or in terms of the probabilities

$$\mu = \sum X P(X) \tag{8.2}$$

where X is the value of the event and $P(X)$ is the probability. The summation is over all values of X.

Suppose one die is thrown; then by Eq. (8.1), the average is

$$\mu = \frac{1 + 2 + 3 + 4 + 5 + 6}{6} = 3.5$$

The mean value 3.5 is unrealizable in a single throw, but what is meant is that in a large number of throws the mean will be 3.5 per throw. For two dice the mean is

$$\mu = \frac{1(2) + 2(3) + 3(4) + \cdots + 3(10) + 2(11) + 1(12)}{36} = 7.0$$

a result which can also be obtained from Eq. (8.1):

$$\mu = 2(0.0278) + 3(0.0556) + \cdots + 11(0.0556) + 12(0.0278) = 7.0$$

For three dice the mean is 10.5.

A very important concept in connection with probability is the *expectation,* which is identical to the average of the occurrences but is used in a very broad sense. The expectation of X is

$$E(X) = \sum X(PX) \tag{8.3}$$

Expected value of X = sum of (value of X)(probability of X occurring)

Example 8.1 One die is thrown, and the player receives twice the face value if odd and one-half the face value if even. Find the expectation.

SOLUTION By Eq. (8.3)

$$E(X) = 2(1)(\tfrac{1}{6}) + \tfrac{1}{2}(2)(\tfrac{1}{6}) + 2(3)(\tfrac{1}{6}) + \tfrac{1}{2}(4)(\tfrac{1}{6}) + 2(5)(\tfrac{1}{6}) + \tfrac{1}{2}(6)(\tfrac{1}{6}) = 4$$

The second difference between distributions is shown by the *variance,* which is a measure of how much the distribution varies from the average. A small variance means that the distribution is peaked, a large variance that it is spread out. Two distributions may have the same average but differ drastically in the variance, or the way they are spread out. The variance σ^2 is defined as

$$\sigma^2 = \frac{\sum\limits_{1}^{N}(X_i - \mu)^2}{N - 1} \tag{8.4}$$

and σ itself is known as the *standard deviation.*

Example 8.2 The life of a tool in hours before breakage is given by the following data obtained for 50 tools. Tabulate the frequency distribution of the tool life and find the average tool life and its variance.

28.1	31.6	29.9	31.4	31.6
29.9	33.7	30.4	33.9	31.5
30.9	26.5	33.6	31.7	36.4
32.6	30.9	29.3	33.7	32.8
27.7	30.5	30.8	34.4	33.5
29.1	37.8	30.4	33.3	30.5
30.5	32.0	26.5	29.6	31.4
29.6	29.3	29.8	34.4	34.4
30.4	31.0	29.8	38.8	38.9
27.7	26.7	31.6	28.3	26.5

The data are grouped and condensed in Table 8.2. The average value is given by Eq. (8.1).

SOLUTION

$$\mu = \frac{4(26.45) + 2(27.45) + \cdots + 1(37.45) + 2(38.45)}{50} = 31.21 \text{ h}$$

and the variance, by Eq. (8.4), is

$$\sigma^2 = \frac{4(26.45 - 31.21)^2 + 2(27.45 - 31.21)^2 + \cdots + 2(38.45 - 31.21)^2}{50 - 1}$$

$$= 8.26$$

8.3 Cumulative-Distribution Function

A frequency distribution function can be shown as a plot of cumulative values known as a *cumulative-distribution function*. The plots are useful in certain applications. Figure 8.2 is a plot of cumulative probability of the value X or less for two dice with data obtained from Table 8.1.

Table 8.2 Tabulation for Example 8.2

Tool life, h	Midpoint X	Number of tools with this life	Probability $P(X)$ of X
26.0–26.9	26.45	4	0.08
27.0–27.9	27.45	2	0.04
28.0–28.9	28.45	2	0.04
29.0–29.9	29.45	9	0.18
30.0–30.9	30.45	9	0.18
31.0–31.9	31.45	8	0.16
32.0–32.9	32.45	3	0.06
33.0–33.9	33.45	6	0.12
34.0–34.9	34.45	3	0.06
35.0–35.9	35.45	0	0.00
36.0–36.9	36.45	1	0.02
37.0–37.9	37.45	1	0.02
38.0–38.9	38.45	2	0.04
		50	

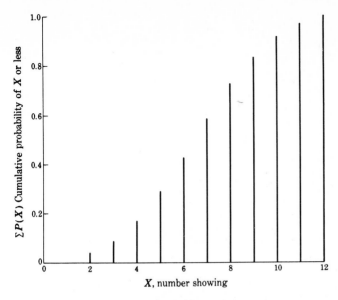

Figure 8.2 Cumulative probability with two dice.

It is emphasized that any frequency function may be changed to a probability function by changing the area under the function equal to unity. The cumulative-probability distribution function is, accordingly, unity when taken in the limit.

Cumulative functions may be used to find the probability for a random variable X lying between two limits A and B. Thus if $A < B$, the probability is

$$P(A < X \le B) = \sum_{i=A+1}^{B} P(X_i) = \sum_{i=0}^{B} P(X_i) - \sum_{i=0}^{A} P(X_i) \qquad (8.5)$$

Example 8.3 Find the probability of throwing a 4, 5, 6, or 7 with two dice. Here the range from above is 3 to 7 inclusive.

SOLUTION By Eq. (8.5) and Fig. 8.2,

$$P(3 < X \le 7) = \sum_{0}^{7} P(X) - \sum_{0}^{3} P(X) = 0.583 - 0.083 = 0.500$$

The same result can be obtained, of course, by adding $P(4)$, $P(5)$, $P(6)$, and $P(7)$.

8.4 Binomial Distribution

The *binomial distribution* applies to events that can take on only two values, such as the head or tail for a tossed coin, or accept or reject for an object. The relationship is

$$P(X) = \frac{N!}{X!(N - X)!} p^X (1 - p)^{N-X} \qquad (8.6)$$

where $P(X)$ is the probability of *exactly* X occurrences in N trials and p is the probability for success in one trial. In the case of an unbiased coin p and $1 - p$ are both 0.5, but in most problems p will not be 0.5. The distribution is symmetrical if and only if $p = 0.5$.

The mean and variance of a binomial distribution, respectively, are

$$\mu = Np \tag{8.7}$$

$$\sigma^2 = Np(1 - p) \tag{8.8}$$

The binomial distribution assumes that the trials are independent. For sampling from a small population without replacement, the binomial distribution should be replaced by the hypergeometric distribution [2], which is

$$P(X) = \frac{\binom{H}{X}\binom{M - H}{W - H}}{\binom{M}{W}} \tag{8.9}$$

where

$$\binom{H}{X} = \frac{H!}{X!(H - X)!}$$

$P(X)$ is the probability of obtaining exactly X objects having a certain property by withdrawing W objects without replacement from a population containing M objects H of which possess the property.

A generalization of the binomial distribution leads to the multinomial distribution. Instead of having two results A and not A with probabilities p and $1 - p$, there are now k results possible each with probability $P_1, P_2, \ldots, P_k$. The probability of getting exactly $X_1, X_2, \ldots, X_k$ occurrence of $A_1, A_2, \ldots, A_k$ is:

$$P(X_1, X_2, \ldots, X_k) = \frac{N!}{X_1!, \ldots, X_k!} P_1^{X_1}, \ldots, P_k^{X_k} \quad \text{with } N = \sum_1^k X_i \tag{8.10}$$

8.5 Poisson Distribution

The Poisson distribution is of great importance and is

$$P(X) = \frac{e^{-\mu}\mu^X}{X!} \tag{8.11}$$

where $P(X)$ = probability of exactly X occurrences
  e = naperian constant $2.71828 \cdots$
  μ = expected or average number of occurrences

The distribution is applicable only when the events occur completely at random and the number that occurs is small compared to the potential number that could occur.

The mean and variance of the Poisson function are both μ. Tables are available for the Poisson function and the accumulated Poisson function.

Example 8.4 At 3 P.M. telephone calls arrive at the company switchboard at the rate of 120/h. Find the probability that exactly 0, 1, and 2 calls arrive between 3:00 and 3:01.

SOLUTION Take 1 min as the period of time. Then

$$\mu = \frac{120}{60} = 2 \text{ calls per minute expected}$$

By Eq. (8.11)

$$P(0) = \frac{e^{-2}2^0}{0!} = 0.1353$$

$$P(1) = \frac{e^{-2}2^1}{1!} = 0.2707$$

$$P(2) = \frac{e^{-2}2^2}{2!} = 0.2707$$

CONTINUOUS FREQUENCY DISTRIBUTIONS

8.6 Continuous Distributions

Continuous distributions arise in practice particularly with measurement as opposed to counting. Although continuous distributions can be reduced to discrete distributions by grouping, much as in Example 8.2, frequently there are advantages in using the continuous property. Essentially this amounts to substituting the integral sign for the summation sign. With this understanding, all the principles that apply for discrete distributions are easily transferred to continuous distributions.

One distribution that must be clearly understood is the use of probability density in place of probability. If a function is continuous, the chance for an event to have exactly the value 3.00 ··· is zero. There is an infinity of possible results, and the chance of exactly 3 is $1/\infty = 0$. However, it is possible to consider a probability density in the neighborhood of 3, with units of probability per unit interval, which when multiplied by an interval length gives the probability for occurrence within the interval.

The symbol $P_d(X)$ will be used for the probability density at the point X, with the understanding that it must be multiplied by some interval of X to convert it to a probability. Generally $P_d(X)$ depends on X and is a function of X.

$$P_d(X) = F(X) \tag{8.12}$$

By analogy with discrete functions, the mean or expected value is

$$\mu = E(X) = \frac{\int X P_d(X)\, dX}{\int P_d(X)\, dX} \tag{8.13}$$

and the variance is

$$\sigma^2 = \frac{\int (X - \mu)^2 P_d(X)\, dX}{\int P_d(X)\, dX} \tag{8.14}$$

All continuous distribution functions used in this chapter are so expressed that the area under the probability-density function is unity and the cumulative probability, as for discrete functions, is unity in the limit. That is,

$$\int_{-\infty}^{+\infty} P_d(X)\, dX = 1 \tag{8.15}$$

8.7 Cumulative-Distribution Function

The cumulative-distribution function for a continuous distribution is obtained by integration and is

$$P_c(A) = \int_{-\infty}^{A} P_d(X)\, dX \tag{8.16}$$

where $P_c(A)$ is the cumulative probability of X having the value A or less. The cumulative probability for some distribution functions has been tabulated in the literature.

The probability that X will lie between A and B, where $A < B$, is

$$P_c(A \leq X \leq B) = \int_{A}^{B} P_d(X)\, dX = \int_{-\infty}^{B} P_d(X)\, dX - \int_{-\infty}^{A} P_d(X)\, dX$$

$$P_c(A \leq X \leq B) = P_c(B) - P_c(A) \tag{8.17}$$

The cumulative-distribution function is the integral of the probability-density function. Hence the derivative of the cumulative-distribution function is the probability-density function at the point.

8.8 Rectangular or Uniform Distribution

Suppose a distribution is uniform over the range 0 to a. Then the probability density is

$$P_d(X) = \begin{cases} \dfrac{1}{a} & 0 \leq X \leq a \\ 0 & \text{elsewhere} \end{cases} \tag{8.18}$$

The condition for Eq. (8.15) is met:

$$\int_{-\infty}^{+\infty} P_d(X)\, dX = \int_{0}^{a} \frac{1}{a}\, dX = 1$$

The cumulative probability for X or less, by Eq. (8.16), is

$$P_c(X) = \int_0^X \frac{1}{a}\, dX = \frac{X}{a} \qquad 0 \le X \le a \qquad (8.19)$$

A plot for the probability-density and the cumulative-probability functions is given in Fig. 8.3.

Example 8.5 A continuous-distribution function is uniform in the range of 0 to 5. Find the average value and the variance.

SOLUTION Here $a = 5$, and by Eq. (8.18),

$$P_d(X) = 0.20 \qquad 0 \le X \le 5 \qquad (8.20)$$

Substitution in Eq. (8.13) gives the mean value

$$\mu = \frac{\int_0^5 X(0.20)\, dX}{\int_0^5 0.20\, dX} = \frac{0.20(\frac{25}{2} - 0)}{0.20(5 - 0)} = 2.5$$

which is obvious. The variance, by Eq. (8.14), is

$$\sigma^2 = \frac{\int_0^5 (X - 2.5)^2 0.20\, dX}{\int_0^5 0.20\, dX} = 0.20\left(\int_0^5 X^2\, dX - 5\int_0^5 X\, dX + 6.25\int_0^5 dX\right)$$

$$\sigma^2 = 0.20[\tfrac{125}{3} - 5(\tfrac{25}{2}) + 31.25] = 2.083$$

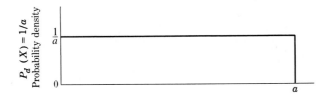

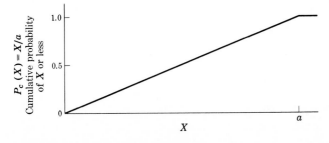

Figure 8.3 Rectangular or uniform distribution.

8.9 Exponential Distribution

The probability-density function for the distribution is

$$P_d(X) = \begin{cases} ae^{-aX} & X \geq 0 \\ 0 & X < 0 \end{cases} \tag{8.21}$$

The condition for Eq. (8.15) is met:

$$\int_0^\infty ae^{-aX}\,dX = -\frac{a}{a}|e^{-aX}|_0^\infty = 1$$

The cumulative probability for X or less, by Eq. (8.16), is

$$P_c(X) = \int_0^X ae^{-aX}\,dX = 1 - e^{-aX} \tag{8.22}$$

The mean of the exponential density distribution is $1/a$ and the variance $1/a^2$. The probability-density function and the cumulative probability of X or less are shown in Fig. 8.4.

There is a connection between the Poisson distribution and the exponential distribution. For example, in queuing problems if the arrival rate, in arrivals per unit time period, follows a Poisson distribution with λ average arrivals per period, then by Eq. (8.11)

$$P(X) = \frac{e^{-\lambda}(\lambda)^X}{X!}$$

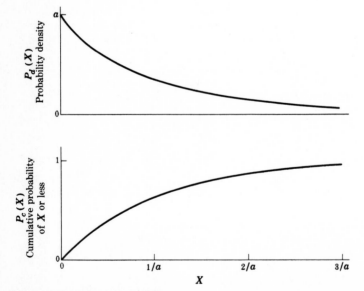

Figure 8.4 Exponential distribution.

It can be shown that the time between arrivals has an exponential distribution with the following probability density:

$$P_d(T_a) = \lambda e^{-\lambda T}$$

where T_a is the time between arrivals measured in periods T. The cumulative probability becomes the time between arrivals of T_a or less and is

$$P_c(T_a) = 1 - e^{-\lambda T}$$

8.10 Normal Distribution in Standard Form

The normal distribution written in standard form using Z for the independent variable has the following probability-density function:

$$P_d(Z) = \frac{1}{\sqrt{2\pi}} e^{-Z^2/2} \tag{8.23}$$

The area under the curve is unity; the curve is centered on $Z = 0$, is symmetrical, and has a mean equal to zero and a variance equal to unity in the standard form.

The cumulative probability of Z or less is given in Table A3.1. The probability-density and cumulative functions are shown in Fig. 8.5.

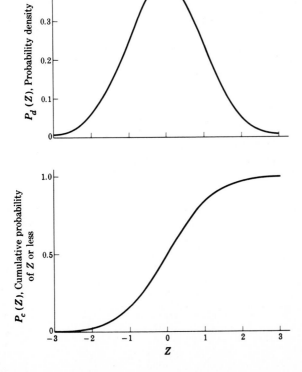

Figure 8.5 Standard normal distribution.

The normal distribution was originally derived from theoretical considerations and has been amply verified by experimental data. It is now recognized that its usefulness is derived from the fact that even when the values in a population are not normally distributed if large groups of samples are drawn from the population, the averages of the groups will be normally distributed. For the same reason, if an experimental measurement is subject to a large number of independent causes for variation, the observations will be normally distributed.

8.11 Normal Distribution with Arbitrary Parameters

Usually a normal distribution has a mean value that is not equal to zero; nor will the variance be unity. The probability density generally will be

$$P_d(X) = \frac{1}{\sigma\sqrt{2\pi}} \exp - (X - \mu)^2/2\sigma^2 \tag{8.24}$$

where μ is the mean and σ^2 is the variance.

If a transformation is made with

$$Z = \frac{X - \mu}{\sigma} \tag{8.25}$$

the normal curve is transformed to a standard normal curve with mean value zero and variance equal to unity. The technique, accordingly, is to transform actual conditions to the standard normal distribution using Eq. (8.25), work out the problem in the standard domain, and revert the answer to the actual domain by Eq. (8.25).

Example 8.6 Measurements on a production-control test give a mean value of 85 with a variance of 17.4. Find the probability for obtaining a value from 80 to 88. Also, find the value which will be exceeded only 10 percent of the time.

SOLUTION Transformation to the standard domain, by Eq. (8.25), gives

$$Z = \begin{cases} \dfrac{80 - 85}{\sqrt{17.4}} = -1.20 & \text{for } X = 80 \\[2mm] \dfrac{88 - 85}{\sqrt{17.4}} = +0.72 & \text{for } X = 88 \end{cases}$$

From Table A3.1,

$$Z = -1.20 \qquad P_c(Z) = 0.1151$$

$$Z = +0.72 \qquad P_c(Z) = 0.7642$$

$$P_c(80 \le X \le 88) = P_c(-1.20 \le Z \le 0.72)$$

By Eq. (8.17),

$$P_c(-1.20 \le Z \le 0.72) = 0.7642 - 0.1151 = 0.6491$$

The probability is 0.649 that Z will lie between -1.20 and $+0.72$, which is the same probability that X will lie between 80 and 88.

For the second part of the example, note that if X is exceeded 10 percent of the time, 90 percent of the time the value will be X or less. By Table A3.1,

$$\text{For } P_c(Z) = 0.900: \quad Z = +1.28$$

By Eq. (8.25),

$$1.28 = \frac{X - 85}{\sqrt{17.4}}$$

$$X = 90.34 = \text{value of } X \text{ which will exceed 10 percent of the time}$$

SIMULATION

8.12 Monte Carlo Method

Problems with uncertainty can be quite complicated, and a solution by direct analytical means may become exceedingly difficult if not impossible. The most general and basic method is to resort to simulation. The number of calculations can become large, but the modern computer is a valuable aid. The application known as the *Monte Carlo method* generates the uncertainty within the problem in conformance with the probabilities involved. A large number of calculations are made to establish averages, distribution within the averages, and any other data sought.

First, it is necessary to understand the use of random numbers in the Monte Carlo method. Numbers are generated at random in order to be as free as possible from any pattern or bias. They can be generated by mechanical or electrical means, but now they are generated mathematically as needed in computers and programmable pocket calculators. Table A3.2 gives 1000 random numbers. The numbers can be read horizontally, vertically, diagonally, at random, and by any method which is not biased or prejudiced.

Any distribution, whether empirical or theoretical, discrete or continuous, can be simulated with random numbers. It is necessary to convert the distribution to a cumulative probability of X or less and then make an assignment of random numbers.

Example 8.7 Show how the Poisson distribution for a mean value of 2 can be simulated from a table of random numbers.

SOLUTION From Eq. (8.11),

$$P(X) = \frac{e^{-2}2^X}{X!}$$

A tabulation is given in Table 8.3. If three random numbers are chosen with 000 representing 1000, then 135 of these numbers can be assigned to $P(0)$, 271 to $P(1)$, etc., to simulate the frequency of the occurrence of these events. The assignment of numbers is easily and systematically made from the cumulative probabilities, as shown in the last column of the tabulation.

The distribution can now be simulated by generating random numbers and converting them to X. Thus, using the first three numbers of each group of five taken from Table A3.2, X is generated as in the tabulation at the top of the following page.

Random no.	X	Random no.	X	Random no.	X
141	1	265	1	897	4
582	2	749	3	592	2
821	3	086	0	328	1
481	2	745	3	841	3
442	2	109	0	665	2
456	2	566	2	346	1
724	3	700	3	063	0

The average X is 1.90, which is reasonably close to the true value 2.0 considering the small number of trials.

Example 8.8 can be evaluated without Monte Carlo simulation but illustrates the method.

Table 8.3 Assignment of random numbers for the Poisson distribution with mean = 2

X	P(X)	Σ P(X)	Numbers assigned, inclusive
0	0.1353	0.135	001–135
1	0.2707	0.406	136–406
2	0.2707	0.677	407–677
3	0.1804	0.857	678–857
4	0.0902	0.947	858–947
5	0.0361	0.983	948–983
6	0.0120	0.995	984–995
7	0.0034	0.999	996–999
8	0.0009	1.000	000–000
9	0.0002		

Example 8.8 Assemblies A and B become one product C. A and B each contain defects on a statistical basis as follows:

Defects	No. of times occurring in A	No. of times occurring in B
0	5	2
1	5	3
2	15	5
3	30	10
4	20	20
5	10	40
6	5	10
7	5	5
8	3	3
9	2	2
	100	100

Table 8.4 Assignment of random numbers in Example 8.8

Defects	A			B		
	$P(X)$	$\Sigma\ P(X)$	Random numbers	$P(X)$	$\Sigma\ P(X)$	Random numbers
0	0.05	0.05	01–05	0.02	0.02	01–02
1	0.05	0.10	06–10	0.03	0.05	03–05
2	0.15	0.25	11–25	0.05	0.10	06–10
3	0.30	0.55	26–55	0.10	0.20	11–20
4	0.20	0.75	56–55	0.20	0.40	21–40
5	0.10	0.85	76–85	0.40	0.80	41–80
6	0.05	0.90	86–90	0.10	0.90	81–90
7	0.05	0.95	91–95	0.05	0.95	91–95
8	0.03	0.98	96–98	0.03	0.98	96–98
9	0.02	1.00	99–00	0.02	1.00	99–00

Table 8.5 Calculation of average defects in C for Example 8.8

(1) Trial no.	(2) Random number	(3) Defects in A	(4) Defects in B	(5) Defects in C (3) × (4)	(6) Cumulative defects in C	(7) Average defects in C
1	14 15	2	3	6	6	6.0
2	58 20	4	3	12	18	9.0
3	82 14	5	3	15	33	11.0
...	....	.	.	...	......	......
19	59 82	4	6	24	256	13.5
20	18 57	2	5	10	266	13.3

The number of defects in C is the product of the defects in A and B. Using the Monte Carlo method, find the expected number of defects in C.

SOLUTION Table 8.4 shows the assignment of random numbers to simulate the defects in A and B. Random numbers are now taken in groups of four, say the first four numbers in column 1 of Table A3.2. The first two numbers are for the simulation of defects in A, the second two numbers for the defects in B. The calculation is given in Table 8.5. The last column is the cumulative average defects in C obtained by dividing the cumulative defects by the cumulative number of trials. The average after 20 trials is 13.3, whereas the correct value is 16.7. Agreement will improve with the number of trials.

UNCERTAINTY AND RISK ANALYSIS

8.13 Normal Distribution and Uncertainty

In practice, a single estimate for a project is desired, along with an analysis of how the actual amount might vary from the expected value. The concept is known by various names, such as *uncertainty analysis* and *risk analysis*. The problem is to combine the uncertainties in the individual components to express uncertainty in the overall project. In this section, it is assumed that the uncertainty in each component is normally distributed. The following section shows a method that can be used if normal distributions do not prevail.

If the components are normally distributed, the variances are additive if they are weighted for their sensitivity to the overall effect. Suppose the estimate for a project is Z and it is made up of estimates for A, B, and C with variances of σ_A^2, σ_B^2, and σ_C^2. The expected value of Z is obtained from a calculation of Z using the expected values for A, B, and C. The sensitivity of Z to the individual components is given by $(\partial Z/\partial A)$, $(\partial Z/\partial B)$, and $(\partial Z/\partial C)$. Thus $(\partial Z/\partial A)$ is a measure of $\Delta Z/\Delta A$, which is the change in Z for a unit change in A, all other uncertainties not included. The variance for Z is given by

$$\sigma_Z^2 = \left(\frac{\partial Z}{\partial A}\right)^2 \sigma_A^2 + \left(\frac{\partial Z}{\partial B}\right)^2 \sigma_B^2 + \left(\frac{\partial Z}{\partial C}\right)^2 \sigma_C^2 \tag{8.26}$$

The partial derivatives squared are the weighting factors for the individual variances.

The variance for a component can be obtained from Eq. (8.4). However, if only a few estimates for a component are available, then a high value and a low value, i.e., the range, can be used. The standard deviation, the square root of the variance, can be taken as

$$\sigma = \frac{\text{high value} - \text{low value}}{8} \tag{8.27}$$

Example 8.9 The total cost C_T for a project in megadollars is given by

$$C_T = 3X^{1.2}Y^{-0.5} + 10Y + F \tag{8.28}$$

X is estimated from six values: 2.5, 1.6, 1.5, 3.0, 1.4, and 2.0. Y is estimated from a high of 2.5 and a low of 0.5. F is certain, with a value of 0.5.
(a) Find the expected value for C_T, its sensitivity to X and Y, and how the actual value for C_T might vary from the expected value. Normal distributions are assumed.
(b) Find the variance for C_T.

SOLUTION
(a) The mean for X is given by Eq. (8.1):

$$\frac{2.5 + 1.6 + 1.5 + 3.0 + 1.4 + 2.0}{6} = 2.0$$

The variance for X is given by Eq. (8.4):

$$\sigma_X^2 = \frac{(2.5 - 2)^2 + (1.6 - 2)^2 + \cdots + (2.0 - 2)^2}{6 - 1} = 0.404$$

The mean for Y is $(2.5 + 0.50) \div 2 = 1.5$ and the variance, by Eq. (8.27), is

$$\sigma_Y^2 = \left(\frac{2.5 - 0.5}{8}\right)^2 = 0.063$$

The mean value for F is 0.5, and the variance is zero.

The expected value for C_T is obtained from Eq. (8.28) using the mean, or expected value, of the components:

$$C_T = 3(2)^{1.2}(1.5)^{-0.5} + 10(1.5) + 0.5 = 21.127$$

The sensitivity of C_T to X and Y is obtained from Eq. (8.28):

$$\frac{\partial C_T}{\partial X} = 3(1.2)X^{0.2}Y^{-0.5} = 3.6(2)^{0.2}(1.5)^{-0.5} = 3.376$$

$$\frac{\partial C_T}{\partial Y} = 3(X)^{1.2}(-0.5)Y^{-1.5} + 10 = -1.5(2)^{1.2}(1.5)^{-1.5} + 10 = 8.124$$

This is equivalent to saying that C_T changes 3.376 times as fast as X for a unit change in X and 8.124 times as fast as Y for a unit change in Y. In other words, a 1 percent change in X ($\Delta X = 0.02$) would cause a change in C_T amounting to $3.376 \times 0.02 = 0.06752$, which on a percentage basis is $(0.06752)(100) \div 21.127$, or 0.320 percent. Similarly for Y, a 1 percent change in Y (at $Y = 1.5$) gives a 0.577 percent change in C_T:

$$\frac{(\partial C_T/\partial Y)Y}{C_T} = \frac{(8.124)(1.5)}{21.127} = 0.577 \text{ percent}$$

The total cost C_T is more sensitive to the estimate for Y than X, since a 1 percent error in X causes a 0.320 percent error in C_T, but a 1 percent error in Y causes a 0.577 percent error in C_T. That completes the sensitivity analysis.

(b) Next, find the variance for C_T, which by Eq. (8.26) is

$$\sigma^2 = (3.376)^2(0.404) + (8.124)^2(0.063) = 8.763$$

$$\sigma = 2.96$$

At this point, it is known that the expected value for C_T is 21.127 and the standard deviation is 2.96. Transform this to a standard normal distribution by Eq. (8.25), which here becomes

$$Z = \frac{C_T - 21.127}{2.96}$$

Assume values for Z, and from Table A3.1 the following tabulation results:

Z	Cumulative probability for Z or less	C_T
-1.65	0.05	16.2
-1.28	0.10	17.3
-0.84	0.20	18.6
-0.52	0.30	19.6
-0.25	0.40	20.4
0	0.50	21.1
$+0.25$	0.60	21.9
$+0.52$	0.70	22.7
$+0.84$	0.80	23.6
$+1.28$	0.90	24.9
$+1.65$	0.95	26.0

A plot of the last two columns gives the cumulative probability of C_T or less and is shown in Fig. 8.6. The plot shows that the probability that the project will cost 20 M$ or less is 0.35 or 35 percent. The probability of a cost of 24 M$ or less is 83 percent. The probability of a cost between 20 M$ and 24 M$ is $83 - 35 = 48$ percent.

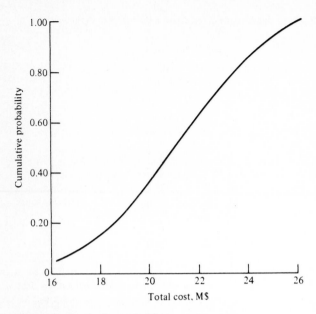

Figure 8.6 Plot for Example 8.9.

8.14 Skewed Distributions and Uncertainty

In practice, estimates and uncertainties will not follow a normal distribution or even be symmetrical about the mean. In practice, a cost for a venture has a greater chance of being 50 percent higher than the estimated cost than it does of being 50 percent lower. In the previous section, normal distributions were assumed for the advantage of using relatively simple mathematical analysis. If the distributions are not normal, the general approach is to employ simulation using the Monte Carlo technique and a computer. Skewed distributions can be represented by mathematical formulations such as the *beta-density function* and the *beta distribution,* but the approach outlined here requires no mathematical representation of the distributions. Indeed, all kinds of distributions can be intermixed in the same problem.

As an illustration, let a net present value, NPV, be represented by the following relationship:

$$\text{NPV} = A + \frac{B}{(1 + r)} + \frac{C}{(1 + r)^2} + \frac{D}{(1 + r)^3} \tag{8.29}$$

Suppose that B, an expected cash flow at the end of year 1, instead of following a normal distribution, as in Fig. 8.5, has a peak at 750 k$ but is unsymmetrical or skewed, with a greater chance for values below 750 k$ than above. The graph of Fig. 8.7 is accepted as fitting the situation. It is peaked at 750 k$ but there is a

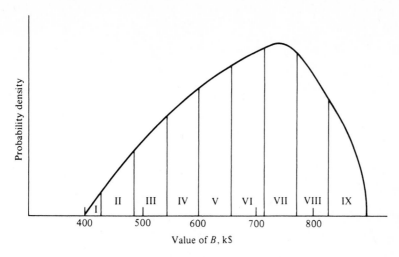

Figure 8.7 Distribution of the value of B.

greater chance for smaller values than for larger values. The ordinate scale here is arbitrary and need not be indicated. The range of B is 400 k\$ to 900 k\$.

First, divide the area into several smaller areas which need not be equally spaced. By some convenient means, which will usually be graphical, estimate the area, width, and center height of each element such that the height multiplied by the base will be the area. The dimensions for the area are arbitrary. The procedure leads to Table 8.6, with assignment of random numbers following the Monte Carlo method of Sec. 8.12. Thus elementary area III is represented by 510 k\$, with a probability of 0.0868 and random numbers 048–134. Similar tabulations are made for other components. The expected value of B is obtained from Eq. (8.3)

Table 8.6 Calculations from Fig. 8.7

Element	Area	Width	Height	Abscissa where height occurs	Fraction of total area	Cumulative fraction of total area	Random numbers
I	3.0	2	1.5	420	0.0128	0.013	001–013
II	8.0	2	4	460	0.0340	0.047	014–047
III	20.4	3	6.8	510	0.0868	0.134	048–134
IV	29.4	3	9.8	570	0.1251	0.259	135–259
V	36.0	3	12.0	630	0.1532	0.412	260–412
VI	41.1	3	13.7	690	0.1749	0.587	413–587
VII	44.1	3	14.7	750	0.1877	0.775	588–775
VIII	34.4	3	11.5	810	0.1464	0.921	776–921
IX	18.6	3	6.2	870	0.0791	1.000	922–000
	235.0				1.0000		

and is

$$E(X) = \sum X(PX) = 420(0.0128) + 460(0.0340)$$
$$+ 510(0.0868) + \cdots + 870(0.0791) = 682$$

The expected value for B is 682, not 750, because the distribution is not symmetrical about 750.

The expected value for the NPV and its distribution is now computed by the Monte Carlo method applied to Eq. (8.29). Three random numbers are generated which by Table 8.6 give a value for B; thus if 264 came up, then B would have the value 630. Single values for A, C, and D would be generated, and Eq. (8.29) would give a value for the NPV. The procedure would be repeated thousands of times and in the end the printout would give the average NPV, which would be the expected NPV. It would also print out how the calculated NPV values occurred in various ranges. For example, it might show that the expected NPV is 15 k$ and that 10 percent of the values were in the range -40 to -20, 25 percent of the values were in the range -20 to -0, and so forth.

A cumulative probability curve such as Fig. 8.6 could then be plotted and would be just as useful. The curve would not be centered on the ordinate value 0.5 since the distributions are skewed and nonsymmetrical.

8.15 Least Squares

Frequently, the problem arises of fitting data or observations to an equation. The only application considered here is the fitting of data to a straight line. The problem is related to probability and the normal distribution applied to the propagation of errors. Not all the data can be made to fall on a straight line. The amount by which each datum fails to fall on a straight line is referred to as the *error*. According to the theory of probability, the best straight line is that for which the variance of the data from the straight line is a minimum. In other words, the sum of the squares of the errors about the line is a minimum; hence the name *least squares* is used. The calculation can now be done using programmable pocket calculators.

Let X_i be a value of the independent variable for which Y_i is a known observation. The equation of a straight line is

$$Y = sX + b \tag{8.30}$$

The error for Y_i, its vertical distance from the straight line, is

$$\text{Error} = Y_i - Y = Y_i - (sX_i + b)$$

For the sum of the squares of the error to be a minimum, it is necessary to minimize the expression

$$\sum \text{error}^2 = \sum (Y - sX - b)^2$$

where the i designation has been dropped, since it is understood that the summation is for all the observed points. For a minimum (see Chap. 10),

$$\frac{\partial \sum}{\partial s} = 2 \sum (Y - sX - b)(-X) = 0$$

$$\frac{\partial \sum}{\partial b} = 2 \sum (Y - sX - b)(-1) = 0$$

The two relationships are

$$\sum XY = s \sum X^2 + b \sum X$$

$$\sum Y = s \sum X + \sum b = s \sum X + Nb$$

where N is the number of observations.

The last two equations are known as the *normal equations*. Solved together, they give

$$s = \frac{N \sum XY - \sum X \sum Y}{N \sum X^2 - \left(\sum X\right)^2} \tag{8.31}$$

$$b = \frac{\sum X^2 \sum Y - \sum X \sum XY}{N \sum X^2 - \left(\sum X\right)^2} \tag{8.32}$$

The last two equations give the value of s and b for the best straight line. The calculations are easily made by tabulation.

An alternative calculation for b is based on summing Eq. (8.30) for the N pairs of X, Y data, giving

$$b = \frac{\sum Y}{N} - s \frac{\sum X}{N} \tag{8.33}$$

If the relationship is in the form

$$Y' = K(X')^s$$

there results, by taking logarithms,

$$\log Y' = s \log X' + \log K$$

By substituting Y for $\log Y'$, X for $\log X'$, and b for $\log K$, the equation reduces to a straight line and can be treated by the method of this section.

A calculation by the method of least squares is given in the next chapter in Example 9.11.

NOMENCLATURE

a	Constant
A	One of k kinds of events in the multinomial distribution
b	Constant in the equation for a straight line
e	Naperian constant 2.71828 $\cdots$
$E(X)$	Expected value of X
H	Number of objects in population with desired property; used in hypergeometric distribution
k	Number of kinds of events in the multinomial distribution
K	Constant
M	Number of objects in the population
N	Number of trials or data
NPV	Net present value
p	Probability for success in a single trial in the binomial distribution
$P(X)$	Probability for event with value X
$P_c(X)$	Cumulative probability of X or less
$P_d(X)$	Probability density at X, probability per unit interval
s	Slope of a straight line
T	Time in periods
T_a	Time between arrivals, periods
W	Number of objects withdrawn from population in a hypergeometric distribution
X	Value of an event
Z	Independent variable in the standard normal distribution measured from the mean value
λ	Average arrivals per period
μ	Mean or average value
σ	Standard deviation
σ^2	Variance

PROBLEMS

8.1 Estimates for a project in megadollars are: A: 1.62, B: 2.84, C: 1.50, D: 3.00, E: 2.05, F: 2.00, G: 2.34, H: 1.67, I: 2.18, J: 2.67. Find the mean value, the expected value, and the variance.

8.2 Repeat Prob. 8.1, weighting the estimates as follows: A: 1.5, B: 1, C: 1, D: 2.0, E: 0.5, F: 0.7, G: 2.0, H: 1.5, I: 0.6, J: 1.5.

8.3 The hours required to perform a certain task by different workers are as follows, with time in hours and minutes:

1:56	1:22	2:16	1:54	2:36	2:04
2:22	2:27	0:50	1:51	2:50	1:11
1:42	2:13	1:20	2:31	1:47	2:09
2:07	1:37	2:22	1:44	2:06	2:06
1:50	1:54	2:02	1:04	2:58	1:42
2:21	2:18	2:43	2:16	1:27	1:32

Take 15-min intervals, such as 0:45 to 1:00, 1:01 to 1:15, etc.
  (a) Make a histogram of the data, i.e., a plot of numbers in the group against work hours.
  (b) Using the midtime for each group, find the mean and the variance.

8.4 For the grouping from Prob. 8.3, make a plot of the cumulative probability of the distribution. If an arbitrary worker is chosen, what is the probability that

(a) The time will be less than 1:30 h?

(b) The time will be more than 2 h but less than 2:30 h?

8.5 The probability of making a successful bid is 10 percent for a construction company. If the company makes 10 bids, find the probability that (a) two will be successful, (b) none will be successful, (c) no more than two will be successful, and (d) at least two will be successful. Use a binomial distribution.

8.6 A maintenance crew consists of nine people, of whom six are workers and three are key-workers. On a certain day, seven people report for work. Find the probability that of the two absentees (a) both are key-workers, (b) one is a key-worker, and (c) neither is a key-worker. Use a hypergeometric distribution.

8.7 A drafting department has 30 persons, 5 of whom are inferior. A job requires three persons chosen at random. Find the probability that one or more of the persons will be inferior. Use a hypergeometric distribution.

8.8 Show that the mean value of the Poisson distribution, Eq. (8.11), is μ.

8.9 A maintenance department receives a call for help on the average of 14 times in an 8-h shift. Find the probability that in a 1.5-h period it will receive (a) no calls, (b) two calls, (c) at least two calls. Use a Poisson distribution.

8.10 Given

$$
F(X) = \begin{cases} 0.25e^{-x/4} & x \geq 0 \\ 0 & x < 0 \end{cases}
$$

(a) Find the mean and variance of the function.

(b) Find the cumulative distribution function.

8.11 Use tables available in handbooks and textbooks on probability and statistics which give the probability, density, and ordinate for the standard normal distribution. Given a normal distribution with a mean value of 1.5 and a variance of 4, tabulate and draw the normal density function.

8.12 A construction company finds that actual costs against its estimates are normally distributed and that for $1 of estimate the variance is 0.04. If it estimates $300,000 on a project, what is the probability that it will cost more than $350,000?

8.13 Actual costs are expected to be normally distributed from an estimate. If the estimate is $100,000 and there is a 20 percent probability that the actual cost is not greater than $115,000, find the probability that the actual cost will not be greater than $125,000.

8.14 A company estimates $100,000 with a standard deviation of $22,000. How much should the company bid in order to be (a) 95 percent sure that the cost will not exceed the bid, (b) 99 percent sure?

8.15 The time needed to complete a project is expected to be 12 months normally distributed with a variance of 2 months². Find the probability that the project will take no longer than 10.5 months.

8.16 A company pays a $10,000 penalty for each day a project exceeds 40 days to complete, up to a limit of 5 days' penalty. The company expects to complete the project in 30 days, normally distributed, with a variance of 100 days². Find the expectation of the penalty for not completing the project in 40 days.

8.17 A company is bidding on repairing machines for a fixed cost per month. It costs $125,000 to repair a machine, and the number of repair jobs required will follow a Poisson distribution, with 0.7 machine per month expected. Find the expected cost per month.

8.18 Using the data of Prob. 8.3 and 15-min intervals, assign random numbers from 01 to 00 for the probabilities of the time required.

8.19 Process A follows a normal distribution, with a mean time of 1 h and a variance of 0.25 h². Plot the cumulative probability of X hours or less against X. Show how to assign two-digit random numbers to processing time X.

8.20 Process B follows an exponential distribution, as in Sec. 8.9, with a mean time of 2 h. Plot the cumulative probability of X hours or less against X. Show how to assign two-digit random numbers to processing time X.

8.21 A product can be manufactured with three sequential independent processes, A, B, and C. Process C has a constant time of 4 h. Processes A and B are described in Probs. 8.19 and 8.20. Using the Monte Carlo method, find the average time required to produce five units.

8.22 At a certain bus stop at the same time each morning, a bus arrives with an average of two empty seats, following a Poisson distribution. The number of passengers at the bus stop averages two, following a Poisson distribution. Using the Monte Carlo method, find the percentage of times that the bus leaves passengers stranded, assuming no passengers are allowed to stand in the bus.

8.23 Solve Prob. 8.21 by using expectancies instead of a Monte Carlo solution.

8.24 The cost C_T of a project in megadollars depends upon two inputs, X and Y, according to the relationship

$$C_T = 7XY^{1.2} + X^2 + 1.4$$

Estimates for X and Y are normally distributed, with X having a mean value of 1 and a variance of 0.5, and Y having a mean value of 2 and a variance of 1. Find the expected cost and the probability that it will lie between the expected cost and 20 percent over the expected cost. Show a cumulative distribution curve for the cost.

8.25 The cash flows for a project in kilodollars are:

End year	Cash flow
0	-1000 A
1	$+750$ B
2	$+390$ C
3	$+180$ D

The cost at zero time is certain since it is a turnkey cost. The cash flow at the end of year 1 is obtained from six estimates: 720, 800, 780, 820, 700, and 680. Cash flow at the end of year 2 is estimated twice as 500 and 280. Cash flow at the end of year 3 is obtained from two estimates: 250 and 110.

    (*a*) Find the expected NPV for the project at an 18 percent rate of return.

    (*b*) How sensitive is the NPV to estimates for B, C, and D?

    (*c*) Find the probability that the NPV will be between $+30$ k\$ and $+50$ k\$.

8.26 The cost in megadollars for a project is composed of two costs, A and B, added together. The costs and probabilities are as follows:

A		B	
Cost	Probability	Cost	Probability
1.0	0.05	2.1	0.05
1.1	0.10	2.2	0.15
1.2	0.20	2.3	0.25
1.3	0.40	2.4	0.35
1.4	0.25	2.5	0.20

Using a Monte Carlo technique, find the expected cost for the project. How would you obtain a distribution of the project cost?

8.27 An electric power company reports to its management for each 30-day period and combines all data by intervals of one 8-h shift. On the average, it goes down one 8-h shift in a 24-h day every 90 days because of failures. Find the distribution for failures of 0 through 6 during a 90-day period if the distribution of failures follows (a) a Poisson distribution, (b) a binomial distribution.

8.28 The following table shows the observed values of two variables, X and Y:

X	6	5	8	8	7	6	10	4	9	7
Y	8	7	7	10	5	8	10	6	8	6

(a) Construct a scatter diagram.

(b) Find a straight-line relationship between Y and X using the method of least squares.

8.29 Solve Prob. 8.26 by using expectancies instead of a Monte Carlo solution.

REFERENCES

1. Blum, J. R., and J. I. Rosenblatt: *Probability and Statistics,* W. B. Saunders Company, Philadelphia, 1972.
2. Brownlee, K. A.: *Statistical Theory and Methodology in Science and Engineering,* 2d ed., John Wiley & Sons, Inc., New York, 1965.
3. Campbell, D. W.: "Risk Analysis," *AACE Bulletin* (now *Cost Engineering*), August–October 1971, pp. 8–11.
4. Gray, H. L., and P. L. Odell: *Probability for Practicing Engineers,* Barnes & Noble, Inc., New York, 1970.
5. Hogg, R. V., and E. A. Tanis: *Probability and Statistical Inference,* The Macmillan Company, New York, 1977.
6. Holland, F. A., F. A. Watson, and J. K. Wilkinson, "Sensitivity Analysis of Project Profitabilities," "Statistical Techniques Improve Decision-Making," "Probability Techniques for Estimates of Profitability," "Estimating Profitability When Uncertainties Exist," and "Numerical Measures of Risk," in H. Popper (ed.), *Modern Cost Engineering,* McGraw-Hill Book Company, New York, 1979, pp. 29, 41, 47, 53, 61.
7. Koehn, E.: "Estimating with Probabilistic Unit Costs and Quantities," *Cost Engineering,* March–April 1978, pp. 63–66.
8. Ostle, B., and R. W. Mensing: *Statistics in Research,* 3d ed., The Iowa State University Press, Ames, 1975.
9. Quirn, W. L.: *Probability and Statistics,* Harper & Row, Publishers, Incorporated, New York, 1978.
10. Wadsworth, G. P., and J. G. Bryan: *Applications of Probability and Random Variables,* 2d ed., McGraw-Hill Book Company, New York, 1974.
11. Woodroofe, M.: *Probability with Applications,* McGraw-Hill Book Company, New York, 1978.

CHAPTER
NINE

PRODUCTIVITY

H. J. Behrens, O. J. Vogl, and
F. C. Jelen

DISPLACEMENT

9.1 Displacement versus Replacement

Replacement will be used to refer to a situation in which an asset is worn out or cannot physically perform its intended use and must be exchanged for another article soon, if not now. *Displacement* will be used to refer to a situation in which an asset can perform its function, at least to some degree, but can be exchanged for another asset for an overall economic gain. One distinguishing difference is that a decision regarding displacement can be ignored or delayed, perhaps for years.

A pure replacement problem, i.e., one based solely on the inability of an existing machine to function, can be analyzed by the methods of Chaps. 2 and 3 for periodic interest and Chap. 4 for continuous interest. The analyst should always regard replacement problems with a broad perspective, for the time when one asset must be replaced may introduce an opportunity for the displacement of other assets.

The possibility of displacement arises from either loss of efficiency or technological advancement. Loss of efficiency has several causes, such as increased maintenance expense. Technological advancement creates new machines that produce a superior product at less cost.

The displacement problem is a key factor in industrial leadership since it is an important part of the productivity problem. Companies which merely confront replacement problems as they occur generally lag behind. Those having an active displacement policy move forward.

9.2 One Year More of the Existent

Let an existing machine be called the *existent* and the new machine the *displacer*. A decision to purchase the displacer depends upon a comparison of the cost of keeping the existent in operation against that of having the displacer in operation. The comparison can be made on the basis of unacost or capitalized cost, and, as for all cost comparisons, only differences need be considered. Costs that are equal in all respects for the two machines can be omitted because they have no effect on the decision.

In general, operating costs and other disadvantages of the existent become progressively greater with time, and only 1 year ahead for the existent need be considered because it will make its best showing in the coming year. Exceptions to this rule will be discussed later.

The cost of keeping the existent in service 1 year more is readily calculated. Suppose the existing machine can be sold now for $C_{sal,0}$, the present realizable salvage value, and next year for $C_{sal,1}$, the realizable salvage value at the end of 1 year. The present realizable salvage value may not be the same as the book value and indeed may be considerably less. The difference between the book value and the salvage value is known as the *sunk cost*. The sunk cost has no influence on the decision except for the tax situation. A sunk cost is revealed but not created by displacement. It represents an inadequate depreciation expense in previous years and cannot be corrected by any policy of future years. The decision insofar as the existent is concerned for 1 year more is to sell the machine now for its realizable salvage value $C_{sal,0}$ or to keep the machine and incur an operating cost C_{e1} as of the end of the year and recover the salvage value at that time, $C_{sal,1}$. These amounts must, of course, be used with the tax situation included.

For the tax situation, let B_0 be the book value now and B_1 that 1 year from now. Suppose that the machine is sold now and that the tax credit is available immediately. The loss on disposal is $B_0 - C_{sal,0}$ and the savings in taxes is $t(B_0 - C_{sal,0})$. If the machine is not sold now, the receipts that are avoided must be regarded as an expense, which on an after-tax basis is

$$C_{sal,0} + t(B_0 - C_{sal,0}) \tag{9.1}$$

At the end of the year, depreciation amounting to D_1 can be taken, $C_{sal,1}$ is received as salvage value, and the loss on disposal is $B_1 - C_{sal,1}$. Altogether the after-tax receipt is

$$tD_1 + C_{sal,1} + t(B_1 - C_{sal,1})$$

Noting that $B_1 = B_0 - D_1$, the last relationship reduces to

$$C_{sal,1} + t(B_0 - C_{sal,1}) \tag{9.2}$$

Let C_{e1} be the operating cost for the coming year as of the end of the year, subject to a tax credit as of the end of the year. Then the after-tax cost is

$$C_{e1}(1 - t) \tag{9.3}$$

Combine Eq. (9.1), referring to the beginning of the year, with Eqs. (9.2) and (9.3), referring to the end of the year. The cost diagram and the present value, basis 1 year, are:

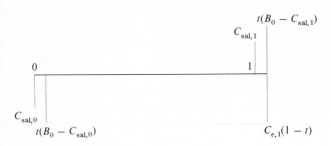

$$P_1 = C_{sal,0} + t(B_0 - C_{sal,0}) - \frac{C_{sal,1} + t(B_0 - C_{sal,1}) - C_{e1}(1 - t)}{1 + r} \quad (9.4)$$

This can be converted to a unacost by Eq. (2.20), with $F_{PR,r,1} = 1 + r$:

$$R = P_1(1 + r)$$

and to a capitalized cost by Eq. (2.32), with $F_{PK,r,1} = (1 + r)/r$:

$$K = P_1 \frac{1 + r}{r}$$

Unacost or capitalized cost for the displacer can be obtained in the conventional manner, as by Table 3.5, and a comparison with the existent will show which is more economical.

Example 9.1 Four years ago a machine cost $23,000. It has been depreciated on a 10-year life with $1000 terminal salvage value using sum-of-the-years-digits depreciation. If sold now, it will bring $2000 with immediate tax credit for any loss on disposal. If sold at the end of the year, it will bring $1500 with immediate tax credit for any loss on disposal. Operating costs for next year are $3800 as of the end of the year with immediate tax credit.

A new machine will cost $50,000 with a 12-year life and has $3000 terminal salvage value. Depreciate for a 10-year life for tax purposes using sum-of-the-years-digits depreciation. The operating cost will be $3000 as of the end of each year with $6000 per year savings due to better quality control, both subject to an immediate tax credit.

If money is worth 10 percent per year after a 50 percent tax rate, should the displacer be purchased?

SOLUTION By item 4, Table 3.2, the book value now of the existent is

$$B_0 = 23,000 - (23,000 - 1000)\frac{4}{10(10 + 1)}(20 + 1 - 4) = \$9400$$

The loss on disposal at zero time is

$$B_0 - C_{sal,0} = 9400 - 2000 = \$7400$$

and the tax credit is

$$t(7400) = 0.5(7400) = \$3700$$

The depreciation plus loss on disposal at the end of the year is

$$B_0 - C_{sal,1} = 9400 - 1500 = \$7900$$

and the tax credit is

$$t(7900) = 0.5(7900) = \$3950$$

The after-tax cost of the end-of-year expense is

$$C_{e1}(1 - t) = 3800(1 - 0.5) = \$1900$$

The cost diagram is

The present value at 10 percent per year is

$$P_1 = 2000 + 3700 - \frac{1500 + 3950 - 1900}{1.10} = \$2473$$

and unacost for 1 year more of the existent is

$$R = 2473(1.10) = \$2720$$

Unacost for the displacer, by Table 3.5, items 1, 2, and 7, is

$$C_d(1 - tF_{SDP,10\%,10})F_{PR,10\%,12} = (50,000 - 3000)[1 - 0.50(0.70099)](0.14676) = \quad 4480$$

$$R(1 - t) = (3000 - 6000)(1 - 0.50) = -1500$$

$$C_{sal}r = 3000(0.10) = \quad \underline{300}$$

$$R \text{ displacer} = \$3280$$

It will be more economical to retain the existent. The unacost is $2720 against $3280 on an after-tax basis.

If the item being displaced is part of a group for depreciation purposes, the tax situation changes. The depreciation will be taken on the group, whether the item is displaced or not. For example, a loss on disposal cannot be taken as an immediate tax advantage.

Example 9.2 Repeat Example 9.1, considering the machine as part of a group for depreciation purposes.

SOLUTION If sold now, the machine would bring $2000 subject to tax. The book value is not a factor since that will remain on the books for future depreciation. The cash flow lost now by retaining the machine is

$$2000(1 - t) = 2000(1 - 0.5) = \$1000$$

At the end of the year it would bring $1500 subject to tax. Any cash flow because of depreciation during the year is the same whether the machine is sold or not. At the end of the year, there will also be a $3800 operating cost subject to tax. The end-of-year cost after taxes is

$$(-1500 + 3800)(1 - 0.5) = \$1150$$

The present value after taxes at 10 percent per year is

$$P_1 = 1000 + \frac{1150}{1.10} = \$2045.5$$

and unacost for 1 more year of the existent is

$$R = 2045.5(1.10) = \$2250$$

As before, the unacost after tax for the displacer is $3280.

Unacost for the existent is less than in the previous example since loss of an immediate tax benefit from the depreciation is not a factor.

9.3 More Than 1 Year of the Existent

It was assumed in the preceding section that 1 year would be the most favorable period for the existent. The treatment of the existent for more than 1 year ahead will now be analyzed, followed by some examples.

Refer to Example 9.1. In addition, let the salvage value at the end of the second year be $900 and the operating cost for the second year as of the end of the year C_{e2} be $1400. The tax timing is the same as before. Now, find the unacost for the existent based on retention for 2 years.

At zero time the expense incurred by retaining the machine is the same as before, and on an after-tax basis is

$$C_{sal,0} + t(B_0 - C_{sal,0}) = 2000 + 3700 = \$5700$$

The end-of-first-year expense is $3800, as before. Depreciation for the year by item 4, Table 3.2, is

$$\tfrac{6}{55}(23,000 - 1000) = \$2400$$

The expense for the end of the first year after taxes is

$$3800(1 - t) - 2400t = 1900 - 1200 = \$700$$

The end-of-second-year expense is $1400. Depreciation for the second year is

$$\tfrac{5}{55}(23,000 - 1000) = \$2000$$

The book value at the end of the second year is

$$B_2 = B_0 - D_1 - D_2 = 9400 - 2400 - 2000 = \$5000$$

where D_1 and D_2 are first- and second-year depreciation. The loss on disposal at the end of the second year is

$$B_2 - C_{sal,2} = 5000 - 900 = \$4100$$

The expense for the end of the second year after a 0.50 decimal tax rate is

$$1400(1 - t) - 2000t - 900 - 4100t = -\$3250$$

The cost diagram on an after-tax basis, by symbols and numbers, is:

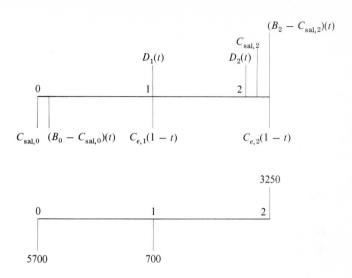

The present value is

$$P_2 = 5700 + \frac{700}{1.10} - \frac{3250}{1.10^2} = \$3650$$

and the unacost is

$$R = P_2 F_{PR, 10\%, 2} = 3650(0.57619) = \$2103$$

This is less than the unacost based on 1 year more, $2720, because there is a sharp drop in operating expenses for the second year. Normally, with operating costs advancing, the unacost for 1 year more will be the minimum.

The example demonstrates how more than 1 year is treated with taxes included. The tax timing must be considered whether 1 or more years ahead is investigated, but the analyst should be able to compute the unacost or capitalized cost for any tax timing and duration of the existent.

Cases in which multiyear service for the existent are more favorable than that for just 1 more year of service are usually evident from inspection. In practice, they arise principally from two causes: (1) net realizable salvage value may decline sharply for the first year, and (2) benefits of an overhaul may extend more than 1 year. Examples of each type follow (on a no-tax basis to simplify the presentation).

Example 9.3 An improperly designed pump has an excessive power cost. The salvage value now is $1000, declining to $250 next year and to $100 the following year. The operating cost for 1 year more is $700 and $800 for the following year as of the end of each year. A displacer costing $3800 will last 5 years with negligible salvage value. The operating expense will be uniform at $400/year as of the end of the year. If money is worth 8 percent per year, should replacement be made now?

SOLUTION Capitalized cost of the existent, basis 1 year more of service, by Table 2.3, is

$$\left(1000 - \frac{250}{1.08}\right)F_{PK,8\%,1} = 768.52(13.500) = 10,375$$

$$700\frac{1}{1.08}F_{PK,8\%,1} = \frac{700}{1.08}13.50 = 8750$$

$$K_1 = \overline{19,125}$$

Capitalized cost of the existent, basis 2 years of service, by Table 2.3, is

$$\left(1000 - \frac{100}{(1.08)^2}\right)F_{PK,8\%,2} = 914.26(7.0096) = 6409$$

$$700\frac{1}{1.08}F_{PK,8\%,2} = \frac{700}{1.08}7.0096 = 4543$$

$$800\frac{1}{1.08^2}F_{PK,8\%,2} = \frac{800}{1.1664}7.0096 = 4808$$

$$K_2 = \overline{15,760}$$

Capitalized cost of the displacer, by Table 2.3,

$$(3800 - 0)F_{PK,8\%,5} = 3800(3.1307) = 11,897$$

$$400\frac{1}{0.08} = 5000$$

$$\text{Displacer } K = \overline{16,897}$$

Based on only 1 more year of service, the existent is more expensive than the displacer: $K_1 = \$19,125$ against $\$16,897$; but it would be incorrect to replace now. If the existent is kept for 2 years, despite its rise in operating cost during the second year, it will be more economical than the displacer: $K_2 = \$15,760$ against $\$16,897$.

Example 9.4 A tank with negligible salvage value now or later can be repaired for $3000 to extend its life 1 year, after which $1000/year at the beginning of each year will extend its life 1 more year. A new tank costs $14,500, lasts 10 years with negligible salvage value, and has no annual expense. If money is worth 8 percent per year, should a new tank be purchased now?

SOLUTION Use unacost and Table 2.3. The cost of the existent, for 1 year of service, is

$$3000F_{PR,8\%,1} = 3000(1.0800) = R_1 = \$3240$$

The unacost of the existent, for 2 years of service, is

$$3000F_{PR,8\%,2} = 3000(0.56077) = 1682.3$$

$$1000\frac{1}{1.08}F_{PR,8\%,2} = \frac{1000}{1.08}(0.56077) = 519.2$$

$$R_2 = \overline{\$2201.5}$$

The unacost of the existent, for 3 years of service is

$$3000F_{PR,8\%,3} = 3000(0.38803) = 1164.1$$

$$1000\frac{1}{1.08}F_{PR,8\%,3} = \frac{1000}{1.08}(0.38803) = 359.3$$

$$1000\frac{1}{(1.08)^2}F_{PR,8\%,3} = \frac{1000}{1.1664}(0.38803) = 332.7$$

$$R_3 = \overline{\$1856.1}$$

The unacost of the displacer, for 10 years of service, is

$$14{,}500 F_{PR,8\%,10} = 14{,}500(0.14903) = \$2160.9$$

Based on 1 or 2 more years of service, the existent compares unfavorably with the displacer, but it is more economical based on a 3-year period.

9.4 Uniform-Gradient Series

In many practical problems the operating expenses of a machine vary from year to year. If the variation from year to year is systematic, the present value of the expenses for a number of years can be computed by an algebraic relationship. A particularly important case arises when the annual cost increases by G as of the end of each year, with the first-year end-of-year cost zero. The costs as of the end of the year form the following sequence for n years:

$$0, G, 2G, 3G, \ldots, (n-1)G$$

The series is known as a *uniform-gradient series,* and G is the uniform gradient. The present value of this series for n years is

$$P = G\left[\frac{0}{1+i} + \frac{1}{(1+i)^2} + \frac{2}{(1+i)^3} + \cdots + \frac{n-1}{(1+i)^n}\right] \qquad (9.5)$$

The series is classified as a recurring series of the second order, which can be summed algebraically by special techniques developed in Chap. 13. The series for $G = 1$ is referred to as the *uniform-gradient-series present-value factor* with the symbol $F_{GP,i,n}$. Mathematically, the sum is

$$F_{GP,i,n} = \frac{1}{i}\left[\frac{(1+i)^n - 1}{i(1+i)^n} - \frac{n}{(1+i)^n}\right] \qquad (9.6)$$

or

$$F_{GP,i,n} = \frac{F_{RP,i,n} - nF_{SP,i,n}}{i} \qquad (9.7)$$

The $F_{GP,i,n}$ factor is tabulated in Table A1.1.

Usually the gradient occurs along with a uniform annual cost, but the two are easily segregated, as shown in the following example.

Example 9.5 A series of costs at the end of each year increases uniformly by $100/year and is $1000 for the first year. If money is worth 8 percent per year, what is the present value of 12 years of costs?

SOLUTION The present value is

$$P = \frac{1000}{1.08} + \frac{1100}{1.08^2} + \frac{1200}{1.08^3} + \cdots + \frac{2100}{1.08^{12}}$$

which can be broken down into two series:

$$P = 1000\left(\frac{1}{1.08} + \frac{1}{1.08^2} + \cdots + \frac{1}{1.08^{12}}\right) + 100\left(\frac{0}{1.08} + \frac{1}{1.08^2} + \frac{2}{1.08^3} + \cdots + \frac{11}{1.08^{12}}\right)$$

or in sigma notation,

$$P = 1000 \sum_{1}^{12} \left(\frac{1}{1.08}\right)^n + 100 \sum_{1}^{12} \frac{x-1}{1.08^x}$$

The two sigma series on the right are, respectively, the present value of a uniform annual series, Eq. (2.17), and a uniform-gradient series, Eq. (9.7). The combined present values are

$$P = 1000F_{RP,8\%,12} + 100F_{GP,8\%,12}$$
$$= 1000(7.5361) + 100(34.634) = \$11,000$$

9.5 Best Policy with a Uniform-Gradient Cost

A practical and instructive case involves an item or machine that incurs an operating inefficiency at a constant rate. Consider the simplified situation for an automobile which costs $12,000 when new and has a salvage value at the end of each year amounting to 75 percent of the salvage value at the beginning of the year; i.e.,

$$C_{sal} = 12,000(0.75^n)$$

where n is any year, and the yearly operating expenses of the automobile increase by $300 per year. The fixed expenses common to all the years need not be considered. The variable expenses are

$$C_{en} = (n-1)(300)$$

where C_{en} is an expense at the end of year n. Money is worth 6 percent per year with no tax.

If an automobile is kept n years, the present value of all the costs is

$$P_n = 12,000 - \frac{C_{sal}}{1.06^n} + 300\left(\frac{0}{1.06} + \frac{1}{1.06^2} + \frac{2}{1.06^3} + \cdots + \frac{n-1}{1.06^n}\right)$$

Substituting the value for C_{sal}, and $F_{GP,6\%,n}$ for the series in parentheses, this is

$$P_n = 12,000 - \frac{12,000(0.75^n)}{1.06^n} + 300F_{GP,6\%,n}$$

Finally, converting to a unacost by Eq. (2.20) gives

$$R = P_n F_{PR} = F_{PR,6\%,n}\left[12,000 - \frac{12,000(0.75^n)}{1.06^n} + 300F_{GP,6\%,n}\right]$$

After obtaining the values in Table A1.1 for $1/1.06^n = F_{PS,6\%,n}$ and for $F_{GP,6\%,n}$, the last relationship yields the following:

Years automobile retained	Unacost, $/year
1	3720
2	3414
3	3186

Years automobile retained	Unacost, $/year
4	3024
5	2910
6	2832
7	2790
8	2769.6
9	2769.9
10	2787

The minimum unacost occurs at 8 years. That is, starting out with a new automobile, the optimum time to trade is at the end of 8 years for the schedule of costs as given.

The same conclusion is reached using the incremental method of Sec. 9.2. The used automobile makes its best showing on the basis of 1 year ahead. Consider the situation at the end of the seventh year. The salvage value then is $1602, dropping 1 year later to $1200. Operating costs for the coming year are $2100 as of the end of the year. The equivalent cost as of the end of the year for just 1 more year of service is

$$1602(1.06) - 1200 + 2100 = \$2598$$

Inasmuch as only 1 year is involved, $2598, being an end-of-year amount, is also a unacost. From the previous tabulation, unacost for keeping a car 7 years is $2790/year. That is, trading at the end of 7 years amounts to a $2790 unacost, whereas keeping the automobile another year amounts to a unacost of $2598. The automobile should be kept another year.

Repeat the calculations at the end of the eighth year. The unacost for 1 year more is

$$1200(1.06) - 900 + 2400 = \$2772$$

whereas from the previous tabulation the unacost for operating for only 8 years is $2769.60. The ninth year will not pay, confirming the previous conclusion.

The preceding analysis was based on the assumption that a new automobile was purchased. Now, consider the situation in which a 2-year-old car is bought, subject to all the previous conditions. The yearly costs will form the series

$$600, \ 900, \ 1200, \ 1500, \ . \ . \ .$$

which can be broken down to a constant amount of 600 and the series

$$0, \ 300, \ 600, \ 900, \ . \ . \ .$$

A tabulation similar to the previous tabulation can be made. The first cost is $6750, and the constant operating cost of $600/year must be included. The tabulation for starting with a 2-year-old automobile is:

Years retained	Unacost, $/year
1	2694
2	2583
3	2496
4	2487
5	2484
6	2499
7	2532

The optimum policy when starting out with a 2-year-old automobile is to trade 5 years later when the automobile is 7 years old. This is not the same conclusion that was reached when starting out with a new automobile, which should be traded when it is 8 years old. Also, the minimum unacost starting with a 2-year-old automobile is $2484, in contrast to $2770 when starting with a new automobile. Proceeding in this manner for all the alternatives gives the data in Table 9.1.

For the figures given, the overall optimum or minimum is $2376/year, obtained by starting out with a 4-year-old automobile and keeping it 1 year. It has been assumed that a used automobile can be purchased for the same amount as its trade-in value. A different set of figures will lead to a different result, but the example is illustrative. If the yearly expenses and salvage value are not systematic, it will be necessary to calculate the present values by discounting each year individually without the benefit of a systematic relationship; but the principles remain the same.

In practice, not everyone can start out with a 4-year-old automobile and follow the optimum policy. In an industrial problem, however, it is not always necessary to start at the beginning of the line. Suppose that a process can use a rough ore, semirefined ore, or refined ore, all of which can be made available in any quantity at different prices. All subsequent processers can choose refined ore if that is the most economical choice.

Table 9.1 Optimum policy for an automobile

Age of automobile when purchased, year	Optimum no. of years retained	Age when traded, years	Unacost, $/year
0	8	8	2769
1	7	8	2598
2	5	7	2484
3	3	6	2412
4	1	5	2376
5	1	6	2382
6	1	7	2466

9.6 Delay Value of the Existent

An existent can be displaced by a new machine merely because it acquires operating inefficiency with time that has nothing to do with technological advancement. The problem is illustrated by the optimum policy for buying an automobile in the preceding section. On the other hand, an existent can suffer no operating inefficiency with time and yet be displaced because of technological advancement of a machine now available. Usually both factors apply. The analyst compares the existent for its unacost now to the unacost of the new machine.

Future technological advancement cannot be purchased in the present. However, the expectation of future technological advancement cannot be ignored. Indeed, its value should be included as a reduction in the unacost or capitalized cost of the existent. By keeping the existent, it becomes possible eventually to obtain a later model of the new machine which presumably will be even more technologically advanced. Technological advancement works in two ways. The technological advancement which has already been made works against the existent, but the possibility of future advancement works for it. This latter factor has generally been overlooked. For new types of assets undergoing rapid technological advancement, the delaying value of the existent may be significant. An exact mathematical evaluation is not possible because the result depends upon the model, but the situation has been mentioned in the literature [3]. However, the delay value of the existent should not become a policy for procrastination since if a company falls too far behind, it may not be in a position to catch up.

9.7 A Mathematical Model for Technological Advancement

Consider a situation in which technological advancement alone makes displacement possible; i.e., the existing machine suffers no loss of efficiency with time. The mathematical model for the machine will be based on two assumptions:

1. Each year a new model will become available, costing $1 - p$ of the cost of the model of the preceding year, where p is the technological advancement rate expressed as a decimal. If the cost of a new machine is C_i now, 1 year from now it will be $C_i(1 - p)$, 2 years from now $C_i(1 - p)^2$, and so forth.
2. Some of the operating costs of the machine will likewise incorporate technological advancement. Let R_p denote the yearly end-of-year operating expense for this year's model that is subject to technological advancement. Next year's model will have a comparable expense $R_p(1 - p)$, the following model $R_p(1 - p)^2$, and so forth. Uniform yearly expenses that are not affected by technological advancement do not vary from model to model and need not be considered. When a machine is bought, the yearly expense R_p remains constant until a new machine is purchased n years later, which will reduce the yearly expense to $R_p(1 - p)^n$.

These two assumptions permit evaluation of the expression for the economic life of a machine for various technological advancement rates. If a machine is bought now and replaced every n years, the pertinent capitalized cost of the machine is the present value after taxes taken to infinity for (1) the first costs occurring every n years and (2) the yearly operating cost occurring every year and changing every n years. This is

$$K = C_i(1 - t\psi_{rn}) + C_i(1 - t\psi_{rn})\frac{(1 - p)^n}{(1 + r)^n}$$

$$+ C_i(1 - t\psi_{rn})\frac{(1 - p)^{2n}}{(1 + r)^{2n}} + \cdots$$

$$+ \sum_{1}^{n} \frac{R_p(1 - t)}{(1 + r)^x} + \sum_{n+1}^{2n} \frac{R_p(1 - t)(1 - p)^n}{(1 + r)^x}$$

$$+ \sum_{2n+1}^{3n} \frac{R_p(1 - t)(1 - p)^{2n}}{(1 + r)^x} + \cdots \tag{9.8}$$

Although Eq. (9.8) appears formidable, it reduces to

$$\frac{K}{C_i} = \frac{(1 + r)^n(1 - t\psi_{rn}) + (R_p/C_i)\{(1 - t)[(1 + r)^n - 1]/r\}}{(1 + r)^n - (1 - p)^n} \tag{9.9}$$

where r = decimal rate of return after taxes
    t = decimal tax rate
    ψ = factor associated with depreciation (see Sec. 3.13)

Multiplication of Eq. (9.9) by r will give an equivalent expression for unacost.

The optimum economic life for a machine subject to the conditions of Eq. (9.9) can be determined by finding the value of n which makes K/C_i a minimum for specified values of the other parameters, r, t, R_p/C_i, p, and the depreciation method. The data in Fig. 9.1, obtained by machine computation, give the optimum service life for a technological advancement rate of 3 percent per year, a realistic figure, and an income tax rate of 50 percent using sum-of-the-years-digits depreciation. From the figure, for a 10 percent per year rate of return after taxes and a ratio for $R_p/C_i = 0.20$, the optimum economic life is 17 years.

The inclusion of all four rates—rate of return, tax rate, inflation rate, and productivity rate—in comparing alternatives was discussed in Sec. 6.5. Under existing conditions, the effect of inflation on increasing costs overwhelms the effect of technological advancement on decreasing costs, so that the mathematical relationships involving technological advancement alone are not realistic.

9.8 The MAPI Model

The Machinery and Applied Products Institute (MAPI) has made a strong effort to promote the comparison of existing machines with current machines as a guide for modernization. The system has undergone several revisions and is offered principally in worksheet form with accompanying charts.

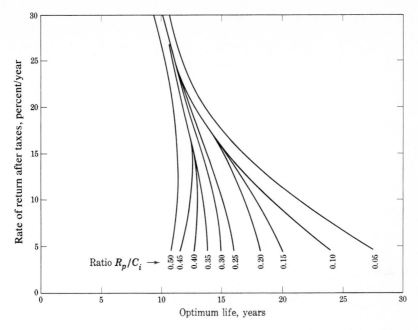

Figure 9.1 Optimum life for an improvement rate of 3 percent per year and a 50 percent tax rate.

The MAPI model is actually based on a new machine that acquires an operating inefficiency at a uniform rate. Applications are made using worksheets and charts. Charts are available for different depreciation methods or for expensing fully in one year, as well as for making 1-year or multiyear comparisons. The charts have several fixed parameters built in, such as tax rate, ratio of borrowed to total capital, and interest rate on borrowed capital. If other values of the parameters are required, algebraic relationships must be used, but they are complicated. The presentation uses picturesque terminology such as *defender, challenger,* and *adverse minimum,* but the method seems to have little current usage.

9.9 Group Displacement

Some assets exist in multiples and fail on an individual basis, light bulbs being a classic example. Such assets can be replaced at a high cost per unit of replacement or displaced as a group at a low cost per unit. There is an optimum policy for the minimum cost obtained by replacing on an individual basis for a certain time and then displacing the entire group.

Suppose that 1000 light bulbs have the distribution of service lives given in Table 9.2. Let 1000 light bulbs be installed; assume that bulbs are replaced at the beginning of each week and that installation on an individual basis is $1 per bulb and on a group basis is $0.45 per bulb. At zero time 1000 bulbs are installed at a cost of $450.

The number of bulbs that must be installed at the beginning of each week on

Table 9.2 Distribution of service lives for 1000 light bulbs

Period week no.	Failures during week	Probability of failure during week
1	150	0.15
2	200	0.20
3	300	0.30
4	250	0.25
5	100	0.10

a replacement basis is given by Table 9.3. An *R* following a number indicates that replacement is required. Thus column 2 shows that in accordance with the probabilities of Table 9.2, 150 replacements will be required at the beginning of period 2, 200 at the beginning of period 3, and so forth. Column 3 of Table 9.3 shows that in accordance with the probabilities of Table 9.2, the 150 bulbs installed at the beginning of period 2 will require 22 replacements at the beginning of period 3, 30 at the beginning of period 4, and so forth. Continuing, column 4 shows that 222 replacements will be required at the beginning of period 3, with replacements as indicated.

Table 9.3 Replacement of light bulbs

(1) Period	(2)	(3)	(4)	(5)	(6)	(7)	(13)	(14)	(15)
1	1000								
2	150R	150							
3	200R	22R	222						
4	300R	30R	33R	363					
5	250R	45R	44R	54R	393				
6	100R	38R	67R	73R	59R	337			
7		15R	56R	109R	79R	51R			
8			22R	91R	118R	67R			
9				36R	98R	101R			
10					39R	84R			
11						34R			
12							339		
13							51R	340	
14							68R	51R	339
15							102R	68R	51R
16							85R	102R	68R
17							34R	85R	102R
18								34R	85R
19									34R

Table 9.4 Optimum displacement schedule for light bulbs

Period	Bulbs installed As a group	As replacements	$ this period	Cumulative $	$ per period average
1	1000		450	450	450
2		150	150	600	300
3		222	222	822	274
4		363	363	1185	296
5		393	393	1578	316
6		337	337	1915	319
Indefinitely		339	339		339

Table 9.3 shows that the replacements per period overshoot, oscillate, and stabilize at 339. The number can be obtained independently. The average life of the bulbs is

$$\frac{1(150) + 2(200) + 3(300) + 4(250) + 5(100)}{150 + 200 + 300 + 250 + 100} = 2.95 \text{ weeks per bulb}$$

The reciprocal is the bulbs per week for 1 bulb installed. For 1000 bulbs installed,

$$\frac{1000}{2.95} = 339 \text{ bulbs per week}$$

Table 9.4 shows how the minimum-cost program is determined. The final column is the cumulative dollars divided by the cumulative periods and is the dollars per period. The minimum cost is obtained by displacing the bulbs as a group every third period with intermediate replacements of 150 and 222 bulbs. If no group displacement is made, the cost is $339 per period, as determined from the equilibrium replacement value of Table 9.3.

THE LEARNING CURVE

9.10 Practice Improves Performance

It is a fundamental human characteristic that a person engaged in a repetitive task will improve his or her performance. This distinguishing quality is one manifestation of optimization, a human trait, to be discussed in Chap. 10. In a broad sense, this quest for improvement is the basis of technological advancement which passes endlessly from generation to generation.

In a large-scale repetitive project, improvement occurs smoothly and continuously. The phenomenon is real and has a specific application in cost analysis, cost estimating, or profitability studies related to examination of future costs and confidence level in an analysis. Improved performance in manufacturing was developed as an analytical tool first by the aircraft industry, which still leads in the application, but the principle is now recognized and used by many other types of manufacturers.

The learning curve as developed in these sections is a powerful tool for predicting, guiding, and encouraging increases in productivity.

9.11 The Learning Curve and Learning-Curve Function

Figure 9.2 shows the effort required, expressed as direct work-hours per pound, plotted against the cumulative plane number for the Century series aircraft from government World War II production data released as part of the Aeronautical Material Planning Report (AMPR) data. The curve is the composite data for eight types of fighter planes, with several series in some types, produced by four manufacturers.

The curve is typical and shows a progressive improvement in productivity, but at a diminishing rate. This suggests an exponential relationship between productivity and cumulative production. Figure 9.3 is a plot of the data on log-log graph paper, where the data plot as a straight line. This suggests a relationship of the form

$$E_N = KN^s \tag{9.10}$$

where E_N = effort per unit of production, such as work-hours, required to produce the Nth unit

K = constant, numerically theoretical effort required to produce the first unit

s = slope constant

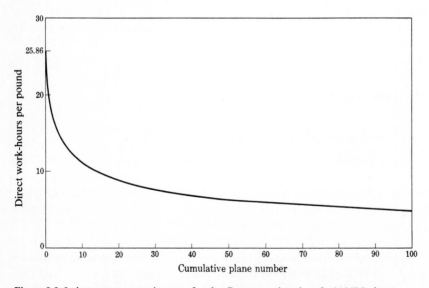

Figure 9.2 Industry average unit curve for the Century series aircraft. (*AMPR data.*)

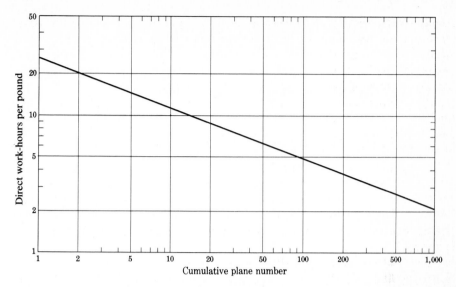

Figure 9.3 Log-log plot of the industry average unit curve for the Century series aircraft. (*AMPR data.*)

The *slope constant s* is negative because the effort per unit decreases with production.

The effort per unit E_N can be expressed in any convenient units. The effort can be converted to dollars, and eventually will be in many cases. Nevertheless, the use of dollars directly in the learning-curve function should be undertaken with care because the dollar is subject to considerable inflation and use of the dollar may mask the true reduction of effort per unit with production.

Relationship (9.10) will plot as a straight line on log-log graph paper. Taking logarithms of both sides,

$$\log E_N = s \log N + \log K$$

which in terms of $Y = \log E_N$, $X = \log N$, and $b = \log K$ has the form

$$Y = sX + b$$

the equation of a straight line.

A *learning curve*, sometimes called a *manufacturing progress curve* or simply *progress curve*, is merely a representation of the decrease in effort per unit required for a repetitive manufacturing operation. Figures 9.2 and 9.3 are both learning curves. Equation (9.10) is a *learning-curve function*. Not all data will fit the simple relationship of Eq. (9.10), and some effort has been made to use more refined relationships in specific cases. Nevertheless, Eq. (9.10) has surprising validity for manufacturing operations and is the only relationship considered in this chapter. Moreover, a log-log plot for the relationship will be used exclusively for the obvious advantage of producing a straight-line plot. A log-log representation also has the advantage of condensing a wide range of variables.

Although Eq. (9.10) will be referred to as the *learning-curve function* and Fig.

9.3 as the *learning curve,* strict usage perhaps requires that they be referred to as the *exponential learning-curve function* and *exponential learning curve,* respectively.

9.12 Properties of the Learning-Curve Function

Equation (9.10) embodies many important properties. It implies a constant fractional or percentage reduction in effort for doubled (or tripled, etc.) production. For example, for any fixed value for s, Eq. (9.10) gives

$$E_1 = K(1^s)$$

$$E_2 = K(2^s)$$

$$\frac{E_2}{E_1} = \frac{K(2^s)}{K(1^s)} = 2^s$$

also

$$E_2 = K(2^s)$$

$$E_4 = K(4^s)$$

$$\frac{E_4}{E_2} = \frac{K(4^s)}{K(2^s)} = 2^s$$

Similarly, $E_8/E_4 = 2^s$, and so forth. Every time production is doubled, the effort per unit required is a constant 2^s of what it was. Inasmuch as s is negative, the effort per unit required decreases with production.

It is common practice to express the learning-curve function in terms of the gain for double production. Thus a 90 percent learning-curve function requires only 90 percent of the effort per unit every time production is doubled.

Let L_D be the decimal ratio of effort, per production unit, required for doubled production. Then, by Eq. (9.10),

$$L_D = \frac{E_{2N}}{E_N} = \frac{K(2N)^s}{K(N)^s} = 2^s$$

Taking logarithms,

$$\log L_D = s \log 2 \qquad (9.11)$$

or

$$s = \frac{\log L_D}{\log 2} \qquad (9.12)$$

Equation (9.12) establishes a relationship between s, the slope constant, and L_D, the *decimal learning ratio.* Table 9.5 gives some corresponding values. Because of the occurrence of negative logarithms, it is sometimes more convenient to use L_P, the *percentage learning ratio,* which by definition is $100L_D$. If $L_P/100$ is substituted for L_D in Eq. (9.12), it becomes

$$s = \frac{\log (L_P/100)}{\log 2} = \frac{\log L_P - 2}{\log 2}$$

Table 9.5 Conversion of the decimal learning ratio to a slope constant

Decimal ratio of effort per unit for doubled production L_D	Slope constant s
1.00	0
0.95	−0.0740
0.90	−0.1520
0.85	−0.2345
0.80	−0.3219
0.75	−0.4150
0.70	−0.5146
0.65	−0.6215
0.60	−0.7370
0.55	−0.8625
0.50	−1.0000

which in terms of minus s, a positive number, is

$$-s = \frac{2 - \log L_P}{\log 2} \tag{9.13}$$

Example 9.6 If 846.2 work-h is required for the third production unit and 783.0 for the fifth unit, find the percentage learning ratio and the work-hours required for the second, fourth, tenth, and twentieth units.

SOLUTION By Eq. (9.10),

$$E_3 = 846.2 = K(3^s)$$
$$E_5 = 783.0 = K(5^s)$$

By division,

$$\frac{846.2}{783.0} = 1.0807 = \left(\frac{3}{5}\right)^s = 0.6^s$$

$$\log 1.0807 = 0.03371 = s \log 0.6 = s(9.77815 - 10) = s(-0.22185)$$

$$s = \frac{0.03371}{-0.22185} = -0.1520$$

By Eq. (9.13),

$$-(-0.1520) = \frac{2 - \log L_P}{0.30103}$$

$$\log L_P = 1.95424$$

$$L_P = 90.0 \text{ percent} = \text{percentage learning ratio}$$

Using the data for $N = 3$ and $s = -0.1520$ in Eq. (9.10),

$$846.2 = K(3^{-0.1520})$$

$$\log 846.2 = \log K - 0.1520 \log 3$$

$$2.9278 = \log K - 0.07252$$

$$\log K = 3.0000$$

$$K = 1000$$

so that the learning-curve function is

$$E_N = 1000N^{-0.1520}$$

The effort required for any unit can now be calculated directly. Thus, for the twentieth unit,

$$E_{20} = 1000(20^{-0.1520})$$

$$\log E_{20} = \log 1000 - 0.1520 \log 20$$

$$= 3.0000 - 0.1520(1.30103) = 2.80224$$

$$E_{20} = 634.2$$

A tabulation for other units is:

E_1	1000.0
E_2	900.0
E_4	810.0
E_5	783.0
E_{10}	704.7
E_{20}	634.2

Note that

$$\frac{E_2}{E_1} = \frac{E_4}{E_2} = \frac{E_{10}}{E_5} = \frac{E_{20}}{E_{10}} = 0.900$$

Example 9.7 Every time production is tripled, the unit work-hours required are reduced by 20 percent. Find the percentage learning ratio.

SOLUTION It will require 80 percent, or 0.80, for the ratio of effort per unit for tripled production. By Eq. (9.10),

$$\frac{E_{3N}}{E_N} = 0.80 = \frac{K(3N)^s}{K(N^s)} = 3^s$$

$$s = \frac{\log 0.80}{\log 3} = \frac{9.90309 - 10}{0.47712} = \frac{-0.09691}{0.47712} = -0.20311$$

By Eq. (9.13), for doubled production,

$$-(-0.20311) = \frac{2 - \log L_P}{\log 2}$$

$$\log L_P = 1.93886$$

$$L_P = 86.86 \text{ percent} = \text{percentage learning ratio}$$

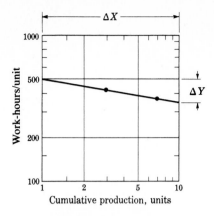

Figure 9.4 Plot for Example 9.8.

Another important property of the learning-curve function is that when plotted on log-log graph paper, the slope of the line is numerically equal to s, the slope constant.

Example 9.8 If 423 work-h is required for the third production unit and 372 for the seventh, find the slope constant by graphical means.

SOLUTION The points are plotted in Fig. 9.4. The slope of the line from measurements, before reproduction, is

$$\text{Slope} = \frac{\Delta Y}{\Delta X} = \frac{-11.4 \text{ mm}}{75.0 \text{ mm}} = -0.152 \text{ slope constant}$$

The modulus of the vertical and horizontal log scales must be the same when using the method, as in the example.

9.13 Cumulative Values

Equation (9.10) gives instantaneous values of effort per unit, and thus the work-hours for the Nth unit.

For the total effort for N units from 1 through N, the *cumulative effort* required, E_T, is

$$E_T = E_1 + E_2 + \cdots + E_N = \sum_{N=1}^{N=N} E_N \qquad (9.14)$$

An exact summation can be made by summing term by term, which can be tedious; from tables, if available; or by approximation relationships.

An approximation is obtained by treating Eq. (9.10) as a continuous function instead of a discrete function and integrating with a suitable change of limits.

$$E_T \cong K \int_{0.5}^{N+0.5} N^s \, dN = K \left[\frac{N^{s+1}}{s+1} \right]_{0.5}^{N+0.5}$$

$$\cong \frac{K}{s+1} [(N+0.5)^{s+1} - 0.5^{s+1}] \qquad (9.15)$$

The approximation improves as N increases. If N is very large compared to 0.5, Eq. (9.15) reduces to

$$E_T = \frac{K}{s+1}N^{s+1} \tag{9.16}$$

The cumulative average unit effort required is the cumulative effort required for the first N units divided by the cumulative number of units N; i.e.,

$$\frac{E_T}{N} = \frac{\displaystyle\sum_{N=1}^{N=N} E_N}{N}$$

which from Eqs. (9.15) and (9.16), respectively, can be approximated by

$$\frac{E_T}{N} \cong \frac{K}{N(s+1)}[(N+0.5)^{s+1} - 0.5^{s+1}] \tag{9.17}$$

$$\frac{E_T}{N} \cong \frac{K}{s+1}N^s \tag{9.18}$$

Example 9.9 If 846.2 work-h is required for the third production unit and 783.0 for the fifth, find the total work-hours required for the fifteenth through twentieth units and the average work-hours for the six units.

SOLUTION From the solution to Example 9.6, the slope constant s is -0.1520, and $K = 1000$. The work-hours required for the first 14 units from Eq. (9.15), with $s + 1 = -0.1520 + 1 = 0.8480$, is

$$E_{T,14} = \frac{1000}{0.8480}[(14 + 0.5)^{0.8480} - 0.5^{0.8480}]$$

$$= \frac{1000}{0.8480}(9.6570 - 0.5556) = 10{,}733 \text{ work-h}$$

For the first 20 units, by the same equation,

$$E_{T,20} = \frac{1000}{0.8480}[(20 + 0.5)^{0.8480} - 0.5^{0.8480}]$$

$$= \frac{1000}{0.8480}(12.953 - 0.556) = 14{,}619 \text{ work-h}$$

Net work-hours for the fifteenth to twentieth units is

$$E_{T,20} - E_{T,14} = 14{,}619 - 10{,}733 = 3886 \text{ work-h}$$

The average for the group of six is

$$\frac{3886}{6} = 647.7 \text{ work-h per unit}$$

The exact answer is 648.0. The approximation equation gives a close answer in this example because N is reasonably large, and moreover the effect of the early units is wiped out.

9.14 The Learning Curve and Economic Evaluation

The learning curve should be considered in making economic evaluations for the future and can be used as a criterion for judging the achievement actually realized. The time value of money becomes an ingredient of the analysis when looking into the future.

Example 9.10 A company has just received an order for four sophisticated space gadgets. The purchaser will take delivery of one unit starting 1 year hence, and one unit at the end of each of the succeeding 3 years. He will pay for each unit immediately upon receipt and will not take delivery ahead of time. The seller, however, can make the units ahead of time, if he prefers, and store them at no cost for later delivery.

The only pertinent cost in making these units is labor at $10/h. All units made in the *same* year are subject to an 80 percent learning ratio for the hours required, and the first unit requires 100,000 h of labor. Learning occurs only within a year and is not carried over from one year to another.

Consider the labor costs to be paid as of the end of the year. If money is worth 16 percent per year after a 52 percent tax, decide which of the following two plans is more economical:
(*a*) Build four units the first year.
(*b*) Build one unit each of the 4 years.

SOLUTION Revenue and its timing are the same for either plan and need not be considered. The object is to choose the plan with the lower present value of cost after taxes. For an 80 percent learning ratio $s = -0.3219$ from Table 9.5. The first unit costs $10(100,000) = \$1,000,000$. The cost of units made in the same year with learning is

First:	$1,000,000(1^{-0.3219}) =$	$1,000,000$
Second:	$1,000,000(2^{-0.3219}) =$	$800,000$
Third:	$1,000,000(3^{-0.3219}) =$	$702,104$
Fourth:	$1,000,000(4^{-0.3219}) =$	$640,000$
	Total $=$	$3,142,104$

The present value of the cost after taxes for all units made in the first year is

$$PV = (1 - 0.52)\frac{3,142,000}{1.16} = \$1,300,200$$

The present value of the cost after taxes for the four units made in successive years is

$$PV = (1 - 0.52)\left(\frac{1,000,000}{1.16} + \frac{1,000,000}{1.16^2} + \frac{1,000,000}{1.16^3} + \frac{1,000,000}{1.16^4}\right) = \$1,343,100$$

It will pay to make the four units the first year. The gain from the learning curve offsets the advantage in deferring the labor costs by the alternate plan. Making four the first year, however, is not necessarily the best possible plan. There are other combinations, such as three the first year and one the last year.

9.15 The Learning-Curve Function from Single-Unit Data

For a given set of data, Eq. (9.10) is established by determining the best values for K and s. If the data are all in the form giving the effort required for single, individual units of production, the data can be plotted on log-log graph paper and the best line drawn by eye. When the best line has been established, any two points on the line can be used to determine, graphically or analytically, the slope of the line (the slope constant) and K, which is numerically the intercept on the ordinate $N = 1$. The graphical method is rapid but depends upon judgment, and can become inaccurate if the points are badly scattered.

The most accurate method for determining the best straight line is application of the method of least squares described in the previous chapter.

9.16 The Learning-Curve Function from Grouped Data

In practice, the effort required to produce single, individual units may not be known, but instead only the average effort to produce a group, or lot, of units. Data for such a circumstance are given in Table 9.6. A total of 35 units has been made, but the precise number of work-hours required for any single unit is not known. Before the work-hours can be plotted, it is necessary to take each group and convert it to a point. This point within a group is some unit number, not necessarily an integer, with which the average work-hour can be associated. It is a balance or focus point and a single value to be paired with the average effort per unit for the group. It is loosely referred to as the *lot midpoint,* but *lot equivalent point* is perhaps more descriptive.

Several methods are available for determining the lot equivalent point. The *lot-midpoint method* uses the arithmetical mean of the first and last unit numbers in the lot. For example, for a lot of 6 units, the equivalent point is at $(1 + 6)/2 = 3.5$, or 3.5 units from the last unit of the preceding lot.

The *rule-of-thumb lot-midpoint method* places the lot equivalent point by means of the following rules:

First lot: Less than 10, take half of the last unit number.

                              10 or more, take one-third of the last unit number.

Subsequent lots: Take one-half of the number of units in the lot.

For the data in Table 9.6,

Lot 1: $\frac{3}{2} = 1.5$ lot equivalent point

Lot 2: $\frac{4}{2} = 2$

Table 9.6 Work-hours for consecutive lots

Lot no.	Quantity in lot	Work-hours for lot	Average work-hours per unit for lot
1	3	621.76	207.24
2	4	357.72	89.43
3	8	428.72	53.43
4	10	449.80	44.98
5	10	368.10	36.81

Table 9.7 Data of Table 9.6 after adjustment by the rule-of-thumb method

Lot no.	Quantity in lot	Assigned to unit	Average work-hours per unit for lot
1	3	1.5	207.24
2	4	5	89.43
3	8	11	53.43
4	10	20	44.98
5	10	30	36.81

or two units from unit 3, the last unit in the preceding lot. Thus $3 + 2 = 5$, or unit 5.

Lot 3: $$\frac{8}{2} = 4$$

or four units from unit 7, the last unit in the preceding lot, thus $7 + 4$, or unit 11. The complete tabulation is given in Table 9.7.

The *geometric-mean lot-midpoint method* uses the geometric mean to determine the equivalent point of each lot. For a lot of five, the geometrical mean is

$$\sqrt[5]{1(2)(3)(4)(5)} = 2.6052$$

which is added to the last unit number of the preceding group. Some geometric means are given in Table 9.8.

The *true lot-midpoint method* is the most accurate method for determining the lot equivalent point but the most difficult to calculate. It involves two separate steps. In the first step, an approximate method is used for the data to establish an approximate value for s, the slope constant. The approximate value is then used in the second step to find the true equivalent point for the lots. The data so generated are then used with the method of least squares to establish a revised but accurate value of s.

As an example, consider the second lot in Table 9.6. It will be shown below that an approximate value for s for the data in Table 9.6 is $s = -0.57616$. The

Table 9.8 Geometric means for lot sizes

No. in lot	Geometric mean	No. in lot	Geometric mean
1	1.0000	9	4.1472
2	1.4142	10	4.5287
3	1.8171	15	6.4235
4	2.2134	20	8.3044
5	2.6052	25	10.177
6	2.9938	30	12.045
7	3.3800	40	15.768
8	3.7644	50	19.483

second lot contains the fourth, fifth, sixth, and seventh units, and for $K = 1$, Eq. (9.1) gives

$$E_4 = 1(4^{-0.57616}) = 0.44990$$
$$E_5 = 1(5^{-0.57616}) = 0.39562$$
$$E_6 = 1(6^{-0.57616}) = 0.35617$$
$$E_7 = 1(7^{-0.57616}) = \underline{0.32590}$$
$$\text{Total} = 1.52759$$

$$\text{Av} = \frac{1.52759}{4} = 0.38190$$

The individual unit which will have an effort of 0.38190 with $K = 1$ and $s = -0.57616$, as before, is

$$0.38190 = 1N^{-0.57616}$$

$$N = 5.3159 = \text{true lot equivalent point}$$

Thus, for the second lot, an average of 89.43 work-h per unit is assigned to unit 5.3159. Other lots are treated similarly. The points so obtained can be subject to a least-squares calculation to determine the best values of s and K.

If tables of cumulative values are available for values of s reasonably close to the approximate value, the numerical work for the true lot-midpoint method can be simplified.

Example 9.11 Using the data in Table 9.6 and the rule-of-thumb lot-midpoint method, determine the learning-curve function that represents the data.

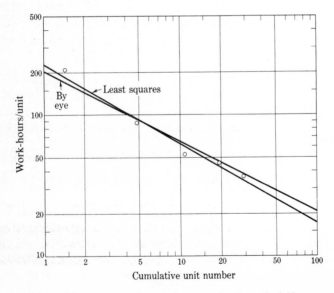

Figure 9.5 Plot of data and learning curves for Example 9.11.

Table 9.9 Least-squares calculation for Example 9.11

N assigned	Effort E_N, work-h	X log N	Y log E	XY	X^2
1.5	207.24	0.17609	2.31647	0.40791	0.03101
5	89.43	0.69897	1.95148	1.36403	0.48856
11	53.39	1.04139	1.72746	1.79896	1.08449
20	44.98	1.30103	1.65302	2.15063	1.69268
30	36.81	1.47712	1.56597	2.31313	2.18188
		4.69460	9.21440	8.03466	5.47862

SOLUTION Application of the rule-of-thumb lot-midpoint method to the group data leads to the five points given by the last two columns of Table 9.7. The points are plotted in Fig. 9.5.

A line put through the points by eye is shown. Using the extreme points on the line, the slope as determined by measurement, before reproduction, is $\Delta Y/\Delta X = -73.6/148.5 = -0.496$.

The percentage learning ratio corresponding to the slope can be obtained from Eq. (9.13):

$$-(-0.496) = \frac{2 - \log L_P}{\log 2}$$

$$\log L_P = 1.8507$$

$$L_P = 70.91 \text{ percent} = \text{percentage learning ratio}$$

and the value for K is the intercept on the axis $N = 1$ and is about 204. Thus the line drawn by eye gives a 70.91 percent learning ratio and the following learning-curve function:

$$E_N = 204N^{-0.496}$$

The data can also be subjected to a least-squares analysis, described in the previous chapter, and shown for the five points in Table 9.9.

$$s = \frac{5\Sigma XY - \Sigma X \Sigma Y}{5\Sigma X^2 - (\Sigma X)^2} = \frac{5(8.03466) - 4.69460(9.21440)}{5(5.47862) - 4.69460^2} = -0.57616$$

$$\log K = \frac{\Sigma Y}{5} - s\frac{\Sigma X}{5} = \frac{9.21440}{5} + 0.57616\frac{4.69460}{5} = 2.38385$$

$$K = 242.02$$

The learning-curve function resulting from the least-squares computation is

$$E_N = 242.02N^{-0.57616}$$

and the slope constant of -0.57616 corresponds to a 67.075 percent learning ratio. The straight line represented by the relationship above is plotted in Fig. 9.5.

Although the slope constant, -0.57616, is not exact because the rule-of-thumb lot-midpoint method was followed, the true lot-midpoint method can now be used, as indicated previously, to establish an even better representation for the learning-curve function.

9.17 Development of Learning-Curve Tables

Tables can be readily constructed for use with the learning-curve function. The decimal or percentage learning ratio is a parameter, and the unit number is a variable. The three most important items to tabulate are:

E_N = effort for the Nth unit taking unity as the effort for the first unit = N^s

$$E_T = \sum_1^N E_N = \text{cumulative effort for units 1 through } N$$

$$\frac{E_T}{N} = \frac{\sum_1^N E_N}{N} = \text{cumulative average unit effort for units 1 through } N$$

Tables can be programmed easily and printed out by a computer for any value of L_D, the decimal learning ratio. An abbreviated tabulation for $L_D = 0.90$, 0.80, and 0.70 is given in Table A4.1.

VALUE ENGINEERING

9.18 General Description

Value engineering is a discipline devoted to increasing productivity and related to cost engineering. Value engineering (VE) applies to hardware and value analysis (VA) to software, but the term *value engineering* alone will be used here. The Society of American Value Engineers is an active organization well known through its appropriate acronym, SAVE.

Value engineering is the systematic use of techniques which identify the required functions of an item, establish values for these functions, and provide the functions at the lowest overall cost without loss of performance. It is widely used by government and industry.

A *function* is the characteristic of an item which meets the need or want of the user. A *basic function* is the primary reason for the existence of an item from the user's point of view, or the purpose or performance feature which must be attained if an item is to work. A *secondary function* is a feature which supports the basic function, such as salability, convenience, or performance.

Value is the measurement of how well an item fulfills its function, considering both performance and cost. The types of value which are important in value engineering are use value, esteem value, cost value, and exchange value.

9.19 Job Plan for Value Engineering

A job plan consists of six major phases:

1. The *feasibility phase* is a type of benefit-cost analysis to determine whether the cost savings or value improvements are worth the expenditures of money and other resources needed to accomplish them.
2. The *information phase* involves gathering the background material, documentation, and other information relevant to the VE study. An important part of this phase is function analysis.
3. The *creativity phase* involves the development of alternative ways of providing the required functions of the item analyzed.

4. The *evaluation phase* involves making cost and other comparisons of the alternatives developed in the creativity phase.
5. The *proposal phase* involves making a plan to describe the value improvement problem and its recommended solution.
6. The *implementation phase* is particularly important to ensure implementation and prevent misrepresentation with resultant higher costs and decreased benefits.

9.20 Applications

An example of a succinct application is the work of a VE team at Picatinny Arsenal, New York, which improved the design for the top charge of a blasting cap, eliminating ferruling in the assembly, as reported by Dobrow [2] and shown in Fig. 9.6. Loading was simplified and safety improved at no reduction in performance. Although unit savings were only $0.08, the yearly savings were $346,000.

VE can be applied to any of the following:

Hardware
Software
Methods and procedures
Services

Before

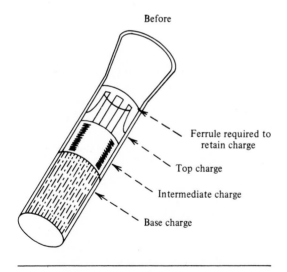

Ferrule required to
retain charge

Top charge

Intermediate charge

Base charge

After
Lower charges now retained by top charge
Ferrule eliminated
Loading procedures simplified
Improved safety of assembly operation

Unit cost reduced by $0.08
Validated savings: $346,900

Figure 9.6 Blast cap before and after modifications of the top charge. (*From Dobrow, 1972.*)

Table 9.10 Checklist for areas to apply value engineering

---

1. Engineering
   a. Design
   b. Development
   c. Product improvement
2. Manufacturing
   a. Methods, material handling, and production
   b. Tool design
   c. Packaging
3. Purchasing
   a. Vendors' alternative proposals
   b. New product information and price against function evaluation
4. Sales
   a. Slow-selling projects
   b. VE service to customers
5. Systems and procedures
   a. Paperwork
   b. Forms processing
   c. Reproduction services
6. Maintenance
   a. Procedures
   b. Materials
   c. Mechanization
   d. Work scheduling

It is not limited by the size of the study or the type of business.

A checklist of areas in which VE can be efficiently and profitably used to increase productivity is given in Table 9.10.

PRODUCTIVITY

9.21 Definition

Productivity has become an important topic in manufacturing and business. The Germans and the Japanese were the first to establish national centers to analyze and increase productivity. It was not until 1977 that the American Productivity Center was established in Houston, Texas, as a private undertaking.

Productivity is basically an engineering topic. The technological elements of productivity are not part of this discussion, but the subject has ancillary aspects that should be recognized by all cost engineers. *Productivity* is readily defined as a ratio of output to input:

$$\text{Productivity} = \frac{\text{measure of output}}{\text{measure of input}}$$

Table 9.11 Activity of workers in public utility construction projects

Activity measured	Time, in %
Late starting, early quitting	3.2
Personal	2.3
Breaks	2.1
Waiting and unexplained idle time	31.9
Receiving instructions and planning work	4.4
Reading drawings	0.7
Traveling	12.8
Transport	4.6
Obtaining tools and materials	3.3
Direct productive work	34.7
	100.0

Source: H. M. Hohns, "Productivity an Updated Concept of Management," *Transactions of the American Association of Cost Engineers,* p. I-1, 1979.

Table 9.11 shows the results on three projects as estimated by a public utility. *Prime productivity,* the direct productive work, as given in the tabulation is 34.7 percent, but this simple conclusion can be challenged. Some classifications not included as productive work, such as obtaining tools and material, may be essential in the opinion of others. Prime productivity is a concept that should be measurable by engineering standards, but the first observation is that even this simple concept is subject to human interpretation.

9.22 Other Aspects

Another consideration is that productivity, as stated above, should occur full time. An assembly line may have to shut down part of the time because of lack of sales, but the shutdown is an underutilization of the equipment. The prime productivity of the assembly line has not changed. Nevertheless, in ordinary parlance, productivity and underutilization are not differentiated. During a recession, productivity is said to fall when actually reduced production is due to underutilization. Then, when sales increase, productivity is said to increase. Not infrequently, demands by labor for a share in the "increased" productivity rise as well.

Even at the prime productivity level, the measure of output and input can be complicated. The investment per worker must be considered. The input of the preceding workers who created the investment must be recognized. The output of a coal miner in tons per shift must include the input of capital for the equipment. Similarly, the introduction of preassembled parts does not necessarily increase the productivity of the individual worker performing the assembly. Indeed, it may actually reduce the productivity of that worker since he or she may become underutilized even though there is a gain in productivity for the project as a whole.

In practice, it is incremental productivity that is measured as follows:

$$\Delta \text{ productivity} = \frac{\Delta \text{ output}}{\Delta \text{ input}}$$

The measurement of output involves many factors. Consider setting telephone poles. The equipment now available is much more efficient than that used in the past, but the gain per work-hour must be corrected for the increased investment. Heavier poles can now be handled, permitting them to be spaced farther apart and thus requiring fewer poles per kilometer. That itself amounts to an increase in productivity. Further, new poles receive improved treatment with preservatives and last longer with less maintenance. That, too, represents increased productivity. Obviously, output should consider the durability and life cycle cost of the poles.

Economists state that capital, labor, and land are the main inputs of production. Land is included since it requires investment. However, natural resources are a major factor in productivity, a fact that has been overlooked. Natural resources include materials, energy, location, climate, and geography, among others. An abundant supply of cheap energy can contribute to apparent productivity. The discovery of a surfeit of crude petroleum and natural gas in the United States in the early twentieth century enhanced the overall productivity of the nation, but the advantage of this cheap energy was not recognized. A similar situation exists today in agriculture; the United States is the world leader in productivity. However, the land is being depleted, and the nation is dissipating its farm land somewhat as it squandered its crude. Failure to include the effect of an exhaustible input such as energy can lead to disaster in the long run since productivity can be misinterpreted.

Conversely, a nonexhaustible resource such as climate can represent a permanent advantage in productivity. Location of the chemical industry in the Gulf Coast region of the United States is a good example. The climate permits equipment to be located outdoors, reducing investment.

All modern industrial nations are based on capital-intensive economies. The gain in productivity comes from increased capitalization. Other sources of increased productivity are scientific and engineering advancement, better use of management techniques, and an improvement in human attitudes.

Scientific and engineering advancement can be a powerful source of increased productivity. A generation ago, a mechanical rotary calculator costing over $1000 was a cumbersome machine that could not perform chain calculations. Today, despite inflation, electronic calculators costing less than $20 can outperform the old mechanical model in every way.

Management is an important factor in productivity since it must make the investment to increase prime productivity, but even more so to prevent the dispersal of productivity gain, discussed below.

Human attitude is a significant factor in productivity but is deemphasized by management, which prefers to avoid personal matters in dealing with its workers.

Japanese workers are employed for life and have a personal relationship with their employer. They are production oriented and will perform multiple tasks. American workers have little inherent loyalty to an employer and regard themselves as skilled specialists with an antagonism to multiple chores. Individuals and even nations differ in their attitude toward work, although it is not prudent to express this fact.

9.23 Levels of Productivity—Diffusion

Although there are problems in determining even prime productivity, the objective is usually obvious—to increase productivity by more investment in better technology. Unfortunately, there are several levels of productivity. Consider an automobile assembly line. Productivity is increased by more automation, but this introduces a chain of events that tends to diffuse the gain in productivity. The number of auxiliary workers increases as more people are needed in supplementary activities. Government, company, and union inspectors proliferate. Special events and activities are required to break the monotony of the new job, and special people are added to orchestrate the special events. The office force expands to keep the more complicated operations in order. More business personnel, supervisors, lawyers, security forces, and psychiatrists are needed. A higher proportion of people are now working as adjuncts to the prime work.

The level of work above the prime productivity level will be referred to as the *local productivity level.* It is made up of a number of prime productivity levels with appropriate adjuncts at the local plant.

In turn, there is a level above the local productivity level that will be called the *organizational productivity level.* The organizational productivity level encompasses the company as a whole and creates a host of new personnel in areas ranging from transportation, distribution, and servicing to a wide spectrum of business specialties, including marketing, advertising, law, and finance. All these new employees add to the diffusion of productivity.

The prime productivity of the workers who make the parts and assemble an automobile is very different from the ultimate productivity at the national level obtained by the purchaser, who must pay for all the diffusion of productivity inherent in the system.

The United States formerly had the best productivity in manufacturing steel and automobiles, but it is no longer first in either. The basic problem does not seem to be understood. Although prime productivity must be improved, it is productivity diffusion that is the serious problem. Unfortunately, the problem must be solved by management and government personnel, who themselves are part of the diffusion but are not inclined to reduce their own numbers. Even when management understands that it must increase prime productivity, its first step is to expand its own diffusive influence by creating various committees and functions to analyze the problem. The parasites increase as the host decreases—an unsatisfactory situation for the host. Diffusion feeds on itself, and the production worker in the United States is becoming an endangered species.

In summary, there are four levels of productivity:

Prime productivity
Local or plant productivity
Organizational productivity
National productivity

The prime productivity level is chiefly an engineering measurement. The other three levels, although necessary to some degree, add to the diffusion of prime productivity. The American enigma is that the nation is overburdened with diffusion. However, that may be the way of the world, with the United States merely leading the parade.

NOMENCLATURE

B	Book value, \$
C_{ex}	Irregular cost at the end of the xth year, \$
C_i	Initial cost, \$
C_{sal}	Salvage value, \$
D_x	Fractional depreciation for the xth year, decimal
E_N	Effort required to produce the Nth unit, e.g., work-hours per unit
E_T	Cumulative effort to produce units 1 through N, e.g., work-hours
E_T/N	Cumulative average unit effort, e.g., work-hours per unit
F_{GP}	Factor to convert G to P; uniform-gradient-series present-value factor, years
F_{PK}	Capitalized cost factor, dimensionless
F_{PR}	Capital recovery factor, year^{-1}
G	Uniform-gradient amount, \$/year
i	Rate of return before taxes, decimal per year
K	Capitalized cost, \$
K	Constant in Eq. (9.10); theoretical effort required for the first unit, e.g., work-hours
L_D	Decimal learning ratio, dimensionless
L_P	Percentage learning ratio, dimensionless
n	A number, such as number of years
N	Cumulative number of units produced, dimensionless
p	Technological advancement rate, decimal per year
P	Present value, \$
r	Rate of return after taxes, decimal per year
R	Unacost, uniform end-of-year annual amount, \$/year
R_p	Yearly operating expense subject to improvement by technological advancement, \$/year
s	Slope constant, decimal, dimensionless
t	Tax rate, decimal
VA	Value analysis
VE	Value engineering
ΣE_N	Same as E_T
$\dfrac{\Sigma E_N}{N}$	Same as E_T/N
ψ	Present value of \$1 of depreciation, decimal

PROBLEMS

9.1 The book value of a machine is $4000, with 4 years of life remaining at straight-line depreciation. Assume that the book value and salvage value are identical at all times. The operating cost for the machine is $10,000/year as of the end of the year. A new machine costs $30,000 and has an 8-year life and an operating cost of $4000/year as of the end of the year. Neglect salvage value and taxes. If money is worth 10 percent per year, does it pay to install the new machine?

9.2 What is the actual rate of return realized with the new machine in the preceding problem?

9.3 Repeat Prob. 9.1 on the basis of a 46 percent tax rate and sum-of-the-years-digits depreciation for the new machine. The rate of return is 10 percent after taxes. Tax benefits can be taken at instant costs occur.

9.4 Assume that 5 years ago a machine cost $35,000 and has been depreciated on a 10-year life with straight-line depreciation and $1000 salvage value. If sold now, it will bring $6000. The machine is part of a group, and depreciation continues at all times, just as if it were in service. If sold at the end of the year, it will bring $5000.

A new machine costs $75,000 with a 10-year life, and has $4000 salvage value and an annual end-of-year savings of $12,000. Use sum-of-the-years-digits depreciation. If money is worth 12 percent per year after a 46 percent tax, should the new machine be bought?

9.5 Repeat Prob. 9.4, assuming that the machine is not part of a group and that any gain or loss on disposal can be taken as an immediate tax consideration.

9.6 A machine costs $20,000 and lasts 6 years, with $1000 salvage value at all times. Past records give the following data, where costs are as of the beginning of the year and include operating and repair costs:

Year	Cost, $
1	3,000
2	3,500
3	8,000
4	15,000
5	5,000
6	15,000

If money is worth 10 percent per year and there is no tax, how long should the machine be kept?

9.7 If money is worth 12 percent per year after taxes, find the unacost and capitalized cost for the following after-tax expenses:

End of year	Expense, $/year
1	1000
2	1200
3	1400
4	1600
...	....
20	4800

9.8 In what respect is sum-of-the-years-digits depreciation related to a uniform-gradient series?

9.9 What is the optimum service life for a machine costing \$100,000 and incurring a \$40,000/year operating rate subject to technological advancement if the technological advancement rate is 3 percent per year, the rate of return is 12 percent per year after a 50 percent tax, and sum-of-the-years-digits depreciation is used?

9.10 Use Eq. (9.9) and find the optimum value for n, the economic life under the following conditions: technological advancement rate 6 percent per year, 20 percent annual expense subject to technological advancement, 15 percent per year rate of return after a 50 percent tax, and sum-of-the-years-digits depreciation.

9.11 If 1000 light bulbs fail statistically according to the following pattern:

Period	Failures
1	50
2	100
3	250
4	500
5	75
6	25

and replacements are made at the beginning of each period if necessary, make a tabulation to find the steady-state number of replacements for an installation of 1000 bulbs and check this against the calculated value.

9.12 In Prob. 9.11, replacement costs are \$1 per bulb but \$0.50 per bulb as a group of 1000. Find the program for minimum cost.

9.13 Find the first unit value when the unit value of unit 100 is 30 h with an 80 percent learning ratio.

9.14 Find the value for unit 300 when the value for unit 3 is 1000 h with a 70 percent learning ratio.

9.15 Find the unit average hours for units 21 to 40 when unit 1 is 250 h with a learning ratio of (a) 90 percent and (b) 83 percent.

9.16 Units 1 through 200 have accumulated 45,000 h. Find the unit time for unit 100 with a learning ratio of (a) 70 percent and (b) 75 percent.

9.17 Using the rule-of-thumb lot-midpoint method, the geometric-mean lot-midpoint method, and the true lot-midpoint method for an 80 percent learning ratio, find the lot midpoints or lot equivalent points for:

Lot 1, units 1 through 14
Lot 2, units 15 through 40
Lot 3, units 41 through 80
Lot 4, units 81 through 100

9.18 Additional information for Prob. 9.17 is:

Lot 1, 15,000 h total time
Lot 2, 20,000 h total time
Lot 3, 25,000 h total time
Lot 4, 10,000 h total time

Continuing the problem, develop the slope constant by plotting on log-log graph paper. Using the method of least squares for the rule-of-thumb lot-midpoint method, establish the learning-curve function.

9.19 Analyze the answers to Prob. 9.18. What can be said about the slope of the curve?

9.20 A company has produced machine parts as a subcontractor for a large military order for a number of years, during which production was consistent with an 89 percent learning ratio. A commercial subcontract for making similar machine parts, but with a rigid schedule, is under consideration. Should you use the same percentage learning ratio for a schedule that is twice as fast as the military schedule? How would you develop the learning curve anticipated for this contract?

REFERENCES

Learning curve

1. Behrens, H. J.: "Improvement Curve, Real or Imaginary?," *AACE Bulletin,* vol. 8, June 1966, p. 75.
2. Fabrycky, W. J., P. M. Ghare, and P. E. Torgersen: *Industrial Operations Research,* 2d ed., Prentice-Hall, Inc., Englewood Cliffs, N.J., 1972.
3. Hirschmann, W. B.: "Profit from the Learning Curve," *Harvard Business Review,* January–February 1964, pp. 125–139.
4. Jordan, R.: *How to Use the Learning Curve,* Material Management Institute, Boston, 1965.
5. Kottler, J. L.: "The Learning Curve: A Case History in Its Application," *Journal of Industrial Engineering,* vol. 15, July–August 1964, pp. 176–180.
6. Noah, L. W., and R. W. Smith: *U.S. Air Force Project Rand Research Memorandum* MR-2786-Pr., January 1962.
7. Paterson, S. F.: "Predicting Learning Time," *Production,* July 1963.

Value engineering

1. Del'Isola, A. J.: *Value Engineering in the Construction Industry,* Construction Publishing Company, New York, 1973.
2. Dobrow, P. V.: "Value Engineering—A Money Saving Tool," *Chemical Engineering,* vol. 79, Aug. 21, 1972, pp. 122–126.
3. EIA Value Management Committee: *Value Engineering Support to Program Management,* Value Engineering Management Bulletin No. 8, Electronic Industries Association, Washington, D.C., 1979.
4. Falcon, W. D., ed.: *Value Analysis, Value Engineering,* American Management Association, New York, 1974.
5. Fallon, C.: *Value Analysis,* 2d ed. Interscience Publishers, a division of John Wiley & Sons, Inc., New York, 1980.
6. Greve, J. W., ed.: *Value Engineering for Manufacturing,* Prentice-Hall, Inc., Englewood Cliffs, N.J., 1967.
7. Heller, E. D.: *Value Management, Value Engineering and Cost Reduction,* Addison-Wesley Publishing Company, Inc., Reading, Mass., 1971.
8. Miles, L. D.: *Techniques of Value Engineering and Analysis,* 2d ed., McGraw-Hill Book Company, New York, 1971.
9. Mudge, A. E.: *Value Engineering,* McGraw-Hill Book Company, New York, 1971.
10. Park, R. J.: *Practical Applications of Function Analysis System Technique "FAST,"* R. J. Park and Associates, Detroit, Mich., 1975.
11. Ridge, W. J.: *Value Analysis for Better Management,* American Management Association, New York, 1969.
12. Parker, D. E.: *Value Engineering Theory,* Value Foundation, Washington, D.C., 1977.
13. *SAVE Encyclopedia of Value,* Society of American Value Engineers, Dallas, Tex., vol. I, 1977, addendum 1, 1978.
14. *Value World,* Society of American Value Engineers, Dallas, Tex.

Productivity

1. *Fortune:* "The Disaster in Productivity," December 1974, p. 24.
2. Grayson, C. J., Jr.: "Eight Ways to Raise Productivity and Profits," *Nation's Business,* November 1972, pp. 30–36.
3. Jelen, F. C., and M. S. Cole: "Can New Equipment Cost Less?," *Hydrocarbon Processing,* July 1971, pp. 97–100.
4. Jelen, F. C., and M. S. Cole: "Methods for Economic Analysis," *Hydrocarbon Processing,* Part 1, July 1974, pp. 133–139; Part 2, September 1974, pp. 227–233; Part 3, October 1974, pp. 161–163.
5. Strandell, M.: "Productivity in the Construction Industry," *Cost Engineering,* March–April 1978, pp. 57–61. (Lists over 100 references.)
6. *Time:* "Productivity, Seeking that Old Magic," Aug. 2, 1971, pp. 56–57.

TEN

OPTIMIZATION

F. C. Jelen

ONE VARIABLE

10.1 Optimization: A Human Trait

Optimization, or the urge for efficiency, has a psychological origin. The human mind can confront a task or problem and recognize more than one course of action. This recognition is followed by a second phase, the selection of what is considered the best action. The second phase is the decision step. The two steps taken together, recognition of alternatives and decision, constitute optimization. Optimization can be qualitative (judged by human preference) or quantitative (detected by exact mathematical means). Optimization permeates society and technology to a far greater degree than is realized. The musical composer struggling for just the right combination of notes, the mathematician seeking an expression for a mathematical or physical principle, and the designer of a bridge are all involved in optimization. In a more prosaic sense, the golfer striving for a minimum score, the student hoping for a high IQ rating, and the newspaper headline announcing the greatest airplane fatality of all time likewise are manifestations of the allure of optimization and extreme values. Optimization is as old as antiquity, and even the ancients suspected that a circle has the greatest ratio of area to perimeter.

As engineering becomes more advanced and the business and industrial world more competitive, the methods for optimizing become more exact and the rewards increasingly great. Detailed consideration of optimization now pervades all successful human endeavor. There are basically three categories of optimization. The first is preferential, as exemplified by the musical composer struggling for just the right combination of notes, a category subject to preference and taste only.The second category is mathematical or physical, such as the demonstration that a sphere contains the maximum volume for a given surface, a fact not subject to human preference or fancy. The third category is economic optimization, which is a combination of these two categories but with the human preference elements introduced in a quantitative way. Economic optimization or engineering optimization, which is the subject of this chapter, has to consider human values, but they must be expressible in quantitative form. Thus the designer of a bridge is concerned with more than an engineering design for maximum strength-to-weight basis. He or she must consider myriad quantitative factors, such as the cost of on-site welding relative to the cost of prefabricated sections, extra construction costs for the safety of the workers, overdesign to meet future demand, and so forth.

10.2 Nature of Optimization

Optimization in a general sense involves the determination of a highest or lowest value over some range. Which is desired is a matter of statement. Thus a problem can be maximized for profit or minimized for loss. *Optimization* will be used as a general term for either case. To the mathematician, maximum and minimum values have special qualities, and it is assumed that the analyst is familiar with these from a study of the calculus. There may be more than one mathematical maximum or minimum in a region, and the highest and lowest values in the region may or may not correspond to mathematical maximum and minimum values. Mathematical optima occur in the case of one independent variable when the slope of the curve is horizontal or zero. This is a necessary but not sufficient condition, as the point may be an inflection point. Certain other conditions must be met; e.g., the function must be continuous at the point in question. Figure 10.1 is a review of the circumstances. Points A and D are mathematical maxima, but the highest value is point F, which is not a mathematical maximum. Point C is a mathematical minimum and also the lowest value. Point B meets the condition that the slope is zero and the curve is horizontal, but B is neither a mathematical maximum nor a mathematical minimum; instead it is a point of inflection. Point E is not a mathematical minimum because the slope or first derivative is not continuous at point E. Maximum and minimum points which are not the highest and lowest values are referred to as *local optimum points*. Thus point A is a local maximum. Although the highest and lowest values in a region may not correspond to mathematical optima, a study of mathematical optima is fundamental to a study of optimization.

In working optimization problems, the analyst must formulate the problem in such a way as to optimize for the proper goal. This is elementary but frequently a source of error because the result depends upon the goal sought. Having obtained a result, the analyst should answer the following questions:

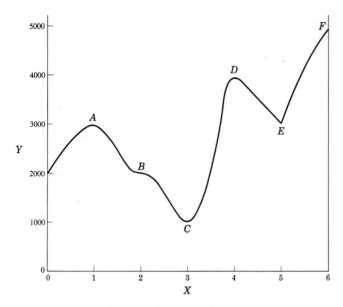

Figure 10.1 Various types of extreme values.

1. Is the answer feasible or realizable?
2. Is the answer optimal?
3. How sensitive is the optimum value to the parameters and how responsive to a variation of the independent variables from their optimum values?

1. The answer may not be feasible for several reasons. A physical barrier may exist. Thus mathematics may find that for optimum conditions, water at $-20\,°C$ should be pumped into the tubes of a heat exchanger, but water at $-20\,°C$ is solid. A computer can print out an optimum answer that requires an undisclosed mixture of benzene and toluene containing 120 percent benzene and -20 percent toluene. The two add up to 100 percent, but the mixture is not realizable. An answer may not be feasible because there are not enough days in the year to reach the yearly production requirement.
2. Is the answer optimum? It may be a minimum instead of a required maximum, or it may be a saddle point, such as point B in Fig. 10.1, or it may be only a local maximum, such as point A.
3. Knowledge of the sensitivity of the optimum value to the parameters is essential. The analyst should know which components are most important and require the most study, and which are insignificant and can be ignored.

10.3 Optimization methods

Many methods have been introduced to determine optimum procedures or policies. This chapter is fundamental and considers three powerful general methods:

1. Analytical
2. Graphical or tabular
3. Incremental

A large part of this book is devoted to optimization, and other chapters describe specialized methods. The search for or calculation of a highest or lowest value in a region can become exceedingly complicated in practice, and the analyst must be capable of originality in setting up solutions. Problems involving manufacturing or processing operations are referred to as *economic balance*. Usually cost data are involved, e.g., operation for maximum profit, but not necessarily so, e.g., operation for maximum output with cost disregarded. The problem may involve a single unit, such as a welding machine, or a combination of units, such as an assembly line. For multiunit systems, optimization is on an overall basis, and it does not follow that optimization of each unit on an individual basis will result in overall optimization.

In practice, many data will be empirical and economic in origin, so that optimization will be based on parameters which are not rigorously fixed or exact numbers. Also, the more variables and parameters involved, the flatter the optimum conditions generally become. Nevertheless optimization as a goal is essential for successful operation even though there are some uncertainties.

Problems reducible to one independent variable will be discussed first, followed by an extension to more than one variable. The literature is replete with special problems leading to mathematical relationships which require only that analysts substitute numerical values. In doing so, they assume all the limitations, assumed values, and other fixed data in the model. Problems are so varied that analysts must be prepared to derive their own relationships.

10.4 Analytical Method for Optimization

Let the problem be to optimize Y, which is a function of the single independent variable X. That is,

$$Y = F(X) \tag{10.1}$$

Generally, an optimum value of Y will be a mathematical maximum or minimum. For this condition, as given by the calculus,

$$\frac{dY}{dX} = \frac{dF(X)}{dX} = 0 \tag{10.2}$$

If the problem can be formulated so that Y can be expressed in terms of X, the analytical method can be used. Equation (10.1) may represent an exact physical or mathematical relationship, an empirical relationship, or a combination. The means of establishing the relationship is not important; only its existence is necessary.

If Y is not a continuous function of X, it can be treated as a series of regions in which it is continuous. Differentiation at a point having a discontinuity is not permissible since the value of Eq. (10.2), the derivative at the point, depends upon from which side of the point $\Delta Y/\Delta X$ is taken. At times, too, X can assume only integral values, e.g., the number of stories in an office building. Only continuous functions can be differentiated rigorously. Discrete variables should be treated by the calculus of finite differences or subjected to a tabular analysis. In many cases, a function of discrete variables can be differentiated just as if it were a continuous function, but solutions so obtained should be investigated carefully.

If Eq. (10.1) can be set up, it is differentiated to satisfy Eq. (10.2), leading to a value or X that is perhaps optimum. The value of X so obtained can be substituted in Eq. (10.1) to give an optimum X, Y. The differentiation of Eq. (10.1) may be difficult, and the resulting Eq. (10.2) may be so complicated that solution by trial and error may be required, but such considerations merely make the solution more tedious.

The solution to Eq. (10.2) will give perhaps several values of X, say X_1, $X_2, \ldots, X_n$. Each value of X, say X_1, can be tested analytically to determine whether it represents a maximum, a minimum, or a point of inflection, which is neither. For maximum and minimum, if

$$\frac{dY}{dX}\bigg|_{X=X_1} = 0 \quad \text{and} \quad \frac{d^2Y}{dX^2}\bigg|_{X=X_1} = +$$

then $\qquad\qquad Y$ at $X = X_1$ is a minimum

If

$$\frac{dY}{dX}\bigg|_{X=X_1} = 0 \quad \text{and} \quad \frac{d^2Y}{dX^2}\bigg|_{X=X_1} = -$$

then $\qquad\qquad Y$ at $X = X_1$ is a maximum

The sign of the second derivative indicates whether the point is a maximum or a minimum. The second derivative may be zero at the point, however, in which case the test is inconclusive. Further testing is necessary, which requires formulating higher-order derivatives. Find the lowest-order derivative that becomes non-zero for $X = X_1$. If this is an odd derivative, then X_1 is an inflection point. If this is an even derivative and positive in value for $X = X_1$, then X_1 is a minimum point, but if it is negative in value, X_1 is a maximum point.

In practice, it is rarely necessary to test beyond the second derivative. Another test is to calculate Y in the neighborhood of $X = X_1$ and from the tabulation or plot to observe the behavior of Y at X_1.

10.5 Graphical Method for Optimization

If $\qquad\qquad\qquad\qquad Y = F(X) \qquad\qquad\qquad\qquad$ (10.3)

then a tabulation and plot of Y against X over the feasible range of X will yield all the information for the optimization of Y. Figure 10.1 shows such a graph.

A plot or tabulation has the advantage of distinguishing between mathematical maxima and minima, inflection points, and highest and lowest values. Also, it discloses how responsive the optimum value is to a variation in the independent variable. The graphical method is widely applicable, and the data and information can be in analytical, tabular, or graphical form.

For one trial in one independent variable, the analytical method has little advantage over the graphical or tabular method, although the analytical method will give the mathematical optima without a search procedure. Even so, the analytical method should be supplemented with some graphical or tabular data in a thorough analysis. The advantage of having an analytical expression for the optimum condition is very real, however, and permits a representation of the problem as mathematical relationships that are particularly valuable for theoretical studies or as expressions in compact, directly usable form.

10.6 Incremental Method for Optimization

The incremental method is instinctive and basic. In essence, it amounts to operating at one value of the independent variable, changing the variable somewhat, and determining by calculation or observation whether the change is advantageous. Referring to Fig. 10.1, operation at $X = 3.5$ gives $Y = 2000$, and incrementing to $X = 3.6$ gives $Y = 2500$, an improvement. In this fashion, small increments will lead to the local optimum D. Some precautions to be observed are evident from Fig. 10.1. It may be necessary to enter at several values of X to avoid determining a local optimum only.

The incremental method is suitable for trial by experiment, including uncertainty in individual observations. With uncertainty, trials are repeated to establish probable values by means of statistical analysis. The incremental method is applicable to analytical or tabular data and can be used for continuous or discrete variables. It is analysis about a point.

10.7 A Cyclic Process

An example in optimization will be developed in detail to illustrate the principles. It is typical of a variety of problems in which at some point in a process it becomes economical to pay a penalty and start over again.

Example 10.1 In drilling wells in a certain type of terrain, an oil company finds that the depth to which it can drill before changing drill bits is given by the relationship

$$Q_c = 300T_d^{0.5}$$

where Q_c = cumulative meters
T_d = drilling time, days

Drilling costs are \$4250/day. It requires 18 h, or 0.75 day, to install a new drill bit at a total cost of \$10,500.

(*a*) Find the optimum cycle time for the minimum monthly cost at a drilling rate of 2000 m/month.

(*b*) Find the optimum cycle time for the minimum monthly cost at a drilling rate of 4750 m/month.

(*c*) Find the optimum cycle time for the maximum drilling depth per month without regard to cost.

(*d*) Find the optimum cycle time for the maximum profit if each meter drilled has a return value to the company of \$60/m.

SOLUTION (*a*) The basis is 2000 m/month. Let N = cycles per month. Then,

$$NQ_c = 2000$$

$$N = \frac{2000}{Q_c} = \frac{2000}{300T_d^{0.5}} = \frac{6.667}{T_d^{0.5}} \tag{10.4}$$

Set up an expression for cost in dollars per month for N cycles per month:

$$\$/\text{month} = N\underbrace{\frac{\text{cycles}}{\text{month}}\,T_d\frac{\text{days}}{\text{cycle}}4250\frac{\$}{\text{day}}}_{\substack{\text{drilling}\\\text{expense}}} + \underbrace{10{,}500\frac{\$}{\text{cycle}}N\frac{\text{cycles}}{\text{month}}}_{\substack{\text{bit}\\\text{expense}}} \tag{10.5}$$

Eliminate N by Eq. (10.4):

$$\$/\text{month} = 28{,}333T_d^{0.5} + 70{,}000T_d^{-0.5} \tag{10.6}$$

Differentiate and set equal to zero:

$$\frac{d(\$/\text{month})}{dT_d} = 0 = 0.5(28{,}333)T_d^{-0.5} - 0.5(70{,}000)T_d^{-1.5}$$

$$T_d = 2.471 \text{ days of drilling per cycle}$$

Check for feasibility:

$$N = \frac{6.667}{T_d^{0.5}} = \frac{6.667}{2.471^{0.5}} = 4.241 \text{ cycles per month}$$

$$\text{Total days/cycle} = 2.471 + \underset{\text{changing}}{0.75} = 3.221$$
$$\phantom{\text{Total days/cycle} = }\underset{\text{drilling}}{}$$

$$3.221\frac{\text{days}}{\text{cycle}}4.241\frac{\text{cycles}}{\text{month}} = 13.66 \text{ days/month}$$

which is feasible inasmuch as $13.66 < 30$.

$$\frac{\$}{\text{m}} = \frac{\$/\text{month}}{\text{m/month}} = \frac{4250NT_d + 10{,}500N}{2000}$$

and for $N = 4.241$ and $T_d = 2.471$, $\$/\text{m} = 44.53$. Check for the optimum by the second derivative. Differentiate Eq. (10.6):

$$\frac{d^2(\$/\text{month})}{dT_d^2} = -0.25(28{,}333)T_d^{-1.5} + 0.75(70{,}000)T_d^{-2.5}$$

which is positive for $T_d = 2.471$, and hence is a minimum cost.

Graphical solution Dollars per month can be calculated from Eq. (10.6). Dollars per meter is obtained from dollars per month by dividing by 2000 m/month.

T_d, days	$\$/\text{month}$	$\$/\text{m}$
1.00	98,333	49.17
1.50	91,851	45.93
2.00	89,568	44.78
2.25	89,167	44.58
2.471	89,069	44.53
2.75	89,198	44.60
3.00	89,489	44.74
4.00	91,666	45.83

A plot of the results is given in Fig. 10.2, confirming that T_d is both a mathematical minimum and a lowest value. It also shows how sensitive the monthly cost is to T_d.

(b) The basis is 4750 m/month.

$$NQ_c = 4750$$

$$N = \frac{4750}{300T_d^{0.5}} = 15.833T_d^{-0.5} \qquad (10.7)$$

$$\$/\text{month} = NT_d(4250) + 10{,}500N$$

Eliminate N by Eq. (10.7):

$$\$/\text{month} = 67{,}290T_d^{0.5} + 166{,}250T_d^{-0.5}$$

$$\frac{d(\$/\text{month})}{dT_d} = 0 = 33{,}645T_d^{-0.5} - 83{,}125T_d^{-1.5}$$

$$T_d = 2.471 \text{ days of drilling per cycle}$$

Check for feasibility:

$$N = 15.833T_d^{-0.5} = 15.833(2.471^{-0.5}) = 10.07 \text{ cycles per month}$$

Total days per cycle $= 2.471 + 0.75 = 3.221$:

$$3.221\frac{\text{days}}{\text{cycle}}10.07\frac{\text{cycles}}{\text{month}} = 32.44 \text{ days/month}$$

which is not feasible as $32.44 > 30$. Collecting results,

$$N = 15.833T_d^{-0.5} \qquad \text{cycles per month necessary}$$

$$\frac{30}{T_d + 0.75} \qquad \text{cycles per month available}$$

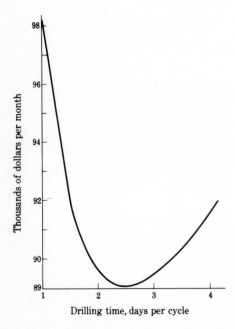

Figure 10.2 Optimum drilling time per cycle for 2000 m/month.

and setting these equal gives

$$15.833T_d^{-0.5} = \frac{30}{T_d + 0.75}$$

or

$$T_d - 1.895T_d^{0.5} + 0.75 = 0$$

which is a quadratic equation in $T_d^{0.5}$. Solve this by the quadratic formula giving $T_d^{0.5}$ and then square to get T_d:

$$T_d = 0.317 \quad \text{or} \quad 1.775 \text{ days}$$

The feasible range for T_d is 0.317 to 1.775 days. The following tabulation shows that the minimum cost is at $T_d = 1.775$ days.

T_d, days	Cycles per month necessary	Cycles per month available	$/month	$/m
0.250	31.67	30.00		
0.317	28.12	28.12	333,160	70.14
1.000	15.83	17.14	233,510	49.16
1.500	12.92	13.33	218,120	45.92
1.775	11.89	11.89	214,510	45.16
2.000	11.20	10.91		

Comparing parts (a) and (b), it should be understood that 2.471 days of drilling per cycle is optimum if production can be met. If not, it becomes necessary to change the drill bit more often. That $T_d = 1.775$ days in part (b) should be evident without the tabulation, since the closer the value is to 2.471, the less the cost per meter drilled. Figure 10.2 gives the same conclusions. For T_d in the range 0.317 to 1.775, the minimum cost per meter drilled is at $T_d = 1.775$. The mathematical minimum at $T_d = 2.471$ is outside the range and is not feasible.

(c) The meters per month drilled is maximized. Set up an expression for meter/month drilled.

$$\text{m/month} = \frac{30 \text{ days/month}}{(T_d + 0.75)\text{days/cycle}} 300T_d^{0.5} \frac{\text{m}}{\text{cycle}} = 9000\frac{T_d^{0.5}}{T_d + 0.75}$$

For the maximum:

$$\frac{d\text{m/month}}{dT_d} = 0 = 9000\frac{(T_d + 0.75)(0.5)T_d^{-0.5} - T_d^{0.5}}{(T_d + 0.75)^2}$$

$$T_d = 0.75 \text{ drilling day/cycle}$$

$$T_{\text{tot}} = 0.75 + 0.75 = 1.50 \text{ days/cycle}$$

$$\text{m/cycle} = 300T_d^{0.5} = 300(0.75^{0.5}) = 260$$

$$\$/\text{cycle} = 4250T_d + 10,500 = 4250(0.75) + 10,500 = \$13,688$$

$$\$/\text{m} = \frac{13,688 \text{ } \$/\text{cycle}}{260 \text{ m/cycle}} = \$52.65$$

$$\$/\text{month} = 13,688\frac{\$}{\text{cycle}} \frac{30}{0.75 + 0.75} \frac{\text{days/month}}{\text{days/cycle}} = \$273,760$$

(d) Maximum profit per month. Set up an expression for profit with

$$\text{Profit} = \text{income} - \text{expenses}$$

$$\underbrace{\frac{\text{Profit \$}}{\text{Cycle}} = 300T_d^{0.5}\frac{\text{m}}{\text{cycle}}60\frac{\$}{\text{m}}}_{\text{income}} - \underbrace{T_d\frac{\text{days}}{\text{cycle}}4250\frac{\$}{\text{day}}}_{\text{drilling}} - \underbrace{10,500\frac{\$}{\text{cycle}}}_{\text{bits}}$$

$$\frac{\text{Profit \$}}{\text{Cycle}} = -4250T_d + 18,000T_d^{0.5} - 10,500$$

$$\frac{\text{Profit \$}}{\text{Month}} = \frac{\text{profit \$}}{\text{cycle}}\frac{\text{cycles}}{\text{month}}$$

$$\frac{\text{Profit \$}}{\text{Month}} = (-4250T_d + 18,000T_d^{0.5} - 10,500)\frac{30}{T_d + 0.75} \tag{10.8}$$

For the maximum:

$$\frac{d(\$/\text{month})}{dT_d} = 0$$

$$= 30\frac{(T_d + 0.75)(-4250 + 9000T_d^{-0.5}) - (-4250 + 18,000T_d^{0.5} - 10,500)}{(T_d + 0.75)^2}$$

which is

$$-9000T_d^{0.5} + 6750T_d^{-0.5} + 7313 = 0$$

By trial and error or by the quadratic formula,

$$T_d = 1.86 \text{ days drilling per cycle}$$

$$\text{m/month} = 9000\frac{T_d^{0.5}}{T_d + 0.75} = 9000\frac{1.86^{0.5}}{1.86 + 0.75} = 4703$$

$$\text{Income (\$/month)} = 60(4703) = \$282,180$$

$$\text{Expenses (\$/month)} = [4250(1.86) + 10,500]\frac{\$}{\text{cycle}}\frac{30}{1.86 + 0.75}\frac{\text{cycles}}{\text{month}} = \$211,550$$

$$\text{Profit (\$/month)} = 282,180 - 211,550 = \$70,630$$

$$\text{Expenses (\$/m drilled)} = \frac{211,550 \text{ \$/month}}{4703 \text{ m/month}} = \$44.98$$

A summary of the results for this example is given in Table 10.1. The optimum policy is different for each of the four conditions specified.

Table 10.1 Summary of Example 10.1

Condition optimized	T_d optimum drilling, days/cycle	m/month	Cost, $/m	Cost, $/month	Profit, $/month
Drilling expense at 2000 m/month	2.471	2000	44.53	89,069	30,931
Drilling expense at 4750 m/month	1.775	4750	45.16	214,510	70,490
Meters drilled per month	0.750	5198	52.65	273,760	37,640
Profit per month at $60/m drilled	1.86	4703	44.98	211,550	70,630

It is instructive to find the maximum profit by the *incremental method*. The profit per month is given by Eq. (10.8). Starting at $T_d = 1.6$ and adding increments to T_d gives the following:

T_d, days	Profit, $/month	Δ Profit, $/month	ΔT_d	Δ Profit$/\Delta T_d$
1.60	69,810			
1.80	70,580	770	0.20	3850
1.86	70,600	20	0.06	333
1.90	70,600	0	0.04	0
2.00	70,430	-170	0.10	-1700

As T_d is incremented, Δ profit$/\Delta T_d$ is positive until at $T_d = 1.86$ there is no further profit as T_d is increased; thus Δ profit$/\Delta T_d$ is zero. As T_d is increased beyond 1.86, the incremental profit is negative, a loss.

10.8 An Example Involving Rate of Return

The time value of money can be included. Thus if the time period is made 1 year and the end of the year is taken as the vantage point, all costs and receipts can be reduced to unacosts and the analysis made on that basis. The following example is instructive because rate of return is used as a criterion.

Example 10.2 An office building can be built 5, 10, 15, or 20 stories high. Investment and revenue representing receipts less expenses as a uniform end-of-year amount are:

	Stories			
	5	10	15	20
Investment, $	5,000,000	7,500,000	11,000,000	15,000,000
Revenue, $/year	662,500	1,450,000	1,782,000	2,209,000

The figures are before a 48 percent tax rate. The life of the building is 25 years for both economic and tax purposes. Sum-of-the-years digits depreciation with no salvage value will be used. Each dollar invested in the building must earn 10 percent per year or more after taxes. Find the maximum number of stories that should be constructed under these circumstances.

SOLUTION If C_d represents the fully depreciable first cost, the unacost by item 1, Table 3.5, is

$$R = C_d(1 - 0.48F_{\text{SDP},r,25})F_{PR,r,25}$$

Using numerical values from Table A1.1 and designating $(1 - 0.48F_{\text{SDP},r,25})F_{PR,r,25}$ by Y, a tabulation is:

r, %	Y	r, %	Y
4	0.0418	12	0.1006
6	0.0547	14	0.1177
8	0.0689	16	0.1354
10	0.0843	20	0.1722

If R_v represents the end-of-year revenue, the unacost after taxes by item 2, Table 3.5, is $R = R_v(1 - t) = 0.52R_v$

For the five-story building with expenses and receipts equal on the basis of unacost,

$$C_d Y = 0.52R_v$$

$$5,000,000Y = 0.52(662,500)$$

$$Y = 0.0689$$

and from the tabulation

$$r = 8 \text{ percent}$$

So, a five-story building cannot be considered. For a 10-story building,

$$7,500,000Y = 0.52(1,450,000)$$

$$Y = 0.1006$$

$$r = 12 \text{ percent}$$

So, a 10-story building will pay. For a 15-story building,

$$11,000,000Y = 0.52(1,782,000)$$

$$Y = 0.0843$$

$$r = 10 \text{ percent}$$

Table 10.2 Summary of Example 10.2

	All stories			Incremental stories		
Stories	Initial investment, $	Yearly revenue, $	Rate of return, %	Initial investment, $	Yearly revenue, $	Rate of return, %
5	5,000,000	662,500	8			
				2,500,000	787,500	19.1
10	7,500,000	1,450,000	12			
				3,500,000	332,000	5.2
15	11,000,000	1,782,000	10			
				4,000,000	427,000	6.1
20	15,000,000	2,209,000	9			

but this is overall and includes the rate of return on the first 10 stories. Based on the incremental upper five stories,

$$(11,000,000 - 7,500,000)Y = 0.52(1,782,000 - 1,450,000)$$

$$Y = 0.0493$$

$$r = 5.2 \text{ percent approx. by interpolation}$$

Hence a 15-story building is not justified because the investment in the upper five stories does not earn 10 percent per year. The calculation is extended to a 20-story building, and the results are given in Table 10.2. The only investment justified is the 10-story building if the investment, including the last dollar invested, is to earn 10 percent per year or more after taxes.

10.9 A Two-Step Example with Recycle

The example developed in this section pertains to a two-step process. The requirement is not to optimize each step on an individual basis but to optimize the two steps together on an overall basis. In addition, the example includes feedback of material to an earlier step, a feature common in the process industries. In this example there is an economic balance between the two steps, the reactor and the separator. If the reactor is made large, the conversion is high and the separation costs are low; but there is an optimum size reactor which results in a lowest overall cost.

Example 10.3 The following homogeneous reaction

$$2A + B = D \qquad \text{(kilogram-moles basis)}$$

is carried out in a liquid-phase reactor in which the conversion of B to D depends upon the residence time according to the relationship

$$r = 0.25T^{0.33} \qquad 0.5 < T < 10 \tag{10.9}$$

where r = mole D formed per mole B fed
T = residence time in reactor, in hours

Fresh feed and recycle are always fed to the reactor in the proportion of 2A to 1B. The density of the feed liquor is 9.6 kg mol/m³.

Operation will be 4000 h/year to produce 3 kg mol of D per onstream hour. Reactors cost \$90,000/m³ of contained liquid and have a 4-year life with no salvage value for economic and tax purposes. Use sum-of-the-years-digits depreciation and a 10 percent per year rate of return after a 46 percent tax. The output from the reactor is run to a separator and is treated at a cost of \$3.65/kg mol of unreacted A and B. The unreacted A and B are recycled back to the reactor.

Find the size reactor in cubic meters that will give the minimum overall cost per year.

SOLUTION First, construct the flowsheet with a basis of 1 h and start with a basis of 2A and 1B fed to the reactor. If T is the residence time in the reactor, then by Eq. (10.9), materials out of the reactor are as follows:

$0.25T^{0.33}$ kilogram-moles of D or kilogram-moles of B reacted per hour

$2 - 2(0.25)T^{0.33}$ kilogram-moles of unreacted A per hour

$1 - 0.25T^{0.33}$ kilogram-moles of unreacted B per hour

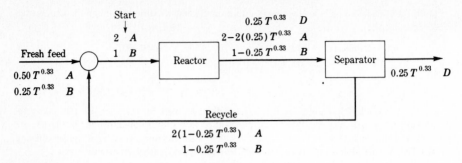

Figure 10.3 Flow rates, in kilogram-moles per hour, basis $2A + 1B$, to a reactor per hour.

Fresh feed, which must be added to maintain $2A + 1B$ to the reactor, is:

$$0.50T^{0.33} \quad \text{kilogram-moles of A per hour}$$

$$0.25T^{0.33} \quad \text{kilogram-moles of B per hour}$$

The flowsheet is given by Fig. 10.3.

If every flow is multiplied by $3/0.25T^{0.33}$, the kilogram-moles per hour of D produced becomes 3, the desired rate. Figure 10.4, so prepared, becomes the flowsheet for the process.

The volume of feed to the reactor in cubic meters per hour is

$$\left(\frac{6}{0.25T^{0.33}} + \frac{3}{0.25T^{0.33}}\right)\left(\frac{\text{kg mol}}{\text{h}} \frac{1}{9.6 \text{ kg mol/m}^3}\right) = \frac{3.75}{T^{0.33}} \text{m}^3/\text{h}$$

The volume V of the reactor in cubic meters and the residence time are related by

$$T = \frac{V, \text{ m}^3}{\text{feed, m}^3/\text{h}}$$

so that, by the last two relationships,

$$V = \frac{3.75 \text{ m}^3}{T^{0.33} \text{ h}} T \frac{\text{h}}{1} = 3.75T^{0.67}$$

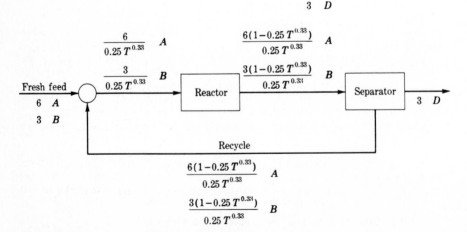

Figure 10.4 Flow rates, in kilogram-moles per hour, basis 3 kg mol/h, for product D.

The cost of reactors, basis unacost after taxes and item 1, Table 3.5, is

$$R = C_d(1 - tF_{SDP, 10\%, 4})F_{PR, 10\%, 4}$$

$$= 90,000(3.75T^{0.67})[1 - 0.46(0.83013)]0.31547$$

$$= 65,814T^{0.67} \quad \$/year$$

The amount of recycle per hour is

$$\frac{6(1 - 0.25T^{0.33})}{0.25T^{0.33}} + \frac{3(1 - 0.25T^{0.33})}{0.25T^{0.33}} = 36T^{-0.33} - 9 \quad \text{kg mol/h}$$

The cost of separation, basis unacost after taxes and item 2, Table 3.5, assuming an end-of-year cost, is

$$R(1 - t) = (36T^{-0.33} - 9)\frac{\text{kg mol}}{\text{h}}4000\frac{\text{h}}{\text{year}}3.65\frac{\$}{\text{kg mol}}(1 - 0.46)$$

$$= 283,820T^{-0.33} - 70,960 \quad \$/year$$

The total cost after taxes, is

$$65,814T^{0.67} + 283,820T^{-0.33} - 70,960 \quad \$/year \tag{10.10}$$

For the minimum:

$$\frac{d \$/year}{dT} = 0 = 44,095T^{-0.33} - 93,660T^{-1.33}$$

$$T = 2.124 \text{ h} \quad \text{optimum retention time}$$

$$V = 3.75T^{0.67} = 6.21 \text{ m}^3 \quad \text{optimum reactor volume}$$

A plot of Eq. (10.10) will verify that a minimum value for dollars per year is actually attained at $T = 2.124$ h.

Although this process involves two steps, it can be resolved in terms of a single variable. Eq. (10.10), of course, can be superseded by a relationship in terms of V as the independent variable. The example illustrates that problems in economic balance demand a good model and care in setting up the expression for the condition to be optimized. In particular, the analyst must guard against errors in units and dimensions.

10.10 Variocyclic Processes

Many processes are cyclic or can be treated as such for convenience. In Example 10.1, concerned with drilling an oil well, the underlying physical relationship was

$$Q_c = 300T_d^{0.5}$$

where Q_c was cumulative meters drilled for a period of T_d days after installing a new drill bit. It was inherent in the problem that each cycle duplicated the others. In some cases, however, there are cycles within cycles. Repetitions within the cycle will be referred to as *rounds*. A cycle comprises a number of rounds. The rounds within a cycle show differences, but each cycle repeats the previous cycle. Such cycles are referred to as *variocyclic* processes to denote the variation of the rounds with a cycle.

Catalytic processes offer some excellent examples of a variocyclic process. Frequently, a catalyst can be rejuvenated but becomes less efficient with each

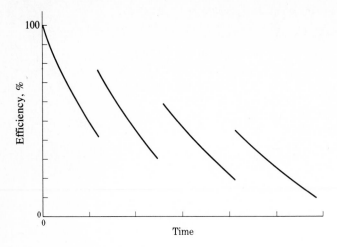

Figure 10.5 Efficiency in a variocyclic process.

rejuvenation, until finally it is discarded and the cycle is repeated with new catalyst. Figure 10.5 is a graph of the situation, showing a series of rounds to complete one cycle. There is a processive degradation from round to round.

In practice, the situation can be very complicated, for more than one variable is usually involved. The temperature of the catalyst can have a pronounced effect, so that the optimization for the time-temperature relationship for all the rounds can become exceedingly intricate. The problem is discussed here in only an elementary and simplified way. In variocyclic processes it is important that the analyst understand the significance and use of such terms as *instantaneous rate, instantaneous efficiency, cumulative throughput, cumulative efficiency, average rate,* and *average efficiency.* An average depends upon how the average is computed; e.g., an average with respect to time can be different from an average based on input.

If there is some systematic relationship between the rounds, an analytic solution is generally possible. Frequently, tabular solutions are convenient, and they

Table 10.3 Optimum use of clay

Round	Treated, kg	Cumulative amount treated, kg	Catalyst cost, $	Activation cost, $	Total cost, $	Cumulative cost, $	Cumulative average, $/kg
1	10,000	10,000	50	25	75	75	0.00750
2	8,500	18,500		25	25	100	0.00541
3	7,225	25,725		25	25	125	0.00486
4	6,141	31,866		25	25	150	0.00471
5	5,220	37,086		25	25	175	0.00472
6	4,436	41,523		25	25	200	0.00481

may be necessary if the data are in tabular form or if the variables are discrete. Dynamic programming can be used successfully in many cases.

Example 10.4 An absorbent clay is used to clarify a liquor. The clay costs $50/1000 kg and $25 for each activation, including the first round. One thousand kilograms of clay will treat 10,000 kg of liquor in the first round, diminishing to 85% of the previous round for subsequent rounds. How many rounds should the clay be used?

SOLUTION The calculation is shown in Table 10.3. If the clay is used for four rounds, a minimum cost of $0.00471/kg treated is obtained.

MULTIVARIABLE OPTIMIZATION

10.11 Analytical Method

The same principles and methods that are useful for one independent variable apply to more than one independent variable but become much more complicated as the number of variables increases. Consider first the analytical method for some criterion W, a function of X and Y, which is to be optimized as follows:

$$W = F(X, Y) \tag{10.11}$$

For a mathematical optimum, two conditions must be met:

$$\frac{\partial W}{\partial X} = 0 \tag{10.12}$$

$$\frac{\partial W}{\partial Y} = 0 \tag{10.13}$$

Generally, each of these relationships involves X and Y; but there will be two equations with two unknowns, and sets of X, Y can be obtained. Each set is a possible optimum for W as given by Eq. (10.11).

An analytical test is available to establish whether a maximum or minimum exists. A maximum for W exists if the three following conditions are satisfied for the X, Y pair considered:

$$\frac{\partial^2 W}{\partial X^2} \text{ negative} \qquad \frac{\partial^2 W}{\partial Y^2} \text{ negative} \qquad \frac{\partial^2 W}{\partial X^2}\frac{\partial^2 W}{\partial Y^2} > \left(\frac{\partial^2 W}{\partial X \partial Y}\right)^2$$

Similarly there will be a minimum if

$$\frac{\partial^2 W}{\partial X^2} \text{ positive} \qquad \frac{\partial^2 W}{\partial Y^2} \text{ positive} \qquad \frac{\partial^2 W}{\partial X^2}\frac{\partial^2 W}{\partial Y^2} > \left(\frac{\partial^2 W}{\partial X \partial Y}\right)^2$$

In either case, if

$$\frac{\partial^2 W}{\partial X^2}\frac{\partial^2 W}{\partial Y^2} < \left(\frac{\partial^2 W}{\partial X \partial Y}\right)^2$$

W is neither maximum nor minimum but a saddle point. If the last relationship is an identity, the nature of the point is inconclusive and further testing of higher-order derivatives is necessary.

If W is a function of three variables, X, Y, and Z:

$$W = F(X, Y, Z)$$

the three following conditions are necessary, but not sufficient, for mathematical optima:

$$\frac{\partial W}{\partial X} = 0 \qquad \frac{\partial W}{\partial Y} = 0 \qquad \frac{\partial W}{\partial Z} = 0$$

In general these will each contain X, Y, Z, but the three equations in three unknowns can be solved for sets of X, Y, Z. The extension to four or more variables is obvious.

10.12 Graphical Method

A graphical method in the form of a three-dimensional space model can be used for two independent variables. For

$$W = F(X, Y) \tag{10.14}$$

X and Y can be measured in a horizontal plane and W measured by the perpendicular distance from the XY plane. In a complicated case the space model will appear like a mountain range. Inasmuch as space is limited to three dimensions, it is not possible to treat higher-order systems by this method.

There are techniques for simplifying multivariable systems which are explained here for the two-variable system of Eq. (10.14). Let Y be given a series of successive fixed values and replace the single relationship by a series of relationships:

$$W_1 = F(X, Y_1)$$

$$W_2 = F(X, Y_2)$$

$$\dots\dots\dots$$

$$W_N = F(X, Y_N)$$

which reduces the two-variable problem to a series of one-variable problems. If the range taken for Y is wide enough, it becomes possible to find the optimum W from a series of tabulations, augmented with a plot if desirable.

Such a plot is shown in Fig. 10.6, where a cost in dollars per year is plotted for two variables, S and P. P is given successive fixed values, and the optimization for each is shown as a series of dotted curves. Connecting these incidental minima leads to the true minimum at $S = 1056$, $P = 300$.

Another method involves calculating W for various points X, Y using only a

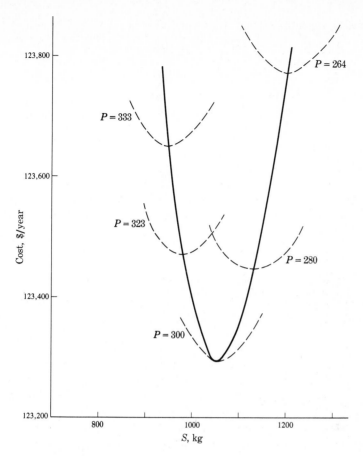

Figure 10.6 Optimum size S for various powers to stirrer P.

two-dimensional plot and connecting points corresponding to equal values of W by contour curves. Such a plot resembles a typographical map. Details are given in Chap. 13.

10.13 Incremental Method

The incremental method involves changing the independent variables singly or in groups. As the number of variables increases, considerable precaution must be exercised to avoid false conclusions. An increase in X alone may lower the value of W, the function to be optimized, yet the optimum may lie in the direction of increasing X and Y simultaneously. Chapter 13 should be consulted for details.

The incremental method with multivariables is suitable for trial by experiment, including uncertainty as mentioned in Sec. 10.6 for one variable. For further study, see Chap. 13.

10.14. A Two-variable Optimization Problem

Example 10.5 A manufacturer prepares a plastic in a special batch reactor-mixer. The time required to process the mixture varies directly as the square root of the capacity of the mixer and inversely as the square of the power to the stirrer. Quantitatively,

$$T = 1000\frac{S^{0.5}}{P^2} \tag{10.15}$$

where T = total time required, h/batch
 S = capacity of reactor-mixer, kg
 P = power to stirrer, kW

The cost for the reactor-mixer is $1265S^{0.5}$ dollars per year. Electricity costs \$0.04/kWh, and there is an overhead or indirect cost expressible in terms of the power requirement as

$$\text{Overhead} = 137.2P \quad \text{\$/year}$$

The full year is available, and 10^7 kg/year must be processed. The time for loading and unloading the reactor-mixer is negligible. Find the optimum size and optimum stirrer power.

SOLUTION Set up an expression for the cost, basis 1 year of operation. Let N be the number of batches per year and S be the number of kilograms per batch.

$$SN = 10^7 \text{ kg/year}$$

$$N = \frac{10^7}{S} \tag{10.16}$$

$$\text{\$/year} = 1265S^{0.5} + \left(T\frac{\text{h}}{\text{batch}}\right)\left(N\frac{\text{batch}}{\text{year}}\right)(P \text{ kW})\left(0.04\frac{\text{\$}}{\text{kWh}}\right) + 137.2P$$

Replace T by Eq. (10.15) and N by Eq. (10.16), yielding

$$\text{\$/year} = 1265S^{0.5} + 4 \times 10^8P^{-1}S^{-0.5} + 137.2P \tag{10.17}$$

For the minimum cost, set the partical derivatives of the above expression equal to zero:

$$\frac{\partial, \text{\$/year}}{\partial S} = 0 = 0.5(1265)S^{-0.5} - 0.5(4 \times 10^8)P^{-1}S^{-1.5}$$

$$PS = 3.162 \times 10^5 \tag{10.18}$$

$$\frac{\partial, \text{\$/year}}{\partial P} = 0 = -4 \times 10^8P^{-2}S^{-0.5} + 137.2$$

$$P^2S^{0.5} = 2.915 \times 10^6 \tag{10.19}$$

Solving for Eqs. (10.18) and (10.19) gives

$$S = 1056\text{-kg capacity}$$

$$P = 300\text{-kW power to stirrer}$$

Check for feasibility:

$$T = 1000\frac{S^{0.5}}{P^2} = 1000\frac{1056^{0.5}}{300^2} = 0.361 \text{ h/batch}$$

Using all the hours in a year, 8760,

$$1056\frac{\text{kg}}{\text{batch}}\frac{8760 \text{ h/year}}{0.361 \text{ h/batch}} = 2.56 \times 10^7 \text{ kg/year possible}$$

Table 10.4 Yearly cost for Example 10.5

S, kg	P, kW	Cost, $/ year	Δ cost, $/year
1056	300	123,298	
1056	315	123,402	104
1056	285	123,400	102
1003	300	123,323	25
1003	315	123,377	79
1003	285	123,481	181
1109	300	123,325	27
1109	315	123,476	178
1109	285	123,374	76

so that 10^7 kg/year is feasible. A check is given in Table 10.4, where increments of ± 5 percent in S or P, or both simultaneously, are considered and the yearly cost is calculated from Eq. (10.17). Regardless how S and P are varied, the cost increases over that for $S = 1056$ and $P = 300$. The cost is not very sensitive to S and P, however, and a 5 percent variation in each simultaneously adds only $181/year to $123,298/year. Note further that the cost is more responsive to a 5 percent change in P than to a 5 percent change in S.

The example can be solved as a one-variable problem by placing successive fixed values for P in Eq. (10.17) and plotting the resulting curves. Figure 10.6 is such a plot, confirming an overall minimum at $S = 1056$ and $P = 300$.

10.15 Lagrange Multipliers

A powerful method for finding mathematical optima in multivariable problems involving supplementary conditions that must be observed, called *restraints*, was introduced by the French mathematician Lagrange (1736–1813). Consider

$$W_1 = F_1(X, Y, Z) \tag{10.20}$$

subject to the two conditional relationships, or restraints,

$$F_2(X, Y, Z) = 0 \tag{10.21}$$

$$F_3(X, Y, Z) = 0 \tag{10.22}$$

In an ideal situation the last two relationships can be used to eliminate, say, Y and Z from Eq. (10.20), leaving a relationship for W in terms of X only. This may not be possible, however, in an actual case. Lagrange's method is applicable to the general case and may make an analytical solution possible.

The method in application requires forming a new function, called the *Lagrange expression*, made up by linear addition of the original relationship and each restraint relationship multiplied by a Lagrange multiplier. There are as many multipliers as there are restraints. The Lagrange expression for the problem at hand is

$$LE = F_1(X, Y, Z) + \lambda_1 F_2(X, Y, Z) + \lambda_2 F_3(X, Y, Z) \tag{10.23}$$

and the optima for Eq. (10.20) are the optima of the Lagrange expression. Equation (10.23) includes the three original variables, X, Y, Z, and the two Lagrange multipliers, λ_1 and λ_2. Five relationships must be satisfied to find the optima for the Lagrange expression in this case:

$$\frac{\partial LE}{\partial X} = 0 \qquad \frac{\partial LE}{\partial Y} = 0 \qquad \frac{\partial LE}{\partial Z} = 0 \qquad \frac{\partial LE}{\partial \lambda_1} = 0 \qquad \frac{\partial LE}{\partial \lambda_2} = 0$$

whereas only three are necessary to find the optima for Eq. (10.20).

$$\frac{\partial W_1}{\partial X} = 0 \qquad \frac{\partial W_1}{\partial Y} = 0 \qquad \frac{\partial W_1}{\partial Z} = 0$$

Nevertheless the Lagrange expression, with its five relationships, has the constraints built in and may be much easier to solve than Eq. (10.20) subject to the restraint equations (10.21) and (10.22).

For an example, return to Example 10.1, part (a), which has already been optimized. The requirement is to minimize

$$\$/month = NT_d(4250) + 10{,}500N \tag{10.24}$$

subject to the condition or restraint

$$N = 6.667T_d^{-0.5} \tag{10.25}$$

Write the restraint function as

$$N - 6.667T_d^{-0.5} = 0$$

The Lagrange expression becomes

$$LE = NT_d(4250) + 10{,}500N + \lambda(N - 6.667T_d^{-0.5}) = 0$$

Set the three partial derivatives equal to zero:

$$\frac{\partial LE}{\partial N} = 0 = 4250T_d + 10{,}500 + \lambda = 0$$

$$\frac{\partial LE}{\partial T_d} = 0 = 4250N - \lambda(6.667)(-0.5)T_d^{-1.5} = 0$$

$$\frac{\partial LE}{\partial \lambda} = N - 6.667T_d^{-0.5} = 0$$

Solving the three equations for N, T_d, and λ gives

$$N = 4.241 \text{ cycles per month}$$

$$T_d = 2.471 \text{ h of drilling per cycle}$$

$$\lambda = -21{,}000$$

The values for N and T_d check those previously found.

In this example, the use of Lagrange multipliers was not as direct as using Eq. (10.25) in Eq. (10.24) to reduce the latter to an equation in one independent variable. Nevertheless, the use of Lagrange multipliers can lead to a solution that otherwise would not be obtained. On the other hand, the use of Lagrange multipliers does not always produce expressions that are solvable by analytical means.

10.16 Sensitivity and Response Analysis

In a detailed analysis, it is desirable to know the optimum and also how sensitive the optimum is to changes in various parameters that make up the problem. For example, suppose Y is expressed by the relationship

$$Y = AX + \frac{B}{X} \qquad (10.26)$$

where X is the independent variable and A and B are parameters. In a given problem A and B may be constant, but they may be subject to change in practice. For example, A and B may be estimates, and it becomes desirable to know how much Y_{opt} is changed by changes in the estimates of A and B. What is desired mathematically is

$$\frac{\partial Y_{opt}}{\partial A} \quad \text{and} \quad \frac{\partial Y_{opt}}{\partial B}$$

The determination of such values is called *sensitivity analysis*.

An analytical solution is easily obtained for the relationship Eq. (10.26). For Y_{opt}

$$\frac{dY}{dX} = 0 = A - \frac{B}{X^2} \quad \text{and} \quad X_{opt} = A^{-0.5}B^{0.5}$$

which, substituted in Eq. (10.26), gives

$$Y_{opt} = 2A^{0.5}B^{0.5}$$

and leads directly to the relationships

$$\frac{\partial Y_{opt}}{\partial A} = A^{-0.5}B^{0.5} \qquad \frac{\partial Y_{opt}}{\partial B} = A^{0.5}B^{-0.5}$$

For the specific case for $A = 10$, $B = 40$,

$$\frac{\partial Y_{opt}}{\partial A} = \sqrt{\frac{40}{10}} = 2$$

The value can be obtained in another way which is particularly useful if the analytical method is too involved or not available. The procedure is to start with Eq. (10.26) or graphical or tabular data and for $A = 10$ and $B = 40$ to find Y optimum. This is found to be

$$Y_{opt} = 40 \qquad A = 10$$
$$B = 40$$

Next, repeat the calculation for, say, $A = 10.1$, $B = 40$. Y_{opt} is found to be

$$Y_{opt} = 40.1995 \qquad A = 10.1$$
$$B = 40$$

From the last two relationships,

$$\frac{\Delta Y_{opt}}{\Delta A} = \frac{40.1995 - 40.000}{10.1 - 10} = 1.995$$

Thus a sensitivity analysis can be made even if analytical expressions are unwieldy or unavailable. A sensitivity analysis is also made in the same manner if there is more than one independent variable.

The sensitivity analysis as such cannot be applied to the independent variable. Thus, for the case under study,

$$\frac{\partial Y_{opt}}{\partial X} = 0$$

That is, Y_{opt} is determined by A and B and not by X. However, it is possible to evaluate the change in Y as a consequence of not operating at the optimum X. This is

$$\frac{Y - Y_{opt}}{X - X_{opt}} = \frac{\Delta Y}{\Delta X}$$

and X can be taken as, say, a 1 percent change. The procedure can be called a *response analysis* to avoid confusion with a true sensitivity analysis. The reader should refer to Table 10.4, which was essentially a response analysis for a reactor-mixer having independent variables S and P.

10.17 Simplification of Multivariable Problems

Optimization in the fields of economic balance can become exceedingly complicated because the problems can be intricately involved with physical, engineering, and cost considerations. This chapter was concerned with general methods, but the analyst must acquire the facility of representing engineering systems in terms of real models, either mathematical or empirical.

As the number of variables mounts, the complexity and scope of the problem increase disproportionately; hence every effort should be made to reduce the number of variables. This can be done by taking fixed values for some variables, particularly if practical considerations will limit the range. If a single value is too restricted, the study might be limited to a few values. Complicated expressions can sometimes be replaced with simpler expressions which might be satisfactory. Thus a linear expression might replace a quadratic expression for which the curvature is small in the feasible range.

As the number of variables and parameters increases and the number of constraints and practical limitations grows larger, the optimum has a propensity to grow flatter, with less likelihood that the realizable optimum will be a mathematical optimum. The tendency toward flatness is real in practical large-scale problems. This does not signify that a problem can be slighted, but it does indicate that some optimization should be considered in deciding upon how much detail a particular problem merits.

Finally, optimization is a continuous study, and old problems take on a fresh challenge as technology, economics, and society change.

10.18 Formulating an Optimization Problem

An optimization problem is an exercise in mathematical modeling that requires great care in setting up the model. Four steps are involved:

1. Decide the exact objective to be optimized. As shown in Example 10.1, many different objectives are possible.
2. Set up the objective function using as many variables as are required. Try for accuracy rather than compactness. Make sure that all the terms have the same dimensional units.
3. Set up all the restraints and relationships between the variables.
4. If possible, reduce the objective function in step 2 to independent variables. Step 2 may contain both dependent and independent variables.
5. The objective function is now ready for solution. If it contains independent variables only, the differentials can be set equal to zero to optimize the expression, or alternatively, a tabulation can be made. If the objective function contains dependent variables in addition to independent variables, a Lagrange expression can be tried. If this fails, the objective function and its restraints must be optimized, using the skill and ingenuity of the analyst.

In formulating problems, it is necessary to recognize the difference between instantaneous rates, cumulative amounts, and various averages that may arise. Careful, detailed labeling is recommended.

Example 10.6 The instantaneous rate of production for a material in kilograms per day with T in days is given by

$$\frac{dP}{dT} = 1000e^{-0.1T}$$

Find the production rate at the end of the tenth day, the total production for 10 days, the average production rate for 10 days, and the average production rate for the last 5 of the 10 days.

SOLUTION The instantaneous production rate at $T = 10$ is

$$1000e^{-0.1T} = 1000e^{-0.1(10)} = 367.88 \text{ kg/day}$$

The total production for 10 days is

$$\int_0^{10} 1000e^{-0.1T} \, dT = \frac{1000}{-0.1}(e^{-0.1(10)} - e^0) = 6321.2 \text{ kg}$$

The average production for 10 days is

$$\frac{6321.2}{10} = 632.12 \text{ kg/day}$$

The average production for the last 5 days is

$$\int_5^{10} \frac{1000e^{-0.1T} \, dT}{10 - 5} = 477.3 \text{ kg/day}$$

NOMENCLATURE

A	Parameter, as in Eq. (10.26)
B	Parameter, as in Eq. (10.26)
C_d	Depreciable cost, $
F_{SDP}	Present value for $1 for sum-of-the-years-digits depreciation, decimal
F_{PR}	Factor to convert a present value to unacost, year^{-1}
LE	Lagrange expression
N	Cycles per month in Example 10.1; batches per year in Example 10.5
P	Power to the stirrer in Example 10.5, kilowatts
Q_c	Cumulative meters drilled from the start of the cycle, Example 10.1
r	Rate of return after taxes, decimal per year
r	Moles of D formed per mole of B fed in Example 10.3; depends on the residence time
R	Unacost, uniform end-of-year amount, $/year
R_v	Revenue, end of year, in Example 10.2, $/year
S	Capacity of the reactor-mixer in Example 10.5, kilograms
t	Tax rate, decimal
T	Residence time in the reactor in Example 10.3, hours
T	Total time required for a batch in Example 10.5, hours
T_d	Drilling time from the start of the cycle in Example 10.1, days
V	Volume of reactor in Example 10.3, cubic meters
W	Dependent variable
Y	Independent variable, generally
Y	$(1 - 0.48F_{\text{SDP},r,25})F_{PR,r,25}$ in Example 10.2
λ	Lagrange multiplier

PROBLEMS

10.1 A causeway is to be built over a long swamp. If X is the distance in meters between piers for the range $20 \leq X \leq 100$, the following cost data apply:

$$\$ \text{ per pier} = 5X^2$$

$$\$/\text{m, span between piers} = 5000X^{-0.5}$$

Find the optimum distance between piers for the minimum cost.

10.2 The cost Y in dollars per hour to produce electrical energy in a unit rated at 20,000 kW, where X is the load generated in megawatts, is given by

$$Y = 12 + 0.2X + 0.27X^2$$

Income in dollars per hour from the sale of electricity is

$$16X - 0.2X^2$$

Find the generated load that gives the maximum profit.

10.3 The fixed cost for a steam line per meter of pipe is

$$350X + 18 \quad \$/\text{year}$$

and the cost for loss of heat from the pipe per meter is

$$\frac{1.6}{X} \quad \$/\text{year}$$

where X is the thickness of insulation in meters. Find the optimum thickness of the insulation.

10.4 A commercial location makes a study to predict added gross income in dollars per year for investment in additional parking space. The survey predicts:

Invested, in parking lot, k$	Added to gross income, k$/year	Invested in parking lot, k$	Added to gross income, k$/year
18	48	153	153
36	69	189	165
54	90	225	174
75	111	270	183
99	126	315	192
126	141		

Added net income can be taken as 33.3 percent of added gross income. Use the incremental method and find how much should be invested in additional parking-lot space if the return on investment is 17 percent or more.

10.5 An office building can be built 5, 10, 15, or 20 stories high. The building will be depreciated uniformly over a 22-year life, and taxes may be neglected. Analysis gives the following, with all amounts in megadollars:

Stories	5	10	15	20
Investment, M$	20	32	44	58
Rent, M$/year	3.6	7	10.2	13.2
Operation, M$/year	1.5	2	2.8	4.0

The minimum acceptable rate of return is 10 percent per year. How high should the building be built?

10.6 Costs associated with an extension to a plant are tabulated below in megadollars. Four production capacities are being considered, and the fixed investment will be written off in 10 years. Neglect taxes. A minimum 13.5 percent return on the original investment is required. What size plant, if any, should be built?

	A	B	C	D
Mkg/year	5	10	15	20
Fixed cost, M$	12	20	32	35
Sales price, $/kg	0.75	0.73	0.63	0.61
Utilities, M$/year	0.21	0.34	0.42	0.50
Raw materials, M$/year	0.2	0.4	0.6	0.8
Labor, M$/year	1.5	1.5	2.4	2.4

10.7 A heat treating furnace must process 1 million parts per year. The furnace can process only one batch per 8-h shift for one shift per day for a maximum of 250 days/year. The cost of the furnace is

$$25X^{0.7} \quad \$/year$$

where X is the number of machine parts processed per batch. The operating cost for the furnace is $200 per shift. Find the optimum size for the furnace to be installed.

10.8 Repeat Prob. 10.7, letting the furnacing time be dependent upon the furnace size:

$$0.04X^{0.5} \quad h$$

The operating cost is a fixed cost of $104 per batch and $12/h for labor. A shift may run over 8 h, but only one shift per day for 250 days/year is allowed.

10.9 The cost for a stamping machine in dollars per year can be expressed as

$$48,000T^{-0.5}$$

where T is the time in hours required to form the stamping. It costs $64/h to operate the machine. Required are 150,000 stampings per year for 8 h/day, not to exceed 300 days/year. Find the size of the stamping machine to be bought and the cost per year for the stamping operation.

10.10 In certain operations, it is necessary to provide one spare machine to ensure continuity of operations at full capacity when one machine is being overhauled. If the cost of machines varies as the 0.8 power of capacity, find the optimum number of machines, including the spare, that should be provided.

10.11 A pressure vessel is to be made from a cylinder of diameter X meters capped with a hemisphere at each end. The cost for the cylindrical section is

$$1800X^{0.5} \quad \$/m \text{ of length}$$

and the cost for one hemispherical end is

$$1400X^{2.4} \quad \$$$

The vessel is to hold 40 m³. Find the optimum dimensions and cost.

10.12 A chemical is produced by a batch process. Chemicals X and Y are used to make Z, with the following relationship for the kilograms of Z produced and the kilograms of X and Y used:

$$Z = 1.5(1.1XZ + 1.3YZ - XY)^{0.5}$$

X costs $0.18/kg and Y $0.08/kg, and Z sells for $1.60/kg. One-half of the selling price for Z is due to costs other than raw materials. Only Z is recovered from the process. Find the maximum profit obtainable per kilogram of Z.

10.13 A machine costs $100,000 and has expenses which increase continuously according to the relationship

$$R = 50,000e^{0.2T}$$

where R is the instantaneous rate of expense in dollars per year and T is the time in years. Use continuously compounded interest with money worth 10 percent per year expressed as the continuous rate. Neglect taxes and find the optimum time to keep the machine in service.

10.14 Suppose that 86,000 kg/day of water for 300 days/year is to be evaporated from a salt solution. One kilogram of steam will evaporate $0.7N$ kg water, where N is the number of effects.

Steam costs $2.50/1000 kg. The cost of the first effect is $75,000 and each additional effect costs $60,000, to be written off over a 10-year period. Repairs, local taxes, and insurance per year cost 10 percent of the first cost. Other costs may be neglected since they are independent of the number of effects.

If the charge for the use of capital is 13.8 percent per year based on the initial investment, find the optimum number of effects to install, neglecting the income tax.

10.15 Water at 3.2×10^6 kg/year is to be obtained from a slurry containing 8 percent solids by filtering. The cake will contain 40 percent solids. The relationship is

$$Q = 13,000T^{0.5}$$

where Q is the kilograms of filtrate water obtained during T hours of filtering time. The cake is not washed. The dumping plus cleaning time is 3 h and costs \$100. The filtration cost is \$40/h. Find (a) the cycle time for the minimum cost, (b) the optimum amount of cake per cycle for the minimum cost, and (c) the cycle time for the maximum production.

10.16 The feed of material A to a process varies with time and is given by

$$F = 1000 - 10T$$

where F is kilograms per day and T is time in days. Material A converts to material B with no change in weight. The instantaneous efficiency E is given by

$$E = \frac{\text{kg B formed}}{\text{kg A fed}}$$

and E varies with time according to

$$E = 1 - 0.05T \qquad T < 20$$

(a) Find the instantaneous rates of A in, A out, B out, and the efficiency, all at $T = 10$ days.
(b) Find the cumulative amounts of A in, A out, and B out for the first 10 days. Also, find their averages and the average efficiency.
(c) Repeat part (b) for the tenth day alone, i.e., from the end of day 9 to the end of day 10.

10.17 An evaporator shows the following variation of U, the overall heat-transfer coefficient, in $J/(m^2 \cdot s \cdot K)$, with T in seconds measured from the starting time with clean tubes.

$$\frac{1}{U^2} = 0.0002 \times 10^{-8}T + 8 \times 10^{-8}$$

A cleanout costs \$330 and causes 10 h downtime. The allocable operating cost is \$15/h; 500,000 kg/day for 250 days/year must be evaporated from a 280-m² area. The latent heat of vaporization is 2.3×10^6 J/kg, and the driving force is 28 K. Find the optimum time for operation before recleaning.

10.18 Jewels, along with recycled jewels, are polished in a special machine for a space project. The retention time in the polisher is inversely proportional to the number of jewels fed per hour and is 1 h when 1000 jewels/h are fed.

The fraction of acceptable jewels produced is $1 - e^{-0.5T}$ where T is the retention time in hours. Rejected jewels are returned in a recycle stream to the polisher, where they perform like raw jewels.

It costs \$1000/h to operate the polisher. The cost of separating good jewels from rejects is proportional to the number of rejects per hour and is \$2/h for one reject per hour. Acceptable jewels are worth \$10 per jewel above the cost for raw jewels.

Find (a) the maximum profit per hour and (b) the number of acceptable jewels produced per hour under maximum profit-per-hour conditions.

10.19 A process uses a catalyst that must be regenerated periodically because of reduction in conversion efficiency. The cost data are:

One regeneration	= \$800
Feed	= \$2.50/kg
Operation	= \$300/day, 300 days/year
Value of product	= \$14/kg
Overhead	= \$100,000/year

The feed rate is maintained constant at a rate of 150 kg/day.

The efficiency of the catalyst expressed as kilograms of product to kilograms of feed is

$$0.87 \quad \text{average} \qquad 0 \leq T = 1$$
$$\frac{0.87 \quad \text{instantaneous}}{T^{0.25}} \qquad T > 1$$

where T is the time in days from the last regeneration.

If no other costs are involved, find the maximum annual profit that can be obtained.

10.20 Lubricating oil is to be treated continuously in a tower using a selective solvent to remove naphthenic material. The parameter R is defined as

$$R = \frac{\text{bbl solvent used}}{\text{bbl oil treated}}$$

and the treatment cost, without recovery of the solvent, in dollars per year, is given by

$$18{,}600R^2 - 111{,}000R + 217{,}000$$

It costs \$0.40 to separate each barrel of solvent circulated through the tower, and 300 bbl/day of oil will be processed for 320 days/year. Find the most economical value of R to be used.

10.21 The savings in dollars per day obtained while operating a heat exchanger is given by the expression

$$(A - C)e^{-kT+C}$$

where T is the time in days from the start of the cycle since cleaning. At the instant the cycle is started, savings are \$90/day, and after a long time they level off at \$30/day. Also, during the first month, savings are \$111/month more than the average savings per month for the first 4 months after cleaning.

It costs \$1000 to clean the heat exchanger, and 3 days are lost with no savings. In this problem, 1 month can be taken as 30 days.

Find the number of days between cleanings to yield the maximum savings.

10.22 Water is treated with an ion-exchange resin which must be regenerated. A large tank is available at no cost for storing softened water to smooth any variations in demand and to supply water while the resin is being regenerated at negligible time.

Demand averages 2.3 m³/h for all the hours in a year. The time T in hours that the resin can be used before it must be regenerated is given by

$$T = \frac{160V}{F}$$

where V is the volume of the resin in cubic meters and F is the flow rate of the water in cubic meters per hour.

Fixed costs are \$9000/m³ for the resin and equipment for a 5-year economic life. Regeneration costs vary as the 0.8 power of the volume of the resin and are \$55 for a 1-m³ unit.

Find the optimum size unit to install and the cost per 1000 m³ of softened water.

10.23 When 100 kg of A reacts, it forms 75 kg of B and 25 kg of C. A batch-processing unit is available for processing 1000 kg of A. The fraction of A that reacts is α:

$$\alpha = \frac{T}{T + 1}$$

where T is the time in hours that a batch is processed. Material A is worth \$1/kg and B \$10/kg. Unreacted A and C formed are not recovered, but B is. The setup time between batches is 2 h.

(a) Find the maximum production rate for B in kilograms per day.

(b) Find the maximum profit in dollars per day.

10.24 An ion-exchange resin is used to make ultrapure water. It is necessary to slow down the process with time according to the relationship

$$q = 75e^{-0.07T}$$

where q is the rate in cubic meters per hour and T is the time in hours since the start of a cycle. Demand is steady at 40 m³/h, and when the softening unit is delivering at a rate higher than that, the excess water is given a credit of $2.50/m³. When the rate is less than 40 m³/h, the deficit below that rate is charged at $3/m³.

The cost for rejuvenating the resin and setting up the equipment is $2500, which includes the charge for no water being produced in the period. The downtime is 5 h. The operating charge while producing water is $12/h. Find the optimum time between rejuvenations.

10.25 A waste material contains a product, P, which can be leached out with water. The water solution is concentrated by evaporation, at a cost, to yield an extract having value.

A batch of waste is leached using 28 m³, 28,000 kg, of water per hour at a constant rate. The concentration of P in the exit water decreases with time and is given by the rate expression

$$\frac{dc}{dT} = -11.8c$$

where c is the concentration of P in kilograms per cubic meter, and at the start is 4.8.

The final solution to be prepared after evaporation is to be 20 percent P. Each 1000 kg of water evaporated is charged at $3, and each 1 kg of P recovered in the 20 percent concentrate receives a credit of $2.

Find how long a batch should be leached to obtain the maximum profit from each batch.

10.26 A process is to be designed to process 5×10^6 kg/year. Processing is by batches, and the processing time is proportional to the square root of the batch size; the processing time is 4 h for a batch of 10,000 kg. The cost of processing units is given by

$$425X^{0.7}\$$$

where X is the size of a batch in kilograms, and the cost will be written off in 5 years. Other costs are a $250 setup cost for each batch, with negligible setup time, and $22/h for labor while the processing unit is in operation. A batch must be finished in 8 h within a shift of 8 h, and only one shift per day for 200 days/year is allowed. Find the optimum batch size.

10.27 One kilogram of A is processed to produce 1 kg of B in a primary converter. At the end of a processing cycle, 1 day is lost in cleaning the converter. The feed rate to the converter is decreased uniformly with time and is 1000 kg/day of A at zero time and would be 500 kg/day at the end of 30 days.

Efficiency in operating the converter is measured by the ratio of kilograms of B produced to kilograms of A fed. The efficiency decreases linearly with time in proportion to the *cumulative* amount of A fed to the converter. The efficiency is 0.9 at the start and 0.75 at the end of 3 days.

Unreacted A from the primary converter can be converted in part to B in a secondary converter which recovers 20 percent of the unreacted A by converting it to B by a kilogram-to-kilogram conversion. The amount of B so formed is added to that obtained in the primary converter.

Determine how long the primary converter should be run before being cleaned to maximize the total production of B.

10.28 It costs $150,000 to overhaul an auxiliary electrical power unit. Immediately after the overhaul, there is a savings of $1000/day by avoiding purchased electricity. The savings per day decreases with time because of scaling, outages, and so forth as it becomes necessary to purchase electricity from outside. The purchased electricity increases linearly with time, and at the end of 1 year it would be purchased at a rate amounting to 25 percent less than the savings rate immediately after the overhaul.

Penalty charges arise with time, and they increase continuously at a rate proportional to the square of the time since the overhaul. For 3 years they would total $300,000.

The time required for the overhaul, during which extra power must be purchased, the time value of money, and taxes may be neglected. Find the optimum time between overhauls.

10.29 A vehicular tunnel is to be built from point A on one side of a river to point C on the other side. The river is 800 m wide at AB, but the tunnel will be dug in a straight line at some angle θ with AB. The

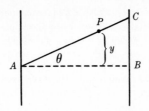

ease of drilling depends upon the distance from line AB. A factor, g, is given by

$$g = \frac{2 \times 10^5}{10^5 + y^2}$$

where y is expressed in meters. The cost for drilling 1 m is proportional to g and would be $33,000/m along line AB.

All other costs, such as those for casing, tile, and ventilation, amount to $25,000/m. Find the optimum angle with respect to line AB at which the tunnel should be built.

10.30 Show how you would obtain the average efficiency in a variocyclic process, such as from the graph of Fig. 10.5. Does it matter if the input to the equipment is not constant?

10.31 The number of items Q that can be made in unit time depends upon the mixture of the number of units of labor and capital and is given by

$$Q = L^{0.6}C^{0.7}$$

where L is the number of units of labor supplied and C is the number of units of capital supplied.

The cost per unit of time for L units of labor is aL, and the cost per unit of time for C units of capital is bC. The sales value of Q items is sQ. Find the optimum mix of labor and capital so that profit is maximized per unit time.

10.32 An evaporator's scales and tests show that the steam condensed in the heating tubes varies with time according to the relationship.

$$S = A(50 - 0.05T) \qquad T < 1000$$

where S is the condensation rate in kilograms per hour, A is the heating area in square meters, and T is the evaporating time in hours since the last cleanout. There will be 0.9 kg of water evaporated per kilogram of steam condensed. The cost for the evaporators is $2100/m^2 heating area and is written off over a 10-year period. The allocable labor cost is $15/h of operation. Each cleaning costs $2500. The water to be evaporated per year is 4.5×10^6 kg, and the operating time is not to exceed 6000 h/year.

Find (a) the optimum evaporating time per cycle before recleaning the tubes and (b) the optimum heating area to be installed.

10.33 A processing unit requires cooling supplied by a chilling liquid. The cost to cool the chilling liquid depends upon ΔT in °C that it is cooled and the flow rate F in kilograms per second at which it is pumped.

The pumping cost is proportional to the cube of the pumping rate, and for a pumping rate of 100 kg/sec it is $12.60/h. The cost for chilling the chilling liquid is proportional to the square of ΔT, and for a drop of 50°C it is 17.5×10^{-5} for 1 kg of liquid. There is a fixed cost of $54/h.

The cooling accomplished in the processing unit depends upon both the flow rate and the ΔT of the chilling liquid and is given by

$$\text{MJ/h removed} = 10.8F(\Delta T)$$

where F is in kilograms per second.

Find the flow rate F and the ΔT for the chilling liquid that will give the minimum cost per unit time per joule per unit time removed in the process.

10.34 A company owns a section of land which is shaped like a right triangle, with legs A and B and hypotenuse C. The company desires to build on the largest rectangular area that can be obtained from the lot. Find the area of that lot (*a*) by using ordinary calculus and (*b*) by using a Lagrange multiplier.

10.35 Maximize the product XY subject to the restraint that X and Y lie on a circle with a radius of 5, that is, $X^2 + Y^2 = 5^2$. (*a*) Use ordinary calculus and (*b*) use a Lagrange multiplier.

10.36 From the Lagrange expression, find the values of X, Y, and Z that minimize the function $X + 2Y^2 + Z^2$, subject to the constraint that $X + Y + Z = 1$.

10.37 For the expression

$$Y = AX^2 - ABX + \frac{BC}{X^2}$$

verify that for $A = B = C = 2$, Y has a minimum value at $X = 1.54$. Find which parameter, A, B, or C, has the greatest effect on the minimum value of Y for the conditions given.

REFERENCES

1. Beightler, C. S., D. T. Phillips, and D. J. Wilde: *Foundations of Optimization,* 2d ed., Prentice-Hall, Inc., Englewood Cliffs, N.J., 1979.
2. Boas, A. H.: "How to Use Lagrange Multipliers," *Chemical Engineering,* Jan. 7, 1963, pp. 95-98.
3. Denn, M. H.: *Optimization by Variational Methods,* McGraw-Hill Book Company, New York, 1969.
4. Fan, L. T.: *The Continuous Maximum Principle,* John Wiley & Sons, Inc., New York, 1966.
5. Hestenes, M. R.: *Optimization Theory—The Finite Dimensional Case,* John Wiley & Sons, Inc., New York, 1975.
6. Iscol, L.: "How to Solve Optimization Problems," *Chemical Engineering,* Feb. 19, 1962, pp. 107-116.
7. Miller, R. E.: *Modern Mathematical Methods for Economics and Business,* Robert E. Krieger Publishing Company, Inc., Huntington, N.Y., reprinted 1978.
8. Peters, M. S., and K. D. Timmerhaus: *Plant Design and Economics for Chemical Engineers,* 3d ed., McGraw-Hill Book Company, New York, 1980.
9. Tao, L. C.: "Process Optimization," *Chemical Engineering,* Apr. 26, 1965, pp. 143-148.

ELEVEN

SUBSETS IN OPTIMIZATION

F. C. Jelen

INVENTORY PROBLEMS

11.1 Introduction

This chapter discusses two types of problems—inventory and queuing—which are treated in operations research but are also important in cost engineering. Inventories are maintained at a cost to gain advantages having monetary value, such as avoiding a shutdown due to a temporary absence of supplies or permitting uniform production for a seasonable supply or demand. There is also an economic balance leading to some optimum policy for minimum cost.

Inventory is simply a stock of physical assets having value, which can be material, money, or labor. Material inventory can be raw material, tools and accessories including spare parts used in production, unfinished or in-process inventory, and finished products. Figure 11.1 illustrates the classes of inventory in the manufacture of steel.

11.2 General Inventory Model

Mathematical models employ the language and symbols of mathematics and may be deterministic, for which all information and values are treated as being definite, or probabilistic (stochastic), for which uncertainty in some of the values is recognized and considered.

The general inventory system is shown in Fig. 11.2. Ordinates represent the number of units available in stock against time as the abscissa. Consider the

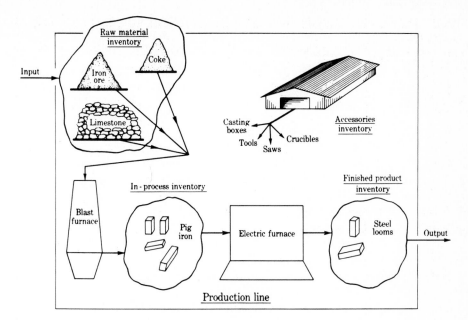

Figure 11.1 Classification of a materials inventory in the manufacture of steel.

quantity Q units received at the beginning of the period T_1. The use during this period U_1 is subtracted from the stock to give a lower stock for the second period, and so forth. At the end of the fourth period T_4 the level of inventory reaches the reorder level R. At that time, an order equal to Q units is placed which is eventually received at the beginning of the tenth period. The procurement time, or lead time T_l, is the time required in periods to receive an order after it is placed. During the procurement time the level of stock may go below zero, creating a shortage for which a penalty can arise.

The object is to minimize the total cost of the system, which is the minimum

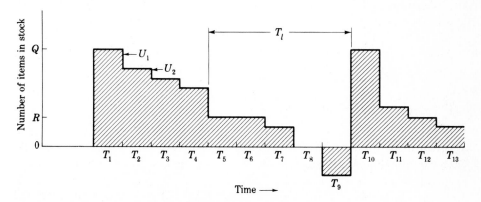

Figure 11.2 Analysis of inventory against time.

cost per period, not per cycle. Some costs are connected with the cycle, such as placing and receiving an order, whereas other costs are connected with a period, such as storage costs. Confusion between a period and a cycle is the most common source of error for a beginner.

11.3 Economic Lot Size

The simplest case makes the following assumptions:

1. All items are delivered simultaneously at the time the stock becomes zero.
2. The cost of placing and receiving an order is constant and is independent of the lot size.
3. Usage is at a constant rate.
4. No safety stock is provided.
5. No shortage is allowed.
6. The warehousing cost is calculated on a continuous basis, and at any instant of time is proportional to the number of items in the inventory.

The schematic diagram is illustrated in Fig. 11.3. Let C_o be the cost for placing and receiving an order, C_h the storage or holding cost *per item per period*, $M = Q$ the maximum inventory where Q is the lot order, and U the rate of use in items per period. Consider a time element T units of time from zero time lasting dT units of time. The inventory at any time is a straight-line function of time and is Q at $T = 0$, and 0 at $T = N$ where N is the duration of the cycle. Then,

$$\text{Inventory} = -\frac{Q}{N}T + Q$$

The number of periods of time N that the lot Q will last is given by

$$N = \frac{\text{items}}{\text{items/period}} = \frac{Q}{U}$$

so that the inventory at any time is

$$\text{Inventory} = -UT + Q$$

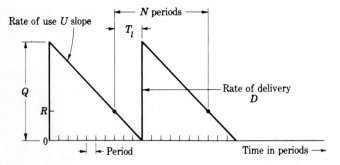

Figure 11.3 Diagram for a simple inventory model.

The total cycle cost is

$$C_{cycle} = C_o + \int_0^{N=Q/U} C_h(-UT + Q)dT$$

$$= C_o + C_h\left(-\frac{U}{2}\frac{Q^2}{U^2} + Q\frac{Q}{U}\right) = C_o + \frac{C_h Q^2}{2U}$$

The function to be optimized is the cost per period C_t and with

$$\text{Cycles/period} = \frac{\text{item/period}}{\text{item/cycle}} = \frac{U}{Q}$$

the cost per period becomes

$$C_t = \frac{\$}{\text{cycle}}\frac{\text{cycle}}{\text{period}} = \left(C_o + \frac{C_h Q^2}{2U}\right)\frac{U}{Q} = C_o\frac{U}{Q} + \frac{C_h Q}{2} \tag{11.1}$$

Differentiating and setting equal to zero,

$$\frac{dC_t}{dQ} = 0 = -C_o\frac{U}{Q^2} + \frac{C_h}{2}$$

Finally,

$$Q_{opt} = \sqrt{\frac{2C_o U}{C_h}} \tag{11.2}$$

The lead time T_l multiplied by the rate of use per period U gives the reorder level R:

$$R = T_l U \tag{11.3}$$

By placing Q_{opt} in Eq. (11.1), there results for the optimum period cost

$$C_t = C_o\frac{U}{\sqrt{2C_o U/C_h}} + C_h\sqrt{\frac{C_o U}{C_h}} = \sqrt{2C_h C_o U} \tag{11.4}$$

Example 11.1 The demand for an item is 100/day. The cost of placing and receiving an order is $50 with an infinite delivery rate. The storage cost is $10 per item per year based on the actual inventory at any time. No shortages are allowed. The reorder level is for an 8-day delivery time. The storage charge per item per day is $10/365 = 0.0274$. Eq. (11.2) is

$$Q_{opt} = \sqrt{\frac{2(50)(100)}{0.0274}} = 604 \text{ items}$$

The minimum cost, by Eq. (11.4), is

$$C_t = \sqrt{2 \times 0.0274 \times 50 \times 100} = \$16.55/\text{day}$$

The reorder level, by Eq. (11.3), is

$$R = 8(100) = 800 \text{ items in stock}$$

11.4 Safety Stock

In some situations the decision maker wants to be sure that a shortage in an item will not occur, and so adds more items to the inventory called *safety stock*. See

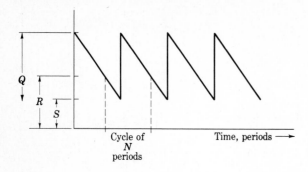

Figure 11.4 Diagram for a simple inventory model with safety stock.

Fig. 11.4, where the safety stock is represented by S. The model has the same assumptions as in the preceding section. The procedure is the same, and the inventory at any time T is

$$\text{Inventory} = -UT + Q + S$$

Individual steps are not shown, but the cost per cycle will be

$$C_{\text{cycle}} = C_o + C_h\left(\tfrac{1}{2}\frac{Q^2}{U} + S\frac{Q}{U}\right)$$

The cost per period will be

$$C_t = C_o\frac{U}{Q} + C_h\frac{Q}{2} + SC_h$$

The optimum order for the minimum cost will be

$$Q_{\text{opt}} = \sqrt{\frac{2C_o U}{C_h}} \qquad (11.5)$$

and the optimum period cost is

$$C_t = \sqrt{2C_h C_o U} + SC_h \qquad (11.6)$$

Equation (11.5) is the same as Eq. (11.2). When $S = 0$, Eq. (11.6) reduces to Eq. (11.4).

11.5 Finite Rate Delivery

The model has important applications since a finite delivery rate is associated with manufacturing. The schematic diagram is shown in Fig. 11.5. Here C_o is a setup cost, and D items per unit of time is a manufacturing rate. The use rate in items per unit of time is U. The value for Q as shown is the total number of items produced and used in the cycle of $(T_1 + T_2)$ periods. The safety stock is S items, and M is the maximum inventory. In time T_1 the inventory builds up by an amount $(M - S)$ at a rate $(D - U)$. Thus,

$$M - S = T_1(D - U)$$

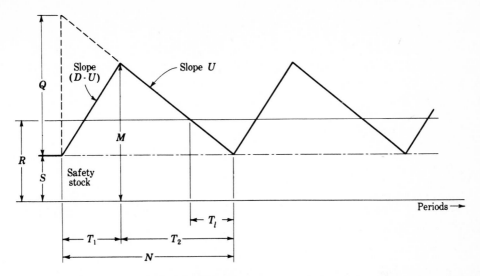

Figure 11.5 Diagram for a model with a finite delivery rate and safety stock.

Also, in time T_1 the amount Q is made at a rate D:

$$Q = T_1 D$$

Eliminating T_1 gives

$$M - S = Q\frac{D - U}{D} \qquad (11.7)$$

but Q is used in N periods at a rate U:

$$Q = UN \qquad (11.8)$$

Placing this value for Q in Eq. (11.7) gives

$$M - S = UN\frac{D - U}{D} \qquad (11.9)$$

In this model $(M - S)$ is linear with time on both sides of the maximum inventory. The average inventory for the total N periods is one-half of $(M - S)$ plus the safety stock.

$$\text{Average inventory} = \frac{UN}{2}\frac{D - U}{D} + S \qquad (11.10)$$

The average inventory is held at a cost C_h per item per period. The cost per period is

$$
\begin{aligned}
C_t &= \left[C_o + \left(\frac{UN}{2}\frac{D - U}{D} + S \right)C_h N \right] \frac{\$}{\text{cycle}} \frac{1}{N} \frac{\text{cycle}}{\text{period}} \\
&= \frac{C_o}{N} + \frac{UN}{2}\frac{D - U}{D}C_h + C_h S \qquad (11.11)
\end{aligned}
$$

For a minimum,

$$\frac{dC_t}{dN} = 0 = -\frac{C_o}{N^2} + \frac{C_h U}{2}\frac{D-U}{D}$$

$$N = \sqrt{\frac{2C_o D}{C_h U(D-U)}} \tag{11.12}$$

Placing N in Eq. (11.8) gives

$$Q_{opt} = \sqrt{\frac{2C_o UD}{C_h(D-U)}} \tag{11.13}$$

and placing N in Eq. (11.11) yields

$$C_t = \sqrt{\frac{2C_h C_o U(D-U)}{D}} + C_h S \tag{11.14}$$

Example 11.2 The setup cost for a small operation is \$600, and the storage cost is \$0.40 per item per day. Usage is uniform at 300 items per day, and the production rate is 500 times per day. A safety stock of 100 units is required. Find the optimum production lot size that will give the minimum cost.

SOLUTION By Eq. (11.13), the optimum lot size is

$$Q_{opt} = \sqrt{\frac{2(600)(300)(500)}{(0.40)(500-300)}} = 1500 \text{ items}$$

The period cost, by Eq. (11.14), is

$$C_t = \sqrt{\frac{2(600)(0.40)(300)(500-300)}{500}} + 0.40(100) = \$280/\text{day}$$

The production time is T_1, given by

$$T_1 = \frac{Q}{D} = \frac{1500}{500} = 3 \text{ days}$$

The total cycle time is N days, given by

$$N = \frac{Q}{U} = \frac{1500}{300} = 5 \text{ days}$$

11.6 Setting Up Inventory Problems

The number of models used to represent inventory problems is infinite, and analysts must learn how to set up their own mathematical models and optimize them by analytical or graphical methods. In some models the inventory may reduce to zero, and unfulfilled demand may be assessed a penalty. The cost of placing or receiving an order may depend on the size. The variations are endless. The general method of setting up a model is illustrated in the following example.

Example 11.3 The demand for an item is uniform at 1000 items per day. Delivery when placing an order is immediate. The cost for placing and receiving an order depends upon the order size and is given by

$$C_o = 13,000 + 5Q^{0.5}$$

It costs \$0.002/day for each item in storage.

In addition, there is an antispoilage cost while in storage which is proportional to the age of the inventory multiplied by the number of units in the inventory at the time. If an inventory is 20 days old and there are 8000 items in stock, the antispoilage cost would be

$$(0.000015)(20)(8000) \qquad \$/\text{day}$$

where 0.00015 is a constant. Find the optimum lot order size.

SOLUTION Let a cycle last N days. The lot size will then be

$$Q = 1000N$$

for which

$$C_o = 13,000 + 5(1000N)^{0.5} = 13,000 + 158.11N^{0.5}$$

The inventory declines linearly from $1000N$ to 0 in N days. The inventory at any time is

$$\text{Inventory} = 1000(N - T)$$

where T is time in days. The cost for antispoilage at that time is

$$(0.000015)T[1000(N - T)] = 0.015(NT - T^2) \qquad \$/\text{day}$$

The cost for one cycle is

$$\$/\text{cycle} = 13,000 + 158.11N^{0.5} + \int_0^N 0.002[1000(N - T)]\,dT + \int_0^N 0.015(NT - T^2)\,dT$$

which after integration reduces to

$$\$/\text{cycle} = 13,000 + 158.11N^{0.5} + N^2 + 0.0025N^3$$

Multiplying by $1/N$ cycles per day gives the dollars per day:

$$\$/\text{day} = 13,000N^{-1} + 158.11N^{-0.5} + N + 0.0025N^2$$

By repeated trials, the minimum occurs at $N = 96.4$, for which

$$Q_{\text{opt}} = 1000N = 96,400 \text{ items per lot}$$

PROBABILISTIC MODELS

11.7 Optimum Stock by Incremental Analysis

When uncertainty is involved in any of the factors, the problem can become complicated. This section and the two following ones illustrate a few examples to acquaint the analyst with some methods.

Example 11.4 The cumulative probability for the sales of a certain item in season is:

Cumulative sales	Cumulative probability of this many sales or less	Cumulative probability of this many sales or more
0	0.00	1.00
1,000	0.03	0.97
2,000	0.09	0.91
3,000	0.18	0.82
4,000	0.29	0.71
5,000	0.44	0.56
6,000	0.60	0.40
7,000	0.80	0.20
8,000	0.94	0.06
9,000	0.98	0.02
10,000	1.00	0.00

An item costs $3, sells for $4 in season, and is disposed of for $1 if not sold in season. There is no storage charge. Determine the optimum number to stock at the start of the season.

SOLUTION The number to stock is such that the last item stocked has an expectation equal to its cost. If another were stocked, its expectation would be less than the cost.
Recall that

$$\text{Expectation} = \Sigma \, (\text{value})(\text{probability})$$

and let $P(X)$ be the probability of selling the last unit and $1 - P(X)$ be the probability of not selling the last unit. If bought, the last unit costs $3 and it brings either $4 or $1. The mathematical relationship for break-even is

$$3.00 = 4.00P(X) + 1.00[1 - P(X)]$$

$$P(X) = 0.67$$

The probability of selling the last unit must be 0.67 and of not selling it $1 - 0.67$, or 0.33. By interpolation in the tabulation, or from a plot, a stock of 4300 meets the requirements. If a stock of 4300 is ordered, the probability of selling the last unit is the probability of selling 4300 or more, that is, 0.67. Analysts must be wary of incorrect reasoning and can assist themselves by recognizing the two complementary cumulative probabilities, as was done in the tabulation of data for the example.

11.8 Optimum Inventory Based on Expectation

The expectation of a variable can be used in many ways to reduce a problem to a tabular or graphical solution. The following example is an illustration.

Example 11.5 The average demand for an item is four per week and follows a Poisson distribution. Find the number of units to be stocked at the beginning of each week so that the number of customers turned away on the average will not exceed 10 percent of demand.

SOLUTION For a Poisson distribution, with an average demand of four the probability of a demand for exactly X, by Eq. (8.11),

$$P(X) = \frac{4^x e^{-4}}{X!}$$

Table 11.1 Example 11.5 with an initial stock of 6

Demand X	P(X)	Lost sales (X − 6)	(X − 6)P(X)
7	0.0594	1	0.0594
8	0.0296	2	0.0592
9	0.0132	3	0.0396
10	0.0053	4	0.0212
11	0.0019	5	0.0095
12	0.0006	6	0.0036
13	0.0002	7	0.0014
14	0.0001	8	0.0008
15	0.0000	9	0.0000
			$\Sigma = 0.195$ expected lost sales per week

and by calculation or from tables:

$$P(0) = 0.0183 \qquad P(1) = 0.0732 \qquad P(2) = 0.1464$$

$$P(3) = 0.1950 \qquad P(4) = 0.1950 \qquad P(5) = 0.1560$$

$$P(6) = 0.1040 \qquad P(7) = 0.0594 \qquad P(8) = 0.0296$$

$$P(9) = 0.0132 \qquad P(10) = 0.0053 \qquad P(11) = 0.0019$$

$$P(12) = 0.0006 \qquad P(13) = 0.0002 \qquad P(14) = 0.0001$$

$$P(15) = 0.0000$$

Let the starting stock be six. Sales will be lost only when the demand is seven or greater. Table 11.1 shows how the expected lost sales are computed using the following underlying relationship:

$$\text{Expected lost sales} = \Sigma \, (\text{lost sales})(\text{probability})$$

If the initial stock is six, the expected lost sales is 0.195, and the average demand is four. Lost sales as a fraction of demand is

$$\frac{0.195}{4} = 0.05$$

Inasmuch as the ratio may be 0.10, a stock of six is more than adequate.

If the calculation is repeated on the basis of a starting stock of five, the expected lost sales is found to be 0.409, which as a fraction of demand is $0.409/4 = 0.102$. The conditions of the example can be very nearly met by stocking five units.

11.9 Monte Carlo Simulation in Inventory

Monte Carlo simulation can be used very effectively to solve probabilistic inventory problems. In this method variables subject to uncertainty are generated by the use of random numbers in such a way that real frequencies of distribution are obeyed on the average by the generated values. A number of cycles are simulated to establish the most probable values. Machine computation makes the method very versatile. Not only can average and expected values be predicted, but frequency distributions are easily established.

The method will be illustrated in a simple form in this section with an inventory problem in which the lead time and demand each follow known probabilistic distributions. The Monte Carlo method itself was described in Chap. 8.

Example 11.6 The mean demand for an article U_m is 4.7 units per day. The distribution of this demand is as follows, with the last column representing the assignment of random rectangular variates:

Demand	$P(U)$	Cumulative relative frequency	Random numbers assigned
2	0.05	0.05	01–05
3	0.15	0.20	06–20
4	0.20	0.40	21–40
5	0.30	0.70	41–70
6	0.20	0.90	71–90
7	0.10	1.00	91–00

The mean lead time T_{lm} is 4 days. The distribution of this time is as follows, with the last column representing the assignment of random rectangular variates:

Lead time T_l	$P(T_l)$	Cumulative relative frequency	Random numbers assigned
3	0.25	0.25	01–25
4	0.50	0.75	26–75
5	0.25	1.00	76–00

It costs $15 to place an order, inventory cost is $1 per unit per day, and shortage cost is $2 per unit per day. Assume that orders are placed and arrive at the end of a day and that charges are based on the end-of-day inventory but before arrival of a new lot.

Find the optimum lot size and reorder level.

SOLUTION Make a Monte Carlo simulation assuming a lot size Q of 30 and a reorder level R of 20. Start the first cycle with 20 units on hand. This is arbitrary, but the effect will be washed out when a great number of cycles is considered.

The tabulation is made in Table 11.2. All random numbers were selected in an unbiased fashion from a table of random rectangular variates. A cycle is assumed to start the day after a new order is placed, which occurs whenever inventory falls to 20 or less at the end of a day.

At zero time the random number 79 indicates a lead time of 5 days for the order placed at that time. During the first period the demand is 3, and the inventory drops to 17 at the end of the

Table 11.2 Monte Carlo method for Example 11.6 with $Q = 30$, $R = 20$

Cycle	Period cumulative	Random no.	Demand	Random no.	Lead time	Stock, end of day	Shortage cost, $	Inventory cost, $	Lot order cost, $	Cumulative cost, $	Average cost per period, $
	0			79	5	20					
1	1	13	3			17	0	17	0	17	17
	2	28	4			13	0	13	0	30	15
	3	77	6			7	0	7	0	37	12
	4	60	5			2	0	2	0	39	10
	5	29	4			−2	4	0	0	43	9
	6	16	3			$30 - 2 - 3 = 25$	0	25	15	83	14
	7	70	5	68	4	20	0	20	0	103	15
2	8	09	3			17	0	17	0	120	15
	9	24	4			13	0	13	0	133	14
	10	16	3			10	0	10	0	143	14
	11	19	3			7	0	7	0	150	14
	12	14	3			$30 + 7 - 3 = 34$	0	34	15	199	17
	13	29	4			30	0	30	0	229	18
	14	49	5			25	0	25	0	254	18
	15	08	3			22	0	22	0	276	18
	16	80	6	45	4	17	0	17	0	293	18

first day. This continues, but between the end of the fifth day and the start of the sixth day, a new lot of 30 arrives and is available for the sixth day. At the end of the sixth day inventory is 25, being the 30 received less the 2 short the previous day less the 3 used during the sixth day. This day is also charged with the lot order cost. The cumulative cost divided by the cumulative days or periods is the average cost per day for $Q = 30$ and $R = 20$. If enough cycles are taken, the simulation will be exceedingly good. Other combinations are tried to find the optimum Q and R.

The method can be used to get averages and distributions of the various variables. Thus by the proper tabulation it is possible to determine the average shortage and distribution of shortage per cycle.

QUEUING PROBLEMS

11.10 Prevalence

A queue is a waiting line, but *queuing* is a term now widely used in a broad sense to cover a wide variety of problems, usually for economic balance and optimization, involving waiting and delay in serving people or servicing machines and equipment. Queuing also covers all facets of the problem, such as the optimum number of long-distance lines required between two cities or the optimum number of repairers and their equipment required to keep an assembly line in operation.

Modern society, with its large population and interdependent functions, has focused considerable attention on queuing problems. There is now strong competition among service industries to gain and retain the customer, and all companies must consider internal queuing for the economic balance of manufacturing and operational efficiency.

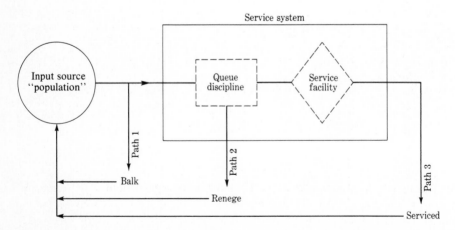

Figure 11.6 Chart of a queuing system.

11.11 Description of a Queuing System

A queuing system is illustrated in Fig. 11.6. The term *machine* will be used in a general way for the object to be serviced. The main elements in the system are the input source and the service system. The latter is characterized by queue discipline and service facility.

A machine needing servicing can follow three possible paths:

1. Return to the population without servicing, called *balking*, path 1.
2. Join the queue but return to population without servicing, called *reneging*, path 2.
3. Return to the population after being serviced, path 3.

11.12 Classification of Queuing Systems

A queuing system can be classified with respect to input source, queue, and service facility.

The input source, or population, is classified as *infinite* or *finite*, which is chiefly the size of the population relative to the number in the queue and being serviced. If the characteristics of the input source (Fig. 11.6) are changed by the number of withdrawals, the population is considered finite and the problem is solved with the changes in the population taken into account.

The queue itself can be classified as *infinite* if it is allowed to grow to any size or *finite* if characterized by a maximum permissible size. Queues can be classified further as *single* or *multiple*. A one-chair barbershop is a good example of a single queue. The multiple-queue case is exemplified by a work center having two input queues from different operations.

11.13 The Simplest Model

The simplest model is a single-channel, single-phase, single queue with an infinite population. Arrivals come at equally spaced intervals and require a constant servicing time. The time periods between arrivals T_a must be greater than the servicing time T_s; otherwise an infinite queue would form. The waiting time or lost time for each item is the servicing time only, since no queue forms.

Consider an interval of time T_a during which one item will enter the service channel and remain there for T_s time periods. If C_w is the cost for one item waiting one time period, the cost for the waiting time is

$$C_w T_s \tag{11.15}$$

In addition, there is the cost of servicing. The size of the servicing channel is optional. It will be assumed that the time for servicing an item T_s is inversely proportional to the size of the channel, and that the period cost of a servicing channel is proportional to its size. Let a service channel which can service one item in one time period cost C_f dollars to operate for one time period. If the servicing time to service one unit is to be T_s and not a unity time period, then a servicing channel of a different size will be required, and its cost for one time period will be

$$\frac{C_f}{T_s} \tag{11.16}$$

The cost for a time interval T_a will be the sum of the cost for the lost time or waiting time for the unit being serviced plus the cost of operating the servicing channel for T_a time periods, which is

$$C_w T_s + \frac{C_f}{T_s} T_a \tag{11.17}$$

Obtain the total cost for one time period C_t by dividing the above by T_a, giving

$$C_t = C_w \frac{T_s}{T_a} + \frac{C_f}{T_s} \tag{11.18}$$

The minimum cost per period is obtained by differentiation with respect to T_s and setting equal to zero:

$$\frac{dC_t}{dT_s} = 0 = \frac{C_w}{T_a} - \frac{C_f}{T_s^2}$$

$$T_{s,\text{opt}} = \sqrt{\frac{C_f T_a}{C_w}} \tag{11.19}$$

and placing this value in Eq. (11.18) gives

$$C_{t,\text{opt}} = 2\sqrt{\frac{C_f C_w}{T_a}} \tag{11.20}$$

It is important to note that C_f is the cost for one time period for a servicing channel which when working full time would service one item.

If there are L channels or service facilities instead of one, as above, then Eq. (11.18) becomes

$$C_t = C_w \frac{T_s}{T_s} + \frac{C_f L}{T_s} \tag{11.21}$$

and Eqs. (11.19) and (11.20) become

$$T_{s,\text{opt}} = \sqrt{\frac{C_f T_a L}{C_w}} \tag{11.22}$$

$$C_{t,\text{opt}} = 2\sqrt{\frac{C_f C_w L}{T_a}} \tag{11.23}$$

Example 11.7 The arrival rate is constant at three items per hour. The cost of providing and maintaining a service facility is $25/h, and it can service four items per hour working full time. If an item waits 1 day, it represents a cost of $2400. Find the optimum time to service one item and the minimum variable cost per item.

SOLUTION Take 1 h as one period.

$$C_f = \frac{25}{4} = \$6.25$$

$$T_a = \frac{1}{3} \text{h between arrivals}$$

$$C_w = 2400/24 = \$100/\text{h} \cdot \text{item}$$

By Eq. (11.19),

$$T_{s,\text{opt}} = \sqrt{\frac{6.25(1/3)}{100}} = 0.1443 \text{ h, optimum servicing time}$$

The total cost per period, by Eq. (11.20), is

$$C_{t,\text{opt}} = 2\sqrt{\frac{6.25 \times 100}{\frac{1}{3}}} = \$86.60$$

but three items enter the system per period. The total cost for servicing one item is $86.60/3 = \$28.87$.

Example 11.8 Redo the previous example. Here the cost of a service facility is proportional to the 0.6 power of its size.

SOLUTION Let S be the size of the service station. Then S is inversely proportional to the time required to service one item.

$$S = \frac{k}{T_s}$$

and the cost of the service facility for 1 h is

$$K'\left(\frac{k}{T_s}\right)^{0.6} = K\left(\frac{1}{T_s}\right)^{0.6}$$

When $T_s = 1/4$, the cost is $25:

$$25 = K\left(\frac{1}{T_s}\right)^{0.6} \qquad K = 10.88$$

and the cost for a service facility in dollars per hour is

$$10.88\left(\frac{1}{T_s}\right)^{0.6} = 10.88T_s^{-0.6} \tag{11.24}$$

For a time interval T_a when one item is serviced, the cost for waiting plus the cost for the service facility is

$$C_w T_s + 10.88T_s^{-0.6}T_a \tag{11.25}$$

Dividing by the time T_a gives the cost per hour:

$$C_t = C_w\frac{T_s}{T_a} + 10.88T_s^{-0.6} \tag{11.26}$$

Differentiate with respect to T_s and set equal to zero:

$$0 = \frac{C_w}{T_a} - 0.6(10.88)T_s^{-1.6}$$

This yields, for the optimum servicing time,

$$T_s = \left(\frac{6.528 T_a}{C_w}\right)^{1/1.6} = \left(\frac{6.528 \times 1/3}{100}\right)^{1/1.6} = 0.0914 \text{ h}$$

The total cost per period, from Eq. (11.26), is

$$C_{t,\text{opt}} = 100\frac{0.0914}{\frac{1}{3}} + 10.88(0.0914)^{-0.6} = \$73.14/\text{h}$$

11.14 Models Involving Probability

In practice, the arrival rate and servicing time will not be constant. The mathematics for these models can become quite complicated, and only the results for some simple cases are given here.

If the number of arrivals per period follows a Poisson distribution with λ average arrivals per period, and the number of items serviced per period follows a Poisson distribution with $(\mu > \lambda)$, then minimum cost per period is obtained at

$$C_{t,\text{opt}} = 3C_w \tag{11.27}$$

for which

$$\lambda'_{\text{opt}} = \frac{C_w}{C_f} \tag{11.28}$$

$$\mu'_{\text{opt}} = 2\frac{C_w}{C_f} \tag{11.29}$$

where λ'_{opt} and μ'_{opt} are specific values of the independent variables λ' and μ' at which the lowest cost of the system occurs.

Example 11.9 The cost of waiting per period is $4, and the cost per hour for servicing an item in a service center which can handle one item in 1 h is $2. If arrivals and servicing rates follow Poisson distributions, find the lowest cost policy.

SOLUTION By Eq. (11.27),

$$C_{t,\text{opt}} = 3C_w = 3(4) = \$12 \text{ per period}$$

for which, by Eqs. (11.28) and (11.29),

$$\lambda'_{\text{opt}} = \frac{C_w}{C_f} = \frac{4}{2} = 2 \text{ items per period}$$

$$\mu'_{\text{opt}} = 2\frac{C_w}{C_f} = 2\frac{4}{2} = 4 \text{ items per period}$$

11.15 Finite Queuing

In practice, the number of items being serviced might have a significant effect on the number of items still in use. The situation becomes one of finite queuing. Analytical solutions for finite queuing problems can become quite complicated and are not considered here. The problem can be treated by the general methods described below, noting that probabilities will be affected by the number of items requiring servicing.

11.16 General Method with Certainty

In many cases an analytical solution is difficult or perhaps impossible. Such cases must be reduced to a tabulation, and considerable ingenuity may be required to represent the problem on paper. The following example is straightforward but would be quite complicated if a reduction to mathematical equations were attempted.

Example 11.10 Five units of a product must be processed at three different stations, A, B, and C. The processes are independent. The process time in days at each station is

A   5 days
B   7 days
C   9 days

The five products must have the same operating sequence. The handling time and cost between stations are negligible.

Stations are all started at zero time and are shut individually when the fifth unit has gone through the station. Idle time for a product is counted whenever a product arrives at a station and must wait for entry. Idle time for a station is counted whenever a station is waiting for work.

Find the most economical sequence of work centers for the following conditions, where C_w is the waiting cost in dollars per part per day and C_e is the cost of idle time in dollars per station per day.

	C_w	C_e
(a)	10	30
(b)	20	20
(c)	30	10

SOLUTION Table 11.3 shows the waiting time of products and the idle time of machines for all possible sequence of work centers. The total time required to produce five parts is 57 days for all the sequences; but the waiting and idle times vary, and in summary are:

Sequence	Part waiting	Station idle
ABC	40	17
ACB	40	27
BAC	20	27
BCA	20	39
CAB	0	47
CBA	0	49

Table 11.4 is the calculation for costs based on the following relationship:

$$C_t = C_w \left(\begin{array}{c} \text{total waiting time} \\ \text{for product} \end{array} \right) + C_e \left(\begin{array}{c} \text{total idle time} \\ \text{for stations} \end{array} \right)$$

The minimum cost for (a) is $910 per period for sequence ABC, for (b) $940 per period for BAC or CAB, and for (c) $470 per period for CAB.

Table 11.3 Time-analysis tabulation for Example 11.10

		Station 1		Station 2				Station 3			
Sequence	Part no.	In	Out	In	Out	Part waiting	Station idle	In	Out	Part waiting	Station idle
ABC	1	0	5	5	12	0	5	12	21	0	12
	2	5	10	12	19	2	0	21	30	2	0
	3	10	15	19	26	4	0	30	39	4	0
	4	15	20	26	33	6	0	39	48	6	0
	5	20	25	33	40	8	0	48	57	8	0
ACB	1	0	5	5	14	0	5	14	21	0	14
	2	5	10	14	23	4	0	23	30	0	2
	3	10	15	23	32	8	0	32	39	0	2
	4	15	20	32	41	12	0	41	48	0	2
	5	20	25	41	50	16	0	50	57	0	2
BAC	1	0	7	7	12	0	7	12	21	0	12
	2	7	14	14	19	0	2	21	30	2	0
	3	14	21	21	26	0	2	30	39	4	0
	4	21	28	28	33	0	2	39	48	6	0
	5	28	35	35	40	0	2	48	57	8	0
BCA	1	0	7	7	16	0	7	16	21	0	16
	2	7	14	16	25	2	0	25	30	0	4
	3	14	21	25	34	4	0	34	39	0	4
	4	21	28	34	43	6	0	43	48	0	4
	5	28	35	43	52	8	0	52	57	0	4
CAB	1	0	9	9	14	0	9	14	21	0	14
	2	9	18	18	23	0	4	23	30	0	2
	3	18	27	27	32	0	4	32	39	0	2
	4	27	36	36	41	0	4	41	48	0	2
	5	36	45	45	50	0	4	50	57	0	2
CBA	1	0	9	9	16	0	9	16	21	0	16
	2	9	18	18	25	0	2	25	30	0	4
	3	18	27	27	34	0	2	34	39	0	4
	4	27	36	36	43	0	2	43	48	0	4
	5	36	45	15	52	0	2	52	57	0	4

11.17 General Method with Uncertainty

The analysis of queuing problems with uncertainty in arrival or servicing time or both can become exceedingly complicated. The most general method is to subject the data to a Monte Carlo computation, which is a simulation process. The distributions are simulated by the use of random numbers, and a large number of periods are taken to get a good average and a good summary of final distributions and costs. The Monte Carlo method was discussed in Chap. 8 and used in Sec.

Table 11.4 Cost tabulation for Example 11.10

Sequence	Total product waiting, product-days	Cost of waiting, $			Total station idle, station-days	Cost of idle stations, $			Total variable cost per period, $		
		(a)	(b)	(c)		(a)	(b)	(c)	(a)	(b)	(c)
ABC	40	400	800	1200	17	510	340	170	910	1140	1370
ACB	40	400	800	1200	27	810	540	270	1210	1340	1470
BAC	20	200	400	600	27	810	540	270	1010	940	870
BCA	20	200	400	600	39	1170	780	390	1370	1180	990
CAB	0	0	0	0	47	1410	940	470	1410	940	470
CBA	0	0	0	0	49	1470	980	490	1470	980	490

11.9 for an inventory problem. The following example is an application to a queuing problem. The example is very simple and can actually be evaluated without the Monte Carlo technique.

Example 11.11 A company has a large number of switches. The number of switches out of service follows a Poisson distribution, with an expectancy of 0.5 switches out of service for a 1-h period. A repair crew makes a round of all the switches and can restore all switches to service in a very short time at negligible cost. A penalty is charged for switches out of service based on the *average* number out of service at a rate of $500/h per switch. The cost for a repair crew depends upon how fast they can make a round, and for an interval of 10 to 60 min it can be expressed as

$$-50\sqrt{T} + 550 \qquad \$/h$$

where T is the time in minutes to make a round.

Assume that a crew makes a round in 36 min, or 0.6 h. For the Poisson distribution of 1 h the expectancy is 0.5 switches down, and for 0.6 h it will be $0.6 \times 0.5 = 0.3$ expected switches down following the Poisson distribution:

$$P(N) = \frac{e^{-0.3}(0.3)^N}{N!}$$

with a cost in dollars per hour for the crew equal to

$$-50\sqrt{36} + 550 = \$250/h$$

Assignment of random numbers is:

N	$P(N)$	Cumulative $P(N)$	Random numbers
0	0.74082	0.74082	0001–7408
1	0.22225	0.96307	7409–9631
2	0.03334	0.99641	9632–9964
3	0.00333	0.99974	9965–9997
4	0.00025	0.99999	9998–0000
5	0.00002	1.00000	

Random numbers are generated with the following tabulation:

Random no.	N switch down	Cost, $/h	Crew, $/h	Total, $/h	Cumulative average, $/h
9804	2	1000	250	1250	1250
0472	0	0	250	250	750
3167	0	0	250	250	583.33
8012	1	500	250	750	625

A large number of trials gave \$400.80/h as the cost for a crew time of 36 min, while the exact cost is $500(0.3) + 250 = \$400$, but generally an exact solution will not be possible. Other crew times can be tried to obtain the dollars per hour against crew time. Some results are:

Crew time, min	$/h
12	427
36	400
60	413

The minimum cost actually occurs for a crew time of 36 min.

NOMENCLATURE

C_f Cost per period for a service facility which can service one item per period when working full time

C_h Storage or holding cost per item per period, \$/(item) (period)

C_o Cost of placing and receiving an order, or setup cost, \$

C_t Total variable cost per period, \$/period

C_T Total cost, \$

C_w Cost for one item waiting one period, \$/(item)(period)

D Production rate, items per period

L Number of channels in the service facility, dimensionless

M Maximum inventory, number of items

N Number of periods per cycle, dimensionless

$P(X)$ Probability of X

Q Lot size, items

R Reorder level, items

T Time, periods

T_a Time between two consecutive arrivals, periods per item

T_l Procurement or lead time, periods

T_s Time in periods to service one item, periods per item

U Usage rate, items per period

PROBLEMS

11.1 The rate of use of an item is 20 units per year. The cost of placing and receiving an order is $40. The cost of storing one item for 1 year is $0.16, and it depends on the average stock. Find (a) the economic lot size, (b) the minimum total variable cost per period, and (c) the reorder level if the lead time is 3 months.

11.2 Repeat Prob. 11.1 with a safety stock of four items.

11.3 For this problem, the storage cost is based on maximum inventory rather than average inventory. Derive expressions for Q_{opt} and $C_{t,opt}$ corresponding to Eqs. (11.2) and (11.4).

11.4 A store has to satisfy a daily demand of 10 items of a product. The holding cost is $0.10 per item per day based on average inventory. The cost of procuring an order is $20, and it takes 3 days for delivery. The minimum number of items per order is 100 and the maximum is 500 with increments of 5 items. The replenishment rate is infinite. Find the optimum lot size.

11.5 An appliance manufacturer can sell 400 per day of an item at $5.99 each with a manufacturing cost of $4.92 each. He can produce them at a rate of 1000 units per day with a setup cost of $1200. The inventory cost is based on carrying the inventory at his cost at 20 percent per year. Find the economic production run and how long the lot will last.

11.6 For this problem, the storage cost is based on maximum inventory rather than average inventory. Derive expressions for Q_{opt} and $C_{t,opt}$ corresponding to Eqs. (11.13) and (11.14).

11.7 The cumulative probability curve for the number of items sold in season can be plotted from the following information:

Cumulative probability of selling this many items or less in season	Number of items
0	0
0.15	2,000
0.52	4,000
0.82	6,000
0.98	8,000
1.00	10,000

Inventory is built up at the start of a season and cannot be replenished during a season. An item costs $3 and can be sold for $6 in season, but if disposed of after the season it will bring only $1. Find the optimum number with which to start the season. The last one ordered should have a sales expectation based on in-season and out-of-season value equal to its cost.

11.8 Redo Example 11.5 using a binomial distribution. Assume that the total demand N might be 40 per week and that with an average of four per week, the probability for demand is $p = 4/40 = 0.1$ in a population of $N = 40$.

11.9 A pharmaceutical company produces 10,000 m³/day of an antibiotic broth. The broth is collected for T days, and then the entire collection is converted to a finished product.

The broth spoils in storage such that the spoilage for each addition is proportional to the number of days that the addition has been held in storage prior to conversion. If a fixed amount of broth is held in storage for 10 days, it loses 0.725 percent of its biological effectiveness.

Broth is held in inventory at $2/m³ at all times. The fractional holding charge for inventory is $0.10/year. Spoilage is taken at the inventory cost. Broth is produced 365 days/year. It costs $1000 to set up the conversion process of broth to the finished product.

How often should the broth be converted to the finished product?

11.10 An item costs $2 and is used at a uniform rate of 10 items per day. No shortage is allowed, and the delivery rate is infinite. It costs $10 to process an order. Inventory charge is $0.02/day for each item in inventory. In addition, there is a flat charge of $3/year per item based on the maximum number of items in inventory. Find the optimum procurement quantity.

11.11 In an emergency repair shop, machinists cost the company $27/h. If a machinist is working on an emergency job, he or she is worth $60/h; otherwise the value is $12/h. The demand for machinists for emergency work is:

Percent of total time	Machinists on emergency work	Percent of total time	Machinists on emergency work
5	1 or less	85	6 or less
15	2 or less	90	7 or less
30	3 or less	95	8 or less
50	4 or less	98	9 or less
70	5 or less	100	10 or less

How many machinists should be kept on hand for emergencies?

11.12 The average demand for a certain item in inventory is 0.7 per week, with the probability for a demand of exactly N items in a week being $P(N)$ and

$$P(0) = 0.500$$

$$P(1) = 0.300$$

$$P(2) = 0.200$$

The time in weeks required to get a new item if it is not in stock averages 1.85 weeks, with the distribution as follows:

$$P(1) = 0.400$$

$$P(2) = 0.350$$

$$P(3) = 0.250$$

If an order is placed when the level on hand is one, find the probability of going short before the order is received.

11.13 A food company makes a flavoring material at the rate of 100 kg/day. It is made batchwise at a cost of $3000 per batch.

Inventory cost is $2/kg/day, since very expensive methods are required to keep the material from spoiling. In addition, there is an extra cost which is proportional to the batch size multiplied by the square of the time that the inventory is held before it is exhausted. If an initial inventory of 1000 kg lasting 10 days is made, this extra cost would be $10,900. Find the optimum batch size.

11.14 Coal is stored and will form a conical pile at all times, with a slope of 45 degrees with the horizontal. The packed coal occupies 1 m³/1600 kg.

The demand for coal is 455,000 kg/day. The cost for ordering and procuring is $1500 per order. The coal is valued at $27.50/1000 kg. Inventory cost is at a rate of 20 percent per year of the value of the coal in the pile at any time. The delivery rate is infinite. There is also a charge of $5.38/m³/year for the maximum ground area that the coal occupies.

Find the optimum amount of coal to be purchased per order.

11.15 It costs $50 to place an order. The cost of putting the order into the warehouse is proportional to the square root of the size of the order and is $25 when the order is for 25 items. A safety stock of 15 items is to be maintained. The cost of maintaining the inventory in storage is proportional to the 1.5

power of the number in storage at any instant and would be $12.50/week for an inventory of 25 items. Usage is continuous at a rate of five items per week, and delivery of an order may be considered instantaneous after it is placed. Find the optimum order size.

11.16 Redo Prob. 11.12 by the Monte Carlo method.

11.17 The demand for a spare part averages two per week and follows a Poisson distribution. The cost for stocking a part is $5/week. If the demand for the part cannot be met due to a shortage, the penalty for each item short is $30. Find the optimum number of parts to stock at the beginning of the week.

11.18 A gasoline station receives an automobile every 10 min. The cost of waiting in the station is $0.10 per automobile per minute. A service station that can service one car in 1 min costs $0.16/min.

(a) Plot the service cost as a function of the service time in minutes.
(b) Plot the waiting cost as a function of the service time in minutes.
(c) Plot the total variable cost as a function of the service time in minutes.
(d) Find the service time for the lowest cost per automobile.

11.19 If in Prob. 11.18 the station has two or three pumps, find the lowest cost situation.

11.20 The demand for a test is four per hour on the average and follows a Poisson distribution. A testing machine can perform one test per hour. The company now has three testing machines. If the demand for a test cannot be made at once, it costs the company $100. How much can the company afford per hour for three additional testing machines?

11.21 An attendant arrives at an inspection point in a plant every 5 h to pick up rejected assemblies. Her pickup truck can accommodate only three assemblies at the most, and the probabilities for vacant space on the truck are:

No. of vacant spaces	Probability
0	0.05
1	0.10
2	0.50
3	0.35

The number of rejected assemblies arriving at the inspection point during the 5 h has the following probabilities:

Number	Probability
0	0.05
1	0.50
2	0.30
3	0.15

Any assemblies that cannot be picked up by the attendant must be picked up by an emergency vehicle. What fraction of the rejected assemblies must be accommodated by the emergency vehicle?

11.22 Redo Prob. 11.21 by the Monte Carlo method.

11.23 A testing station runs tests on motors. Twenty percent of the motors are class A, 30 percent are class B, and 50 percent are class C. A motor arrives every 10 h with negligible variation. While a motor is being serviced, the cost to the company for waiting in dollars per hour is $100 for class A motors, $150 for class B, and $180 for class C.

Any size inspection station can be installed, and the inspection time per motor for one test will be inversely proportional to the size of the station. A station that can make 200 tests per hour will cost $47.60/h.

Class A motors require 100 tests, class B 200, and class C 300. Find the optimum size inspection station.

11.24 A company has automatic machines which require overhauls on the average of one per week with a Poisson distribution. A single crew can overhaul a machine in 4 days at a cost of $1000/week, two crews can overhaul a machine in 2 days at $2000/week, and so forth. Lost time is charged against a machine only while it is actually being overhauled, i.e., no waiting in line occurs, at a charge of $2000/day. Find the optimum number of crews to be hired and the cost per week.

11.25 The cost in dollars per day for an inspection crew is given by

$$150M^{0.9}$$

Where M is the number of persons in the crew. The time in days to complete an inspection is given by

$$\frac{2}{M^{0.8}}$$

An inspection is made every 3 days, and the loss for the operating item while it is being inspected is $1000/h. Find the optimum number of persons for the inspection crew.

11.26 Every 10 min a bus arrives at a stop with a distribution of empty seats as follows:

No. of empty seats	Times occurring
0	1
1	3
2	5
3	6
4	4
5	1
6	0

The number of persons arriving at the stop in a 10-min period has a distribution as follows:

No. arriving	Times occurring
0	5
1	4
2	3
3	3
4	1
5	2
6	2
7	0

No standees are allowed in the bus. What fraction of the total persons arriving will have to wait for the next bus? Use a Monte Carlo technique.

11.27 Ten items of product X are to be processed on machines A, B, C, and D. Machine A must be used first, but the other machines can be selected in any order. The process times in days are:

A	B	C	D
3	2	7	5

The waiting costs are $100 per item per day and $150 per machine per day.

The same sequence of operations must be used for all the items. A machine must finish one item before starting another. The handling time between machines is negligible.

Find the optimum sequence of operations.

REFERENCES

Inventory Problems

1. Arrow, K. J., S. Karlin, and H. Scarf: *Studies in the Mathematical Theory of Inventory and Production,* Stanford University Press, Stanford, Calif., 1958.
2. Fabrycky, W. J., P. M. Ghare, and P. E. Torgersen: *Industrial Operations Research,* rev. ed., Prentice-Hall, Inc., Englewood Cliffs, N.J., 1972.
3. Hadley, G., and T. M. Whitin: *Analysis of Inventory Systems,* Prentice-Hall, Inc., Englewood Cliffs, N.J., 1963.
4. Larson, S. E.: *Inventory Systems and Controls Handbook,* Prentice-Hall, Inc., Englewood Cliffs, N.J., 1976.
5. Lewis, C. D.: *Demand, Analysis, and Inventory Control,* Lexington Books, Lexington, Mass., 1975.
6. Love, S. F.: *Inventory Control,* McGraw-Hill Book Company, New York, 1979.
7. Starr, M. K., and D. W. Miller: *Inventory Control, Theory and Practice,* Prentice-Hall, Inc., Englewood Cliffs, N.J., 1962.
8. Welch, W. E.: *Scientific Inventory Control,* Management Publishing Company, Greenwich, Conn., 1956.

Queuing problems

1. Allen, A. O.: *Probability and Statistics, and Queuing Theory,* Academic Press, Inc., New York, 1978.
2. Cooper, R. B.: *Introduction to Queueing Theory,* The Macmillan Company, New York, 1972.
3. Fabrycky, W. J., P. M. Ghare, and P. E. Torgersen: *Industrial Operations Research,* rev. ed., Prentice-Hall, Inc., Englewood Cliffs, N.J., 1972.
4. Gross, D., and C. M. Harris: *Fundamentals of Queuing Theory,* John Wiley & Sons, Inc., New York, 1974.
5. Kleinrock, L.: *Queuing Systems,* vol. 1, *Theory,* John Wiley & Sons, Inc., New York, 1975.
6. Page, E.: *Queuing Theory in OR,* C. Russak and Company, Inc., New York, 1972.
7. Peck, L. G., and R. N. Hazelwood: *Finite Queuing Tables,* John Wiley & Sons, Inc., New York, 1958.
8. Prabhu, N. U.: *Queues and Inventories,* John Wiley & Sons, Inc., New York, 1965.
9. White, J. A., J. W. Schmidt, and G. K. Bennett: *Analysis of Queueing Systems,* Academic Press, Inc., New York, 1975.

TWELVE

LINEAR, DYNAMIC, AND GEOMETRIC PROGRAMMING

A. H. Boas

LINEAR PROGRAMMING

12.1 Optimization of an Objective Function

Linear programming (*LP*) is a mathematical tool for finding the optimum solution to a certain class of problems. The word *linear* implies that the relations involved will be linear, i.e., in all relationships every term can be expressed in only one variable raised to the first power. Thus,

$$A_1 X_1 + A_2 X_2 + A_3 X_3 = B_1$$

is linear, while the following are not:

$$A_1 X_1 + A_2 X_2^2 + A_3 X_3 = B_1$$

$$A_1 \ln X_1 + A_2 X_2 + A_3 X_3 = B_1$$

$$A_1 X_1 + A_2 X_1 X_2 + A_3 X_3 = B_1$$

The term *programming* in this context means planning of activities.

In any type of optimization problem, linear or nonlinear, the problem can be represented as indicated in Fig. 12.1. The black box constitutes the mathematical model, i.e., relations that describe some physical situation. The input to the black box is composed of two quantities, fixed and variable. The fixed quantities are those which are inherent in the problem; the formulator has no control over these.

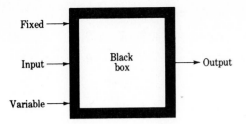

Figure 12.1 A mathematical model.

On the other hand, the variable inputs are the unknowns in the problem; the values of these variables determine the optimum solution to the problem. Hence, the interest is in the values of the variables that will give the optimum value to a term called the *objective function*, which is the output from the black box.

12.2 Development of Linear-programming Equations

Define each of the variable inputs by X_j and assume that there are n of them. These variables are also called *activity levels*. In LP problems a restriction that is imposed on these variables is that they be nonnegative; i.e.,

$$X_j \geq 0 \tag{12.1}$$

The objective function, the output from the black box, will be called Z and defined as

$$Z = C_1 X_1 + C_2 X_2 + \cdots + C_n X_n \tag{12.2}$$

where the C_j values, $j = 1, 2, \ldots, n$, are unit costs. Associated with each variable X_j is a cost coefficient C_j, which may be zero, positive, or negative. Some examples of C_j values are:

X_j	C_j	$C_j X_j$
kg/h	$/kg	$/h
Btu/h	$/Btu	$/h
work-h/h	$/work-h	$/h

In each case, the product of X_j and C_j represents a cost in dollars per hour. These may be added together to represent the total cost, in dollars per hour, and could therefore represent a meaningful objective function. Mathematically, the objective function may be represented as

$$Z = \sum_{j=1}^{n} C_j X_j \tag{12.3}$$

Note that this equation is linear, as required in LP models.

The mathematical model itself may be represented as

$$A_{11}X_1 + A_{12}X_2 + \cdots + A_{1j}X_j + \cdots + A_{1n}X_n = B_1 \tag{12.4}$$

$$A_{21}X_1 + A_{22}X_2 + \cdots + A_{2j}X_j + \cdots + A_{2n}X_n = B_2 \tag{12.5}$$

$$\cdots\cdots\cdots\cdots\cdots\cdots\cdots\cdots\cdots\cdots\cdots\cdots\cdots\cdots\cdots\cdots\cdots$$

$$A_{m1}X_1 + A_{m2}X_2 + \cdots + A_{mj}X_j + \cdots + A_{mn}X_n = B_m \tag{12.6}$$

Each equation in an LP model takes this form and is called a *row* or *constraint*. The right-hand sides are the B_i values. Each A_{ij} is a coefficient with the usual mathematical convention of rows before columns; thus i represents the row and j the column. The coefficients may be zero, positive, or negative. Equations (12.4) to (12.6) can be expressed mathematically:

$$\sum_{j=1}^{n} A_{ij}X_j = B_i \qquad i = 1, 2, \ldots, m \tag{12.7}$$

12.3 Slack Variables

Suppose that a process unit can handle a maximum of 100 kg/h. Mathematically,

$$X_1 \leq 100 \tag{12.8}$$

It is necessary to express this fact within the LP model. Any solution to the problem which violates this capacity limit is not a valid solution because the process unit cannot physically handle such a quantity of stream 1. Equation (12.8) is not in the form of a linear equation, and it becomes necessary to convert the inequality equation to an equality relationship. This is done through the introduction of a *slack variable*. Express Eq. (12.8) as

$$X_1 + X_4 = 100 \tag{12.9}$$

The variable X_4, the slack variable, is treated like any other variable with the restriction of nonnegativity imposed on it. When X_4 assumes its minimum value of 0, then $X_1 = 100$, and for the maximum value $X_4 = 100$, $X_1 = 0$.

In contrast, suppose the following inequality appears:

$$X_2 \geq 10 \tag{12.10}$$

This equation states that X_2 must equal *at least* 10 kg/h. In a manner similar to the foregoing illustration, a *negative* slack variable X_5 is introduced, so that Eq. (12.10) becomes

$$X_2 - X_5 = 10 \tag{12.11}$$

The slack variable X_5 must meet the nonnegativity restriction, of course. Thus, a method is available of converting inequalities to equalities and satisfying the form of the standard LP constraint equations.

12.4 Quality Constraints

Another type of constraint encountered in LP work is the *quality* constraint. Normally, some type of specification or quality of product must be met. As an example, let us consider the blending of gasoline components to meet a certain octane number. The assumption of linear blending will be made so that the following relation holds:

$$Q = \frac{X_1 Q_1 + X_2 Q_2 + X_3 Q_3}{X_1 + X_2 + X_3} \tag{12.12}$$

where X_i = quantity of blending stock i, bbl
$\quad\quad Q_i$ = quality of ith blending stock, octane numbers/bbl

Equation (12.12) then represents the octane number of a three-component mixture. In general,

$$Q = \frac{\Sigma X_i Q_i}{\Sigma X_i} \tag{12.13}$$

Example 12.1 Three components with octane numbers 108, 93, and 100 are to be linearly blended such that the final octane number of the mixture is at least 98. Express this condition as a standard equation in an LP model.

SOLUTION The final octane number of the mixture can be obtained from Eq. (12.12), and it must be a minimum of 98. That is,

$$\frac{108X_1 + 93X_2 + 100X_3}{X_1 + X_2 + X_3} \geq 98$$

or
$$108X_1 + 93X_2 + 100X_3 \geq 98X_1 + 98X_2 + 98X_3$$

or
$$10X_1 - 5X_2 + 2X_3 \geq 0$$

Introducing the slack variable X_4, the required relationship is

$$10X_1 - 5X_2 + 2X_3 - X_4 = 0$$

which is in standard LP form and can be used as one of the equations in the model to represent the quality constraint on octane blending.

12.5 Method of Solution

Refer to Fig. 12.2 and identify the various inputs to a standard LP problem. The fixed inputs would be all the constants in the problem, A_{ij}, B_i, and C_j; the variables would be the X_j values. The model within the black box would be all the constraint equations, given by Eq. (12.7). The output from the black box is represented by the objective function, Eq. (12.3). Figure 12.2 is a representation of the standard LP problem.

The formulation of the problem, the variables, the constraint equations, and the objective function have all been discussed. The problem of finding the optimal

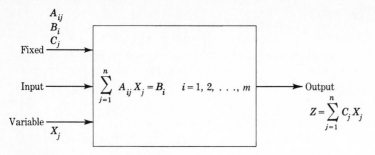

Figure 12.2 A linear-programming model.

solution remains. Examine the constraint Eqs. (12.7). Where the number of variables n is equal to the number of equations m, one can solve this set of independent linear equations. One unique solution will be obtained. There is no problem in optimization here because there is only one solution. The values obtained for the variables from the constraint equations are substituted in the equation for the objective function, and the solution is obtained immediately. It is when there are more variables than equations, i.e., $n > m$, that a problem arises. Now, there is not one unique solution but an infinite number of solutions. From all these solutions, one wishes to find the one which optimizes the objective function.

The method used is to set $n - m$ variables equal to zero, i.e., the excess number of variables. Then there will be m variables left and m equations to solve. For example, if $n = 23$ variables and $m = 10$ equations, set $23 - 10$, or 13, variables equal to zero and solve the 10 equations for the 10 unknowns. The object is to determine which variables to set equal to zero. By examining every possible way of selecting $n - m$ variables from n variables, setting these equal to zero, and solving for the m remaining variables, it can be shown that the optimum solution lies among these many solutions [3].

The $n - m$ variables set equal to zero are called *nonbasic variables;* the remaining m variables are known as *basic variables,* and the collection of them is called a *basis.* The solution of the equations involving the basic variables is termed a *basic solution.* If the values of the basic variables are nonnegative and do not violate any of the constraint equations, the solution is called a *basic feasible solution.* The optimal solution is to be found among the basic feasible solutions. Therefore, once a basic feasible solution is obtained, the object is to find a more optimal solution. This process is continued until a point is reached at which no further improvement is possible.

Example 12.2 A production facility manufactures two products, A and B. Each product must pass through a bank of type 1 machines and a bank of type 2 machines serially, as shown in Fig. 12.3. Each bank consists of many machines of its particular type. Product A requires 2 h on type 1 machines and 1 h on a type 2 machine, while product B requires 1 h on type 1 and 4 h on type 2 machines. The total time available per week on machines of type 1 is 6000 h, while the time available for type 2 machines is 10,000 h. The net profit is $5 per unit of product B and $3.5 per unit of product A. Find the optimal solution, the production schedule which will maximize profit per week.

Figure 12.3 A model for Example 12.2.

SOLUTION First, present all the data in tabular form.

Machine type	Time required, h		Maximum h/week
	Product A	Product B	
1	2	1	6,000
2	1	4	10,000
	Profit, $ per unit		
	3.5	5.0	

The first task is to represent all the data in the form of standard LP equations. Let X_1 be the quantity of product A produced per week, and X_2 correspondingly will be the quantity of product B produced per week. For each unit of product A, $2X_1$ h will be required on type 1 machines; similarly, X_2 hours will be required on type 1 machines for product B. The total time required on type 1 machines is $2X_1 + X_2$, and the maximum weekly time is 6000 h; hence

$$2X_1 + X_2 \leq 6000 \tag{12.14}$$

which becomes, upon introduction of slack variable X_3,

$$2X_1 + X_2 + X_3 = 6000 \tag{12.15}$$

Similarly, for type 2 machine with slack variable X_4,

$$X_1 + 4X_2 + X_4 = 10,000 \tag{12.16}$$

Equations (12.15) and (12.16) are the constraint equations of the LP model. The variables are X_1, X_2, X_3, and X_4. It remains to define the objective function and solve the problem. Based on the unit profit figures, the objective function is

$$Z = 3.5X_1 + 5X_2 \tag{12.17}$$

At this point, it might be instructive to see a geometric interpretation of the problem. Equation (12.14) is plotted in Fig. 12.4 as line *BD*. Any point (X_1, X_2) that lies above this line violates the constraint on maximum available time on machine type 1. The solution must, therefore, lie in the region *OBD*. However, there is another restriction on the solution. Equation (12.16) is plotted as line *AC*. A feasible solution that does not violate the maximum available time on machine 2 must lie in area *OCA*. Therefore, in order to satisfy both constraints of the problem, the solution must lie in area *OBEA*. Thanks to the nonnegativity restriction on the variables, it was necessary to consider only the first quadrant. The problem, therefore, is to seek a point (X_1, X_2) that lies within the region *OBEA* and maximizes the profit function. Equation (12.17) represents the profit function and can be plotted on Fig. 12.4 for various values of Z. Four of these lines have been plotted for illustrative purposes. Any point along the line $Z = 10,500$ indicates a profit of

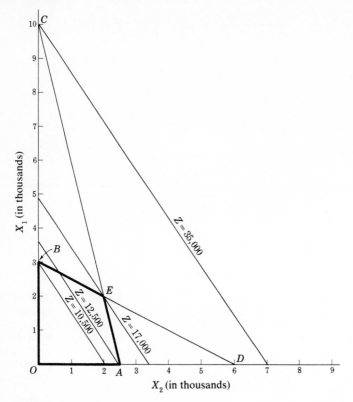

Figure 12.4 A graph for Example 12.2.

$10,500. There are an infinite number of these points along the line. However, the line for $Z = 12,500$ is a line of higher profit and thus indicates a more favorable direction. The line for $Z = 17,000$ has only one point within the feasible region, the intersection of the two constraint lines. Any line to the right of this line, such as $Z = 35,000$, contains points that yield higher profits, but the constraints of the problem are violated and hence these solutions are not feasible. By inspection, therefore, the optimal solution is

$$X_1 = 2000 \text{ units per week of product A}$$

$$X_2 = 2000 \text{ units per week of product B}$$

$$X_3 = 0$$

$$X_4 = 0$$

$$Z = \$17,000/\text{week net profit}$$

In describing the method of solution, it was stated that the excess number of variables should be set equal to zero and the remaining variables solved for. If every possible way of doing this is investigated, one should be able to find the optimal solution. Let us now illustrate this. Equations (12.15) and (12.16) indicate that there are four variables and two equations. Two of these four variables must be set equal to zero and the other two solved for. For example, setting X_1 and X_2 equal to zero and solving for X_3 and X_4, one obtains $X_3 = 6000$ and $X_4 = 10,000$. How many ways are there to set two variables equal from a total of four? This is the well-known problem for the number of combinations of P things taken Q at a time, which is

$$\frac{P!}{Q!\,(P-Q)!} \tag{12.18}$$

which for $P = 4$ and $Q = 2$ gives

$$\frac{4!}{2!\,(4-2)!} = \frac{4(3)(2)(1)}{2(1)(2)(1)} = 6 \text{ combinations}$$

Tabulating all six solutions:

X_1	X_2	X_3	X_4	Z	Fig. 12.4
0	0	6,000	10,000	0	0
0	6,000	0	−14,000	NF	
0	2,500	3,500	0	12,500	A
3,000	0	0	7,000	10,500	B
10,000	0	−14,000	0	NF	
2,000	2,000	0	0	17,000	E

where NF indicates a solution that is nonfeasible. It will be noted that the four feasible solutions correspond to O, A, B, and E of Fig. 12.4. These points are called *extreme points* or *vertices*. In all LP problems these extreme points correspond to basic feasible solutions. In examining the six solutions, the nonfeasible ones were eliminated immediately; of the remaining basic feasible solutions, the optimal solution was found.

12.6 Algebraic Method

The technique used will follow four steps:

1. Select the basic variables.
2. Solve the constraint equations and the objective function for the basic variables in terms of the nonbasic variables.
3. Inspect the objective function equation. Decide which variable will leave the basis and which will enter in order to improve the value of the objective function.
4. Continue until no further improvement is possible.

Example 12.3 Repeat Example 12.2 using the algebraic technique.

Step 1 Consider a feasible mixture that violates no restraints. In this problem it is feasible to have a mixture of X_1 and X_2 only. It may be the worst mixture possible for a maximization, but it is a feasible mixture. As per Sec. 12.5, X_1 and X_2 can be set equal to zero, becoming nonbasic variables, and X_3 and X_4 become the basic variables.

Step 2 Solve the constraint Eqs. (12.15) and (12.16) in terms of the nonbasic variables X_1 and X_2.

$$X_3 = 6000 - 2X_1 - X_2 \tag{12.19}$$

$$X_4 = 10{,}000 - X_1 - 4X_2 \tag{12.20}$$

The objective function was given by Eq. (12.17):

$$Z = 3.5X_1 + 5X_2 \tag{12.21}$$

For $X_1 = X_2 = 0$, the initial solution from above is

$$X_1 = 0 \qquad X_2 = 0 \qquad X_3 = 6000 \qquad X_4 = 10{,}000 \qquad Z = 0$$

This basic feasible solution corresponds to point O in Fig. 12.4.

Step 3 By inspecting Eq. (12.21), it is noted that it is better to introduce X_2 than X_1 since X_2 has the larger coefficient and will improve the objective function more than X_1. How much X_2 should be introduced? Obviously, as much as possible. Equation (12.20) indicates that when X_2 exceeds 2500, X_4 becomes negative. Therefore, the most that X_2 can be is 2500. The variable X_2 enters the *basis;* it becomes a *basic variable,* and X_4 leaves the basis and becomes a *nonbasic* variable. When one variable enters the basis, another must leave to make room for it. The variable that is driven to zero is the one that leaves the basis. Now, repeat steps 1, 2, and 3 for the next iteration.

   Solving the constraint equations and the objective-function equation for X_2 and X_3 in terms of X_1 and X_4 gives, from Eqs. (12.20) and (12.21), respectively,

$$X_2 = 2500 - \tfrac{1}{4}X_1 - \tfrac{1}{4}X_4 \tag{12.22}$$

$$X_3 = 6000 - 2X_1 - (2500 - \tfrac{1}{4}X_1 - \tfrac{1}{4}X_4) = 3500 - \tfrac{7}{4}X_1 + \tfrac{1}{4}X_4 \tag{12.23}$$

and
$$Z = 3.5X_1 + 5(2500 - \tfrac{1}{4}X_1 - \tfrac{1}{4}X_4) = 12{,}500 + \tfrac{9}{4}X_1 - \tfrac{5}{4}x_4 \tag{12.24}$$

For $X_1 = 0$ and $X_4 = 0$, the solution is:

$$X_1 = 0 \qquad X_2 = 2500 \qquad X_3 = 3500 \qquad X_4 = 0 \qquad Z = 12{,}500$$

This basic feasible solution corresponds to point A in Fig. 12.4. The improvement in the objective function has been from 0 to 12,500. Upon examination of Eq. (12.24), it is seen that X_1 should be introduced since it has a positive coefficient. Inspection of Eqs. (12.22) and (12.23) shows that Eq. (12.23) limits the value of X_1 to 2000, at which point X_3 goes to zero. Therefore, X_1 enters the basis and x_3 leaves. Repeating the technique of solving for the basic variables in terms of the nonbasic variables,

$$X_1 = 2000 - \tfrac{4}{7}X_3 + \tfrac{1}{7}X_4 \tag{12.25}$$

$$X_2 = 2000 + \tfrac{1}{7}X_3 - \tfrac{2}{7}X_4 \tag{12.26}$$

$$Z = 17{,}000 - \tfrac{9}{7}X_3 - \tfrac{13}{14}X_4 \tag{12.27}$$

For $X_3 = 0$ and $X_4 = 0$ the solution is:

$$X_1 = 2000 \qquad X_2 = 2000 \qquad X_3 = 0 \qquad X_4 = 0 \qquad Z = 17{,}000$$

This basic solution corresponds to point E in Fig. 12.4. It is the optimal solution because both coefficients of the nonbasic variables in Eq. (12.27) are negative; introducing any of these nonbasic variables would only lower the value of Z and lead to a less optimal solution.

12.7 Simplex Method

The method of solution most widely used is the simplex method developed by Dantzig [2]. It will be seen that this method facilitates the actual calculations of the algebraic method just described. The development will make many references to the equations just derived in the algebraic method.

   The first step requires a representation of the problem in matrix form. Start by listing the variables X_1, X_2, X_3, and X_4 with the respective cost coefficients directly above. Note that the slack variables X_3 and X_4 have zero cost coefficients because they do not appear in the objective function. The original Eqs. (12.15) and (12.16) are now entered, with the right-hand sides given under B_i.

i \ j	3.5 X_1	5 X_2	0 X_3	0 X_4	B_i
	2	1	1	0	6,000
	1	4	0	1	10,000

At this point, the basic variables must be selected. This requires a feasible mixture since the computation must start with a feasible mixture. As explained in Example 12.3, step 1, X_3 and X_4 become the basic variables and are entered in the first column of the tabulation, with the cost coefficients alongside, as shown in Table 12.1. The rest of the manipulation is the solution of the constraint equations and the objective function for the basic variables in terms of the nonbasic variables. The following calculations are required:

$$Z_j = \sum_{i=1}^{m} C_i A_{ij} \tag{12.28}$$

$$\Delta_j = Z_j - C_j \tag{12.29}$$

where C_i refers to the cost coefficient of the ith basic variable. In the first iteration, X_3 and X_4 have been chosen as basic variables. Therefore, $C_1 = C_2 = 0$ since these are the values of the cost coefficients for X_3 and X_4 in the objective function. For the first basic variable X_3, use the first line of the matrix for $i = 1$ and read

$$A_{11} = 2 \qquad A_{12} = 1 \qquad A_{13} = 1 \qquad A_{14} = 0$$

For the second basic variable X_4, use the second line of the matrix for $i = 2$ and read

$$A_{21} = 1 \qquad A_{22} = 4 \qquad A_{23} = 0 \qquad A_{24} = 1$$

Table 12.1 First tableau of the simplex method

		3.5 X_1	5 X_2	0 X_3	0 X_4	B_i	Corresponding equation	
X_3	0	2	1	1	0	6,000	(12.15) $\frac{6,000}{1} = 6,000$	
X_4	0	1	4	0	1	10,000	(12.16) $\frac{10,000}{4} = 2,500$ ← Smallest positive
Z_j		0	0	0	0	0		
Δ_j		-3.5	-5	0	0			

Largest
negative

Therefore

$$Z_1 = C_1 A_{11} + C_2 A_{21} = 0(2) + 0(1) = 0$$

$$Z_2 = C_1 A_{12} + C_2 A_{22} = 0(1) + 0(4) = 0$$

$$Z_3 = C_1 A_{13} + C_2 A_{23} = 0(1) + 0(0) = 0$$

$$Z_4 = C_1 A_{14} + C_2 A_{24} = 0(0) + 0(1) = 0$$

The Δj values can now be calculated.

$$\Delta_1 = Z_1 - C_1 = 0 - 3.5 = -3.5$$

$$\Delta_2 = Z_2 - C_2 = 0 - 5.0 = -5.0$$

$$\Delta_3 = Z_3 - C_3 = 0 - 0 = 0$$

$$\Delta_4 = Z_4 - C_4 = 0 - 0 = 0$$

Note that the Δj values are the coefficients of the nonbasic variables with the sign reversed.

These values are entered in the appropriate spaces, giving the first tableau in Table 12.1. The value of Z, the objective function, is also listed in the lower right-hand corner. It is obtained from the relationship

$$Z = \sum_{i=1}^{m} C_i B_i \tag{12.30}$$

The bottom row is examined for the most negative coefficient. This is seen to be under column X_2, which indicates that it is desirable to introduce this variable into the basis. Each right-hand side is divided by the element in the X_2 column to determine which is limiting. It is that calculation which gives the smallest *positive* number. For this example,

$$\frac{6000}{1} = 6000$$

$$\frac{10,000}{4} = 2500$$

The second row controls since X_4 cannot be larger than 2500 (lest X_4 become a negative number). Hence, X_2 enters the basis and X_4 leaves. Element $A_{22} = 4$ is called the *pivot element*. The variable that enters the basis is on the *pivot column*, and the variable leaving lies on the *pivot row*. Their intersection is the pivot element.

The method now prepares a new tableau without violating any restraint. The procedure is based on a matrix manipulation in which each row in the first tableau is an algebraic equation, and algebraic manipulation is permissible.

The old pivot row is operated on to make the pivot element unity. In this case, row 2 of tableau 1 is divided by 4, giving

X_1	X_2	X_3	X_4	B_i	
$\frac{1}{4}$	1	0	$\frac{1}{4}$	2500	(12.31)

which becomes row 2 of tableau 2.

Next, each of the other rows (there is only one other row here) is manipulated so as to make the element in the pivot column equal to zero. Proceed exactly as follows. Write the row. Below it, write Eq. (12.31) multiplied by a factor such that addition will give zero in the pivot column. Here, multiply Eq. (12.31) by minus unity, giving

X_1	X_2	X_3	X_4	B_i
2	1	1	0	6000
$-\frac{1}{4}$	-1	0	$-\frac{1}{4}$	-2500
$\frac{7}{4}$	0	1	$-\frac{1}{4}$	3500

which becomes row 1 in the second tableau, Table 12.2. Note also that X_2 with a value of 5 has replaced X_4.

The Z_j and Δ_j calculations are then carried out, with the results given in Table 12.2.

The final iteration is carried out in the same way. Briefly, in tableau 2 the first column is the pivot column and the first row is the pivot row, with $\frac{7}{4}$ the pivot element. Multiplying the first row by $\frac{4}{7}$ makes the pivot element unity and gives

X_1	X_2	X_3	X_4	B_i	
1	0	$\frac{4}{7}$	$-\frac{1}{7}$	2000	(12.32)

which becomes row 1 of the new tableau. Write row 2 and below it Eq. (12.32) multiplied by $-\frac{1}{4}$, giving

X_1	X_2	X_3	X_4	B_i
$\frac{1}{4}$	1	0	$\frac{1}{4}$	2500
$-\frac{1}{4}$	0	$-\frac{1}{7}$	$\frac{1}{28}$	-500
0	1	$-\frac{1}{7}$	$\frac{2}{7}$	2000

which is row 2 for the third tableau shown completed in Table 12.3. The bottom row has no negative coefficients, and hence no further improvement is possible. The best mixture is $X_1 = X_2 = 2000$ for $Z = \$17,000$.

Table 12.2 Second tableau of the simplex method

		3.5	5	0	0		Corresponding
		X_1	X_2	X_3	X_4	B_i	equation
X_3	0	$\frac{7}{4}$	0	1	$-\frac{1}{4}$	3,500	(12.23)
X_2	5	$\frac{1}{4}$	1	0	$\frac{1}{4}$	2,500	(12.22)
Z_j		$\frac{5}{4}$	5	0	$\frac{5}{4}$	12,500	
Δ_j		$-\frac{9}{4}$	0	0	$\frac{5}{4}$		

Table 12.3 Third tableau of the simplex method

		3.5 X_1	5 X_2	0 X_3	0 X_4	B_i	Corresponding equation
X_1	3.5	1	0	$\frac{4}{7}$	$-\frac{1}{7}$	2,000	(12.25)
X_2	5	0	1	$-\frac{1}{7}$	$\frac{2}{7}$	2,000	(12.26)
Z_j		3.5	5	$\frac{9}{7}$	$\frac{13}{14}$	17,000	
Δ_j		0	0	$\frac{9}{7}$	$\frac{13}{14}$		

12.8 Applications

The previous development of the simplex method was based on maximization. However, it can be applied to a minimization problem by changing the sign of the objective function and treating the problem as one in maximization. (This is equivalent to changing the signs of all cost coefficients from positive to negative.)

Computer programs are available which can accommodate thousands of restraints and variables. In practice, an LP problem requires setting up all the restraint equations to realistically represent the simulation. Restraints may arise from limited resources (raw materials), equipment capacities, quality specifications on products, physical constraints (e.g., freezing of water at 0°C to prevent pumping), etc. An implied constraint of nonnegativity is assumed for all the variables so that all variables will have values of 0.0 or positive numbers.

Once all the restraint equations have been formulated, a mathematical model is available. The LP computer program can now be applied. It will find a feasible solution that will satisfy all the constraints. The procedure will then be to improve this initial solution to obtain the optimal solution using the objective function as the criterion of optimality. As a by-product of the final solution, a sensitivity analysis is also obtained.

DYNAMIC PROGRAMMING

12.9 An Allocation Example

Dynamic programming (DP) is an optimization technique that is especially applicable to the solution of multistage problems. *Programming* is used in the same context as in linear programming, i.e., the planning of activities. Bellman introduced the concept of dynamic programming through the principle of optimality. "An optimal policy has the property that whatever the initial state and initial decision are, the remaining decisions must constitute an optimal policy with regard to the state resulting from the first decision."†

Consider the following optimization problem, consisting of three black boxes,

†R. E. Bellman, *Dynamic Programming,* Princeton University Press, Princeton, N.J., 1957, p. 83.

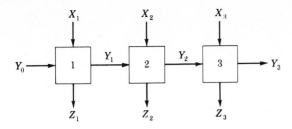

Figure 12.5 Model for a multistage problem.

shown in Fig. 12.5. The output Y_3 from black box 3 is related to the input Y_2 and the *decision variable* X_3. The profit function Z_3 from black box 3 is also related to Y_2 and X_3. Once Y_2 is known, the optimization problem is to find the best value of X_3. The path taken to arrive at point Y_2 is no longer important. The object is to do the best possible from this point to Y_3. If the procedure is repeated for all possible values of Y_2, then the optimal values of X_3 corresponding to every possible value of Y_2 become known.

Now, consider black box 2. Assume that Y_1 is known and the best value of X_2 must be chosen. By selecting a value of X_2, Y_2 can be calculated from the model for black box 2, as well as the profit function Z_2. Once Y_2 is calculated, X_3 has already been determined, as well as Z_3 from the study made on black box 3. It will be assumed that the profit functions are additive so that the total profit function at this point is $Z_2 + Z_3$. Various values of X_2 are tried for an assumed value of Y_1 until that value is found which optimizes the objective function $Z_2 + Z_3$. We therefore have found the best values of X_2 and X_3 for a given Y_1. This process is repeated for all possible values of Y_1. We are now in a position to specify the optimal policy for any Y_1. The quantities Y_1 and Y_2 are called *state variables*, being the outputs from one stage and the inputs to another stage. Then, for any state variables we have determined the best decision variables. The process is continued until Y_0 is reached. This is the input variable, and it is assumed that this quantity is known. The optimal policy can actually be determined for any input. The technique of dynamic programming can best be illustrated by examples.

Example 12.4 A decision must be made with respect to the allocation of a fixed sum of money, $4 million. Three types of research and development are shown graphically in Figs. 12.6, 12.7, and 12.8, respectively, and in Table 12.4. In project A, returns do not materialize until appreciable money is invested; in project B, the law of diminishing returns applies; and in project C, the return is linear with respect to the investment.

If $4 million is available for the three projects, what is the optimal way to allocate these funds assuming integral million dollars investments in each project?

SOLUTION Consider projects B and C only. Table 12.5 gives all possible combinations for investing $4 million or less among the two projects. A summary of the optimal conditions for projects B and C together is given in Table 12.6. The table was prepared from the optimum conditions found in Table 12.5. Thus if $3 million is the total investment available for projects A and B, the maximum profit, $5.05 million, is obtained by investing $2 million in project A and $1 million in project C. It should be noted that the 14 calculations of Table 12.5 can now be replaced by the five calculations of Table 12.6.

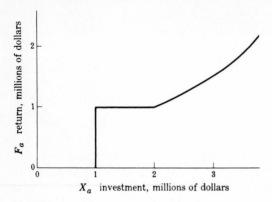

Figure 12.6 Project A.

Projects A, B, and C can now be considered together, but this requires consideration only of project A in conjunction with the optimum policies for projects B and C together. A tabulation is given in Table 12.7. In the tabulation the investment in projects A, B, and C all together is $4 million, inasmuch as there are no other projects to invest in.

The optimal policy is to invest $0 million in project A, and $4 million in projects B and C together, which in turn should be $2 million in project B and $2 million in project C for a maximum total profit of $6.10 million.

12.10 A Transportation Example

Example 12.5 A sales representative must travel from city A to city B. The territory is divided into three districts, and she is required to visit at least one city in each district. Travel costs between cities are indicated in Fig. 12.9. The object is to meet the objective of visiting at least one city in each district at the minimum total travel cost. Use the technique of dynamic programming to solve this problem.

SOLUTION The solution is given in Tables 12.8 to 12.13, obtained as follows. Consider destination city B. There are three routes from district 1 to city B: J–B, K–B, L–B. Having arrived in district 1, therefore, there is no decision to be made. Only one route is available from any city in district 1 to city B as given in Table 12.8.

Now go back one stage, one district, and examine the situation from district 2 to city B. All the combinations are given in Table 12.9 and the optimal routes in Table 12.10. Thus if the salesperson is in city H, her best route is H–K–B and she need not consider other routes originating from H.

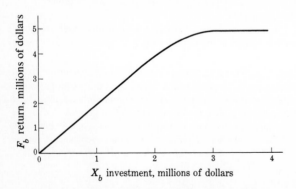

Figure 12.7 Project B.

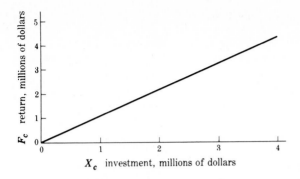

Figure 12.8 Project C.

Next go back another stage to district 3. All the *pertinent* combinations are given in Table 12.11 prepared from Table 12.10 and the costs from district 3 to district 2. Table 12.12 gives the optimal routes from district 3 to city B.

Finally, Table 12.13 gives all the pertinent combinations from city A to city B. The optimal path is route A–D–I–J–B at a cost of 14.

There may of course be ties, so that more than one optimal path can exist. The problem is subject to many variations. If no path exists between two cities, it can be included by placing a very high cost on the route so that it will eliminate itself but not break up the systematic formulation of the solution.

12.11 Pros and Cons of Dynamic Programming

In the dynamic programming examples discussed, it has been assumed tacitly that the manner in which a stage has been reached has no influence on the stages ahead. The existence and importance of this limitation should be appreciated. Thus in Example 12.5 the road ahead is not influenced by the road behind, and use of the optimality principle was very effective in reducing the number of calculations. However, the later stages can be influenced by the manner in which the earlier stages have occurred. For example, suppose that in Example 12.5 the fatigue of the sales representative must be considered and that her fatigue contributes to the cost of travel. That is, the more tired she becomes, the more she will require extra service, and the greater the travel cost. It then becomes necessary to consider the degree of fatigue of the salesperson at each city, which in turn depends upon the manner in which the previous steps were executed.

Table 12.4 Investments for Example 12.4

Project A		Project B		Project C	
Investment X_a, M\$	Return F_a, M\$	Investment X_b, M\$	Return F_b, M\$	Investment X_c, M\$	Return F_c, M\$
0	0	0	0	0	0
1.0	1.0	1.0	2.0	1.0	1.05
2.0	1.0	2.0	4.0	2.0	2.10
3.0	1.5	3.0	5.0	3.0	3.15
4.0	2.5	4.0	5.0	4.0	4.20

Table 12.5 Investment in projects B and C only

Total investment, M$	Invest project B, M$	Return F_b, from B, M$	Invest project C, M$	Return F_c, from C, M$	Return from B and C, M$
4	4.0	5.0	0.0	0.0	5.0
	3.0	5.0	1.0	1.05	6.05
	2.0	4.0	2.0	2.10	6.10
	1.0	2.0	3.0	3.15	5.15
	0.0	0.0	4.0	4.20	4.20
3	3.0	5.0	0.0	0.0	5.0
	2.0	4.0	1.0	1.05	5.05
	1.0	2.0	2.0	2.10	4.10
	0.0	0.0	3.0	3.15	3.15
2	2.0	4.0	0.0	0.0	4.0
	1.0	2.0	1.0	1.05	3.05
	0.0	0.0	2.0	2.10	2.10
1	1.0	2.0	0.0	0.0	2.0
	0.0	0.0	1.0	1.05	1.05

Some of the pros and cons of dynamic programming are as follows:

1. At any stage, the optimization technique to use is left up to the user. The most efficient method can therefore be used for any particular problem.
2. The problem of constraints usually hinders most optimization techniques. In dynamic programming, however, the constraints are actually helpful because they limit the range to be investigated.
3. Discontinuities in the objective function can be handled by dynamic programming because no analytical function need be used; tables and curves are adequate.
4. Dynamic programming greatly reduces the number of optimization problems to be examined. In general, where there are N stages and K decisions to be made at each stage, the overall optimization problem involves K^N possible

Table 12.6 Optimum profit from projects B and C only

Total investment in B and C, M$	Invest in project B, M$	Invest in project C, M$	Optimal profit from B + C, M$
0	0	0	0
1.0	1.0	0	2.0
2.0	2.0	0	4.0
3.0	2.0	1.0	5.05
4.0	2.0	2.0	6.10

Table 12.7 A 4 M$ investment in projects A, B, and C

Total investment, M$	Invest project A, M$	Return from A, F_a, M$	Invest B + C, M$	Optimal return B + C, M$	Return from A + B + C, M$
4	0.0	0.0	4.0	6.10	6.10
	1.0	1.0	3.0	5.05	6.05
	2.0	1.0	2.0	4.0	5.0
	3.0	1.5	1.0	2.0	3.5
	4.0	2.5	0.0	0.0	2.5

solutions. For a five-stage problem with three decisions at each stage, this means 3^5, or 243, possible combinations. By dynamic programming, only 3×5, or 15, one-stage optimization problems would have to be solved. The main disadvantage of dynamic programming is the problem of dimensionality (many variables), which arises in the optimization at each stage. For example, when many decisions are to be made at one time, the single-stage optimization itself may become quite complex.

GEOMETRIC PROGRAMMING

12.12 The Format

Geometric programming (GP) is an optimization technique that can handle a certain form of functions with fractional and negative exponents in an efficient man-

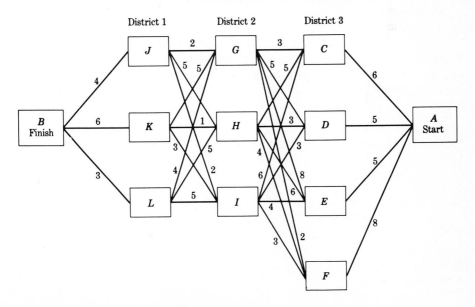

Figure 12.9 Travel costs between cities.

Table 12.8 District 1 to city B

Route	Cost
J–B	4
K–B	6
L–B	3

Table 12.9 District 2 to city B

Route	Cost
G–J–B	$2 + 4 = 6$
G–K–B	$5 + 6 = 11$
G–L–B	$4 + 3 = 7$
H–J–B	$5 + 4 = 9$
H–K–B	$1 + 6 = 7$
H–L–B	$5 + 3 = 8$
I–J–B	$2 + 4 = 6$
I–K–B	$3 + 6 = 9$
I–L–B	$5 + 3 = 8$

Table 12.10 Optimal routes, district 2 to city B

Route	Cost
G–J–B	6
H–K–B	7
I–J–B	6

ner. The technique was introduced by Duffin, Paterson, and Zener [R1] and reviewed by a number of others [2, 3, 4].

A polynomial of the form

$$g = u_1 + u_2 + \cdots + u_n \tag{12.33}$$

is known as a *posynomial* when each term on the right has the form

$$u_i = c_i X_1^{a_1} X_2^{a_2} X_3^{a_3} \ldots X_n^{a_n}$$

where c_i must be a positive constant and the exponents can be positive, negative, or zero. The variables $X_1 \ldots X_n$ are also positive.

Example 12.6 Which of the following are permissible terms in a posynomial?
(a) $4X_1^2 X_2^{-1.5}$
(b) $-4X_1^2 X_2^3$

Table 12.11 District 3 to city B

Route	Cost
C–(G–J–B)	$3 + 6 = 9$
C–(H–K–B)	$5 + 7 = 12$
C–(I–J–B)	$6 + 6 = 12$
D–(G–J–B)	$5 + 6 = 11$
D–(H–K–B)	$3 + 7 = 10$
D–(I–J–B)	$3 + 6 = 9$
E–(G–J–B)	$8 + 6 = 14$
E–(H–K–B)	$6 + 7 = 13$
E–(I–J–B)	$4 + 6 = 10$
F–(G–J–B)	$2 + 6 = 8$
F–(H–K–B)	$4 + 7 = 11$
F–(I–J–B)	$3 + 6 = 9$

Table 12.12 Optimal routes, district 3 to city B

Route	Cost
C–G–J–B	9
D–I–J–B	9
E–I–J–B	10
F–G–J–B	8

Table 12.13 City A to B

Route	Cost
A–(C–G–J–B)	$6 + 9 = 15$
A–(D–I–J–B)	$5 + 9 = 14$
A–(E–I–J–B)	$5 + 10 = 15$
A–(F–G–J–B)	$8 + 8 = 16$

(c) $4X_1 \sqrt{2}$
(d) $4X_1 \log X_2$
(e) 12

SOLUTION Only (a), (c), and (e) are permissible; (b) is negative, and (d) does not have the correct form.

12.13 A Simple Application

If the function to be optimized is a posynomial, geometric programming is an efficient method for optimization. Only the application of the method is described here.

Example 12.7 Reconsider the two-variable optimization given by Eq. (10.17).

$$ Z = 1265S^{0.5} + 4 \times 10^8 P^{-1}S^{-0.5} + 137.2P \tag{12.34} $$

Let α denote a weighting factor and consider the following expression for v, known as the *dual function:*

$$ v = \left(\frac{1265S^{0.5}}{\alpha_1}\right)^{\alpha_1} \left(\frac{4 \times 10^8 P^{-1}S^{-0.5}}{\alpha_2}\right)^{\alpha_2} \left(\frac{137.2P}{\alpha_3}\right)^{\alpha_3} \tag{12.35} $$

Now, find relationships between the α factors which will make the expression dimensionless.

SOLUTION This dimensionless expression is obtained when both S and P are reduced to the zero power. For the terms in S,

$$ S^{0.5\alpha_1}S^{-0.5\alpha_2} = S^0 $$

$$ 0.5\alpha_1 - 0.5\alpha_2 = 0 $$

and similarly, from the terms in P,

$$ -\alpha_1 + \alpha_3 = 0 $$

Since the α factors are only weighting factors, one α can be chosen arbitrarily, but it is most convenient to set their sum equal to unity. Thus there are three relationships:

$$ \alpha_1 + \alpha_2 + \alpha_3 = 1 \tag{12.36} $$

$$ 0.5\alpha_1 - 0.5\alpha_2 \quad\quad = 0 \tag{12.37} $$

$$ -\alpha_1 \quad\quad + \alpha_3 = 0 \tag{12.38} $$

The first is known as the *normality condition;* the others are called the *orthogonality conditions.*
The three equations are easily solved:

$$ \alpha_1 = \alpha_2 = \alpha_3 = \tfrac{1}{3} \tag{12.39} $$

but in general the weighting factors will not be equal.
With α_1, α_2, and α_3 now known and with the dual function dimensionless in S and P, Eq. (12.35) becomes

$$ v = \left(\frac{1265}{\alpha_1}\right)^{\alpha_1}\left(\frac{4 \times 10^8}{\alpha_2}\right)^{\alpha_2}\left(\frac{137.2}{\alpha_3}\right)^{\alpha_3} = \$123{,}298 $$

Here v is the optimum sought, and it checks the value given previously in Table 10.4. The optimum value has been obtained without knowing the individual values for S and P that lead to it.

The optimum values for S and P at the optimum are easily obtained from v, the dual function, and the weighting α factors:

$$1265S^{0.5} = \alpha_1 v = \tfrac{1}{3}(123{,}298) = 41{,}099$$

$$S = 1056$$

$$137.2P = \alpha_3 v = \tfrac{1}{3}(123{,}298) = 41{,}099$$

$$P = 300$$

which check the optimum values for S and P found previously in Example 10.5.

12.14 Some Comments on Geometric Programming

Equation (12.33) is known as the *primal function* in geometric programming. The basis of geometric programming work is the inequality which states that the arithmetic mean is equal to or greater than the geometric mean. The dual function, the optimum, can be calculated without finding the values of the problem variables, a unique feature of geometric programming.

The degree of difficulty D in geometric programming is defined as the total number of terms T in the posynomial minus one more than the number of independent variables N:

$$D = T - (N + 1)$$

In Example 12.7 there are three terms and two variables, and therefore zero degree of difficulty.

Geometric programming has been expanded to overcome the limitations of a simple problem such as that expressed by Eq. (12.34). The removal of each limitation is obtained at the expense of increased mathematical difficulty. Restraints can be included using Lagrange's method of undetermined multipliers, and even inequality constraints can be accommodated. Also, the severe requirement that all terms in the objective function be positive can be circumvented.

The advantages of geometric programming are that functions with fractional and negative exponents can be handled efficiently, the value of each variable need not be found, and constraints may be incorporated in the solution techniques.

The disadvantages are that if the degree of difficulty is more than zero, the problem formulated can be more complicated than the original problem, all variables must be continuous or at least treated as such, and it is not applicable to large, complex design problems since the equations will not necessarily be in the proper format.

NOMENCLATURE

A_{ij}	Coefficient in a restraint equation ith row, jth column, LP
a_n	Exponent in a GP term
B_i	Right side of the ith restraint equation, LP
c_i	Constant in a GP term

C_i	Unit cost associated with a basic variable in the ith row, LP
C_j	Unit cost associated with X_j, LP
D	Degree of difficulty, GP
DP	Dynamic programming
F	Objective function
F_a	Return from project A, DP
g	A posynomial, GP
GP	Geometric programming
i	The ith row
j	The jth column
LP	Linear programming
N	The Nth stage, DP
N	Number of variables, GP
Q	Quality coefficient
T	Number of terms in a posynomial, GP
u	A term in a posynomial
v	Dual function, GP
X_a	Investment in project A, DP
X_j	Variable input, LP
X_N	Decision variable associated with the Nth stage, DP
Y_N	Output from the Nth stage, input to the $(N + 1)$ stage, DP
Z	Objective function
Z_g	$\sum_{i=1}^{m} C_i A_{ij}$, LP
α	Weighting factor, GP
Δ_j	$Z_j - C_i$, LP

PROBLEMS

12.1 Verify all terms in the first and second rows of the second and third tableaus in Tables 12.2 and 12.3.

12.2 Verify the Z_j and Δ_j values in Table 12.3.

12.3 Set up the equations for the following diet problem given by Garvin [3]. One unit of hog's liver contains 1 unit of carbohydrates, 3 units of vitamins, 3 units of proteins, and costs 50 units. Suppose 1 unit of castor oil contains 3, 4, and 1 units of these, respectively, and costs 25 units. Minimum daily requirements are 8 units of carbohydrates, 19 units of vitamins, and 7 units of proteins. Obtain the minimum-cost diet subject to the constraints of the problem.

12.4 Express Prob. 12.3 geometrically, indicating the feasible region.

12.5 Solve Prob. 12.3 by the enumeration of all solutions.

12.6 Solve Prob. 12.3 by the algebraic method.

12.7 Verify the equations in the algebraic method of Prob. 12.6 by solving the problem using the simplex method.

12.8 A student is given an examination and is to solve no more than 100 problems, with the problems divided into three groups.

First group—moderately difficult (X_1), worth five points each, with an average time of 3 min per problem.

Second group—least difficult (X_2), worth four points each, with an average time of 2 min per problem.

Third group—most difficult (X_3), worth six points each, with an average time of 4 min per problem.

The total time available to the student is 3.5 h. The problems in groups 1 and 2 are the most tedious, and the student can take only 2.5 h of this type of work. How many problems of each type should the student solve to obtain a maximum grade on the examination?

12.9 An advertising budget is to be prepared which aims to reach 32 million potential customers, of whom 21 million are required to have incomes over $18,000/year. The following data are given:

Medium	Cost, $/thousand	Audience, million per ad	Audience with income over $18,000, millions
Magazine	28	1	0.6
Television	400	9	2
Radio	20	0.8	0.7

Set up the model and find the optimal solution by the enumeration method.

12.10 A small machine shop has capabilities in turning, milling, drilling, as follows:

Turning,   140 h/day
Milling,   80 h/day
Drilling,   120 h/day

Two products, A and B, are made, yielding a net profit of $2.20 and $3.80 per item, respectively. Each will require the following amount of machine time in hours:

	A	B
Turning	2.4	3.2
Milling	0.0	2.6
Drilling	4.1	0.0

Use the simplex method and find the policy that gives the maximum profit per day.

12.11 A customer needs 1000 kg of a mixture made up from three different raw materials with the following costs:

X_1,   2 $/kg
X_2,   3 $/kg
X_3,   4 $/kg

The mixture must not contain more than 400 kg of X_1, at least 200 kg of X_2, and at least 100 kg of X_3. Use the simplex method and find the least-cost mixture that will satisfy all the requirements.

12.12 Crude A is limited to 15,000 bbl/day and if sent to process C gives a profit of $1.10/bbl. If sent to process D, it gives a profit of $1.20/bbl.

Crude B is limited to 18,000 bbl/day and if sent to process C gives a profit of $1.60/bbl. If sent to process D, it gives a profit of $1.40/bbl.

Process C has a limit of 12,000 bbl/day and process D a limit of 20,000 bbl/day.

Find the policy that maximizes the profit per day.

12.13 Refer to Example 12.4. What is the optimal policy if 3 M$ instead of 4 M$ were available for all three projects?

12.14 Repeat Prob. 12.13 for (a) 2 M$ and (b) 1 M$. Plot the optimal return against available funds.

12.15 Repeat Example 12.5, traveling from city B to city A.

12.16 Describe some application of DP with which you are familiar, e.g., investment decisions, staged engineering calculations, or network problems.

12.17 Find the number of each of three items to include in a package so that the value of the package will be a maximum subject to the limitation that the mass must not exceed 6 kg.

Item	Mass, kg	Value, $
1	3	6
2	2	4
3	4	7

12.18 The following type of industrial operation is being evaluated:

$$\longrightarrow \text{Unit 1} \longrightarrow \text{unit 2} \longrightarrow \text{unit 3} \longrightarrow$$

Two types of units, A and B, costing $10,000 and $20,000, respectively, are being considered for unit 1. The operating costs for unit 1 are:

Efficiency, %	Operating cost, $/year	
	A	B
60	3,000	2,000
80	6,000	5,000
95	12,000	8,000

The operation of unit 2 is related to the efficiency of unit 1. One type of unit is being considered which may operate at one of three different temperatures, 370°C, 400°C, or 430°C, with the following data:

% efficiency of unit 1	Cost, $/year		
	370°C	400°C	430°C
60	2,000	10,000	16,000
80	1,800	8,000	12,000
95	1,000	6,000	10,000
Fixed cost, $	5,000	20,000	20,000

Unit 3 is related to the temperature of unit 2:

Temp. °C, unit 2	Income unit 3, $ for 5 years
370	$530,000
400	630,000
430	680,000

Assume a 5-year basis for costs. Find the optimal operation using DP.

12.19 Three types of purification units are available, as indicated in the following scheme:

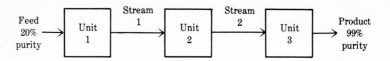

Each of the three units may be operated in any one of three modes of operation, A, B, and C, with the following data:

Unit 1		
Mode of operation	Purity stream 1, %	Cost, $/year
A_1	99	150,000
B_1	60	70,000
C_1	50	50,000

Unit 2			
Purity stream 1, %	Operation unit 2	Purity stream 2, %	Cost, $/year
50	A_2	99	60,000
	B_2	60	10,000
	C_2	50	0
60	A_2	99	40,000
	B_2	60	0
99	A_2	99	0

Unit 3		
Purity stream 2, %	Operation unit 3	Cost, $/year
50	A_3	60,000
60	A_3	40,000
99	A_3	0

It is not required that all three units be used. Find the optimal operation policy.

12.20 The total cost C_t in dollars per year for a piping installation is given by

$$C_t = \frac{A}{D^{4.8}} + BD^{1.5}$$

where D is the diameter of the pipe in centimeters and A and B are constants. The first term represents the annual operating cost, and the second term represents the fixed cost. Find the percentage of optimal total costs that arise from the operating costs and the fixed costs, respectively, using GP

(*a*) For the equation as written
(*b*) For the equation with A doubled
(*c*) For the equation as written, with the exponent on D changed from 4.8 to 5.2
(*d*) For part (*c*) with A doubled

12.21 A cost function with A, B, and C constants is given by

$$Z = AX_1 + \frac{B}{X_1^2 X_2^3} + CX_2^2$$

(*a*) Find a minimum value for Z using GP.
(*b*) Find the value for X_1 and X_2 in terms of A, B, and C at minimum Z.

REFERENCES

Linear programming

1. Charnes, A., and W. W. Cooper: *Management Models and Industrial Application of Linear Programming,* John Wiley & Sons, Inc., New York, 1960.
2. Dantzig, G. B.: *The Simplex Method,* Rand Corporation Report P-891, 1956.
3. Garvin, W. W.: *Introduction to Linear Programming,* McGraw-Hill Book Company, New York, 1960.
4. Gass, S. I.: *Linear Programming: Methods and Applications,* 3d ed., McGraw-Hill Book Company, New York, 1969.
5. Hadley, G.: *Linear Programming,* Addison-Wesley Publishing Company, Inc., Reading, Mass., 1962.
6. Vajda, S.: *Readings in Linear Programming,* John Wiley & Sons, Inc., New York, 1958.

Dynamic programming

1. Bellman, Richard E.: *Dynamic Programming,* Princeton University Press, Princeton, N.J., 1957.
2. Bellman, R., and S. Dreyfus: *Applied Dynamic Programming,* Princeton University Press, Princeton, N.J., 1962.
3. Hadley, G. *Nonlinear and Dynamic Programming,* Addison-Wesley Publishing Company, Inc., Reading, Mass., 1964.
4. Nemhauser, G. L.: *Introduction to Dynamic Programming,* John Wiley & Sons, Inc., New York, 1966.

Geometric programming

1. Duffin, R. J., S. F. Paterson, and C. Zener: *Geometric Programming,* John Wiley & Sons, Inc., New York, 1967.
2. Kermode, R. I.: "Geometric Programming—A Simple, Efficient Optimization Technique," *Chemical Engineering,* vol. 74, 1967, p. 26.
3. LaMonte, R. R., and P. B. Lederman: "The Uses and Limitations of Geometric Programming," *British Chemical Engineering and Process Technology,* vol. 17, no. 1, 1972, p. 34.
4. Wilde, D. J., and C. S. Beightler: *Foundations of Optimization,* 2d ed., Prentice-Hall, Inc., Englewood Cliffs, N.J., 1979.

THIRTEEN

SPECIAL MATHEMATICAL TECHNIQUES

A. H. Boas

UNIVARIABLE SEARCH METHODS

13.1 Search Techniques

Very often, a problem is not defined in specific analytical form. An objective function, such as a maximum or minimum value, may be evaluated, however, by trial. This may entail direct laboratory measurement or experimentation, a computer run, an iterative procedure involving graphs and tables, and so forth. The problem is to select the proper values of input data to optimize a specified objective function by using search techniques. A base point is given or assumed, the objective function is evaluated, and the process is repeated until no further improvement is possible.

After a certain number of experiments are performed, the interval within which the optimum value must lie is a measure of the efficiency.

13.2 Uniform Search

Assume that we are concerned only with one independent variable and that the experimental error of measuring or computing is negligible. Furthermore, the important assumption of unimodality (one optimum value only) is made.

This section develops the uniform-search method in which experimental or trial points are spaced equally within the allowable range.

Example 13.1 A function F is given by the relationship

$$F = 1.3X^2 - 4X + 3 \qquad (13.1)$$

It is known that the function is unimodal and that the range of the answer lies between 0 and 5. Four experiments, or computations, are permitted. Find the minimum value of F.

SOLUTION It should be emphasized that the form of the function F as given by Eq. (13.1) may or may not be known to the analyst. If known, the solution proceeds by computation; if unknown, by experimentation. The method of analysis is the same, however: For each value of X, the corresponding value of F can be determined by some means.

Spacing four experimental points equidistantly within the interval leads to a choice $X = 1$, 2, 3, 4, and experiment gives

X	F
1	0.3
2	0.2
3	2.7
4	7.8

The results are plotted in Fig. 13.1. The minimum may lie between 1 and 2, as indicated by curve A, or between 2 and 3, as indicated by curve B. It must lie between 1 and 3. Originally the range covered five units, and after four experiments the range has been narrowed down to two units, or 40 percent of the original range.

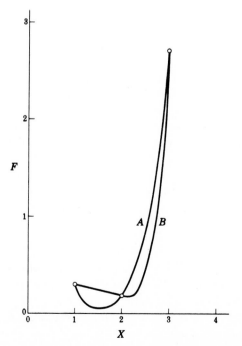

Figure 13.1 The unimodal search of $F = 1.3X^2 - 4X + 3$.

If N is the number of experiments allowed in an entire interval L, there are $N + 1$ intervals and the final answer spans two intervals. If F_r is the fraction of the original interval in which the optimum lies, then

$$F_r = \frac{2}{N + 1} \tag{13.2}$$

For Example 13.1, $N = 4$ and $F_r = 0.40$.

13.3 Uniform Dichotomous Search

In this procedure, experiments are performed in pairs to establish whether the function is increasing or decreasing at a point.

Example 13.2 Repeat Example 13.1 using a dichotomous search.

SOLUTION Four experiments, or two pairs, are allowed. Divide the range 0 to 5 into three equal intervals. Thus, let

$$X = \tfrac{1}{3}(5) = 1.667 \qquad X = \tfrac{2}{3}(5) = 3.333$$

One pair of experiments is used in the neighborhood of $X = 1.667$ to establish the trend of F at that point. Similarly the second pair of experiments establishes the trend at $X = 3.333$. For the latter pair,

X	F
3.30	3.96
3.36	4.24

At $X = 3.333$ the function is increasing; hence the minimum must be below that value of X. The other pair of experiments gives

X	F
1.60	−0.072
1.70	−0.043

so that the function is increasing with increasing X. The minimum must be below $X = 1.667$. The four experiments have established, at best, that the minimum must lie in the range 0 to 1.667, which is one-third of the total range 0 to 5.

The procedure will now be generalized. For N experiments there will be $N/2$ pairs, and the entire interval is divided into $(N/2) + 1$ intervals, each of width $L/(N/2 + 1)$. The optimum is located over the width of one interval, $L/(N/2 + 1)$. Therefore,

$$F_r = \frac{L/(N/2 + 1)}{L} = \frac{2}{N + 2} \tag{13.3}$$

For the example $N = 4$, $F_r = \tfrac{2}{6} = 0.33$.

This method has reduced the original interval to 33 percent and is seen to be superior to a uniform search, which reduced the original interval to 40 percent.

13.4 Sequential Dichotomous Search

In a sequential search, the information available from previous experiments is used before the next one is performed. A sequential dichotomous search involves two experiments run very closely together (closely enough to distinguish between the two outcomes) near the middle of the region under consideration. In Example 13.1, two experiments close to $X = 2.5$ show it to be increasing at $X = 2.5$; therefore the region above $X = 2.5$ is eliminated. Another pair of experiments near the middle of the remaining region at $X = 1.25$ shows that the function is decreasing at $X = 1.25$. The minimum lies between $X = 1.25$ and 2.50, or one-quarter of the range 0 to 5.

Generalizing this procedure, it is seen that each pair of experiments bisects the previous interval. At the end of the first pair of experiments, the remaining interval is one-half of the original interval; after the next pair, it is one-quarter of the original interval. After $N/2$ pairs of experiments, the remaining interval is $\frac{1}{2}^{N/2}$ of the original interval.

$$F_r = \frac{1}{2^{N/2}} \tag{13.4}$$

Referring again to the example, $N = 4$, $F_r = \frac{1}{2}^2 = 0.25$. The original interval has now been reduced to 25 percent, which is the lowest value obtained thus far.

13.5 Fibonacci Search Techniques

It has been shown in the articles by Johnson [13] and Keifer [14] that the most efficient sequential technique to follow for the case of one variable and unimodality is the Fibonacci search. Fibonacci numbers F_i corresponding to an index N are related by the recurring relationship

$$F_{i,N+2} = F_{i,N} + F_{i,N+1}$$

where $F_{i0} = 1$ and $F_{i1} = 1$.

Table 13.1 gives some Fibonacci numbers. Each number is the sum of two previous Fibonacci numbers. The use of Fibonacci numbers is illustrated in the next example.

Example 13.3 Repeat Example 13.1 using a Fibonacci search technique.

SOLUTION First, a pair of experiments is run equidistant from each end of the interval. The distance D_1 is determined from the expression

$$D_1 = \frac{F_{i,N-2}}{F_{i,N}} L \tag{13.5}$$

where $F_{i,N} = N$th Fibonacci number
$N =$ number of experiments
$L =$ interval length

Table 13.1 Some Fibonacci numbers

N	$F_{i,N}$	N	$F_{i,N}$
0	1	8	34
1	1	9	55
2	2	10	89
3	3	11	144
4	5	12	233
5	8	13	377
6	13	14	610
7	21	15	987

$$F_{i,N+2} = F_{i,N} + F_{i,N+1}$$
$$F_{i0} = 1$$
$$F_{i1} = 1$$

For the example, $N = 4$, $L = 5$, $F_{i,N-2} = F_{i2} = 2$, and $F_{i,N} = F_{i4} = 5$. From Eq. (13.5),

$$D_1 = \tfrac{2}{5}(5) = 2$$

Therefore, two experiments are run two units from each end of the interval; i.e.,

$$X = 0 + 2 = 2$$
$$X = 5 - 2 = 3$$

The experiments give

X	F
2	0.200
3	2.700

The region for X greater than 3 is eliminated.

Second, the procedure is repeated for the remaining interval, changing N to $N - 1$ because one experiment less remains. Equation (13.5) becomes

$$D_2 = \frac{F_{i,N-2-1}}{F_{i,N-1}}L = \frac{F_{i1}}{F_{i3}}L = \tfrac{1}{3}(3) = 1$$

The next two experiments should be run 1 unit from each end of the remaining interval.

$$X = 0 + 1 = 1$$
$$X = 3 - 1 = 2$$

However, one of these two new experiments has already been run (at $X = 2$). Hence, only one new experiment is added, at $X = 1$. It will always turn out that one of the previous experiments is a Fibonacci experiment for the next run. Performing the new experiment gives the pair

X	F
1	0.300
2	0.200

The region for X less than 1 is eliminated.

Third, the procedure is repeated again with $L = 3 - 1 = 2$, changing N to $N - 2$. Equation (13.5) becomes

$$D_3 = \frac{F_{i,N-2-2}}{F_{i,N-2}}L = \frac{F_{i0}}{F_{i2}}L = \tfrac{1}{2}(2) = 1$$

This indicates that the last experiment should be run at

$$X = 1 + 1 = 2$$

$$X = 3 - 1 = 2$$

This experiment is run around $X = 2$ to determine in which half of the remaining region the answer lies. Choose $X = 2 \pm \varepsilon$, where ε is a small number but large enough to detect a difference in F. Taking $\varepsilon = 0.05$,

X	F	
2.05	0.263	New experiment
2.00	0.200	Old experiment

The function is increasing in the neighborhood of $X = 2$; thus the minimum is below $X = 2$. The fraction of the original interval is $\tfrac{1}{5} = 0.20$, or 20 percent, which is the lowest obtained. For the general case the result is

$$F_r = \frac{1}{F_{i,N}} \tag{13.6}$$

For the example, $N = 4$, $F_{i4} = 5$, $F_r = \tfrac{1}{5} = 0.20$.

Table 13.1 provides a method of determining the number of experiments needed to obtain a certain value of F_r, the fractional ratio of interval left to the original interval. For example, to narrow the original interval to at least 1 percent,

$$\frac{1}{F_{i,N}} \le 0.01 \qquad \text{or} \qquad F_{i,N} \ge 100$$

At least 11 Fibonacci experiments would have to be performed since F_{i11} is the first Fibonacci number above 100.

13.6 Comparison of Methods

Table 13.2 summarizes the results obtained by the various methods described. As N gets larger, the advantage of the Fibonacci technique over all the others increases.

Table 13.2 Comparison of methods for univariable search techniques

Search technique	Interval left after N experiments / Original interval	For $N = 4$
Uniform	$\dfrac{2}{N + 1}$	0.40
Uniform dichotomous	$\dfrac{2}{N + 2}$	0.33
Sequential dichotomous	$\dfrac{1}{2^{N/2}}$	0.25
Fibonacci	$\dfrac{1}{F_{i,N}}$	0.20

MULTIVARIABLE FUNCTIONS

13.7 One-at-a-time Method

More common and much more complex is the multivariable optimization problem. A review of some widely used methods follows.

Friedman and Savage [11] described the elementary one-at-a-time method. All variables except one are kept constant, and that one is varied to obtain an improvement in the objective function. Referring to Fig. 13.2, an initial search is started, changing Y only (constant X) along the line AB until the lowest value of the objective function is obtained at point C. A new search is started along line CD until point D is reached, and the process is continued until the optimum is

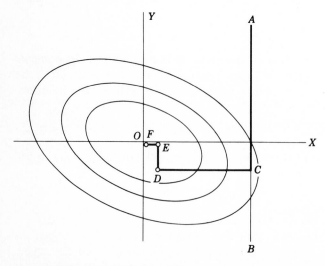

Figure 13.2 The one-at-a-time method.

approached very closely. When the contours come to a sharp point, this method might not work.

Example 13.4 Solve the two-variable problem given by Eq. (10.17).

$$Z = 1265S^{0.5} + 4 \times 10^8 P^{-1}S^{-0.5} + 137.2P$$

minimizing Z by the one-at-a-time method. Start at $P = 250$ and $S = 1000$.

SOLUTION With $P = 250$ and not varied, find the best value of S.

$$\frac{\partial Z}{\partial S} = 0 = \frac{0.5(1265)}{S^{0.5}} - \frac{0.5 \times 4 \times 10^8}{PS^{1.5}}$$

$$S = \frac{3.1621 \times 10^5}{P} = \frac{3.1621 \times 10^5}{250} = 1265$$

Now, for $S = 1265$, find the best value of P.

$$\frac{\partial Z}{\partial P} = 0 = \frac{-4 \times 10^8}{P^2 S^{0.5}} + 137.2$$

$$P = \left(\frac{4 \times 10^8}{137.2S^{0.5}}\right)^{0.5} = \left(\frac{4 \times 10^8}{137.2(1265)^{0.5}}\right)^{0.5} = 286$$

Continue until no further improvement results.

P	S
250 $\longrightarrow$	1265
286 $\longleftarrow$	1265
286 $\longrightarrow$	1106
296 $\longleftarrow$	1106
. . .	. . .
299.5 $\longleftarrow$	1057
299.5 $\longrightarrow$	1056
299.5 $\longleftarrow$	1056

Thus the minimum is at $P = 299.5$ and $S = 1056$, which checks the answer in Sec. 10.14.

13.8 Method of Steepest Ascent (or Descent)

The direction of the gradient is the direction that gives the greatest response of the objective function per unit length of the independent variable. The incremental change in each variable is taken proportional to its partial derivative, which determines the gradient direction. This direction is called the *direction of steepest ascent* (or *descent*).

Example 13.5 Given the following objective function:

$$Z = X^2 + 2Y^2 + XY \tag{13.7}$$

find the minimum value of this function and the values of X and Y at the minimum, starting at $X = 2$, $Y = 2$.

SOLUTION The starting point is represented by M in Fig. 13.3. The partial derivatives must first be calculated and evaluated at the starting point.

$$\frac{\partial Z}{\partial X} = 2X + Y \qquad (13.8)$$

$$\frac{\partial Z}{\partial Y} = 4Y + X \qquad (13.9)$$

At the starting point, these partial derivatives become

$$\frac{\partial Z}{\partial X} = 6 \quad \text{and} \quad \frac{\partial Z}{\partial Y} = 10 \qquad (13.10)$$

Because both partial derivatives are positive, the objective function varies in the same direction as X and Y. Therefore, as this is a minimization problem, both X and Y should be decreased in order to decrease the value of Z. The ratio of the decrease in X to the decrease in Y is taken to be proportional to their respective partial derivatives, that is, $\frac{6}{10}$. An arbitrary step size ΔY of -0.5 is chosen initially; the corresponding ΔX would then be $0.60(-0.50) = -0.30$.

Initially $X_0 = 2$; $Y_0 = 2$; $Z_0 = 16$. After the first step,

$$X_1 = X_0 + \Delta X = 2 - 0.30 = 1.70$$

$$Y_1 = Y_0 + \Delta Y = 2 - 0.50 = 1.50$$

$$Z_1 = 1.70^2 + 2(1.50^2) + 1.70(1.50) = 9.94$$

As long as the objective function is decreasing, the procedure is continued along the same gradient line. Table 13.3 summarizes the results of the calculation.

Referring to Fig. 13.3, point P corresponds to (X_5, Y_5) and point N to (X_6, Y_6). Table 13.3 indicates that at point N the objective function is no longer decreasing. Point P, in fact, represents the last successful calculation. A new gradient direction is now calculated at this point and the entire procedure repeated. At point P, the gradient direction is

$$\frac{\partial Z}{\partial X} = 2X + Y = 2(0.5) - 0.5 = 0.50$$

$$\frac{\partial Z}{\partial Y} = 4Y + X = 4(-0.5) + 0.5 = -1.50$$

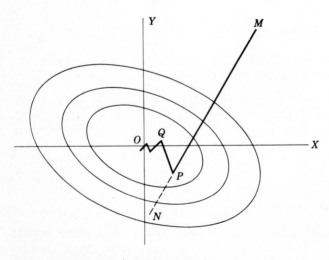

Figure 13.3 The method of steepest ascent.

Table 13.3 Calculations along the first gradient

i	X_i	Y_i	Z_i
0	2.00	2.00	16.00
1	1.70	1.50	9.94
2	1.40	1.00	5.36
3	1.10	0.50	2.26
4	0.80	0.00	0.64
5	0.50	−0.50	0.50
6	0.20	−1.00	1.84

It is noted that $\partial Z/\partial X$ is positive but $\partial Z/\partial Y$ is negative. Therefore, in order to decrease the value of Z, X should be decreased and Y should be increased. Inasmuch as the solution should now be close at hand, the step size in Y will be arbitrarily reduced from 0.5 to 0.3. From the recently calculated partial derivatives, the ratio of $\Delta X/\Delta Y = -\frac{1}{3}$ so that $\Delta X = -0.1$.

Step 6 is calculated as follows:

$$X_6 = X_5 + \Delta X = 0.50 - 0.1 = 0.40$$

$$Y_6 = Y_5 + \Delta Y = -0.50 + 0.3 = -0.20$$

$$Z_6 = 0.40^2 + 2(-0.20^2) + 0.40(-0.20) = 0.16$$

Continuing along this new gradient line, the following points are obtained:

i	X_i	Y_i	Z_i
6	0.40	−0.20	0.16
7	0.30	0.10	0.14
8	0.20	0.40	0.44

Since Z_8 is greater than Z_7, it is obvious that the extrapolation has gone too far along this line. The last successful step, point Q (X_7, Y_7), is indicated on Fig. 13.3. The direction is changed once again, the step size increased, etc., until the optimum point is finally approached.

The method of steepest ascent requires two decisions—when to calculate a new gradient direction, and what step size to use. Experience and familiarity with the problem increase the efficiency of the solution.

If an analytical expression is not available, it is necessary to explore by experimentation the value of Z, the objective function, in various directions from M (Fig. 13.3) and to establish the steepest gradient experimentally.

The method can be applied to more than two variables by vector analysis techniques.

13.9 Constrained Optimization

Suppose now that there are bounds or constraints imposed on problems, such as restrictions on the independent variables or on some calculated values. Equality constraints can be handled by the technique of Lagrange multipliers (Sec. 10.15), while the Kuhn-Tucker conditions can be used for handling inequality constraints.

Constraints can cause serious mathematical difficulties, and various methods have been proposed. The work of Fiacco and McCormick [6–10] is recommended, along with that of Box [5].

SUMMATION OF SERIES

13.10 Recurring Power Series

Frequently it is necessary to sum a discrete power series, such as a present-value series:

$$P = \frac{A_0}{(1 + i)^0} + \frac{A_1}{(1 + i)^1} + \cdots + \frac{A_m}{(1 + i)^m} + \cdots + \frac{A_n}{(1 + i)^n} \quad (13.11)$$

The series generally is more complicated than a geometric series, and it is desirable to develop a method for summing these more complicated types. The subject of series summation is involved, but one method, summation of recurring power series, depends upon algebraic methods and is easily understood upon demonstration. Moreover, the method will accommodate most series that arise in economic analysis.

If $U = 1/(1 + i)$, then Eq. (13.11) can be written more compactly:

$$\sum = A_0 + A_1 U^1 + \cdots + A_m U^m + \cdots + A_n U^n \quad (13.12)$$

Each A is a coefficient, and the series is a *recurring power series* if and only if successive coefficients are connected by a recurring relationship.

For example, consider the series

$$\sum = 1 + 3U + 7U^2 + 15U^3 + 31U^4 + \cdots \quad (13.13)$$

Every three successive coefficients in the series are related by the equation (where the constant multiplying the highest coefficient has been made equal to unity)

$$2A_m - 3A_{m+1} + 1A_{m+2} = 0 \quad (13.14)$$

Thus

$$2(1) - 3(3) + 1(7) = 0$$

$$2(3) - 3(7) + 1(15) = 0$$

Equation (13.14) has the algebraic form

$$B_2 A_m + B_1 A_{m+1} + 1A_{m+2} = 0 \quad (13.15)$$

where each B is constant and independent of m and the right side is zero. Equation (13.15) is known as the *scale of relation*, and the B multipliers are referred to

Table 13.4 Summing a recurring power series

$$\sum = A_0 \qquad\quad + A_1 U + A_2 U^2 + \cdots + \quad A_n U^n$$

$$- 3U\sum = \qquad - 3A_0 U - 3A_1 U^2 + \cdots - 3A_{n-1}U^n \qquad - 3A_n U^{n+1}$$

$$+ 2U^2 \sum = \qquad\qquad\qquad + 2A_0 U^2 + \cdots + 2A_{n-2}U^n \qquad + 2A_{n-1}U^{n+1} + 2A_n U^{n+2}$$

$$(1 - 3U + 2U^2)\sum = A_0 + (A_1 - 3A_0)U + 0 \qquad + \cdots + 0 \qquad + (2A_{n-1} - 3A_n)U^{n+1} + 2A_n U^{n+2}$$

with $A_0 = 1 \qquad A_1 = 3 \qquad A_n = 2^{n+1} - 1$

$$\sum = \frac{1 + [2(2^n - 1) - 3(2^{n+1} - 1)]U^{n+1} + 2(2^{n+1} - 1)U^{n+2}}{(1 - 2U)(1 - U)}$$

$$\sum = 2\frac{1 - (2U)^{n+1}}{1 - 2U} - \frac{1 - U^{n+1}}{1 - U}$$

$$\sum = 1 + 3U + 7U^2 + \cdots + (2^{m+1} - 1)U^m + \cdots + A_n U^n$$

Scale of relation $= 2A_m - 3A_{m+1} + 1A_{m+2} = 0$

as the *constants of the scale*. Equation (13.14) has two independent constants of the scale; hence Eq. (13.13) is a recurring series of the second order.

Recurring power series were first investigated by the French mathematician de Moivre (1667–1754). A recurring power series is readily summed when its scale of relation and an expression for the general coefficient are known. The general term for Eq. (13.13), as will be shown, is

$$(2^{m+1} - 1)U^m \tag{13.16}$$

The summation is accomplished by adding the original series and others derived from it in a way that will utilize the scale of relation and compel most of the terms in the addition to become zero and vanish. For example, to find the sum Σ for Eq. (13.13), use the numbers $+1$, -3, and $+2$, taken by reading from right to left in the scale of relation, form the series

$$+1 \sum \quad -3U \sum \quad +2U^2 \sum$$

and add the series so formed. The work is shown in Table 13.4.

13.11 General Term from a Generating Function

In summing Eq. (13.13) by Table 13.4, an expression for the general term coefficient A_m was needed. Consider the following fraction taken from Table 13.4:

$$\frac{A_0 + (A_1 - 3A_0)U}{1 - 3U + 2U^2} = \frac{1}{1 - 3U + 2U^2}$$

$$= 1 + 3U + 7U^2 + 15U^3 + 31U^4 + \cdots$$

The fraction is known as the *generating function,* which can be used to find the general term, inasmuch as it generates the series. By the use of partial fractions developed in standard algebra textbooks with C and D constants,

$$\frac{1}{1 - 3U + U^2} = \frac{1}{(1 - 2U)(1 - U)} = \frac{C}{1 - 2U} + \frac{D}{1 - U}$$

The above is true for all values of U. Placing $U = 0$ gives $1 = C + D$, and placing $U = -1$ gives $1/6 = C/3 + D/2$. Solved simultaneously, it yields $C = 2$ and $D = -1$. Thus,

$$\frac{1}{1 - 3U + 2U^2} = \frac{2}{1 - 2U} - \frac{1}{1 - U}$$

and

$$\frac{2}{1 - 2U} = 2 + 4U + \cdots + 2^{m+1}U^m + \cdots$$

$$-\frac{1}{1 + U} = -1 - U + \cdots \qquad -1U^m + \cdots$$

The general term is $(2^{m+1} - 1)U^m$, as used previously.

13.12 Some Recurring Power Series

The following series all have the scale of relation $A_m - 2A_{m+1} + A_{m+2} = 0$:

$$\sum = U + 2U^2 + 3U^3 + \cdots + nU^n \qquad (13.17)$$

$$\sum = U^2 + 2U^3 + 3U^4 + \cdots + (n - 1)U^n \qquad (13.18)$$

$$\sum = BU + (B + A)U^2 + (B + 2A)U^3 + \cdots + [B + (n - 1)A]U^n \qquad (13.19)$$

Summations, up to and including the term in U^n, are, respectively, where A or B can be negative:

$$\sum = U\frac{1 - (n + 1)U^n + nU^{n+1}}{(1 + U)^2} \qquad (13.20)$$

$$\sum = U^2\frac{1 - nU^{n-1} + (n - 1)U^n}{(1 - U)^2} \qquad (13.21)$$

$$\sum = U\frac{B - (B - A)U - (B + A_n)U^n + (B - A + A_n)U^{n+1}}{(1 - U)^2} \qquad (13.22)$$

The series

$$\sum = GU^2 + 2GU^3 + \cdots + (n - 1)GU^n$$

is particularly important because it arises in connection with a machine for which the annual expense increases at a constant rate, G per year. The summation is given by Eq. (13.21), which when multiplied by G and with $1/(1 + i)$ replacing U reduces to

$$\sum = \frac{G}{i(1 + i)^n} \left[\frac{(1 + i)^{n-1}}{i} - n \right]$$

The uniform-gradient series was discussed in Sec. 9.4.

Another series of considerable importance is the present value of \$1 of depreciation taken with the sum-of-the-years-digits method of depreciation. Sum-of-the-years-digits is algebraically $0.5n(n + 1)$, and the series is

$$\sum = \frac{n}{0.5n(n + 1)} U + \frac{n - 1}{0.5n(n + 1)} U^2 + \cdots + \frac{1}{0.5n(n + 1)} U^n$$

The series can be summed by Eq. (13.22), which after multiplication by $1/[0.5n(n + 1)]$ and with $B = n$, $A = -1$, and $U = 1/(1 + i)$, becomes

$$\sum = \frac{1}{0.5n(n + 1)i} \left[n - \frac{(1 + i)^n - 1}{i(1 + i)^n} \right] \tag{13.23}$$

or

$$\sum = \frac{n - F_{RP,i,n}}{0.5n(n + 1)i} \tag{13.24}$$

NOMENCLATURE

B	Constant in scale of relation, Sec. 13.10
D	Distance from the end of the interval in the Fibonacci search
F	Objective function to be optimized
F_i	Fibonacci number
F_r	Fractional ratio of the interval left to the original interval
$F_{RP,i,n}$	The factor $[(1 + i)^n - 1]/i(1 + i)^n$; see Sec. 2.2
G	Gradient amount, or constant increase or decrease
i	Decimal rate of return in Secs. 13.10 to 13.12
L	Range of a variable
m	The mth or general term
n	The nth or last term
N	Number of experiments or calculations allowed
P	Present value, \$
U	The factor $1/(1 + i)$

PROBLEMS

13.1 Assume that the model is really the evaluation of the square root of 3. Assume that the solution is known to lie in the range 0 to 5. Evaluate $\sqrt{3}$ to an interval that is 0.05 wide. Use the Fibonacci search technique. Hint: $|X^2 - 3|$ is a minimum if $X = \sqrt{3}$.

13.2 Repeat Prob. 13.1 using a uniform search method. For a condensation, tabulate only in the neighborhood of the minimum.

13.3 Repeat Prob. 13.1 using a sequential dichotomous search.

13.4 Suppose

$$Z = 2X^3 - 3X^2 - 14X + 2$$

Using six trials or experiments, find a minimum value for Z in the range $X = 1$ to $X = 3$ using (a) a uniform search, (b) a uniform dichotomous search, (c) a sequential dichotomous search, and (d) a Fibonacci search. Also, compare the fractional ratio of the interval left to the original interval for these methods.

13.5 Using the method of steepest ascent, find the minimum value of

$$Z = X^2 + 2Y^2 + XY$$

starting at the point $X = 2$ and $Y = 3$. Use an initial step on Y of 0.5 until no further improvement results; then shift to a step size on Y of 0.05.

13.6 Repeat Prob. 13.5 using the one-at-a-time method.

13.7 In the method of steepest ascent, surfaces with spherical contours give the fastest convergence rates. Given the objective function $Z = X^2 + 25Y^2$, how can this problem be transformed into one giving spherical contours?

13.8 The one-at-a-time method can be applied to a three-variable problem by the technique used in Example 13.4. Given the function

$$f = X^2 + Y^2 + Z^2 - XY - YZ - 4X$$

find the values of X, Y, and Z that make f a minimum. Hint: Start at $X = Y = 0$ and find the best value of $Z = 0$. Then, for $Y = 0$ and $Z = 0$, find the best value of $X = 2$. Then, for $X = 2$ and $Z = 0$, find the best value of $Y = 1$. Then, for $X = 2$ and $Y = 1$, find the best value of $Z = 0.5$. Continue until no further improvement results.

13.9 Confirm the answer to Prob. 13.8 by solving analytically.

13.10 Repeat Prob. 13.8 using the method of steepest ascent. Start at point $X, Y, Z = (0, 0, 0)$.

13.11 For the series

$$\sum = -1 + X - 3X^2 + 5X^3 - 11X^4 + 21X^5 + \cdots$$

find (a) the scale of relation, (b) the generating function, (c) the coefficient for the term in X^n, and (d) the sum of the terms through X^n.

13.12 Find the scale of relation and the next term of the series

$$\sum = 1 + 3U + 6U^2 + 10U^3 + 15U^4 + 21U^5 + 28U^6 + \cdots$$

13.13 Find the generating function for Prob. 13.12.

13.14 Verify Eq. (13.21).

13.15 Verify Eq. (13.23).

13.16 Let F_N denote the nth Fibonacci number and form the following series, where $U = 1$:

$$\sum = F_0 + F_1 U + F_2 U^2 + \cdots + F_N U^N$$

Knowing the scale of relation given by

$$-F_N - F_{N+1} + F_{N+2} = 0$$

find (a) a generating function for the series, and (b) an expression for the nth Fibonacci number.

13.17 Find an expression for the limit of the ratio of two successive Fibonacci numbers as N becomes very large.

13.18 Find an expression for the sum of the first N Fibonacci numbers.

REFERENCES

Search methods

1. Abadie, J.: *Integer and Non-Linear Programming*, North-Holland Publishing Company, New York, 1970.
2. Beveridge, G. S. G., and R. S. Schecter: *Optimization: Theory and Practice*, McGraw-Hill Book Company, New York, 1970.
3. Boas, A. H.: "How Search Methods Locate Optimum in Univariable Problems," *Chemical Engineering,* Feb. 4, 1963, p. 105.
4. Boas, A. H.: "Optimizing Multivariable Functions," *Chemical Engineering,* Mar. 4, 1963, p. 97.
5. Box, M. J.: "A New Method of Constrained Optimization and a Comparison with Other Methods," *Computer Journal,* vol. 8, April 1965, pp. 42–52.
6. Fiacco, A. V., and G. P. McCormick: *SUMT without Parameters,* Northwestern University Technical Institute Systems Research Memo 121, April 1965.
7. Fiacco, A. V., and G. P. McCormick: Programming under Nonlinear Constraints by Unconstrained Minimization: A Primal-Dual Method, Research Analysis Corporation Technical Paper RAC-TP-96, September 1963.
8. Fiacco, A. V., and G. P. McCormick: "Computational Algorithm for the Sequential Unconstrained Minimization Technique for Non-linear Programming," *Management Science,* vol. 10, no. 4, 1964, pp. 601–617.
9. Fiacco, A. V., and G. P. McCormick: "The Sequential Unconstrained Minimization Technique for Nonlinear Programming, a Primal-Dual Method," *Management Science,* vol. 10, no. 2, 1964, pp. 360–366.
10. Fiacco, A. V., and G. P. McCormick: "Extensions of SUMT for Non-linear Programming: Equality Constraints and Extrapolation," *Management Science,* vol. 12, no. 11, 1966, pp. 816–828.
11. Friedman, M., and L. S. Savage: *Selected Techniques of Statistical Analysis*, McGraw-Hill Book Company, New York, 1947.
12. Himmelblau, D. M.: *Applied Non-Linear Programming*, McGraw-Hill Book Company, New York, 1972.
13. Johnson, S. M.: *Best Exploration for Maximum Is Fibonaccian,* U.S. Air Force Project Rand Research Memorandum, RM-1590, Nov. 18, 1955.
14. Keifer, J.: "Sequential Minimax Search for a Maximum," *Proceedings of the American Mathematics Society,* vol. 4, no. 3, 1953.
15. Koester, J. L., and J. H. Mize: *Optimization Techniques with Fortran,* McGraw-Hill Book Company, New York, 1973.
16. Umeda, T., and A. Ichikawa: "Modified Complex Method for Optimization," Ind & Eng Chem, *Process Design and Development,* vol. 10, no. 229, 1971.
17. Wilde, D. J.: *Optimum Seeking Methods,* Prentice-Hall, Inc., Englewood Cliffs, N.J., 1964.
18. Wilde, D. J., and C. S. Beightler: *Foundations of Optimization,* 2d ed., Prentice-Hall, Inc., Englewood Cliffs, N.J., 1979.

Series summation

1. Chrystal, G.: *Algebra,* Dover Publications, Inc., New York, 1961, pt. 2, chap. 31.
2. Fine, H. B.: *College Algebra,* Dover Publications, Inc., New York, 1961, chap. 35.
3. Jelen, F. C.: "Series Summation for Economic Analysis," *Hydrocarbon Processing,* December 1967, pp. 153–157.

THREE

COST ESTIMATION AND CONTROL

FOURTEEN

CAPITAL INVESTMENT COST ESTIMATION

H. C. Nelson, O. P. Kharbanda,
W. A. Janda, and J. H. Black

14.1 Definitions

Decisions made by management pertaining to capital investments affect the financial stability and future earnings of the firm. These earnings are usually obtained by converting capital into facilities or plants that yield a return exceeding the amount of the capital investment. This capital investment is composed of two parts:

1. The fixed capital investment to provide the physical facilities.
2. The working capital investment, a revolving fund to keep the facilities operating. An investment cost is a one-time cost, as opposed to operating costs, which are continuous or repeated.

The *fixed capital* investment is the capital required to provide *all* the facilities needed for the project regardless of how remote they might be to manufacturing operations, e.g., a rolling mill or a refrigerator in the employees' cafeteria. Fixed capital investment can be divided into various categories, such as design and engineering, land purchase and improvement, manufacturing and allied buildings, equipment, utilities and services, receiving and shipping, and start-up.

Although land does involve a capital investment, some prefer to include in fixed capital only those items for which depreciation is allowed by the Internal Revenue Service [75], thus exluding land. However, if land costs are included in the total investment, it makes little difference in what subcategory they appear.

Working capital in finance and accounting is the difference between current assets and current liabilities. Current assets are chiefly cash, accounts receivable, inventories, and negotiable securities. Current liabilities are chiefly accounts pay-

Table 14.1 Components of working capital

Raw materials inventory
Work-in-progress inventory (semifinished goods)
Finished-products inventory
Supplies for product manufacture
Prepaid expenses
Taxes payable
Accounts receivable
Cash or equivalent on hand for salaries, wages, etc.
Accounts payable

able, short-term debts, and the portion of long-term debts due within 1 year. In this book, working capital comprises the value of funds, in addition to fixed capital and start-up costs, needed to get a project started and meet subsequent obligations [75]. Working capital is regarded as funds frozen in the project that will be recovered in full at the end of the project or even if the project is terminated at any time. A list of the components of working capital is given in Table 14.1. Inventory, from raw material to finished goods, comprises one major group. Accounts receivable usually allows a 30-day payment and thus ties up funds. On the other hand, accounts payable can be delayed about 30 days, works contrary to accounts receivable, and reduces the working capital requirements.

In general, working capital is based on requirements for a 1-month period. In manufacturing industries, working capital is about 10 to 20 percent of the fixed capital investment. In some industries with very high inventories of finished goods, the figure can exceed 40 percent. As a rough estimate, working capital may be taken as 20 percent of the total investment.

Before a new investment is made, a capital cost estimate is usually developed as part of calculating the economic return for the venture. This return is used as partial justification for the project. Cost estimates are required before and after a decision is made to go ahead with the investment. During the course of a capital project, various types of cost estimates will be necessary for a variety of reasons and purposes.

Cost estimates for capital projects consist of two major cost elements:

1. *Direct costs* are the costs of all permanent equipment, materials, labor, and other resources involved in the fabrication, erection, and installation of the permanent facilities [2].
2. *Indirect costs* are all costs other than direct costs which do not become a permanent part of the facilities but are required for the orderly completion of the project. These may include, but are not limited to, construction management, start-up costs, fees, insurance, and taxes [2].

14.2 Types of Estimates

Capital cost estimating is essentially an intuitive process which attempts to predict the final outcome of a future capital expenditure program even though not all parameters and conditions concerning a project are known or not fully defined when the cost estimate is prepared.

Estimate types vary considerably depending upon the available information, time demands, and purposes of the estimate. Lists of estimate types prepared by various authors, companies, professional and industrial organizations, and government agencies are available [7, 52, 56, 68]. These lists identify the various types, or classifications, of estimates most commonly prepared. The four lists presented below are typical examples.

Edmunds [14] gives six basic types of estimates for building construction used by architects and engineers:

1. Quickie
2. Original
3. Preliminary
4. Official or budget
5. Final or definitive
6. Revised

The AACE [1] has proposed five classifications of estimating types:

1. Order of magnitude (ratio estimate)
2. Study (factored estimate)
3. Preliminary (budget authorization estimate)
4. Definitive (project control estimate)
5. Detailed (firm estimate)

A more recent and more condensed classification by AACE [33] shows the following types, with their probable accuracies and their method of preparation:

1. Order of magnitude: −30 to +50 percent. Cost-capacity curves and cost-capacity ratios are used.
2. Budget: −15 to +30 percent. Made for the owner's budget. Flowsheets, layouts, and equipment details are used.
3. Definitive: −5 to +15 percent. Defined engineering data such as site data, specifications, basic drawings, and detailed sketches are used.

Another classification of estimate types [27] based on probable accuracy is:

1. Detailed unit cost: ±3 percent
2. Material takeoff: ±6 percent
3. Defined equipment ratio: ±12 percent
4. Preliminary equipment ratio: ±25 percent
5. Cost-capacity curve: ±40 percent

The various types of estimates are designed to identify the method of preparation and the range of accuracy. However, even with this diversity of names and

Table 14.2 Accuracy of cost estimates

Types of estimates	% accuracy
Preliminary	
Order of magnitude	+50 to −25
Conceptual	+30 to −15
Detailed	
Semidefinitive	+25 to −15
Definitive	+15 to −10
Check	+10 to −5

classifications, cost estimates generally fall into two major categories, preliminary and detailed, as shown in Table 14.2.

Preliminary estimates are used during the formative stages of a capital expenditure program when there is a lack of firm or verifiable information for the initial evaluation of the project. In this situation various estimating techniques are used, ranging from experience and judgment as the dominant characteristics to techniques using historical cost charts, rule of thumb, and simple mathematical calculations to arrive quickly at a relatively inexpensive cost estimate. The accuracy of a preliminary estimate is usually very low. Obviously, accuracy is dependent upon the amount and quality of information and the time available to prepare the estimate. Other terms commonly used in practice to identify preliminary estimates include: *order of magnitude, conceptual, factored, quickie,* and *feasibility estimates* [52].

Detailed estimates are used after the scope and schedule of a project have been stabilized. Methods used for developing detailed estimates are based on more quantitative information in conjunction with formal, exacting procedures in arriving at estimated values. Techniques for detailed cost estimates vary with circumstances and the industry. However, emphasis is on factual data and mathematical modeling, the latter ranging from simple manual calculations to complex computer simulation programs. These cost estimates are used for many purposes: bid proposals, establishment of budgets, fair-price estimates for bid evaluations, contract change orders, extra work orders, legal claims, appropriation of funds, and obtaining government permits and approvals. Generally, these estimates consist of firm prices for materials and equipment. Quantitative data are obtained from design drawings and specifications; engineering, procurement, and construction execution plans are developed to arrive at the estimated costs for those functions. Most practitioners claim that detailed estimates have an accuracy of ±10 percent. Whether the accuracy of a particular estimate is ±5 percent or ±50 percent is not significant. If the amount of detail and the quality of information are increased, greater accuracy can be achieved. However, increased accuracy of a cost estimate requires more time, effort, and expense. Other terms commonly used in practice to identify detailed estimates include: *definitive, semidefinitive, budget, check, final, official,* and *defined* [24, 52].

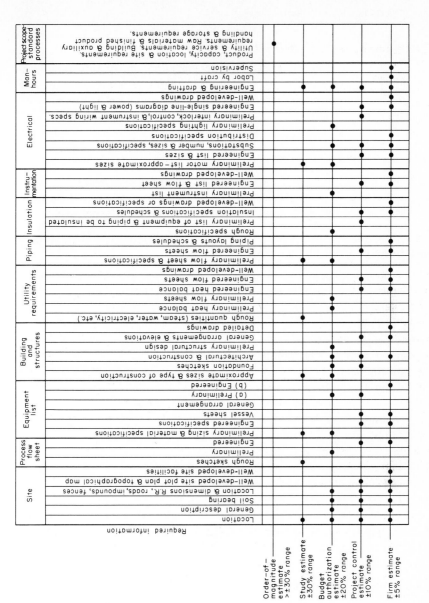

Figure 14.1 A guide for capital investment estimating. (*R. H. Perry and C. H. Chilton: Chemical Engineers' Handbook, 5th ed., McGraw-Hill Book Company, New York, 1973.*)

325

14.3 Available Information against Accuracy

As stated earlier, the accuracy of an estimate depends upon the quality and quantity of available project information. A guide for judging estimate accuracy and the type of estimate to be used, based on available information, is presented in Fig. 14.1.

14.4 Cost of Making Estimates

The cost of making an estimate of a given accuracy can vary over a large range [8], p. 41; [58]. The figures in Table 14.3 show that an estimate of 50 percent accuracy costs very little, while one of 5 percent accuracy can be quite expensive.

14.5 Functions of Capital-Cost Estimates

Estimates of capital costs are made for various purposes, the most common of which are:

Preliminary estimates:
1. To study project feasibility
2. To obtain designs and economic studies
3. To obtain financing
4. To obtain government permits or appraisals

Detailed estimates:
1. To seek funding
2. To prepare bid proposals
3. To prepare fair-price estimates for bid evaluations
4. To prepare changes and extra work orders for contracts
5. To bolster legal claims

Before private funds are appropriated, feasibility studies must show that the expected profit is sufficient to justify the risk of capital expenditure. Order-of-

Table 14.3 Typical costs for preparing estimates in kilodollars, December 1980

Type of estimate	Accuracy range, %	Cost of project in M$		
		Less than 1.0	1.0–5.0	5.0–50.0
Order of magnitude	−30 to +50	7.5 to 20	17.5 to 45	30 to 60
Budget	−15 to +30	20 to 50	45 to 85	70 to 130
Definitive	−5 to +15	35 to 85	85 to 175	150 to 330

Source: Based on A. Pikulik and H. E. Diaz, "Cost Estimating for Major Process Equipment," *Chemical Engineering,* vol. 84, Oct. 10, 1977, p. 106.

magnitude studies are used for screening purposes. Further studies can optimize the project design and the plan of execution. Cost estimates play a very important role in obtaining financing from lending institutions, if such arrangements are required.

After the decision has been made to go ahead with a project, execution of the capital expenditure program begins. Contractors to perform services in management, engineering, procurement, and construction must be selected. Usually such services are obtained through competitive bidding, which requires bid estimates. If the bid is too high, the contractor may not be awarded the job; but if the bid is too low, the contractor may be awarded the job and take a substantial loss. During the course of the project, changes in the scope of the work may be required; estimates are necessary to determine these costs. Upon completion of the project, legal claims may arise from conflicts that occurred during the project's execution. Cost estimates may be involved in presenting the claim and agreeing to a settlement.

For the allocation of public funds, e.g., to explore space or improve military preparedness, there are available measures of feasibility rather than measures of profitability. Such funds are limited, so some kind of feasibility study is important, as discussed in Chap. 5.

14.6 Estimating Techniques

Cost estimates are prepared by a number of techniques, based on cost of similar projects or the cost of project components. Many companies have extensive files of cost data derived from previous projects. Published cost information is voluminous, but its accuracy is not always known.

The conference estimating technique is commonly used for developing quick preliminary estimates. This method draws on the collective judgment of a group of individuals with experience in cost estimating for similar projects in the past. The goal is to arrive at an estimate based on a group consensus [52].

The comparison estimating technique is a common method for estimating new capital projects and determining appraisals for existing facilities. This technique is similar to the conference method except that a more formal logic is used in arriving at a final figure. Estimating by comparing, weighing, and relating the costs of similar past projects requires adjusting for the new location and time period, which usually requires judgment or well-defined adjustment factors [52].

Graphic relationships and tabular relationships are two more estimating techniques. These methods are more quantitative and more accurate than the conference and comparison techniques, and hence are used in conceptual cost estimates. In these techniques, limited historical data are used to develop cost relationships. Examples of these simple techniques are tabulations of data such as Table 14.4, exclusion charts such as Fig. 14.2, band charts such as Fig. 14.3, upper- and lower-limit charts such as Fig. 14.4, and simple charts such as Fig. 14.5. Unfortunately, when using published cost data, it is not always clear what is included in the cost. Some equipment costs refer to purchased prices, while others are in-

Table 14.4 Example of tabular estimating data
Manhours for fabrication and erection of 3/4-in carbon steel piping

Operation	Source A	Source B (8)	Source C (9)	Source D	Source E	Source F (10)
Butt weld—std. wt.	1.1	0.8	0.4	....	0.5	
Butt weld—X.H.	1.3	0.9	1.2	0.7	0.6	
Lin. ft. std. pipe	0.13	0.07	0.2	....	0.19	
Lin. ft. X.H. pipe	0.15	0.08	0.25	0.16	0.20	
Std. weld ells	2.2	1.6	0.9	2.1	1.3	
Std. weld tees	3.3	2.4	1.2	3.0	1.8	
X.H. weld ells	2.6	2.7	1.0	2.1	1.5	
X.H. weld tees	3.9	2.7	1.5	3.0	2.1	
Std. weld reducer	2.0	1.4	0.5	2.1	1.30	
X.H. weld reducer	2.2	1.5	0.9	2.1	1.5	
150# slip-on flange	1.1	....	0.9	....	0.80	
300# slip-on flange	1.2	....	1.0	....	0.90	
600# slip-on flange	1.3	....	1.1	....		
150# weld neck flange	1.1	1.1	0.4	....	0.80	
300# weld neck flange	1.2	1.1	1.3	....	0.95	
600# weld neck flange	1.3	1.1	1.4	....		
150# flange bolt-up	....	0.6	1.5	....		
300# flange bolt-up	....	0.7	1.6	....		
600# flange bolt-up	....	0.7	....	....		
Handle 150# valve	....	0.3	1.5	0.8		
Handle 300# valve	....	....	1.6	....		
Handle 600# valve	....	....	....	1.1		

Source: AACE Cost Engineers' Notebook, "Data for Estimating Piping Cost," Paper 64–31, June 1964.

stalled costs, which may or may not include auxiliary equipment. A similar difficulty exists with plant cost data. Some costs are for complete new plants, including land, utilities, and auxiliary facilities. Others are restricted to battery-limits costs. Ordinarily, all manufacturing facilities are within the battery limits, but administrative offices, storage areas, utilities, and other auxiliary facilities are outside these limits. The term *battery limits* is a geographical designation for the process area of a proposed project. Published cost data should be used only for order-of-magnitude estimates since their accuracy is seldom known, and discrepancies of 50 percent are not uncommon. Some cost data are based on specific information and are not generally applicable. Other cost data are average values, and an average is questionable when applied to a particular case.

The unit rate technique is one of the most widely used approaches in preliminary and detailed estimating. This method takes the form of either the so-called bare unit cost or the complete unit cost. Examples of bare unit rates are wage cost per work-hour, purchased material cost per unit of weight or volume, and machinery cost per piece. Examples of complete unit cost are delivered price of fabricated structural steel per unit of weight, fully erected storage tank cost per unit of capacity, and installed cost of pipeline per unit of diameter and length.

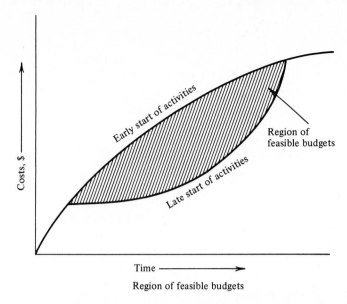

Figure 14.2 Example of an exclusion chart. (*H. Kerzner: A System Approach to Planning, Scheduling and Controlling, Van Nostrand Reinhold Company, New York, 1979. Used with permission.*)

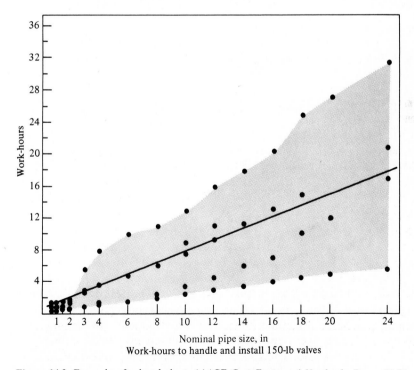

Figure 14.3 Example of a band chart. (*AACE Cost Engineers' Notebook. Paper 64–31, June 1964.*)

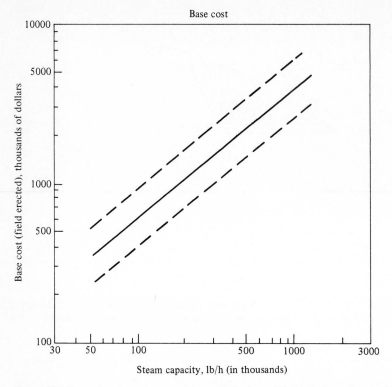

Base cost

Base cost (field erected), thousands of dollars

Steam capacity, lb/h (in thousands)

Figure 14.4 Example of an upper- and lower-limit chart. (*K. M. Guthrie: Process Plant Estimating, Evaluation, and Control, Craftsman Book Company, Solana Beach, Calif., 1974. Used with permission.*)

Table 14.4 is an example of bare unit work-hours needed to fabricate and erect piping elements.

Example 14.1 Using Table 14.4, source C, estimate the work-hours required to install 1000 ft of standard pipe, eight standard-weld ells, and one 150-lb valve.

SOLUTION

$$
\begin{array}{r}
1000 \times 0.2 = 200 \\
8 \times 0.9 = \ \ 7.2 \\
1 \times 1.5 = \underline{\ \ 1.5} \\
209 \ \ \text{work-h}
\end{array}
$$

The unit rate technique is used in building construction. It is generally used in heavy civil construction, such as tunneling and pipeline projects.

The resource allocation method is popular in developing cost estimates for less tangible or nonrepetitive items, such as applications of new technology, research and development projects, technical services, management services, or consulting or advisory services. This technique is used in conceptual and detailed

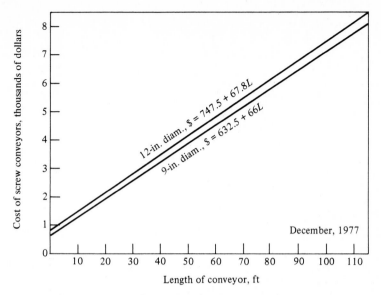

Figure 14.5 Example of a simple cost chart. (*W. M. Vatavuk and R. B. Neveril: "Estimating Costs of Dust-Removal and Water-Handling Equipment, Part VI," Chemical Engineering, vol. 88, Mar. 23, 1981, p. 223.*)

estimates when it is difficult to quantify the end product or output. Therefore, a plan is developed in terms of the times required for the various scheduled activities. The available resources—manpower, equipment, and materials—are allocated against each planned activity at the appropriate unit cost to arrive at an estimated cost for the project [35]. An example of manpower resource allocation is shown in Table 14.5.

Example 14.2 Estimate the labor cost for equipment construction during the month of April 1977 from Table 14.5.

SOLUTION
Cost = (work-h)(rate)(1 + overhead + indirect costs + corporate costs + profit)
Program management = 20(11)(1 + 1.2 + 0.14 + 0.01 + 0.12)   =   543.40
Engineering = 154(10)(1 + 1.2 + 0.14 + 0.01 + 0.12)   =   3803.80
Fabrication = 154(10)(1 + 1.25 + 0.14 + 0.01 + 0.12)   =   3880.80
                                                                         ─────────
                                                                         $8228.00

Factor estimating is a technique generally used to arrive at a preliminary capital cost estimate inexpensively and quickly with a reasonable accuracy of 15 to 30 percent. In this method, a cost is estimated by multiplying one cost factor to get another cost. Thus, the cost of a building complete with electrical, plumbing, heating, ventilation, and air-conditioning systems might be estimated by multiplying the cost of the shell by 1.6. Such factors are called *ratio cost factors* [81]. A common use of ratio cost factors is the estimation of the installed cost of equipment from the purchase cost, applying an *equipment installation cost ratio*. The

Table 14.5 Manpower resource allocation, cost engineering method example

Projects/tasks	1977 Jan.	Feb.	Mar.	Apr.	May	June	July	Aug.	Sept.	Oct.	Nov.	Dec.	1978 Jan.	Feb.	Mar.	Apr.	May	June	July	Aug.	Sept.
I. Program plans and building design																					
1. Program management	102	50	48																		
2. Engineering	20	174	283																		
II. Equipment purchase																					
1. Program management		52	40	40	30	30	30	30	30	30	20	20	20	20	20						
2. Engineering		267	232	200	165	131	95	60	40												
3. Testing and inspection					154	154	154	154	154	154	154	154	154								
4. Shipping										266	266	266	154	112							
5. Procurement								77	77	308	308			308	308						
III. Equipment construction																					
1. Program management			20	20	30	30	30	30	30	30	20	20	20	20	20						
2. Engineering			154	154	154	154	154	154	154	154	77										
3. Procurement					154	308	308	308	308	308	308	308	308	308	308						
4. Shipping													47	100	154						
5. Fabrication				154	308	462	616	770	924	1078	1386	1848	1848	1848	154						
IV. Training																					
1. Program management											20	20	20	20							
2. Engineering											60	60	60	60							
3. Training											154	154	154	154							
V. Plant start-up																					
1. Program management													20	20	5	5	5	5	5	20	
2. Field engineering													154	154	154	154	308	154	154	154	

II. Other costs

Purchased goods	$121,981
Freight	1,988
Other	3,049
Overseas packing	6,242
	$133,260

III. Other costs

Purchased mat'ls	39,527
Subcontracts	28,082
Overseas packing	3,520
Freight	2,598
	$73,727

IV. Other costs

Supplies	$980

	Rate	Oh
Program management	11.00	120%
Engineering	10.00	120%
Testing	8.00	117% ⎫
Procurement	8.00	110% ⎬ 125%
Shipping	5.70	100%
Fabrication	10.00	125%
Training	10.50	120% ⎫ 12%
Field engineering	9.00	80% ⎭

Indirect costs	14%
Corporate costs	1%
Profit	12%

Raw materials escalation	10%
Demanding ratio	10%
Termination liability	0%
Salary increases	6%

Note: One-man month = 154 Manhours

Source: H. Kerzner: *Project Management, a System Approach to Planning, Scheduling, and Controlling,* Van Nostrand Reinhold Company, New York, 1979. Used with permission.

installed cost is the sum of the FOB cost, transportation, cost of foundations, erection, and making connections to service facilities. Although each of these costs can be estimated separately, it is much easier to multiply the FOB cost by a factor to give the installed cost. On the average, the installed cost is about 1.5 times the purchase cost, but it can vary considerably. Factor estimating has numerous synonyms, such as *ratio estimating, parameter estimating, module estimating, Lang factor estimating,* and *percentage estimating*. These methods involve substantially the same technique, with refinements added to increase the accuracy and identification of the various cost elements comprising the estimate [25].

An exponent estimating technique is sometimes used to prepare preliminary cost estimates. The costs of similar plants or pieces of equipment of different sizes vary with the size raised to some power. This relationship can be expressed mathematically as follows:

$$\frac{C_2}{C_1} = \left(\frac{Q_2}{Q_1}\right)^X \tag{14.1}$$

Here, C_2 is the desired cost of capacity Q_2 and C_1 is the known cost of capacity Q_1. A frequent value for X is 0.6, so this relationship is often referred to the *six-tenths (0.6) factor rule* [11, 76]. The exponent X can be determined by plotting actual historical costs for the equipment or plant as the ordinate on log-log paper and the equipment or plant size as the abscissa. The slope of the resulting line through the data will be the cost-capacity factor.

Example 14.3 An ethylene plant with a capacity of 100,000 t/year costs 16 M$. Estimate the cost of a 200,000 t/year plant.

SOLUTION Using the 0.6 factor rule for Eq. (14.1),

$$C_2 = 16\left(\frac{200,000}{100,000}\right)^{0.6} = 24.2 \text{ M\$}$$

A standard unit estimating technique is frequently used in preparing detailed cost estimates. There are various forms for these standard estimating units, such as cost units, work-hour units, crew size units, construction equipment-use units, manpower in units per day, and manpower per units of operation. Standard unit data are available as a list, by industry, of standard units required to accomplish a specified task. Such data are particularly useful in systematizing procedures to give consistent estimates. The following equation explains the method:

Estimated value = (standard estimating unit)(quantity)(adjustment)   (14.2)

The adjustment in this equation takes care of variations in the units with such factors as location, time frames, and productivity.

14.7 Computer-Aided Estimating

The application of computer technology to the preparation of cost estimates has become commonplace. Considerable investment has been made by firms in developing computer hardware and software for proprietary use. With the advent of lower-cost computer systems, more estimating programs and computer estimating services have become available commercially [34, 57, 59, 60].

Cost-estimating computer systems have a wide variety of applications, ranging from aids for quantity takeoffs to sophisticated parametric modeling systems. Examples of computer-aided estimating applications are as follows [52]:

1. Quantity takeoff
2. Factor cost estimating
3. Historical cost data retrieval
4. Probabilistic simulation
5. Parametric modeling
6. Unit cost estimating
7. Free-form computer estimating as part of an integrated data base project control system
8. Regression analysis
9. Compilation of estimating data

Computer-aided estimating is a technique used to increase accuracy and to reduce the time and cost of preparing capital-cost estimates. Care is required in using computer programs for estimating purposes. It is easy to lose the judgment of experienced estimators by relying solely on mechanically derived, computer-generated estimates. The computer should be viewed only as an aid in the cost estimating process.

14.8 Cost Indexes

A *cost index* is a dimensionless number used to adjust the cost of an item from one time period to another. The adjustment is necessary because of the changing value of money with time. The change in value is shown in Figs. 14.6 and 14.7, in which the Producers Price Index for Capital Goods and the Consumer Price Index are shown for the years 1970–1979. It is apparent from these plots that costs used in 1975 would not be valid for estimating purposes in 1979.

To convert costs from one time period to another, the following relationship is used:

$$C_2 = C_1 \frac{\text{index}_2}{\text{index}_1} \tag{14.3}$$

Example 14.4 A construction project was built in 1976 at a cost of 750 k\$. The index for this project in 1976 was 238. What would the project cost in 1979 if the index in 1979 is 320?

SOLUTION From Eq. (14.3),

$$750 \frac{320}{238} = 1008 \text{ k\$}$$

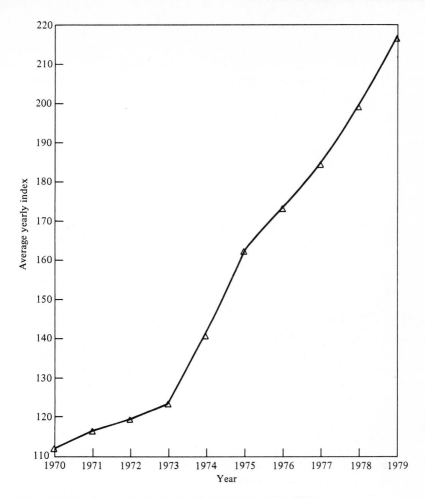

Figure 14.6 Producers Price Index for Capital Goods, 1970–1979.

There are numerous cost indexes covering many types of construction and many related disciplines. Some of the more common indexes are the *Engineering News-Record* (ENR) *Building Cost Index, Engineering News-Record Construction Cost Index, Chemical Engineering Plant Construction Cost Index, Marshall and Swift Installed Equipment Cost Index,* and the *Nelson Refinery Construction Cost Index.* Values for these indexes for the years 1970 to 1980 are shown in Table 14.6, where the values have been rounded off.

Table 14.7 presents recent values for some of the specialized indexes available. These values are all related to the year 1972 = 100, and are based mainly on data supplied by the U.S. Department of Commerce.

All indexes are calculated from a base year and vary depending upon the index and when it was begun. To convert all indexes to a common base year, the

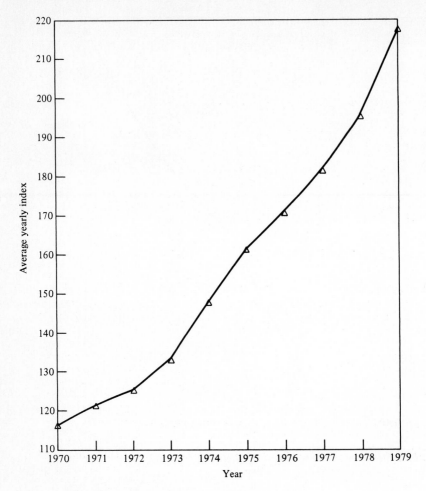

Figure 14.7 Consumer Price Index, 1970–1979.

following relationship may be used [16]:

$$\text{Index value on new base} = \frac{\text{index value (old base) to be converted}}{\text{index value (old base) for new base year}}(100)$$

$$(14.4)$$

Example 14.5 Convert the ENR Building Cost Index of March 1979 based on 1913 = 100 to a base year of 1967. The index for March 1979 is 1750 based on 1913 = 100.

SOLUTION The index value for the new base year (1967 average value) on the old base (1913 = 100) is 672. From Eq. (14.4),

$\frac{1750}{672}(100) = 260.4$, the ENR Building Cost Index for March, 1979 based on 1967 = 100

Table 14.6 Some major cost indexes

Year	Engineering News-Record Construction 1967 = 100	Engineering News-Record Building 1967 = 100	Marshall and Swift Installed Equipment 1926 = 100	Nelson Refinery Construction 1946 = 100	Chemical Engineering Plant Construction 1957-1959 = 100
1970	130	124	303	127	126
1971	148	141	321	141	132
1972	164	156	332	153	137
1973	177	169	344	163	144
1974	188	178	398	182	165
1975	206	193	444	200	182
1976	223	211	472	215	192
1977	240	229	505	223	204
1978	250	248	545	244	219
1979	280	269	599	264	239
1980	301	288	660	288	261
Dec. 1980	316	301	688	299	277

Source: See *Engineering News-Record,* vol. 180, Mar. 21, 1968, pp. 77–86, for descriptions of a number of indexes; also includes data on *Chemical Engineering Plant Construction Cost Index.*

The Bureau of Labor Statistics of the Department of Labor publishes much data on price changes in all industries. Some of these data, such as the Producers Price Index, formerly the Wholesale Price Index, are components of many cost indexes, and others are useful for the construction of highly specialized indexes.

The ENR indexes [15] are usually based on 1913 = 100, although the ENR

Table 14.7 Some specialized cost indexes

	Sewer Treatment Plants*	Sewer Line Construction†	Federal Highways Composite†	Water and Power Resource	Electric Light and Power‡	Telephone Outside Plant§
1975	145	140	148	139	149	130
1976	152	148	144	149	158	141
1977	162	158	157	158	169	148
1978	177	173	192	167	179	155
1979	195	194	223	184	197	171

*Environmental Protection Agency.
†Federal Highway Administration.
‡Hanoy-Whiteman public utility.
§Bell System telephone plant.

Table 14.8 Components of the ENR indexes

Component	Construction Index	Building Index
Structural steel shapes, base mill price, cwt	25	25
Portland cement 20-cities average, bulk, bbl	6	6
2 × 4, S4S lumber, 20-city average, thousand board-ft (MBF)	1.088	1.088
Common labor, 20-city average, h	200	
Skilled labor, 20-city average, h		68.38

and Nelson indexes in Table 14.6 have been converted to a 1967 base because the values have become so large over the years. The components of the ENR indexes are shown in Table 14.8.

It is apparent from this breakdown that the only difference between the two indexes is the use of common labor, 200 h, in the Construction Index, and skilled labor, 68.38 h, in the Building Index.

The Marshall and Swift Equipment Cost Index is the average of installed costs in 47 industries and is obtained by a complex procedure involving equipment appraisals, modifying factors, and a judgment about current economic conditions [65]. The Nelson Refinery Construction Index is based on a 40 percent material and 60 percent labor distribution, as explained in the literature [51]. The Chemical Engineering Plant Cost Index is a complex index consisting of four major components [6], as shown in Table 14.9. The equipment component is made up of seven subcomponents, as shown in Table 14.10. As with many indexes, the Chemical Engineering Index allows for a productivity factor of 2.5 percent per year compounded. The various components are calculated from Producers Price Indexes published by the Bureau of Labor Statistics [70].

In addition to indexes published in the literature, there are special indexes

Table 14.9 Components in the Chemical Engineering Plant Cost Index

Component	Percent
Equipment machinery and supports	61
Erection and installation labor	22
Buildings, material, and labor	7
Engineering and supervision manpower	10
Total	100

Table 14.10 Equipment subcomponents in the Chemical Engineering Plant Cost Index

Component	Percent
Fabricated equipment	37
Process machinery	14
Process instruments	7
Pumps and compressors	7
Electrical equipment	5
Pipes, valves, and fittings	20
Structural supports, insulation, and paint	10
	100

published by econometric companies. These are client-oriented indexes and are available on an annual-fee basis. Some of the more notable are:

Stanford Research Institute (SRI) Index
Data Resources Inc. (DRI) Index
Chase Econometric Inflation Index

When all cost indexes are converted to a common base year and plotted, the curves are very irregular. It is imperative that cost indexes be used with caution and an index selected that most nearly suits the project being adjusted. One of the major pitfalls of indexes is the tendency to use them over too long a period. A rule of thumb is to use indexes only to adjust costs for a 4- to 5-year time span. Using them over longer time spans greatly reduces the accuracy of the results.

Cost indexes can reflect only average costs and must be used with judgment for the specific purpose intended. The results can vary with the index selected, as shown in Example 14.6.

Example 14.6 An air conditioning system was purchased in 1973 for $85,000. Estimate its cost in 1979 using both the Marshall and Swift Index and the Chemical Engineering Plant Cost Index.

SOLUTION Marshall and Swift Index:

$$85,000 \frac{599 \ (1979 \ \text{index})}{344 \ (1973 \ \text{index})} = \$148,000$$

Chemical Engineering Index:

$$85,000 \frac{239 \ (1979 \ \text{index})}{144 \ (1973 \ \text{index})} = \$141,000$$

The error for the 6-year period using two different indexes is approximately 5 percent. Using indexes over a larger time span would sharply increase the margin of error.

The American construction industry is changing from a strictly domestic business to an international one. With this change in emphasis, more indexes are being established by foreign countries. This is more specifically true for the Chemical Process Industry (CPI). One of the broadest indexes has been developed in Canada by Statistics Canada [64]. It is updated quarterly for both residential and nonresidential construction. The Chemical Plant Index is patterned after the Chemical Engineering Plant Cost Index, as shown below.

Annual plant cost		100%
Machinery and equipment		54.2
Fabrication and equipment	23.6	
Process machinery	26.7	
Pumps and compressors	7.5	
Pipe, valves, and fittings	10.2	
Process instruments	6.2	
Electrical equipment	9.6	
Other utilities equipment	7.8	
Structural supports	8.4	
	100.0	
Field erection		16.6
Buildings		13.5
Engineering and administration		15.7
Total annual plant cost		100.0

The base year is 1971 = 100, and no credit is taken for labor productivity.

Table 14.11 Selected foreign cost indexes

Year	Canadian Plant Cost Index* 1971 = 100	U.K. Plant Cost Index† 1970 = 100	West German Plant Cost Index‡ 1962 = 100
1971	100	108	136
1972	105	122	140
1973	112	136	146
1974	133	170	157
1975	156	205	167
1976	168	238	176
1977	182	300	185
1978	196	329	192
1979	217	380	202

*Used with permission of the Minister of Supply Services, Canada.
†*The Cost Engineer* (London).
‡Used with permission of *Chemische Industrie* (Germany).

Other foreign indexes published on a regular basis are:

West Germany Plant Cost Index [37]
U.K. Chemical Plant Cost Index [66]

Table 14.11 presents values for these indexes.

14.9 Planning Estimates

Planning the preparation of cost estimates helps to establish the purpose of the estimate, the time available for its completion, the type of estimate required, the estimating technique to use, and the range of accuracy desired. When these elements are determined, a time schedule can be developed to help decide whether a pre-estimate survey is needed and whether manual or computer methods are to be used. It is important to have a written description of the proposed facilities, the bases for the costs, and any underlying assumptions. The final estimate must be reviewed by management to check the results against similar or past project experiences.

The planning of capital estimates provides data for gauging the cost of estimate preparation for future projects. Because the cost of preparing an estimate of a given type and accuracy can vary over a large range (see Table 14.3), standard cost factors are not usually applicable. Currently, contractors have been giving considerable attention to lowering estimate preparation costs through the use of advanced estimating techniques and computer applications. Figure 14.8 shows an example of an estimate preparation schedule or plan.

14.10 Detailed Estimates against Preliminary Estimates

As pointed out in Sec. 14.2, detailed estimates are expensive and time-consuming to prepare, although they are the most accurate. The checklist in Table 14.12 gives some idea of the scope of items involved in a detailed estimate. Each builder or contractor has lists of this type to help minimize omissions. Because of the expense, time, and lack of information in the early project stages, many more preliminary estimates than detailed estimates are prepared. Because not all projects pass through the screening procedure successively, much time, effort, and expense are saved by making preliminary estimates. The next several sections will discuss and provide examples of the common preliminary estimate types.

14.11 Equipment Costs for Preliminary Estimates

Preliminary estimates can range from very quick, inaccurate ones to reasonably detailed, accurate ones, as shown in Fig. 14.1. In many preliminary estimating methods, the term *purchased equipment* or *delivered equipment* is used. These terms refer to the cost of the uninstalled equipment delivered to the construction site.

The most accurate method of determining equipment costs is to obtain firm

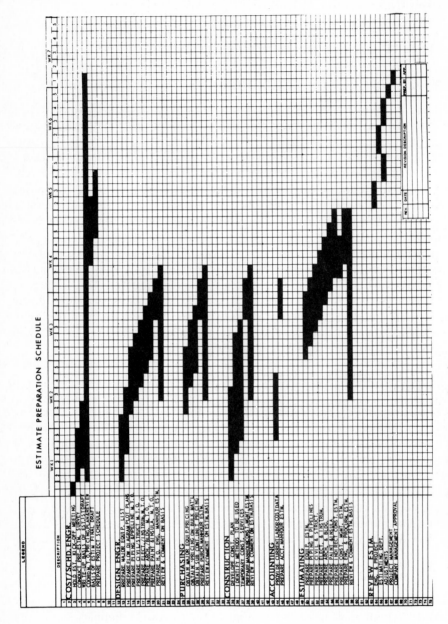

Figure 14.8 Example of an estimate preparation plan.

Table 14.12 Checklist of items for fixed capital cost estimating

Land
  Surveys
  Fees
  Property cost
Site development
  Site clearing
  Grading
  Roads, access and on-site
  Walkways
  Railroads
  Fence
  Parking areas
  Other paved areas
  Wharves and piers
  Recreational facilities
  Landscaping
Process buildings
  (List as required) Include in each as required substructure, superstructures, platforms, supports, stairways, ladders, access ways, cranes, monorails, hoists, elevators
Auxiliary buildings
  Administration and office
  Medical or dispensary
  Cafeteria
  Garage
  Product warehouse(s)
  Parts or stores warehouse
  Maintenance shops—electric, piping, sheet metal, machine, welding, carpenters, instrument
  Guard and safety
  Hose houses
  Change houses
  Smoking stations (in hazardous plants)
  Personnel building
  Shipping office and platforms
  Research laboratory
  Control laboratories
Building services
  Plumbing
  Heating
  Ventilation
  Dust collection
  Air conditioning
  Sprinkler systems
  Elevators, escalators
  Building lighting
  Telephones
  Fire alarm
  Paging
  Intercommunication systems
  Painting

Process equipment
  (List carefully from checked flow sheets)
Non-process equipment
  Office furniture and equipment
  Cafeteria equipment
  Safety and medical equipment
  Shop equipment
  Automotive heavy maintenance and yard material-handling equipment
  Laboratory equipment
  Lockers and locker-room benches
  Garage equipment
  Shelves, bins, pallets, hand trucks
  Housekeeping equipment
  Fire extinguishers, hoses, fire engines
Process appurtenances
  Piping—carbon steel, alloy, cast iron, lead-lined, aluminum, copper, asbestos-cement, ceramic, plastic, rubber, reinforced concrete
  Pipe hangers, fittings, valves
  Insulation—piping, equipment
  Instruments
  Instrument panels
  Electrical—panels, switches, motors, conduit, wire, fittings, feeders, grounding, instrument and control wiring
Utilities
  Boiler plant
  Incinerator
  Ash disposal
  Boiler feed-water treatment
  Electric generation
  Electrical substations
  Refrigeration plant
  Air plant
  Wells
  River intake
  Primary water treatment—filtration, coagulation, aeration
  Secondary water treatment—deionization, demineralization, pH and hardness control
  Cooling towers
  Water storage
  Effluent outfall
  Process-waste sewers
  Process-waste pumping stations
  Sanitary-waste sewers
  Sanitary-waste pumping stations
  Impounders, collection basins
  Waste treatment, including gases
  Storm sewers

Table 14.12 (*Continued*)

Yard distribution and facilities (outside battery limits)	Process, project, and general engineering
Process pipe lines—steam, condensate, water, gas, fuel oil, air, fire, instrument, and electric lines	Drafting
	Cost engineering
	Procurement, expediting, and inspection
Raw-material and finished-product handling equipment—elevators, hoists, conveyors, airveyors, cranes	Travel and living expense
	Reproductions
	Communications
	Scale model
Raw-material and finished-product storage—tanks, spheres, drums, bins, silos	Outside architect and engineering fees
Fuel receiving, blending, and storage	Construction expense
Product loading stations	Construction, operation, and maintenance of temporary sheds, offices, roads, parking lots, railroads, electrical, piping, communication, and fencing
Track and truck scales	
Miscellaneous	
Demolition and alteration work	
Catalysts	Construction tools and equipment
Chemicals (initial charge only)	Warehouse personnel and expense
Spare parts and non-installed equipment spares	Construction supervision
	Accounting and timekeeping
Surplus equipment, supplies and equipment allowance	Purchasing, expediting, and traffic
	Safety and medical
Equipment rentals (for construction)	Guards and watchmen
Premium time (for construction)	Travel and transportation allowance for craft labor
Inflation cost allowance	
Freight charges	Fringe benefits
Taxes and insurance	Housekeeping
Duties	Weather protection
Allowance for modifications and extra construction work during startup	Permits, special licenses, field tests
	Rental of off-site space
Engineering costs	Contractor's home office expense and fees
Administrative	Taxes and insurance, interest

Source: R. H. Perry and C. H. Chilton: *Chemical Engineers' Handbook,* 5th ed., McGraw-Hill Book Company, New York, 1973, p. 25–13.

bids from fabricators or suppliers [56]. Often these same suppliers can provide quick equipment prices orally which are close to a bid price, although not to be taken as such. Next in accuracy are equipment costs from the owner's files of past purchases, adjusted by cost indexes to current costs. Last in accuracy are the equipment costs published in the literature [7, 17, 18, 50, 55, 56, 58, 61, 72, 80]. One is never certain, however, about what is included in published costs.

14.12 Equipment Costs by Scaling

The most common method for obtaining delivered equipment costs for equipment of a desired size is to scale the cost from data of another size, such as the exponential estimating technique discussed in Sec. 14.6. Table 14.13 gives exponents for various types of equipment [56]. An exponent of 0.6 should be used only in the absence of more specific information [44, 46]. Table 14.14 gives the magni-

Table 14.13 Exponent values for typical process equipment

Equipment	Size range	Exponent
Blender, double cone rotary, c.s.	50–250 ft^3	0.49
Blower, centrifugal	10^3–10^4 ft^3/min	0.59
Centrifuge, solid bowl, c.s.	10–10^2 hp drive	0.67
Crystallizer, vacuum batch, c.s.	500–7000 ft^3	0.37
Compressor, reciprocating, air-cooled, two-stage, 150 psi discharge	10–400 ft^3/min	0.69
Compressor, rotary, single-stage, sliding vane, 150 psi discharge	10^2–10^3 ft^3/min	0.79
Dryer, drum, single vacuum	10–10^2 ft^2	0.76
Dryer, drum, single atmospheric	10–10^2 ft^2	0.40
Evaporator (installed), horizontal tank	10^2–10^4 ft^2	0.54
Fan, centrifugal	10^3–10^4 ft^3/min	0.44
Fan, centrifugal	2×10^4–7×10^4 ft^3/min	1.17
Heat exchanger, shell and tube, floating head, c.s.	100–400 ft^2	0.60
Heat exchanger, shell and tube, fixed sheet, c.s.	100–400 ft^2	0.44
Kettle, cast iron, jacketed	250–800 gal	0.27
Kettle, glass lined, jacketed	200–800 gal	0.31
Motor, squirrel cage, induction, 440 volts, explosion proof	5–20 hp	0.69
Motor, squirrel cage, induction, 440 volts, explosion proof	20–200 hp	0.99
Pump, reciprocating, horizontal, cast iron (includes motor)	2–100 gpm	0.34
Pump, centrifugal, horizontal, cast steel (includes motor)	10^4–10^5 gpm $\times$ psi	0.33
Reactor, glass lined, jacketed (without drive)	50–600 gal	0.54
Reactor, S.S., 300 psi	10^2–10^3 gal	0.56
Separator, centrifugal, c.s.	50–250 ft^3	0.49
Tank, flat head, c.s.	10^2–10^4 gal	0.57
Tank, c.s., glass lined	10^2–10^3 gal	0.49
Tower, c.s.	10^3–2×10^6 lb	0.62
Tray, bubble cup, c.s.	3–10 ft diameter	1.20
Tray, sieve, c.s.	3–10 ft diameter	0.86

Source: M. S. Peters and K. D. Timmerhaus: *Plant Design and Economics for Chemical Engineers,* 3d ed., McGraw-Hill Book Company, New York, 1980, p. 167.

Table 14.14 Error introduced by use of the 0.6 factor

Ratio of Q_2/Q_1	Actual cost-capacity factor								
	0.2	0.3	0.4	0.5	0.6	0.7	0.8	0.9	1.0
	Percent error								
5:1	+89	+61	+37	+17	0	−16	−28	−39	−48
10:1	+150	+100	+59	+26	0	−21	−37	−50	−60

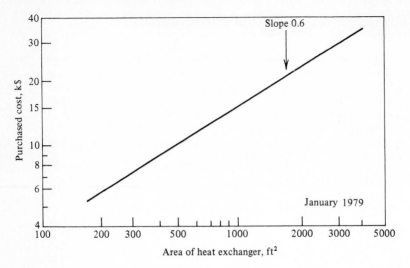

Figure 14.9 Application of the six-tenths factor rule to heat exchanger costs. (M.S. Peters and K. D. Timmerhaus: "Plant Design and Economics for Chemical Engineers," 3d ed., McGraw-Hill Book Company, New York, 1980, p. 166.)

tude of the error by using the 0.6 factor in place of various actual cost-capacity factors.

Figure 14.9 shows the technique used to determine an exponential value. The slope of a log-log plot of this nature is defined as the *cost-capacity factor.*

Example 14.7 The purchased cost of a 400-ft^2 heat exchanger was \$9000 in January 1979, when the cost index was 561. What cost would you expect for a 1000-ft^2 exchanger in December 1980, when the index was 688?

SOLUTION From Eq. (14.1), Eq. (14.3), and Fig. 14.9, the new exchanger cost will be

$$9000\left(\frac{1000}{400}\right)^{0.6}\frac{688}{561} = \$19,130$$

14.13 Turnover Ratios

Probably the least accurate but most rapid method of estimating plant costs is the use of the turnover ratio, which is defined as follows:

$$\text{Turnover ratio} = \frac{\text{gross annual sales}}{\text{fixed capital investment}} \qquad (14.5)$$

Gross annual sales is the product of unit selling price and annual production rate.

Turnover ratios vary from a low of about 0.2 to a high of 8.0 [68]. A rough figure for the chemical industry would be in the range 1.00 to 1.25. Values less than 1.00 are generally found in large-volume, capital-intensive industries which use raw basic materials, such as steel making or power generation [68]. The reciprocal of the turnover ratio, or capital ratio, is sometimes used [41].

Example 14.8 Estimate the capital investment for a plant to produce 12 Mlb/year of chemical Y, which sells for $0.40/lb. M is mega-.

SOLUTION Assume a turnover ratio of 1.00 as typical of the chemical industry. From Eq. (14.5),

$$\text{Fixed capital investment} = \frac{12 \times 10^6 \times 0.40}{1.00} = \$4.8 \times 10^6$$

14.14 Investment Cost per Unit of Capacity

It is possible to make an order-of-magnitude estimate from data expressed as the fixed capital investment per unit of annual production capacity. These costs are for average capacities, and hence there will be inaccuracies since the plant capacities deviate from average. Table 14.15 shows some typical costs per unit data for chemical plants and refinery units. One should note that the cost-capacity factor, or six-tenths rule, applies to total plants as well as to items of equipment [11]. However, some authorities recommend a seven-tenths factor for entire plants.

Example 14.9 Estimate the capital cost in December 1980 for a plant to produce 80 Mlb/year of butanol. M is mega-.

SOLUTION Plant size is:

$$\frac{80,000,000}{2000} = 40,000 \text{ t/year}$$

From Table 14.15,

$$\text{Capital cost} = 40,000 \times 600 = 24 \text{ M\$}$$

14.15 Power Factors Applied to Plant Costs

The use of cost-capacity factors, as in Eq. (14.1), along with cost indexes to adjust for inflation, is probably the easiest way of estimating plant costs [7]. There are numerous literature references for plant costs of various types and for various capacities [22, 42, 50, 67]. Adjustments for the desired plant size can be made with cost-capacity factors of the type shown in Table 14.15.

Example 14.10 Estimate the capital cost as it would have been in 1979 for the butanol plant in Example 14.9 if the capacity were 20 Mlb/year instead of 80 Mlb/year. Use 277 for the Chemical Engineering Plant Construction Index for 1980. M is mega-.

SOLUTION From Eq. (14.1) and Table 14.6,

$$\text{Capital cost} = 24 \left(\frac{20}{80} \right)^{0.4} \left(\frac{239}{277} \right) = 11.9 \text{ M\$}$$

The method is improved by separating out the direct and indirect costs and applying the cost-capacity relationship to just the direct-cost portion of the fixed capital investment. The fixed cost thus becomes

$$C_F = f[D(R)^X + I] \tag{14.6}$$

Table 14.15 Typical unit capital cost data for process plants, December 1980

Product or process	Units of production	Typical plant size	Fixed capital investment, $ per unit	Cost-capacity factor
Chemical plants	1000 ton/year		Per ton/year	
Acetic acid		10	480	0.68
Acetone		100	240	0.45
Ammonia		100	180	0.53
Ammonium nitrate		100	35	0.65
Butanol		50	600	0.40
Chlorine		50	410	0.45
Ethylene		50	195	0.83
Ethylene oxide		50	745	0.78
Formaldehyde (37%)		10	1200	0.55
Glycol		5	2170	0.75
Methanol		60	160	0.60
Nitric acid		100	50	0.60
Polyethylene (high density)		5	2400	0.65
Propylene		10	240	0.70
Sulfuric acid		100	25	0.65
Urea		60	95	0.70
Refinery units	1000 bbl/day		Per bbl/day	
Alkylation		10	1450	0.60
Coking (delayed)		10	1925	0.38
Coking (fluid)		10	1200	0.42
Cracking (fluid)		10	1200	0.70
Cracking		10	360	0.70
Distillation (atm.)		100	240	0.90
Distillation (vac.)		100	145	0.70
Hydrotreating		10	240	0.65
Reforming		10	2170	0.60
Polymerization		10	360	0.58

Source: Based on M. S. Peters and K. D. Timmerhaus: *Plant Design and Economics for Chemical Engineers,* 3d ed., McGraw-Hill Book Company, New York, 1980, pp. 184–185:

where C_F is the fixed cost, f is the ratio of cost indexes, D is the direct plant cost, R is the ratio of plant sizes, X is the cost-capacity factor, and I is the indirect-cost portion of the fixed capital cost [56].

14.16 Factor Methods of Estimating

The factor method as a technique was mentioned in Sec. 14.6. The fixed capital investment, or in some cases the total capital investment, is estimated by multiplying the delivered or purchased equipment cost by a factor. The use of a single constant factor cannot possibly allow for all the variations among projects, geographical locations, or site-specific conditions that a detailed estimate would consider.

The earliest of these methods was the Lang factor method [39]. Improvements and refinements have been made to increase the accuracy without complicating the procedure unduly. The improvements introduce flexible factors and allow the exercise of the estimator's judgment in place of an inflexible constant factor. A discussion of the more frequently used methods follows.

14.17 The Lang Factor Method

This technique, originally proposed by Lang, has been used in the past for quick order-of-magnitude cost estimates. Lang suggested multiplying the delivered cost of equipment by the following factors to obtain total process plant costs [39]:

3.10 for solid process plants
3.63 for solid-fluid plants
4.74 for fluid process plants

It is obvious that such factors do not allow the estimator to exercise any judgment.

Example 14.11 Estimate the total process plant cost for a refinery addition for which the delivered cost is 8 M$. The refinery is a fluids-processing plant.

SOLUTION Using the appropriate Lang factor,

$$\text{Total plant cost} = 8 \times 4.74 = 38 \text{ M}\$$$

14.18 Hand Factors

Greater accuracy can be achieved by using factors based on different types of equipment. Hand [28, 29] recommends the following factors, applied to delivered equipment costs, to obtain the battery-limits cost:

4   for fractionating columns, pressure vessels, pumps, and instruments
3.5   for heat exchangers
2.5   for compressors
2   for fired heaters

These factors are very close to those in a more complete list developed later by Wroth [78] and shown in Table 14.16.

Example 14.12 Estimate the battery-limits cost of the refinery in Example 14.11 if the equipment distribution and costs are as follows:

Equipment	Cost, k$
Blowers and fans	117
Compressors, reciprocating	584
Furnaces	1168
Heat exchangers	934
Instruments	584
Motors, electric	700
Pumps, centrifugal, motor driven	234
Tanks:	
Process	292
Field erected	1051
Towers	2336
	8000

Using the individual cost ratios from Table 14.16:

Equipment	Cost, k$
Blowers and fans	$117 \times 2.5 = 290$
Compressors, reciprocating	$584 \times 2.3 = 1{,}340$
Furnaces	$1{,}168 \times 2.0 = 2{,}340$
Heat exchangers	$934 \times 4.8 = 4{,}480$
Instruments	$584 \times 4.1 = 2{,}390$
Motors, electric	$700 \times 8.5 = 5{,}950$
Pumps, centrifugal, motor driven	$234 \times 7.0 = 1{,}640$
Tanks:	
Process	$292 \times 4.1 = 1{,}200$
Field erected	$1{,}051 \times 2.0 = 2{,}100$
Towers	$2{,}336 \times 4.0 = 9{,}340$
	31,070

The cost is approximately 31 M$, not too different from the 38 M$ estimated by the Lang factor method.

A recent improvement in equipment factors has been proposed by Cran [12]. The breakdown in types of equipment is more extensive, a distinction is made in factors for materials of construction, and the instrument and indirect costs are factored separately. In addition, Cran has introduced accuracy determinations into his estimates. His cost relationship is:

$$\text{Total plant cost} = (\Sigma \, EF_D + IF_I)(1 + F_0) \tag{14.7}$$

where E is the delivered equipment cost, F_D is a direct-cost factor that varies with the type of equipment and its material of construction, I is the sum of the costs of all instruments, F_I is the direct-cost factor for instruments, and F_0 is the indirect cost factor. By this variation on Lang factors, Cran is able to increase his accuracy considerably, resulting in an error of only 6 percent [12].

Table 14.16 Process-plant cost ratio from individual equipment

Equipment	Factor*
Blender	2.0
Blowers and fans (including motor)	2.5
Centrifuges (process)	2.0
Compressors:	
Centrifugals, motor-driven (less motor)	2.0
Steam turbine (including turbine)	2.0
Reciprocating, steam and gas	2.3
Motor-driven (less motor)	2.3
Ejectors (vacuum units)	2.5
Furnaces (package units)	2.0
Heat exchangers	4.8
Instruments	4.1
Motors, electric	8.5
Pumps:	
Centrifugal, motor-driven (less motor)	7.0
Steam turbine (including turbine)	6.5
Positive displacement (less motor)	5.0
Reactors—factor as approximate equivalent type of equipment	
Refrigeration (package unit)	2.5
Tanks:	
Process	4.1
Storage	3.5
Fabricated and field-erected (50,000 + gal)	2.0
Towers (columns)	4.0

Source: W. F. Wroth, "Factors in Cost Estimation," *Chemical Engineering,* vol. 67, October 1960, p. 204.

*Multiply the purchase cost by a factor to obtain the installed cost, including the cost of site development, buildings, electrical installations, carpentry, painting, contractor's fee and rentals, foundations, structures, piping, installation, engineering, overhead, and supervision.

14.19 Chilton Method

Using this method, the estimator must exercise some judgment because the factors are given as ranges. In contrast to the Lang and Hand methods, this method begins with equipment installation costs. As pointed out in Sec. 14.6, these costs are about 50 percent of the delivered equipment costs. Aries and Newton [5] give 43 percent, Woods gives a range of 40 to 120 percent [77], and Peters and Timmerhaus give an average of 43.7 percent [56]. This factor changes, of course, as the relative costs of installation labor and materials vary with changes in the economy. A figure of 43 percent is common, however. Table 14.17 is a summary of the factors used in the Chilton method. Again, these factors are designed to cover the items in a detailed estimate, as shown in Table 14.12, to give a total plant cost. The use of ranges for the factors should give more accurate estimates.

Example 14.13 Estimate the cost of the refinery addition in Examples 14.11 and 14.12 by the Chilton method.

Table 14.17 Chilton factors for estimating total plant costs†

Item no.	Item	Multiplying factor	Operating on item no.
1.	Delivered equipment cost	1.00	1
2.	Installed equipment cost	1.40–2.20	1
3.	Process piping		
	Type of plant—solid	0.07–0.10	2
	solid/fluid	0.10–0.30	2
	fluid	0.30–0.60	2
4.	Instrumentation		
	Amount—little or none	0.02–0.05	2
	some	0.05–0.10	2
	extensive	0.10–0.15	2
5.	Buildings and site development		
	Type of plant—existing	0.00	2
	outdoor	0.05–0.20	2
	outdoor/indoor	0.20–0.60	2
	indoor	0.60–1.00	2
6.	Auxiliaries (power, steam, water)		
	Extent—none	0.00	2
	minor addition	0.00–0.05	2
	major addition	0.05–0.25	2
	new facilities	0.25–1.00	2
7.	Outside lines		
	Unit—closely integrated	0.00–0.05	2
	separated	0.05–0.15	2
	scattered	0.15–0.25	2
8.	Total physical plant costs (Σ cost of items 2–7)		
9.	Engineering and construction		
	Complexity—simple	0.20–0.35	8
	difficult	0.35–0.50	8
10.	Contingency and contractor's fee		
	Process—firm	0.10–0.20	8
	subject to change	0.20–0.30	8
	speculative	0.30–0.50	8
11.	Size factor		
	Unit—large	0.00–0.05	8
	small	0.05–0.15	8
	experimental	0.15–0.35	8
12.	Total plant cost (Σ cost of items 8–11)		

†Based on C. H. Chilton: "Cost Data Correlated," *Chemical Engineering,* vol. 56, June 1949, pp. 97–106.

SOLUTION This fluids plant is to be closely integrated with existing facilities and is to be highly instrumented with automatic controls. The refinery is located in the Gulf Coast area, where outdoor construction exists, and the addition involves well-known processes. There need be only a minor addition—a large commercial unit—to the auxiliary facilities.

The Chilton method is tabulated in Table 14.18. As one might expect, the estimate is lower than that obtained with the Lang and Hand methods because this is a plant addition rather than a new, or "grass-roots" plant.

Table 14.18 Example of the use of Chilton factors for capital cost estimation of a refinery addition

Item no.	Item	Multiplying factor	Operating on item no.	Cost of item
1.	Equipment costs, delivered	1.0	1	ΣE
2.	Installed equipment cost	1.43	1	$1.430 \, \Sigma E$
3.	Piping (includes insulation)	0.40	2	$0.572 \, \Sigma E$
4.	Instrumentation	0.15	2	$0.215 \, \Sigma E$
5.	Buildings and site development	0.10	2	$0.143 \, \Sigma E$
6.	Auxiliaries	0.05	2	$0.072 \, \Sigma E$
7.	Outside lines and site development	0.05	2	$0.071 \, \Sigma E$
8.	Total physical cost (items 2–7)	....	....	$2.503 \, \Sigma E$
9.	Engineering and construction	0.25	8	$0.626 \, \Sigma E$
10.	Contingencies and contractor's fee	0.15	8	$0.376 \, \Sigma E$
11.	Size factor	0.03	8	$0.075 \, \Sigma E$
12.	Total plant cost (items 8–11)	....	....	$3.580 \, \Sigma E$

Another method, proposed by Peters and Timmerhaus [56], is a combination of the Lang and Chilton methods. The method is based on *delivered* equipment costs but retains Lang's classification of solid, solid-fluid, and fluid processing plants. Table 14.19 shows the factors to be used in estimating major process plant additions to an existing site.

The text accompanying this table gives ranges for each item in the table so that the estimator can modify the factors to fit special project conditions. The table gives average factor values, and the end results with average values are reasonably close to those obtained using Lang factors.

Another variation on factor methods is one proposed by Rudd and Watson [62]. In this method the fixed capital cost is again calculated from the delivered equipment cost, but the multiplying factors are all greater than 1. The method can be expressed mathematically as

$$\text{Fixed cost} = \phi_1 \phi_2 \phi_3 \cdots \phi_n \tag{14.8}$$

in which the ϕ's are the factors for the various items, such as in Tables 14.17 or 14.19. Note that in the Rudd-Watson method the factors are multiplied together, whereas in the Chilton and Peters-Timmerhaus methods the factors are added together [33].

14.20 Plant Cost by Analytical Procedure

Today, with the emphasis on computers, there is interest in the use of equations, instead of tables and charts, for cost estimation. Hirsch and Glazier [31] developed the following equation, suitable for computer calculation:

$$I = E[A(1 + F_L + F_P + F_M) + B + C] \tag{14.9}$$

Table 14.19 Peters-Timmerhaus ratio factors for capital-cost estimation

	Percent of delivered-equipment cost for		
Item	Solid-processing plant	Solid-fluid-processing plant	Fluid-processing plant
Direct costs			
Purchased equipment delivered (including fabricated equipment and process machinery)	100	100	100
Purchased equipment installation	45	39	47
Instrumentation and controls (installed)	9	13	18
Piping (installed)	16	31	66
Electrical (installed)	10	10	11
Buildings (including services)	25	29	18
Yard improvements	13	10	10
Service facilities (installed)	40	55	70
Land (if purchase is required)	6	6	6
Total direct plant cost	264	293	346
Indirect costs			
Engineering and supervision	33	32	33
Construction expenses	39	34	41
Total direct and indirect plant costs	336	359	420
Contractor's fee (about 5% of direct and indirect plant costs)	17	18	21
Contingency (about 10% of direct and indirect plant costs)	34	36	42
Fixed capital investment	387	413	483
Working capital (about 15% of total capital investment)	68	74	86
Total capital investment	455	487	569

Source: M. S. Peters and K. D. Timmerhaus: *Plant Design and Economics for Chemical Engineers,* 3d ed., McGraw-Hill Book Company, New York, 1980, p. 180.

where I = total battery-limits investment, \$

A = total purchased equipment cost on an FOB basis less the incremental cost for corrosion-resistant alloys, \$

B = installed equipment cost, \$†

C = incremental cost of alloy materials used only for their corrosion-resisting properties, \$

E = indirect cost factor representing contractors' overhead and profit, engineering, supervision, and contingencies; E is normally assumed to be 1.4

†All vessels larger than 12 ft in diameter should be considered as field erected unless specifically designated as shop fabricated.

F_L = cost factor for field labor; $F_L A$ is the total cost for field labor, less supervision, and excluding the labor charges in item B

F_M = cost factor for miscellaneous items; $F_M A$ includes the materials cost for insulation, instruments, foundations, structural steel, buildings, wiring, painting, and the cost of freight and field supervision

F_P = cost factor for piping materials; $F_P A$ is the total cost of piping materials, including pipe, fittings, valves, hangers, and supports but excluding insulation and installation charges

The three factors F_L, F_P, and F_M are not simple ratios, but are defined by the equations

$$\log F_L = 0.635 - 0.154 \log A_0 - 0.992\frac{e}{A} + 0.506\frac{f}{A} \qquad (14.10)$$

$$\log F_P = -0.266 - 0.014 \log A_0 - 0.156\left(\frac{e}{A}\right) + 0.556\frac{p}{A} \qquad (14.11)$$

$$F_M = 0.344 + 0.033 \log A_0 + 1.194\frac{t}{A} \qquad (14.12)$$

where $A_0 = A/1000$, k\$

e = total heat exchanger cost, less the incremental cost of the alloy, \$

f = total cost of field-fabricated vessels, less the incremental cost of the alloy, \$; ordinarily all vessels larger than 12 ft in diameter are field erected

p = total pump plus the cost for the driver less the incremental cost of the alloy, \$

t = total cost of the tower shells less the incremental cost of the alloy, \$

These equations are easily solved, but the effort can be reduced by a nomograph prepared by Walas [74].

Example 14.14 Estimate the total battery-limits investment for a process plant using the estimated equipment costs shown in Table 14.20.

SOLUTION From the data given, the following factors are derived:

$$A = \$4,310,000$$

$$A_0 = \$4310$$

$$B = \$371,000$$

$$C = \$280,000 + \$100,000 + \$20,000 = \$400,000$$

$$\frac{e}{A} = \frac{\$1,200,000}{\$4,310,000} = 0.278$$

$$\frac{f}{A} = \frac{\$60,000 + \$105,000 + \$350,000}{\$4,310,000} = 0.119$$

Table 14.20 Estimated major equipment costs for Example 14.14

Item	Cost calculated on carbon steel basis k$	Total calculated cost, including incremental alloy, k$	Incremental cost of alloy, k$
Towers:			
Shells, 12 ft diam. or less	$200		
Over 12 ft diam.	60	340	280
Trays and internals	400	500	100
Drums, 12 ft diam. or less	80		
Over 12 ft diam.	105		
Miscellaneous vessels:			
Shells, 12 ft diam. or less	40		
Over 12 ft diam.	350		
Total internals	55	75	20
Heat exchangers	1200		
Pumps	140		
Pump drives	120		
Compressors	1500		
Miscellaneous FOB equipment	60		
Total FOB equipment cost	$4310		
Erected equipment costs:			
Fired heaters	360		
Tanks	11		
Total erected equipment costs	$371		

$$\frac{p}{A} = \frac{\$140{,}000 + \$120{,}000}{\$4{,}310{,}000} = 0.060$$

$$\frac{t}{A} = \frac{\$200{,}000 + \$60{,}000}{\$4{,}310{,}000} = 0.060$$

from which Eqs. (14.10) to (14.12) give F_L, F_P, and F_M.

$$\log F_L = 0.635 - 0.154 \log A_0 - 0.992\frac{e}{A} + 0.506\frac{f}{A}$$

$$= 0.635 - 0.154 \log 4310 - 0.992(0.278) + 0.506(0.119) = -0.140$$

$$F_L = 0.724$$

$$\log F_P = -0.266 - 0.014 \log A_0 - 0.156\frac{e}{A} + 0.556\frac{p}{A}$$

$$= -0.266 - 0.014 \log 4310 - 0.156(0.278) + 0.556(0.060) = -0.327$$

$$F_P = 0.471$$

$$F_M = 0.344 + 0.033 \log A_0 + 1.194\frac{t}{A}$$

$$= 0.344 + 0.033 \log 4310 + 1.194(0.060)$$

$$= 0.536$$

Finally, from Eq. (14.9),

$$I = 1.4[4,310,000(1 + 0.724 + 0.471 + 0.536) + 371,000 + 400,000]$$

$$= \$17,600,000$$

or approximately \$18 million.

It is interesting to note that the ratio of total investment to major equipment costs is

$$\frac{17,600,000}{4,310,000 + 371,000 + 280,000 + 100,000 + 20,000} = 3.5$$

and the ratio of total investment to FOB costs of equipment is

$$\frac{17,600,000}{4,310,000 + 280,000 + 100,000 + 20,000} = 3.7$$

These factors are very close to overall plant investment factors given previously.

In the absence of other information, multiplying the cost of major equipment items by 4 seems to be the easiest way to get a reasonable figure for the total process plant investment cost.

14.21 Compartmentalization Methods of Estimating

It has been possible to improve cost-estimating techniques by building up the estimate by parts to which separate factors are applied. In these methods the equipment is grouped, or compartmentalized, by being broken down into its components. A factor is applied to each component, and then the factored components are summed. The method proposed by Cran and described in Sec. 14.18 is a combination of the Lang method and a compartmentalized method.

Compartmentalization methods can be grouped into four types of estimates. These are (1) the average unit-cost estimate, (2) the functional-unit cost estimate, (3) the unit-operations estimate, and (4) the modular estimate [56].

Each of these approaches involves the same breakdown into components. In the average unit-cost method, in choosing the factors, the size of the equipment, the materials of construction, the operating pressures, and the process are considered. The factors are then applied to the basic equipment cost to get the battery-limits cost, from which the total plant cost is derived by using additional factors [47-49].

In the functional-unit cost method, the equipment is grouped according to function, such as distillation, evaporation, or filtration. The auxiliary equipment, such as pumps and exchangers, is grouped with the functional equipment, and a factor is applied to the group to calculate the final capital investment. A method of this type has been proposed by Bridgewater [10].

In the unit-operations estimate, it is assumed that similar units have similar accessory equipment and should therefore have similar costs [7]. As an example, nearly all distillation columns have a condenser, a reflux drum, a reboiler, and reflux and bottom pumps. Instead of treating this distillation operation as consisting of six pieces of equipment, it is treated as a single-unit operation. In this method the cost for a specific operation is a function of size, materials of construction, pressure, and temperature. The method was originally proposed by Zevnik and Buchanan [79], and more recently, improvements in the method have been proposed by Hensley [30] and Viola [73]. Each of these authors gives examples of the method, but so far, this procedure suffers from insufficient suitable data because of the wide range of possible process operations modules [68].

In the modular estimating method, the focus is on individual modules in the total system, with each module consisting of a group of similar items [56]. For example, all furnaces would be included in one module, all process vessels in another, all pumps in a third, and so on. The total module cost is made up of the sum of the direct material cost M (itself a sum of equipment items E and field installation materials m), and field labor costs L for each piece of equipment, then multiplied by indirect cost factors to give the total module cost. To find the fixed capital investment, the extra costs for adjuncts and auxiliaries are added to the sum of the total module costs. The module estimating method was first proposed by Dodge et al. [13] but has been expanded and improved upon by Guthrie [23, 26]. The method as proposed by Guthrie is the most complicated of the factor methods and so should yield more reliable results because of the sounder basis for the installation charges. It is a popular method, but the types of equipment for which the necessary cost data are correlated are limited [68].

14.22 The Miller Method

Because it seems to be the most complete of the compartmentalized methods, an example will be worked out by the Miller average unit-cost method.

To make a factored estimate more accurate, the Lang factor can be broken down into parts, permitting a different factor for each project [7]. In Miller's method, this idea is expanded by suggesting that it is difficult to choose a single value for each subfactor; hence the estimator should determine a low, a probable, and a high value for each factor. From these factors would come a minimum, a most likely, and a maximum plant investment cost.

Miller also proposes that because large equipment, high-pressure equipment, and alloy equipment are all very expensive per piece, the installation, piping, foundation, and other such factors should be lower than the corresponding factors for less expensive equipment. For this reason, Miller developed a sliding scale for factors depending upon the average cost of the "main plant items" (MPI) [49]. The average unit cost is calculated by adding all the costs of all the MPIs, through pumps, and dividing by the total number of such items. All equipment costs are based on delivered prices in 1958 in order to choose the proper factors, but these factors are applied to current costs.

In this method, the plant is divided into four areas: battery limits (B/L), storage and handling (S&H), utilities (U), and services (S). The last three areas can be estimated separately because they are a function of the size of the facility, rather than of the type of product being made [7].

The battery limits cost is calculated from the factors given in Table 14.21. The costs for storage and handling, utilities, and services are calculated from the factors given in Table 14.22. The following example illustrates the use of these tables.

Table 14.21 Ratio factors for estimating battery-limits costs

| | | Range of factors as percent of basic equipment | | | | | | |
| | | Average unit cost of MPI in 1958 dollars | | | | | | |
		Under $3000	3000–5000	5000–7000	7000–10,000	10,000–13,000	13,000–17,000	Over 17,000
Basic equipment (delivered to site, excluding sales taxes and catalyst)	MPI (Main plant items)	X	X	X	X	X	X	X
	MUE (Miscellaneous unlisted items) Early flowsheet stage Scope of work well defined Note: Top of ranges: Complicated processes Many process steps Bottom of ranges: Simple processes Few process steps	20–10% of MPIs in all categories 10–1% of MPIs in all categories						
	Basic equipment = MPI + MUE	100	100	100	100	100	100	100
Field erection of basic equipment	High percentage of equipment involving high field labor Average (Mild steel equipment)	23/18 18/12.5	21/17 17/11.5	19.5/16 16/10.8	18.5/15 15/10	17.5/14.2 14.2/9.2	16.5/13.5 13.5/8.5	15.5/13 13/8
	High percentage of corrosion materials and other high unit-cost equipment involving little field erection	12.5/7.5	11.5/6.7	10.8/6	10/5.5	9.2/5.2	8.5/5	8/4.8

Table 14.21 (*Continued*)

	Average unit cost of MPI in 1958 dollars						
	Under $3000	3000–5000	5000–7000	7000–10,000	10,000–13,000	13,000–17,000	Over 17,000
Equipment foundations and structural supports							
High—Predominance of compressors or mild steel equipment requiring heavy fdns.			17/12	15/10	14/9	12/8	10.5/6
Average—For mild steel fabricated equipment			12.5/7	11/6	9.5/5	8/4	7/3
Average—For predominance of alloy and other high unit-price fabricated equipment	7/3	8/3	8.5/3	7.5/3	6.5/2.5	5.5/2	4.5/1.5
Low—Equipment more or less siting on floor	5/0	4/0	3/0	2.5/0	2/0	1.5/0	1/0
Piling or rock excavation	Increase above values by 25–100%						
Piping (includes ductwork excludes insulation)							
High—Gases and liquids, petrochemicals, plants with substantial ductwork	105/65	90/58	80/48	70/40	58/34	50/30	42/25
Average for chemical plants—Liquids, electrolytic plants	65/33	58/27	48/22	40/16	34/12	30/10	25/9
Liquids and solids	33/13	27/10	22/8	16/6	12/5	10/4	9/3
Low—Solids	13/5	10/4	8/3	6/2	5/1	4/0	3/0
Insulation of equipment only							
Very high—Substantial mild steel equipment requiring lagging and very low temperatures	13/10	11.5/8.5	10/7.4	9/6.2	7.8/5.3	6.8/4.5	5.8/3.5
High—Substantial equipment requiring lagging and high temperatures (petrochemicals)	10.3/7.5	9/6.3	7.8/5.2	6.7/4.2	5.7/3.4	4.7/2.8	4.8/2.5
Average for chemical plants	7.8/3.4	6.5/2.6	5.5/2.1	4.5/1.7	3.6/1.4	2.9/1.1	2.2/0.8
Low	3.5/0	2.7/0	2.2/0	1.8/0	1.5/0	1.2/0	1/0

Insulation of piping only	Very high—Substantial mild steel piping requiring lagging and very low temperatures	22/16	19/13	16/11	14/9	12/7	9/5	6/3.5
	High—Substantial piping requiring lagging and high temperatures (petrochemicals)	18/14	15/12	13/10	11/8	9/6	7/4	4.5/2.5
	Average for chemical plants	16/12	14/10	12/8	10/6	8/4	6/2	4/2
	Low	14/8	12/6	10/5	8/4	6/3	4/2	2/1
All electrical (except building lighting and instrumentation)	Electrolytic plants (includes rectification equipment)		55/42	50/38	45/33	40/30	35/26	
	Plants with mild steel equipment, heavy drives, solids	26/17	22.5/15	19.5/12.5	17/10	14/8.5	12/7	10/6
	Plants with alloy or high unit-cost equipment, chemical and petrochemical plants	18/9.5	15.5/8.5	13/6.5	11/5.5	9/4.5	7.3/3.5	6/2.5
	Note: Above figures include 1 to 3% for B/L outside lighting which is not covered in Building Services							
Instrumentation	Substantial instrumentation, central control panels, petrochemicals		58/31	46/24	37/18	29/13	23/10	18/7
	Miscellaneous chemical plants		32/13	26/10	20/7	15/5	11/3	8/2
	Little instrumentation, solids		21/9	17/7	13/5	10/3	7/2	5/1
	Note: Total instrumentation cost does not vary a great deal with size and hence is not readily calculated as a percent of Basic Equipment. This is particularly true for distillation systems. If in doubt, detailed estimates should be made.							

Table 14.21 (*Continued*)

Miscellaneous (includes site preparation, painting and other items not accounted for above)	Top of range: large complicated processes Bottom of range: smaller, simple processes	Range for all values of basic equipment 6–1%
Buildings—architectual and structural (excludes building services)	Note: When building specifications and dimensions are known, a high-speed building cost estimate is recommended especially if buildings are a significant item of cost. If a separate estimate is not possible, evaluate the buildings as follows before selecting the factors.	

Building evaluation
when most of process units are located inside buildings

	High: Brick and steel	Medium	Low: Economical		Evaluation
Quality of const.	+4	+2	0		
Type of equipment	Very high unit cost equipment −3	Mostly alloy steel −2	Mixed materials −1	Mostly carbon steel 0	
Operating pressures	Very high −2	Intermed. −1	Atmos. 0		

Building class = algebraic sum =

Average unit cost of MPI in 1958 dollars

	Bldg. class	Under $3000	3000–5000	5000–7000	7000–10,000	10,000–13,000	13,000–17,000	Over 17,000
Most of process units inside buildings	+2	92/68	82/61	74/56	67/49	59/44	52/39	46/33
	+1 to −1	72/49	62/43	56/38	51/33	45/29	41/26	36/21
	−2	50/37	44/33	40/29	35/25	30/21	27/18	23/15
Open-air plants with minor buildings		37/16	32/13	28/11	24/8	20/6	17/4	14/2

Note: The following factors are for Battery Limit (process) buildings only and are expressed in percent of the Building-Architectual & Structural cost. They are not related to the Basic Equipment cost.

	High	Normal	Low
Compressed air for general service only	4	1½	5
Electrical lighting	18	9	5
Sprinklers	10	6	3
Plumbing	20	12	3
Heating	25	16	8
Ventilation: without air conditioning	18	8	0
with air conditioning	45	35	25
Total overall average*	85	55	20

The above factors apply to those items normally classified as building services. They do not include:

1. Services located outside the building such as sub-stations, outside sewers, outside water lines, etc., all of which are considered to be outside the Battery Limit, as well as outside the building.
2. Process services.

*The totals provide the ranges for the type of buildings involved and are useful when the individual service requirements are not known. Note that the over-all averages are not the sum of the individual columns.

Building services

Source: C. A. Miller, "Factor Estimating Refined for Appropriation of Funds," AACE Bulletin, vol. 7, September 1965, pp. 92–118.

Table 14.22 Ratio factors for auxiliary costs

Item	Grass-roots plant	Battery-limit addition to existing site
Storage and handling (S&H) (% of B/L cost)		
Low—Raw material by pipeline, little warehouse space	2	0
Average—Average raw material storage and finished product warehousing	15–20	2–6
High—Tank farm for raw material, substantial warehousing	70	20
Utilities (U) (% of B/L cost)		
Low	10	3
Average	20–30	6–14
High	50	30
Services (S) (% of [B/L + (S&H) + U)]		
Low	5	0
Average	10–16	2–6
High	20	15

Source: C. A. Miller: "Factor Estimating Refined for Appropriation of Funds," *AACE Bulletin,* vol. 7, September 1965, pp. 92–118.

Example 14.15 Repeat Examples 14.11, 14.12, and 14.13. The refinery addition for these examples is to be made up predominantly from mild steel equipment, and the processes involved in the addition are simple, involving few petrochemical-type process steps. The delivered equipment cost for 250 pieces of process equipment is 8 M$. Estimate the cost of the refinery addition by the Miller method.

SOLUTION The first step is to calculate the average unit cost in 1958 dollars. The Marshall and Swift index is 688 for December 1980 and 229 for 1958. The average unit cost in 1958 is

$$\frac{8 \times 10^6}{250} \frac{229}{688} = \$10,700$$

and therefore the factors in the fifth column of Table 14.21 are used.

Table 14.23 shows the development of the estimate. In the Miller method, miscellaneous unlisted items (MUE) are added to the MPI, and low, average, and high figures are obtained by taking ±10 percent on each side of the average. In addition, an adjustment of ±10 percent is made in the factored items just before the B/L costs are calculated. This is done because it is unlikely that all costs would be at their low or high estimates. One should note that storage, utilities, and services are based on a factor times the average B/L cost and the average (B/L + S + U) cost because these auxiliaries should not vary widely.

From the results of this example, the refinery addition would probably cost 31.2 M$. It should not cost more than 37.7 M$ and is not likely to cost less than 24.4 M$.

Table 14.24 is a summary of the four example estimates for the refinery addition. The estimates group about a cost of 31 M$. If 31 M$ is assumed correct, the estimates are within about ±20 percent.

14.23 Building Cost

Preliminary cost estimates for building costs can be made by multiplying the square feet of floor area or the cubic feet of volume by a unit cost. Unit costs for several building types are shown in Table 14.25. The data in the table are based on average unit costs for the major cities in the United States. Variations in labor costs, local efficiency, and materials prices will result in local fluctuations from these values [43].

A more accurate method of estimating is to use unit costs for each component of a building, such as walls, floors, and foundations; however, such figures can still be in error for a specific case. Table 14.26 shows a detailed breakdown of costs for a particular type of building and gives the unit costs for various components.

14.24 Pipeline Costs

The weighted average cost per mile for 49 onshore pipeline projects and 30 off-shore pipeline projects is shown in Fig. 14.10. The total cost per mile for placing an onshore pipeline in the ground ranges from about $100,000 for a 6-in-diameter line to about $500,000 for a 30-in-diameter line.

14.25 Costs of Port Facilities

Port facilities encompass a wide range of design types, functions and locations. Some examples of structures for port facilities include breakwaters, finger piers, marginal and free-standing wharves, small-craft piers, bulk cargo (dry) piers, bulk liquid and gas-handling facilities, container wharf systems, and conventional general cargo facilities [4].

Because of the broad differences between projects, there appears to be no valid correlation of costs between one project and another. This is in contrast to plant project cost data. The building of ports and facilities is construction-equipment intensive. It is not uncommon to see construction expenses exceed 60 percent of the cost for construction work. Therefore, construction planning, equipment selection, and the method of construction become significant factors in developing capital estimates for ports and facilities. The costs for four port projects are given in Table 14.27. The original source contains details of the structures.

14.26 Power Plant Costs

Construction costs for nuclear and fossil fuel power plants in the United States have been difficult to estimate because of long project delays in an inflationary economy. Most power project delays result from design changes and added regu-

Table 14.23 Total plant cost for Example 14.15

Average unit cost in 1958 = $10,700

Date: December 1980

No. of MPIs 250	M & S indexes 1958 229 Current 688	Factors or adjustments	Thousands of dollars		
			Low	Average	High
MPI (main plant items)		—	—	$8,000	—
MUE (misc. unlisted items)		±2%		160	
Basic equipment		±10%	7,344	8,160	8,976
Field erection—average		9.2/11.7/14.2			
Foundations and supports—average steel		5.0/7.3/9.5			
Piping—high, petrochemicals		34.0/46.0/58.0			
Insulation of equipment—high		3.4/4.6/5.7			
Insulation of pipe—high		6.0/7.5/9.0			
Electrical—low, petrochemical		4.5/6.8/9.0			
Instrumentation—substantial		13.0/21.0/29.0			
Miscellaneous—simple		1.0/2.0/3.0			
Buildings—minor, open air		6.0/13.0/20.0			
Building services—low—20% buildings		1.2/2.6/4.0			
Subtotal, factored items		83.3/122.5/161.4			
Adjustments, ±10%		+8.3/0.0/−16.1			
Total factored items		91.6/122.5/145.3	6,727	9,996	13,042

		Col 1	Col 2	Col 3
B/L cost		14,071	18,156	22,018
Storage—high, addition	10.0/15.0/20.0			
Utilities—average, addition	6.0/10.0/14.0			
Subtotal—S + U – %(average B/L)	16.0/25.0/34.0	2,905	4,539	6,173
Subtotal—B/L + S + U		16,976	22,695	28,191
Services—average, addition	2.0/4.0/6.0	454	908	1,362
Total direct cost		17,430	23,603	29,553
Indirect costs				
Engineering and construction	20%	4,721	4,721	4,721
Total direct and indirect costs		22,151	28,324	34,274
Contingency (including 3% construction fee)	10%	2,215	2,832	3,427
Total plant cost		24,366	31,156	37,701

Table 14.24 Results of selected estimates for a refinery addition

Method	Cost estimate, M$
Lang	38
Hand	31
Chilton	28.6
Miller—average	31.2
High	37.7
Low	24.4

Table 14.25 Typical unit costs for several classes of buildings

Type of building	$/ft^2$ of floor area	$/ft^3$
Apartments, mid-rise	32.25	3.42
Auditoriums	56.80	2.27
Banks	69.20	5.10
Churches	46.40	2.92
Colleges, classrooms	55.85	3.73
Colleges, laboratories	60.60	4.43
Courthouses	62.35	4.77
Department stores	26.85	1.28
Dormitories, mid-rise	55.25	4.75
Factories	21.35	1.52
Fire stations	52.20	3.32
Garages (commercial)	28.35	1.78
Gymnasiums	38.95	1.89
Hospitals	77.60	5.95
Housing (public)	31.55	2.80
Jails	76.30	5.85
Libraries	50.60	3.35
Motels	34.60	3.08
Offices, mid-rise	51.60	3.97
Police stations	69.25	4.15
Restaurants	54.30	4.96
Retail stores	27.50	2.00
Schools, jr. high and middle	42.00	2.66
Supermarkets	27.80	1.54
Telephone exchanges	69.25	4.05
Terminals (bus)	35.85	2.46
Theatres	43.70	2.19
Town halls	52.70	3.88
Warehouses	16.90	1.01

Source: Selected from information copyrighted by the Robert Snow Means Company, Inc., Kingston, Mass. It is reproduced from *Building Construction Cost Data,* 39th ed., 1981, and used with permission.

Table 14.26 Detailed breakdown of mid-rise apartment building costs

			Unit costs			% of total		
		Unit	1/4	Median	3/4	1/4	Median	3/4
17.1	S.F., C.F., and % of total costs							
950	Total: Mechanical and electrical	Apt.	4,330	5,490	7,250			
02	**Apartments, Mid-rise**							
002	Total project costs	S.F.	26.75	32.25	48.15			
		C.F.	2.37	3.42	4.09			
180	Equipment	S.F.	0.35	0.71	1.34	2.50%	2.50%	3.20%
272	Plumbing		2.22	2.41	2.91	8.40%	8.90%	9.50%
277	Heating, ventilating, air conditioning		1.09	1.13	1.40	4.10%	4.30%	10%
290	Electrical		1.80	2.19	4.13	7.10%	7.90%	11.40%
310	Total: Mechanical and electrical		5.75	7.60	11.60	21%	26.60%	30%
900	Per apartment unit, total cost	Apt.	21,170	31,970	45,430			
950	Total: Mechanical and electrical	Apt.	6,160	10,330	13,040			

Source: This information is copyrighted by the Robert Snow Means Company, Inc., Kingston, Mass. It is reproduced from *Building Construction Cost Data*, 39th ed., 1981, and used with permission.

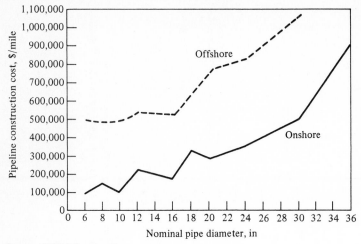

Source: FERC Construction
Permit applications, 1979–1980.

Figure 14.10 Average pipeline construction costs. (*Earl Seaton: "Pipeline Economics," Oil and Gas Journal, vol. 78, Aug. 11, 1980, p. 59. Used with permission.*)

latory requirements. The delays account for 64 percent of the plant costs for an average nuclear unit built between 1969 and 1978. The same comments apply to coal-fired units, with delays and changes adding 62 percent to the total project cost.

A 1-month delay for a nuclear power project with a 2200-MW capacity would cost 20 M$, while the same delay for a coal-fired plant would cost about 15 M$. It is not uncommon for power projects to cost close to $1 billion and take over 10 years to complete.

Delays increase power project costs more than the costs of most other types of

Table 14.27 Port structure costs

Location	Year completed	Size of structure, ft	Overall cost, M$	$/linear ft
Morehead City, N.C.	1969	1000	1.65	1650
Savannah, Ga.	1969	300	1.3	4300
North coast of Tasmania	1969	5860	3.6	614
Oakland, Calif.	1967	1282	1.6	1250

Source: American Society of Civil Engineers, Task Committee on Port Structure Cost: *Port Structure Cost,* New York, 1974, p. 1. Used with permission.

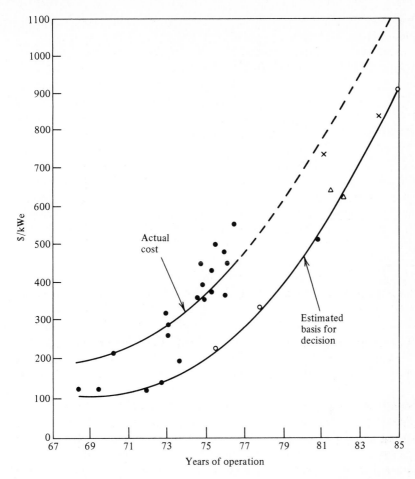

Figure 14.11 Trends in nuclear power plant costs. (*W. C. Finch and F. D. Postula: "Benefits of Probabilistic Cost Estimating for Nuclear Power Plant Projects," AACE Transactions, Paper B4, 1979.*)

projects because of the long construction periods and large capital outlays involved. Therefore, probabilistic cost-estimating techniques are generally used to help define these project costs. The techniques aid in identifying the associated risks and improve the accuracy of and confidence in the capital estimates [19].

Figure 14.11 shows how actual costs have escalated above the estimated costs over the years. The deviations are apparently the result of design changes and project delays.

14.27 International Costs

The operations of multinational corporations cut across national boundaries, and many major contractors work worldwide. As a result of fast and inexpensive air travel, the world is shrinking. For these and other political and social reasons, comparative international costs assume great signficance.

In a completely free world economy, competition should tend to equalize prices throughout the world. Such, however, is not the case; in fact, price may not be related to cost. The domestic and export prices of the same item may differ widely, e.g., as a result of fiscal policies intended to protect new industries. This is particularly true of developing countries, such as India, that want to encourage local industries and also maintain an increasing share of world exports. These policies have not always been successful.

A good portion of future capital expenditure programs will be related to natural resources or forms of energy supply. However, most of the supplies of natural resources are located in remote areas, such as the Arctic, mountainous areas, deserts, offshore areas, or underdeveloped countries. Underdeveloped countries are eager to reap the benefits of their mineral wealth and become part of the modern world.

It is not feasible for estimators to prepare capital-cost estimates for projects located in areas of the world with which they are unfamiliar. Preparing cost estimates for such projects is extremely difficult, especially if the organization has no experience in the prospective geographical area. The cost estimator, together with other engineers, is responsible for developing capital estimates as accurately as possible. To do so requires the evaluation of all cost and scheduling factors that will affect the capital estimate.

Obviously, the estimator can rely on published location adjustment factors supplied by estimating services and trade journals, which consider the effects of labor, materials, construction equipment, and indirect costs applied to a particular location. This method is commonly used for preliminary estimates. Table 14.28 presents such a list of international adjustment factors [9]. This list is the result of literature searches and discussions with informed people from many countries. Miller [45] has published location factors for the United States, the United Kingdom, and Canada. A wider range of countries is represented in an earlier list by Kharbanda [36].

Example 14.16 The cost of a 1000 ton/day ammonia plant, based on natural gas, was 30 M$ in the United States in 1976. Estimate the cost of the same plant in Britain in the same year.

SOLUTION From Table 14.28, the location adjustment factors are 1.0 for the United States and 0.9 for Britain. The cost for the plant in Britain would be

$$30\frac{0.9}{1.0} = 27 \text{ M\$}$$

The costs of plants in different countries vary with the exchange rate, productivity, and inflation. If inflation must be taken into account, Kharbanda [36] lists exchange rates and inflation indexes for various currencies in different years. This situation arises when the base plant and the new plant are built in different years. Fluctuating currency rates add a new dimension to overseas plant costs. An objective discussion appears in an article by Kumar [38].

Table 14.28 International adjustment factors

Location factors for chemical plants of similar function

		U.K. = 1.0	U.S. = 1.0
Australia		1.4	1.3
Austria		1.1	1.0
Belgium		1.1	1.0
Canada		1.25	1.15
Central Africa		~2.0	~2.0
Central America		1.1	1.0
China	(imported element)	1.2	1.1
	(indigenous element)	0.6	0.55
Denmark		1.1	1.0
Eire		0.9	0.8
Finland		1.3	1.2
France		1.05	0.95
Germany (West)		1.1	1.0
Greece		1.0	0.9
Holland		1.1	1.0
India	(imported element)	2.0	1.8
	(indigenous element)	0.7	0.65
Italy		1.0	0.9
Japan		1.0	0.9
Malaysia		0.9	0.8
Middle East		1.2	1.1
Newfoundland		1.3	1.2
New Zealand		1.4	1.3
North Africa	(imported element)	1.2	1.1
	(indigenous element)	0.8	0.75
Norway		1.2	1.1
Portugal		0.8	0.75
South Africa		1.25	1.15
South America (North)		1.5	1.35
South America (South)		2.5	2.25
Spain	(imported element)	1.3	1.2
	(indigenous element)	0.8	0.75
Sweden		1.2	1.1
Switzerland		1.2	1.1
Turkey		1.1	1.0
U.K.		1.0	0.9
U.S.		1.1	1.0
Yugoslavia		1.0	0.9

Notes:
1. Increase a factor by 10% for each 1000 miles, or part of 1000 miles, that the new plant location is distant from a major manufacturing or import center, or both.
2. When materials or labor, or both, are obtained from more than a single source, prorate the appropriate factors.
3. Investment incentives have been ignored.

Source: A. V. Bridgewater: "International Construction Cost Location Factors," *Chemical Engineering*, vol. 86, Nov. 5, 1979, p. 5.

The major economic factors affecting overseas costs are:

1. Political system
2. Financial system
3. Legal system
4. Social system
5. Geography
6. Type of industry

These factors must be considered from one or more of the following points of view:

1. International
2. National
3. Local
4. Company
5. Project

No generalizations are possible, so each case must be considered individually and in depth.

If capital estimates are required for projects in a location for which there are no published adjustment factors or in-house historical information, the estimator must obtain estimating data by making a preestimate survey. The survey requires careful study of local market conditions and the many components of a capital estimate. Estimating data may be obtained by talking with local or foreign contractors who have worked that area. A prepared checklist containing questions on major materials, labor, and contracting cost elements is an excellent guide. By relating these questions to the normal estimating base, the estimator will understand local costs and any special conditions which may exist. Table 14.29 is a list of factors important in framing questions for a preestimate survey [9, 54].

There is another important reason for a preestimate survey. Conditions are constantly changing, and international cost comparisons are far from simple. Climatic conditions, for example, can have a pronounced effect both directly (concrete must be heated in Canada but cooled in Iran) and indirectly through productivity. Until recently, the United States was noted for its high productivity in various trades, but today certain trades abroad may be equally productive, or perhaps more so, in their own environments.

Productivity, defined as the ratio of output to input, enters into estimating overseas costs. Productivity is an abstract and controversial subject, and there is no mathematically correct way of measuring it. Productivity applies to the use of land, labor, and capital, but cost estimating generally is more concerned with labor productivity.

Statistical data on labor productivty are usually based on how much actual work is performed for a given amount of time, effort, or resources expended compared with a standard amount of work accomplished. In process plant esti-

Table 14.29 Preestimate survey checklist

Social, economical, and political
    Local culture and customs
    Language problems
    Religion and its holidays
    Attitude of community toward project
    Political climate
    Type of government
    Government agencies
    General economical climate
    Taxes and duties
    Currency fluctuations
    Inflation rate
    Cost of living

Site and environmental
    Soil conditions
    Terrain
    Site access
    Site elevation
    Earthquake factors
    Offshore wind, wave, and water depth
      data
    Climate
    Weather storms
    Environmental requirements

Material and equipment
    Local available materials
    Materials required to import
    Special selected material suppliers
    Freight and logistic cost
    Export preparation cost
    Local fabrication capability
    Acceptable foreign fabricators and
      suppliers

Contracting
    Prime contractors
    Joint ventures
    Subcontractors availability
    Contracting through agents or brokers
    Types of contracts in common use

Legal and financial
    Legal assistance
    Cost of doing business
    Multiple taxation
    Insurance
    Letters of credit
    Bid bonds
    Financing cost
    Overseas premium

Labor
    Labor cost
    Working hours practiced
    Work week practiced
    Labor availability
    Amount of trained labor
    Labor productivity
    Unionized labor
    Labor contracts
    Training requirements

Procurement
    Local practices
    Worldwide procurement
    Public bid opening practices
    Trading company participation
    Expediting and inspection practices

Support
    Construction facilities
    Temporary facilities
    Utilities
    Housing
    Communications
    Security
    Warehousing
    Medical facilities
    Sanitary facilities
    Food and catering
    Recreational facilities
    Education facilities
    Permanent community

Source: A. Patrascu, *Construction Cost Engineering,* Craftsman Book Company of America, Solana Beach, Calif., 1978. Used with permission.

mating, the standard base generally used is the U.S. Gulf Coast area. It is usually difficult to get statistical data on productivity because bidding is a competitive situation and contractors are reluctant to share this type of information. If in-house productivity data are not available, the estimator must consult the limited published data, such as those shown in Fig. 14.12, or obtain assistance from local contractors or labor unions. For accuracy in evaluating productivity data, one should consider individual crafts rather than the entire work force. Coupling ac-

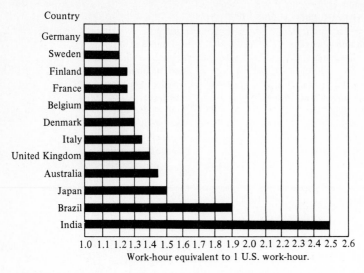

Figure 14.12 Typical worldwide construction-labor productivities. (*J. T. Gallagher: "Analyzing Field Construction Costs," in H. Popper (ed.): Modern Cost Engineering Techniques, McGraw-Hill Book Company, New York, 1970, p. 227.*)

tual productivity data with sound judgment is the best approach to use in determining productivity factors.

Many forces affect labor productivity, such as the lack of experienced craft labor, restrictive labor regulations, other projects in the area, bad weather, poor construction planning, inefficient use of manpower, lack of supervision, and low morale. Some factors can be controlled and others cannot, but all must be evaluated during the project and after its completion. This information will then serve as a historical basis for future estimates [54].

14.28 Staying within Limitations

Cost estimation requires aptitude, but experience is necessary for professionalism. No estimator should work in unfamiliar fields. What may be an easy estimation for one analyst is difficult, if not impossible, for another. Even competent estimators can overextend themselves. Inflation has greatly increased the estimator's task, and information must be continuously updated. Costs may be changing continuously, but estimators must limit their work to the area they know.

NOMENCLATURE

Nomenclature for Chap. 14 is restricted for use within the chapter and conflicts, in part, with nomenclature used in other chapters.

A Total cost of all battery-limits investments, Eqs. (14.9) to (14.12), \$

A_0 $A/1000$, Eqs. (14.10) to (14.12), k\$

B Cost of all erected equipment, Eq. (14.9), \$

C Cost, Eqs. (14.1) and (14.6), \$

C Incremental cost of alloys used for corrosion resistance, Eq. (14.9), \$

D	Direct plant cost, Eq. (14.6), $
e	Total heat-exchanger cost less the incremental cost of the alloy, Eqs. (14.10) and (14.11), $
E	Delivered-equipment cost, Eqs. (14.7) and (14.8), $
E	Indirect-cost factor, Eq. (14.9), dimensionless, normally assumed to be 1.4
f	Ratio of cost indexes, Eq. (14.6), dimensionless
f	Total cost of field-fabricated vessels, Eq. (14.10), $
F_I	Direct-cost factor for instruments, Eq. (14.7), dimensionless
F_L	Cost factor for field labor, Eqs. (14.9) and (14.10), dimensionless
F_M	Cost factor for miscellaneous items, Eqs. (14.9) and (14.12), dimensionless
F_0	Indirect-cost factor, Eq. (14.7), dimensionless
F_P	Cost factor for piping materials, Eqs. (14.9) and (14.11), dimensionless
I	Total battery-limits investment, Eq. (14.9), $
I	Indirect costs, Eq. (14.9), $
I	Instrument costs, Eq. (14.7), $
p	Total pump plus driver cost less the incremental cost of the alloy, Eq. (14.11), $
Q	Capacity, Eq. (14.1), any convenient dimensions
R	Ratio of plant sizes, Eq. (14.6), dimensionless
t	Total cost of tower shells less incremental cost of alloys, Eq. (14.12), $
X	Total cost-capacity factor, Eqs. (14.1) and (14.6), dimensionless
ϕ	Cost factors, Eq. (14.8), dimensionless

PROBLEMS

14.1 A construction project cost 600 k$ in 1972. Estimate the cost in 1977.

14.2 Equipment for a process plant cost 150 k$ in 1973. Estimate the cost in 1977.

14.3 The Chemical Engineering Plant Cost Index shown in Table 14.6 is based on 1957–1959 = 100. Convert this index to a base year 1970 = 100.

14.4 A construction project cost 1 M$ in 1971. Find the cost in 1976.

14.5 Equipment installed in a project in 1972 cost 800 k$. Find the cost in 1979.

14.6 Find the cost in December 1976 or a 1000-hp reciprocal compressor designed to deliver air at 6000 psi. Use the data in the figure below.

14.7 From the following figure showing costs for centrifugal and reciprocating air compressors, deter-

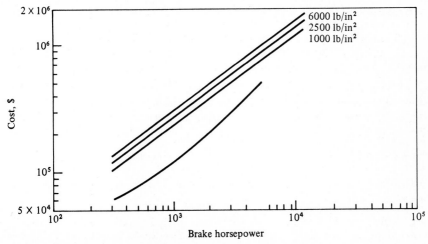

From "Modern Cost Engineering," compiled by *Chemical Engineering*, McGraw-Hill Book Company, New York, 1979, p. 314.

mine the cost-capacity factor for motor-driven reciprocal compressors designed to deliver air at 1000 psi.

14.8 A 300-gal jacketed, glass-lined kettle cost $8500 in 1979. Estimate the cost for a 700-gal kettle of the same type in 1970.

14.9 The delivered cost of the major equipment for a solid-fluid process plant is 2 M$. Estimate the total cost of the plant.

14.10 Estimate the total battery-limits investment for a process plant with an estimated equipment cost based on carbon steel as follows:

	k$	Incremental cost of corrosion-resisting alloy, k$
Tower shells over 12 ft diameter	200	500
Trays and internals	350	600
Drums over 12 ft diameter	150	250
Miscellaneous vessels over 12 ft diameter	400	
Heat exchangers	600	
Pumps and drivers	165	300
Miscellaneous FOB equipment	135	
Erected equipment cost	225	

14.11 The delivered equipment for a plant to beneficiate iron ore costs 6 M$ and consists of 140 individual pieces of equipment. The plant is a "grass-roots" plant to be built in northern Minnesota. The equipment involves little field erection and is subject to erosive action. The process is relatively simple and does not involve very high or very low temperatures. Pressures are atmospheric, and the quality of building construction is medium. Estimate the total plant cost by the Lang factor method.

14.12 Repeat Prob. 14.11 using the Chilton method.

14.13 Repeat Prob. 14.11 using the Peters and Timmerhaus factors.

14.14 Repeat Prob. 14.11 using the Miller method.

14.15 If the capital cost of a process plant in Britain was 33 M$ in 1979, estimate the cost of an identical plant in India in 1979.

14.16 A 1600 ton/day sulfur-burning sulfuric acid plant was reported to cost 20 M$ in the United States in 1978. Estimate the cost of an identically designed, 3500 ton/day plant in Italy in 1979.

14.17 A high-density polyethylene plant built in Norway was reported to cost $38 M$. Estimate the cost if it had been built in Italy.

REFERENCES

1. *AACE Bulletin,* vol. 1, November 1958, p. 12.
2. *AACE Cost Engineers' Notebook,* "Cost Engineering Terminology," May 1971, pp. 5 and 7.
3. *AACE Cost Engineers' Notebook,* "Data for Estimating Piping Cost," Paper 64–31, June 1964.
4. American Society of Civil Engineers, Task Committee on Port Structure Cost, *Port Structure Cost,* New York, 1974, p. 1.
5. Aries, R. S., and R. D. Newton: *Chemical Engineering Cost Estimation,* McGraw-Hill Book Company, New York, 1955.
6. Arnold, T. H., and C. H. Chilton: "New Index Shows Plant Cost Trends," *Chemical Engineering,* vol. 70, Feb. 18, 1963, pp. 143–152.
7. Baasel, W. D.: *Preliminary Chemical Engineering Plant Design,* American Elsevier Publishing Company, Inc., New York, 1976.
8. Bauman, H. C.: *Fundamentals of Cost Engineering in the Chemical Industry,* Reinhold Publishing Corporation, New York, 1964.

9. Bridgewater, A. V.: "International Construction Cost Location Factors," *Chemical Engineering,* vol. 86, Nov. 5, 1979, p. 5.

10. ———:"The Functional-Unit Approach to Rapid Cost Estimation," *AACE Bulletin,* vol. 18, September–October 1976, p. 153.

11. Chilton, C. H.: "Six-tenths Factor Applies to Complete Plant Costs," *Chemical Engineering,* vol. 57, April 1950, pp. 112–114.

12. Cran, J.: "Improved Factored Method Gives Better Preliminary Cost Estimates," *Chemical Engineering,* vol. 88, Apr. 6, 1981, pp. 65–79.

13. Dodge, W. J., et al.: The Module Estimating Technique as an Aid in Developing Plant Capital Costs," *AACE Transactions,* June 1962.

14. Edmunds, K. B.: "Cost Estimating Building Service System, 1," *ACHV,* vol. 62, February 1965, pp. 93–94.

15. ———: "Engineering News-Record Indexes of Cost Trends 1913–68," *Engineering News-Record,"* vol. 180, Mar. 12, 1968, pp. 77–80.

16. ———: "ENR Indexes Reflect Material and Labor Trends," *Engineering News-Record,* vol. 202, Mar. 22, 1979, p. 74.

17. Enyedy, G.: "Cost Data for Major Equipment," *Chemical Engineering Progress,* vol. 67, May 1971, p. 73.

18. ———: "Generate Your Own Cost Curves with the COME Computer Program," *AACE Bulletin,* vol. 14, November–December 1972, p. 171.

19. Finch, W. C., and F. D. Postula: "Benefits of Probabilistic Cost Estimating for Nuclear Power Plant Projects," *AACE Transactions,* Paper B-4, 1979.

20. Gallagher, J. T.: "Analyzing Field Construction Costs," in H. Popper (ed.), *Modern Cost Engineering Techniques,* McGraw-Hill Book Company, New York, 1970, p. 227.

21. Gorman, J. E.: *Simplified Guide to Construction Management for Architects and Engineers,* Cahners Books, Inc., Boston, Mass., 1976, pp. ix–x.

22. Guthrie, K. M.: "Capital and Operating Costs for 54 Chemical Processes," *Chemical Engineering,* vol. 77, June 15, 1970, p. 140.

23. ———: "Capital Cost Estimating," *Chemical Engineering,* vol. 76, Mar. 24, 1969, pp. 114–142.

24. ———, et al.:*Managing Capital Expenditures for Construction Projects,* Craftsman Book Company, Solana Beach, Calif., 1977, p. 80.

25. ———: *Process Plant Estimating, Evaluation, and Control,* Craftsman Book Company, Solana Beach, Calif., 1974, pp. 115, 334.

26. ———: *Process Plant Estimating, Evaluation, and Control,* Craftsman Book Company, Solana Beach Calif., 1974, p. 23.

27. Hackney, J. W.: *Control and Management of Capital Projects,* John Wiley & Sons, Inc., New York, 1965, p. 94.

28. Hand, W. E.: "Estimating Capital Costs from Process Flow Sheets," *AACE Cost Engineers' Notebook,* Paper A3, January 1964.

29. ———: "From Flow Sheet to Cost Estimate," *Petroleum Refiner,* vol. 37, September 1958, pp. 331–334.

30. Hensley, E. F.: "The Unit Operations Approach," *AACE Transactions,* June, 1967.

31. Hirsch, J. H., and E. M. Glazier: "Estimating Plant Investment Costs," *Chemical Engineering Progress,* vol. 60, December 1964, pp. 23–25.

32. *How to Estimate Electrical Work, Electrical Construction and Maintenance,* New York, p. 17.

33. Jelen, F. C.: *Project and Cost Engineers' Handbook,* American Association of Cost Engineers, Morgantown, W. Va., 1979.

34. Johnson, P. W., and F. A. Peters: *A Computer Program for Calculating Capital and Operating Costs,* Bureau of Mines Information Circular 8426, U.S. Department of the Interior, Washington, D.C., 1969.

35. Kerzner, H.: *Project Management: A System Approach to Planning, Scheduling and Controlling,* Van Nostrand Reinhold Company, New York, 1979, pp. 355, 418.

36. Kharbanda, O. P.: *Process Plant and Equipment Cost Estimation,* Vivek Enterprises, Bombay, 1977.

37. Kolbel, H., and J. Schulze: "The Price Index of Chemical Plants," *Chemical Industries,* vol. 27, May 1975.

38. Kumar, P.: "Fluctuating Currency Rate Effects on an Internation Project," Engineering and Process Economics, vol. 3, 1978, pp. 115–121.

39. Lang, H. J.: "Simplified Approach to Preliminary Cost Estimates," *Chemical Engineering,* vol. 55, June 1948, pp. 112–113.

40. LeBlanc, L.: "Platform Price Tag Climbs," *Offshore,* vol. 24, September 1978, pp. 86–88.

41. Lynn, L., and R. F. Howland: "Use Capital Ratio," in H. Popper (ed.), *Modern Cost Engineering Techniques,* McGraw-Hill Book Company, New York, 1970.

42. McGlamery, G. G., R. L. Torstrick, W. J. Broadfoot, J. P. Simpson, S. V. Tomlinson, and J. F. Young: *Detailed Cost Estimates for Advanced Effluent Desulfurization Processes,* EPA-600/2-75-006, IERL, U.S. Environmental Protection Agency, Research Triangle Park, N.C., January 1975.

43. Means, R. S.: *Building Construction Cost Data,* 39th ed., Robert S. Means Company, Inc., Kingston, Mass., 1981, p. 276.

44. Miller, C. A.: "Capital Cost Estimation—A Science Rather Than an Art," *Cost Engineers' Notebook,* June 1968, p. A-1.000.

45. ———: "Converting Construction Costs from One Country to Another," *Chemical Engineering,* vol. 80, July 2, 1979, pp. 89–93.

46. ———: "Current Concepts in Capital Cost Forecasting," *Chemical Engineering Progress,* vol. 69, May 1973, p. 77.

47. ———: "Factor Estimating Refined for Appropriation of Funds," *AACE Bulletin,* vol. 7, September 1965, pp. 92–118

48. ———: "New Cost Factors Give Quick, Accurate Estimates," *Chemical Engineering,* vol. 72, Sept. 13, 1965, p. 226.

49. ———: New Cost Factors Give Quick, Accurate Estimates, in H. Popper (ed.), *Modern Cost Engineering Techniques,* McGraw-Hill Book Company, New York, 1970, p. 58.

50. ———: *Modern Cost Engineering: Methods and Data,* McGraw-Hill Book Company, New York, 1979.

51. Nelson, W. L.: "Refinery Construction Cost Indexes," *Oil and Gas Journal,* vol. 54, Oct. 1, 1956, vol. 54.

52. Ostwald, P. F.: *Cost Estimating for Engineering and Management,* Prentice-Hall, Inc., Englewood Cliffs, N.J., 1974, chaps. 6, 7.

53. Page, J. S.: *Cost Estimating for Pipelines and Marine Structures,* Gulf Publishing Company, Houston, Tex., 1977, p. 15.

54. Patrascu, A.: *Construction Cost Engineering,* Craftsman Book Company, Solana Beach, Calif., 1978, p. 55.

55. Perry, R. H., and C. H. Chilton: *Chemical Engineers' Handbook,* 5th ed., McGraw-Hill Book Company, New York, 1973.

56. Peters, M. S., and K. D. Timmerhaus: *Plant Design and Economics for Chemical Engineers,* 3d ed. McGraw-Hill Book Company, New York, 1980.

57. Peterson, J. N., C. C. Chen, and L. B. Evans: "Computer Programs for Chemical Engineers," *Chemical Engineering,* vol. 85, July 31, 1978, p. 79.

58. Pikulik, A., and H. E. Diaz: "Cost Estimating for Major Process Equipment," *Chemical Engineering,* vol. 84, Oct. 10, 1977, p. 106.

59. ———: "Process Designers Get More Software," *Chemical and Engineering News,* vol. 48, Mar. 30, 1970, p. 38.

60. ———: *Project Economic Analysis,* Program # 198049E, Software Catalog, Texas Instruments, Inc., Lubbock, Tex., September 1980.

61. Popper, H.: *Modern Cost Engineering Techniques,* McGraw-Hill Book Company, New York, 1970.

62. Rudd, D. F., and C. C. Watson, *Strategy of Process Engineering,* John Wiley & Sons, Inc., New York, 1968.

63. Seaton, E.: "1980 Pipeline Economics," *Oil and Gas Journal,* vol. 78, August 11, 1980.

64. ———: *Statistics Canada Bulletin,* vol. 2, no. 7, June 1973, p. 59.

65. Stevens, R. W.: "Equipment Cost Indexes for Process Industries," *Chemical Engineering,* vol. 54, November 1957, pp. 124–126.
66. ———: *The Cost Engineer,* March 1974, pp. 152–157.
67. Tomlinson, S. V., F. M. Kennedy, F. A. Sudhoff, and R. L. Torstrick: *Definitive SO_x Control Process Evaluations: Limestone, Double Alkali, and Citrate FGD Processes,* EPA-600/7-79-177, IERL, U.S. Environmental Protection Agency, Research Triangle Park, N.C., August 1979.
68. Uhl, V. W.: *Standard Procedure for Cost Analysis of Pollution Control Operations—Vol. 1,* EPA-600/8-79-018a, IERL, U.S. Environmental Protection Agency, Research Triangle Park, N.C., June 1979.
69. U.S. Department of Labor, Bureau of Labor Statistics: *Consumer Price Index,* Washington, D.C.
70. U.S. Department of Labor, Bureau of Labor Statistics, *Monthly Labor Review,* Washington, D.C.
71. U.S. Department of Labor, Bureau of Labor Statistics, *Producer Prices and Price Indexes,* Washington, D.C.
72. Vatavuk, W. M., and R. B. Neveril: "Estimating Costs of Dust-Removal and Water Handling Equipment—Part VI," *Chemical Engineering,* vol. 88, Mar. 23, 1981, p. 223.
73. Viola, J. L., Jr.: "Estimate Capital Costs Via a New, Shortcut Method," *Chemical Engineering,* vol. 88, Apr. 6, 1981, pp. 80–86.
74. Walas, S. M.: "Plant Investment Costs by the Factor Methods," *Chemical Engineering Progress,* vol. 57, June 1961, pp. 68–69.
75. Weaver, J. B., and H. C. Bauman: "Glossary of Cost Estimating Terms, I," *Industrial and Engineering Chemistry,* vol. 52, June 1960, pp. 69A–72A.
76. Williams, R., Jr.: "Six-tenth Factor Aids in Approximating Costs," *Chemical Engineering,* vol. 54, December 1947, pp. 124–125.
77. Woods, D. R.: *Financial Decision Making in the Process Industry,* Prentice-Hall, Inc., Englewood Cliffs, N.J., 1975.
78. Wroth, W. F.: "Factors in Cost Estimation," *Chemical Engineering,* vol. 67, October 1960, p. 204.
79. Zevnik, F. C., and R. F. Buchanan: "Generalized Correlation of Process Investment," *Chemical Engineering Progress,* vol. 59, February 1963, p. 70.
80. Zimmerman, O. T.: "Process Equipment Cost Data," *Cost Engineering,* vol. 18, February 1973, p. 12.
81. Zimmerman, O. T.: "Use of Ratio Cost Factors in Estimating," *Cost Engineering,* vol. 10, October 1965, pp. 13–17.

FIFTEEN

OPERATING-COST ESTIMATION

J. H. Black

15.1 Definitions

Operating costs differ from capital costs mainly in that operating costs must be paid repeatedly, in contrast to *capital costs,* which involve a one-time payment. Because they recur repeatedly, operating costs are often called *repeated costs,* whereas capital costs are often referred to as *first costs.* It is important to understand these two types of costs. The operating cost, or manufacturing cost, is the expense involved in keeping a project, operation, or piece of equipment running and producing. The terminology and the connection between the component cost items are shown in Table 15.1.

In this chapter the terms *operating cost* and *manufacturing cost* are used synonymously. In a broad sense, this cost is incurred at the factory level. In addition, a company incurs costs above the factory level associated more closely with management; here it is termed *general expense,* but it is also known as *administrative, sales,* and *general expense.* The sum of the operating cost, or manufacturing cost, and the general expense is designated here as the *total product cost.*

The arrangement shown in Table 15.1 is subject to variation depending upon accounting procedures and circumstances. Thus royalty payments may be directly proportional to throughput and treated as a direct production cost. In some cases, however, the purchase of a patent in connection with a project may be regarded as an expense in lieu of research and development and may appear as part of general expense. Alternatively, the royalty can be a paid-up royalty and hence will be included in the capital investment. Nevertheless, Table 15.1 illustrates the classification of operating costs and will serve well for defining the terms and showing their relationship.

Operating costs fall into two major classifications: direct and indirect. *Direct costs* tend to be proportional to throughput, such as raw materials, and are called *variable costs*. *Indirect costs* tend to be independent of throughput, such as local property taxes, and are called *fixed costs*. Some costs are neither fixed nor directly proportional to output and are known as *semivariable costs* or *regulated costs*. These concepts were developed in Chap. 7 in connection with break-even and minimum-cost analysis.

15.2 Operating-Cost Estimation—General

Operating-cost estimates are made for a variety of reasons. Preliminary estimates act as a screening procedure to minimize the expenditure of time, effort, and money on uneconomical projects and to choose the more economical route among alternatives to the same end result. The use of preliminary estimates to guide research efforts is common practice. Detailed cost estimates are made after screening estimates when the choice has been narrowed to a relatively few alternatives.

Some judgment is required in determining how far to investigate individual operating-cost items. The estimate will indicate which costs dominate and deserve more study and which costs, even if drastically misjudged, will not produce significant changes in the estimate.

It is important to calculate costs at full and reduced levels inasmuch as in practice operating costs are not a linear function of the production rate. Calculations should be made and reported in such a way that they can be reviewed in the future without ambiguity.

Frequently, the operating cost to be calculated is connected with changes in and additions to existing equipment and processing, and an incremental analysis can be made. Costs based on an incremental analysis should be used only in connection with the specific project for which they were intended. Even so, it might be well to make an integral evaluation for the entire project. The increment might pay off, yet it may be wise to abandon the entire project.

Operating costs are commonly calculated on one of three bases: daily, unit-of-product, or annual. Unit-of-product cost is usually expressed as dollars per unit of end product. The annual basis has some advantages over the daily basis because (1) the effect of seasonable variations is evened out, (2) the on-stream time or equipment operating factor is considered, (3) it is more adaptable to less than full-capacity operation, (4) it provides an expeditious way of considering infrequently occurring large expenses, such as the regularly scheduled replacement of parts or equipment or the periodic turnaround costs in a refinery, and (5) the calculations are more directly usable in a profitability analysis.

It is best to use a prepared form for estimating operating costs, and it should conform as closely as possible to that used for recording and controlling costs. Ideally the form should provide such supplementary information as the date, the capital investment, a cost index, identification of the product or service, plant

Table 15.1 Components of total product cost

I. Operating cost or manufacturing cost
  A. Direct production costs
    1. Materials
      a. Raw materials
      b. Processing materials
      c. By-product and scrap credit
      d. Utilities
      e. Maintenance materials
      f. Operating supplies
      g. Royalties and rentals
    2. Labor
      a. Direct operating labor
      b. Operating supervision
      c. Direct maintenance labor
      d. Maintenance supervision
      e. Payroll burden on all labor charges
        i. FICA tax
        ii. Workers' compensation coverage
        iii. Contributions to pensions, life insurance, hospitalization, and dental plans
        iv. Vacations, holidays, sick leave, overtime premium
        v. Company contribution to profit sharing
  B. Indirect production costs
    1. Plant overhead or burden
      a. Administration
      b. Indirect labor
        i. Laboratory
        ii. Technical service and engineering
        iii. Shops and repair facilities
        iv. Shipping department
      c. Purchasing, receiving, and warehousing
      d. Personnel and industrial relations
      e. Inspection, safety, and fire protection
      f. Automotive and rail switching
      g. Accounting, clerical, and stenographic
      h. Communications—telephone, mail, and teletype
      i. Plant custodial and protective
      j. Plant hospital and dispensary
      k. Cafeteria and clubrooms
      l. Recreational activities
      m. Local contributions and memberships
      n. Taxes on property and operating licenses
      o. Insurance—property, liability
      p. Nuisance elimination—waste disposal and pollution control
    2. Depreciation
  C. Contingencies
  D. Distribution costs
    1. Containers and packages
    2. Freight
    3. Operation of terminals and warehouses
      a. Wages and salaries—plus payroll burden
      b. Operating materials and utilities
      c. Rental or depreciation

Table 15.1 (*Continued*)

II. General expense
  A. Marketing or sales costs
    1. Direct
      a. Salespersons' salaries and commissions
      b. Advertising and promotional literature
      c. Technical sales service
      d. Samples and displays
    2. Indirect
      a. Sales supervision
      b. Travel and entertainment
      c. Market research and sales analysis
      d. District office expenses
  B. Administrative expense
    1. Salaries and expenses of officers and staff
    2. General accounting, clerical, and auditing
    3. Central engineering and technical
    4. Legal and patent
      a. Inside company
      b. Outside company
      c. Payment and collection of royalties
    5. Research and development
      a. Own operations
      b. Sponsored, consultant, and contract work
    6. Contributions and dues to associations
    7. Public relations
    8. Financial
      a. Debt management
      b. Maintenance of working capital
      c. Credit functions
    9. Communications and traffic management
    10. Central purchasing activities
    11. Taxes and insurance

Source: Adapted from R. H. Perry et al., *Chemical Engineers' Handbook,* 5th ed., McGraw-Hill Book Company, New York, 1973, pp. 25–27.

location, plant capacity, and operating rate or operating days. The supplementary information is important because these factors affect the component costs, and hence the finished estimate.

Table 15.10 illustrates a prepared form and Table 15.11 an outline or checklist to match it. Other suggested forms appear elsewhere in the literature [20, 34, 47].

The best source of information for use in operating-cost estimates is the data from similar or identical projects within the company. Most companies have extensive records of their operations, so that reliable estimates of operating costs for new projects can be made from the existing records if the new project is similar to installations now operated by the company. Adjustments for increased costs as a result of inflation must be made, and differences in plant site and geographical location must also be considered. Company experience, however, probably yields the most accurate information, particularly for preliminary estimates.

Literature references also provide mcuh information on costs [11, 37]. Many engineering journals and magazines also publish cost information [1, 2, 3, 10, 18, 24, 26].

Methods for estimating cost elements in the absence of specific information are discussed in the following sections. The elements of cost are presented essentially in the order shown in Table 15.11.

Perhaps the most serious sources of error in estimating operating costs result from overlooking elements of cost. A tabular form and checklist cannot always guide the analyst to every cost component. Detailed cost estimation requires that the cost engineer be much more the engineer than a cost accountant. Technical familiarity with the project is essential, and no analyst should undertake detailed cost estimation in fields not directly within his or her specialty.

Many computer programs have been developed for performing operating-cost calculations [27, 36, 38]. While the initial program may be expensive, its use is justified when a number of variations are to be explored by repetitive calculations. Manual calculation is still used, however, because appraising the data takes more effort than making the actual calculations. Care must be exercised in using computer programs so that the resulting estimates are not so mechanically derived that judgment, so necessary in evaluating alternatives, is lost. Programs are also available for hand calculators [39].

15.3 Raw Materials

Raw materials may constitute one of the largest items in the manufacturing cost, varying from 10 to 50 percent of the total product cost [35]. The operating-cost estimate should list raw materials, the units in which the materials are purchased, unit cost, and cost per unit of time per unit of production. Credit should be given for by-products and scrap.

Information on the quantities of the raw materials required can come from published articles, from experiments, or from calculations to which reasonable yields or waste allowances are applied.

Prices of raw materials purchased from outside sources are available on request from the sources of supply. Quality is an important consideration when buying raw materials, such as concentration of an ore, surface finish of a metal, and so forth. Generally, unit price decreases with quantity. However, if the material is available at a low price as a by-product, it may be available in large quantities, if at all, only at a substantially higher unit price.

Prices suitable for estimating purposes can be obtained from catalogs, price lists, or sources such as spot prices published in *Iron Age, Chemical Marketing Reporter, Chemical and Engineering News,* and *European Chemical News* for European prices. Actual prices are negotiated and will vary from the quotations depending on quantity, quality, and such factors as the state of competition, the duration of a contract, and penalties for contract violations. If the raw material is in a competitive market, the published price could be discounted by 20 percent [20].

Raw materials produced captively present a special situation. If the product is now sold, the transfer price is normally the going market price, sales price, less any direct sales costs not incurred, plus freight and other transportation costs for the transfer. If the transferred material is at an intermediate stage that has no going market price, the transfer price can be taken as the going price of the nearest downstream converted product having a going price, less any direct sales cost not incurred, less standard manufacturing costs for any operations which are avoided by the transfer, plus the distribution cost necessary for the transfer.

Periodic makeup of losses, such as for a catalyst in a chemical process, is included as a raw material. If the initial fill of these materials has a life of more than 1 year, its cost is treated as nondepreciable capital investment or working capital. If the life is less than 1 year, the cost is included as a start-up expense. In any case, the periodic makeup is an operating cost.

Fuels used by a project, such as natural gas for a heat-treating furnace, may be considered as a raw material or a utility, depending upon the accounting procedure and circumstances. Costs of fuels are discussed in the following section under utilities. In some chemical processes fuels may be used as a bona fide raw material, e.g., cracking refinery gases for the production of ethylene, and also as a fuel. The two uses may be combined as one raw material or segregated into two costs as desired.

The quantities of by-products are estimated in much the same fashion as the raw-material quantities. The prices of by-products can be estimated from market or selling prices of salable materials less purification, packaging, selling, and transportation costs. If the by-products are intermediate products for which no market exists, they may be credited at their net value to downstream or subsequent operations at a cost value equal to their replacement value.

Many price quotations for raw materials will be FOB the vendor's plant or some basing point, and freight to the plant and local handling costs must be added. Shipping rates are a complicated topic. The company's traffic department can help, or the carriers can be consulted. A valuable source of basic transportation information is Perry [33, sec. 9].

Large-volume, long-haul (over 250 mi) rates in dollars per ton-mile are shown in Table 15.2. The lower values in each range apply to longer distances and larger shipments, while the higher values apply to shorter distances and smaller ship-

Table 15.2 Approximate shipping costs

Transportation mode	Approximate shipping costs, $/ton · mi
Pipeline	0.004–0.006
Barge	0.009–0.020
Tanker	0.003–0.008
Rail	0.025–0.075
Truck	0.09 –0.110

Source: Adapted from J. M. Winton, "Plant Sites: Spreading Out from Houston," *Chemical Week,* vol. 127, Dec. 3, 1980, p. 54.

ments [5]. Rates for distances less than 250 mi may be two to five times higher. Small-volume (less than carload lots) rates may be two to three times higher.

15.4 Utilities

Utility prices can be obtained from company cost records for nominal utility increments. Otherwise they will have to be determined by inquiry from outside sources of a study of the company-owned system if the utility is self-generated. Unit costs often decrease substantially as demand goes up. Rates for most utilities vary considerably with location. Bauman [6] gives a comprehensive study of rates. Table 15.3 gives typical values for common utilities [7, 35, 48, 51].

Table 15.3 Rates for industrial utilities, December 1980

Utility	Cost, $ (Dec. 1980)	Unit
Steam		
500 psig	2.60–3.00	1000 lb
100 psig	1.30–2.70	1000 lb
Exhaust	0.70–1.10	1000 lb
Electricity		
Purchased	0.020–0.050	kWh*
Self-generated	0.016–0.045	kWh
Cooling water		
Well	0.065–0.35	1000 gal
River or salt	0.045–0.13	1000 gal
Tower	0.045–0.175	1000 gal
Process water		
City	0.25–1.00	1000 gal
Filtered and		
softened	0.35–0.90	1000 gal
Distilled	1.70–2.70	1000 gal
Compressed air		
Process air	0.045–0.135	1000 ft^3
Instrument	0.09–0.27	1000 ft^3
Natural gas	2.00–4.50	1000 ft^3
Manufactured gas	1.20–3.60	1000 ft^3
Fuel oil	0.45–1.05	gal
Coal	22.00–50.00	ton
Refrigeration,	1.40	ton/day (288,000 Btu
(ammonia) to		removed)
34°F		

Source: Adapted from M. S. Peters and K. D. Timmerhaus, *Plant Design and Economics for Chemical Engineers,* 3d ed., McGraw-Hill Book Company, New York, 1980, p. 199; J. M. Winton, "Plant Sites Spreading Out from Houston," *Chemical Week,* vol. 127, Dec. 3, 1980, p. 54; and B. R. Beattie and H. S. Foster, Jr., "Can Prices Tame the Inflationary Tiger?" *American Water Works Association Journal,* vol. 72, August 1980, p. 441.

*Highly dependent upon the load factor and the location.

Steam can be used at several pressure levels. High-pressure steam can be reduced to lower pressures by the use of steam turbines that are used for the production of motive power. This is an economical practice if use at various pressure levels can be kept in balance. The allocation of costs between the different pressure levels can become complicated, and the balance can be upset.

Steam is generally charged on the basis of thousand pounds used but can be charged on the basis of million Btus. Credit should be given for the heating value of condensate returned to the boiler. If the condensate is frequently contaminated by processing leaks, a penalty may be assessed.

The cost of steam depends principally upon the cost of the fuel and the capacity and pressure level of the generating units. The full cost of steam includes fuel, boiler-water treatment, operating labor, depreciation, and maintenance. A rough estimate is two to three times the cost of fuel.

Water costs vary widely depending upon the quality and quantity required. If the water can become contaminated, some provision for purification cost before disposal should be provided. Table 15.3 gives some representative water costs. Water rates may be obtained from the local water supplier or through the American Water Works Association. River water on a once-through basis without treatment costs only a few cents per thousand gallons, while most industrial water falls in the range of $0.25 to $1.00/1000 gal.

All major contracts for electrical energy contain a demand component, an energy component, and a variable-fuel-cost component. The demand charge is based on the maximum power requirement drawn by a plant during a measurable period, usually 15 or 30 min in any 1 month. The energy charge is based on kilowatt-hours used. The ratio of average usage to demand is termed the *load factor*. Most electric-power rate schedules are designed to yield greater savings preferably by improving the load factor more than by increasing the load. The fuel-adjustment charge ties the cost, say, to the price of coal, but it is an escalation clause and is used even for electricity generated from water power.

Electric rates may be obtained from the utilities. The Federal Power Commission (FPC) publishes and distributes at a nominal charge the published rates of all public utilities in the United States, but other literature sources are available [51]. Bauman [6] illustrates how rates vary with the location, load, and load factor. Table 15.3 shows some typical power costs.

Cost of fuels varies with the type of fuel and the number of Btus per unit of fuel. Special preheating and firing equipment is required to burn the cheaper but heavier fuel oils. An analysis of fuel economy is sometimes necessary to select the type and quality of fuel for a particular location and situation. Choice of fuel must allow for the required storage, which can make a difference in coal versus oil versus gas.

The estimator must check carefully to make sure that proper provisions are made for delivery of utilities to the site. A common estimating error is the assumption that utilities are available at the site when, in fact, they may not be.

The required utilities can sometimes be estimated in the early stages of project development from available information about similar operations [15, 18, 42]

Table 15.4 Utility factors for losses and contingencies

Utility	Steam	Water	Fuel	Electricity
Factor	1.25–2.00	1.10–1.50	1.10–1.50	1.10–1.25

or estimated from a preliminary design. Utility consumption does not vary directly with the production rate, and variation to the 0.9 power of the capacity ratio might be a better relationship. A factor should be applied to the estimated quantities of utilities to allow for equipment efficiencies, other losses, and contingencies. Appropriate factors for the various utilities are suggested in the literature [14], as shown in Table 15.4. The exact value to use in each range depends upon the load and the site conditions.

15.5 Operating Labor

The most accurate means of establishing operating labor requirements is with a complete manning table, but shortcut methods are satisfactory for most cost estimates.

If the project is a novel one, work-hour requirements can be estimated using elemental time values as developed by Haines [19] or using labor requirements as given in the literature [15, 18, 35]. Table 15.5 shows typical labor requirements for some types of process equipment [35]. Ostwald [32] gives examples and references for estimating labor for such nonprocess industries as metal working, sheet metal operations, craft work, construction, and machine shop operations. Wessell [50] relates labor requirements to process plant capacity by the equation

$$\frac{\text{Operating work-h}}{\text{Tons of product}} = t\frac{\text{no. of process steps}}{(\text{capacity in tons/day})^{0.76}} \tag{15.1}$$

where t is determined by the type of process as follows:

Batch operations with a maximum of labor, 23
Operations with average labor requirements, 17
Well-instrumented continuous-process operations, 10

The equation recognizes the improvement in labor productivity as plants increase in throughput, and can be used to extrapolate known work-hour requirements from one plant to another of different capacity.

It has been suggested that labor requirements vary to about the 0.2 to 0.25 power of the capacity ratio when processing-plant capacities are scaled up or down [23, 31].

Example 15.1 A plant to produce 50 tons/day of product A has been designed. The plant is a modern, highly automated fluid-processing plant. The raw material for product A goes through the four processing steps of heat transfer, chemical reaction, solvent extraction, and distillation. Estimate the annual labor requirement for continuous operation of the plant for 330 days per year.

Table 15.5 Typical labor requirements for process equipment

Type of equipment	No. of workers per unit shift
Dryer, rotary	0.5
Dryer, spray	1
Dryer, tray	0.5
Centrifugal separator	0.25–0.5
Crystallizer, mechanical	1/6
Filter, vacuum	1/8–1/4
Evaporator	0.25
Reactor, batch	1
Reactor, continuous	0.5
Steam plant (100,000 lb/h)	3

Source: M. S. Peters and K. D. Timmerhaus, *Plant Design and Economics for Chemical Engineers,* 3d ed., McGraw-Hill Book Company, New York, 1980, p. 195.

SOLUTION From Eq. (15.1),

$$\frac{\text{Operating work-h}}{\text{Tons of product}} = t\frac{\text{no. of process steps}}{(\text{capacity, tons/day})^{0.76}}$$

$$= 10\frac{4}{(50)^{0.76}} = 2.0457$$

$$\text{Operating work-h/year} = 50 \times 330 \times 2.0457 = 33,750$$

The average hourly rate for labor can be obtained from company records or data in such sources as *Engineering News-Record,* the *Cost Engineers' Notebook* of

Table 15.6 Labor rates in selected industries, September 1980

Industry	Earnings, $/h
Mining	9.26
Construction	10.17
Metals	9.98
Miscellaneous manufacturing	5.53
Food	6.97
Paper	8.02
Chemicals	8.48
Petroleum	10.25

Source: Survey of Current Business, Table S-13, U.S. Department of Commerce, Washington, D.C., October 1980.

the AACE, the *Monthly Labor Review* from the U.S. Department of Labor, and the *Survey of Current Business* from the U.S. Department of Commerce. Table 15.6 gives labor rates in selected industries as of September 1980 [48]. Labor rates, such as those shown in Table 15.6, can be adjusted for geographical differences in rates and productivity and for rate escalation using the methods and information given in Peters and Timmerhaus [35].

Labor rates have been increasing, seemingly on an accelerated basis; hence, care must be taken to get the current figures. Average rates should include the effect of shift and scheduled overtime premiums, but care is required to avoid the overlap with indirect payroll items which are added separately.

Supervision of labor is generally estimated as a percentage of operating labor, a typical value being 20 percent. The range is usually 10 to 25 percent.

15.6 Payroll Charges

Payroll charges involve the cost of workers' compensation, pensions, group insurance, paid vacations and holidays, social security and unemployment taxes, contributions to profit sharing, and a growing list of other fringe benefits. Payroll charges are computed as a percentage of the labor and supervision cost, and generally the same rate is used for both labor and supervision. The best figures are from the company's records, but an accounting definition is important to avoid counting some items that may be included under general expense. A rough estimate is 30 to 45 percent of the labor-plus-supervision cost [51]. This figure is rising and may eventually reach 60 percent or more, as it does in parts of Europe today.

15.7 Maintenance

Records for the company's existing plants are the only reliable source of maintenance cost. Little has been published on the subject, but with experience, maintenance can be estimated as a function of investment. Maintenance cost as a percentage of the investment per year ranges from 11 percent for a complex plant with severe corrosive conditions to 2 percent for a relatively simple plant with mild, noncorrosive conditions. For example, average maintenance in a petroleum refinery is about 5 percent of the investment per year, but for individual units it ranges from about 3 percent for dewaxing units to about 7 percent for sulfuric acid alkylation units.

Generally, maintenance can be taken as 60 percent labor cost and 40 percent material cost. When a unit is operating at 75 percent of capacity, the maintenance cost will be about 85 percent the maintenance cost at 100 percent of capacity. When operating at 50 percent of capacity, the maintenance cost will be about 75 percent of the maintenance cost at 100 percent of capacity.

Maintenance is known to increase with the age of equipment, but the estimation must use an average figure. The procedure is equalized, in part, by the use of average value for depreciation, whereas accelerated depreciation may, in fact, be used for tax purposes. Thus the overall cost estimate is balanced because average values for maintenance and depreciation are used.

A project may create the need for additional maintenance supervisors and foremen. For major projects, maintenance manning and cost tables may be required to establish the additional maintenance supervision cost; but for small additions no additional maintenance supervision may be required.

15.8 Operating Supplies

Operating supplies include such things as lubricating oil, instrument charts, custodial supplies, and other items normally used by the project exclusive of those included as raw, repair, or packaging materials. Company records should be used for this cost when available. If they are not, operating supplies may be assumed to be about 6 percent of operating costs. Some authorities base such costs on a percentage of investment, such as 0.5 to 1.0 percent of the plant investment [35].

15.9 Laboratory and Other Service Costs

Depending on company practice and the type of project, operating costs may include service charges by other units of the company. Charges made by a central control laboratory are one example.

Laboratory costs can be estimated (1) on the basis of past experience; (2) on the basis of work-hours required, using an ample allowance for the overhead associated with a laboratory technician's pay, i.e., $20 to $30/h [47]; (3) as a percentage of the operating labor cost, in the range of 10 to 20 percent; and (4) from the literature [11, 19, 35].

15.10 Waste Disposal and Avoidance of Nuisances

An increasingly important item is the cost for the reduction of various nuisances. Common nuisances are the pollution of air and water, but any such problem, such as odor or noise, must be considered. Even if the nuisance is confined to company property, it must not be overlooked. For example, a moderately high noise level may appear to be harmless, but it may become a costly liability as a result of lawsuits brought by employees for alleged impairment of health. There has been much publicity about chemical-waste disposal problems at sites such as Love Canal and the Chemical Control Corporation's dump in Elizabeth, New Jersey [44]. These too are serious nuisances which have become costly liabilities to the owners and the public.

There are excellent discussions of the laws, regulations, pollutants, and methods of pollution control in the literature [13]. Peters and Timmerhaus [35] give an excellent presentation of waste disposal of all types, including the sources of regulations, key aspects of environmental regulations, and pollution-abatement methods for various types of pollutants, including, air, water, land, noise, and thermal pollution sources. Baasel [5] presents an excellent discussion of the various pollution-abatement problems and gives some costs.

The costs associated with pollution control have been published widely in the literature by the Environmental Protection Agency, Tennessee Valley Authority, research institutes, and private corporations [12, 29, 30, 40, 43, 46]. Table 15.7 shows some typical incremental costs for selected control technologies in power stations, and Example 15.2 illustrates the use of these incremental data.

> **Example 15.2** Suppose we wish to get a rough approximation for the increase in capital costs from adding SO_x (limestone), NO_x (combustion modifications), and particulate matter (cold ESP) controls to a power plant. The plant has a 1000-MW boiler, and the costs are to be based on 1980 costs.
>
> SOLUTION To solve this simple problem, one might use average costs. The boiler is to burn high-sulfur eastern coal. Thermal discharge control is also to be installed.

Average cost of conventional boiler, eastern coal		$635/kW
Average cost of limestone SO_x control facilities	$105/kW	
Average cost of cold ESP control facilities	35/kW	
Average cost of boiler combustion modifications	5/kW	
Average cost of fossil thermal discharge control facilities	22/kW	
Average cost of control facilities		$167/kW

> From these figures, we get
>
> $$\text{Average percent increase in capital cost} = \frac{\text{environmental}}{\text{base}}(100)$$
>
> $$= \frac{167}{635}(100) = 26.3 \text{ percent, or a 26 percent increase}$$

Solid residues from some operations must be disposed of by impounding if they are initially slurries, or by hauling to industrial dumps if they are in solid form. The expense of disposal is considered part of the main operation and is included in the operating expense, or as a service charge for hauling and further processing by another organization if this is required [47]. Costs in 1976 for landfilling or ponding of sludges were estimated to range from $3.50/dry ton for natural clay-lined ponds to $7.80/dry ton for Hypalon-lined ponds, depending upon land costs and the auxiliary equipment required [47].

Removal of organic contaminants from water is also important in keeping rivers and streams clean. Costs for such removal have been estimated to range from $0.005 to $0.05/1000 gal in 1980 for disinfecting the water and adsorbing the contaminants on granular activated carbon [12].

15.11 Royalties and Rentals

Royalties and licensing costs are generally an operating expense. If the charge is proportional to production, or a running royalty, it can be treated as a direct cost.

A single-sum payment, or a paid-up royalty, will become part of the project's capital investment, although it is not part of the depreciable investment. A tax specialist may be required to handle specific situations.

Royalty payments may range from 1 to 5 percent of the product's sales price, but the agreement may involve so many combinations that such an approximation should be used only when no data are available.

Certain royalty payments and patent purchase costs may be treated by the accountants under general expense, since they can be regarded as a replacement for research and development expense.

The cost of rentals can be obtained through the purchasing department. Payments for rentals, along with taxes and insurance on the items, are a current expense, and should not be overlooked in estimating operating expense.

15.12 Contingencies

An allowance for contingencies should be made for both the direct and indirect costs. Contingencies for the direct costs can be applied as a percentage of the direct cost, in the range of 1 to 5 percent.

15.13 Indirect Costs

As pointed out earlier, costs independent of the production rate are called *indirect costs*. Such costs include depreciation, real estate taxes, insurance costs, and general plant overhead. These costs must be paid whether there is any plant output or not.

Depreciation was discussed in detail in Chap. 3, along with its various methods of calculation. Suffice it to say that depreciation is generally taken as straight-line depreciation over the useful life of the project, although accelerated depreciation would very likely be used in determining taxes. Balancing of the estimate by using straight-line depreciation and level maintenance cost over the life of the project was discussed under maintenance. The useful life for cost estimation is not necessarily the life required for tax purposes. The depreciation charge does not include any return on investment.

Real estate taxes depend upon the region of the country and the local laws. A figure of 2 to 4 percent of the fixed capital investment could be used for heavily populated areas, while a figure of 1 to 2 percent of the fixed capital investment could be used for less populated areas [35].

Insurance rates depend upon the type of product being manufactured and the availability of protection facilities. A figure of 1 percent of the fixed capital investment could be used to cover the costs of fire, explosion, tornado, and product liability insurance [9].

Plant overhead costs cover most of the items shown under plant overhead in Table 15.1. If data are not available from company records, a figure often used is 50 to 70 percent of the total expense for operating labor, supervision, and maintenance [35].

A separate contingency charge for indirect costs may not be necessary.

Table 15.7 Typical incremental costs for selected control technologies, mid-1980 dollars

	500 MW		1000 MW	
	Capital, $/kW	Annualized, mills/kWh	Capital, $/kW	Annualized, mills/kWh
Base generation system (typical heat rate, Btu/kWh)				
Conventional fossil fuel-fired boilers				
High-sulfur eastern coal (9800)	640–775	20–24	575–700	17–21
Low-sulfur western coal (9200)	670–820	20–25	605–740	18–22
Liquid fuel (9200)	395–490	12–15	360–440	11–13
Conventional nuclear light-water reactor (10,400)	1000–1220	30–36	900–1100	26–33
Combined cycle				
Liquid fuel (7500)	360–440	12–15	320–395	11–13
Low-Btu gasification (8400)	870–1175	28–36	780–1060	24–32
Medium-Btu gasification (8200)	880–1190	28–36	790–1070	24–33
Fluidized bed combustion (FBC)				
Atmospheric FBC (9500)	690–930	21–28	625–840	20–25
Pressurized FBC (8800)	835–1260	26–38	750–1130	24–34
Incremental costs				
Fuel*				
High-sulfur eastern coal		10–16		11–16
Low-sulfur western coal		15–20		14–20
Physically cleaned coal		13–18		13–18
Chemically cleaned coal		18–29		18–29
Solvent-refined coal		33–40		33–40
Liquefied coal		40–46		40–46
Liquid fuel		20–33		20–33
Uranium		7–8		7–8

Environmental control technology				
Sulfur control				
Flue gas desulfurization				
Limestone	100–135	4.6–5.5	90–120	4.1–5.0
Wellman Lord	115–155	4.6–5.8	105–140	4.2–5.3
Magnesia	120–180	4.7–6.5	110–165	4.3–5.8
Dual alkali	125–185	5.9–7.2	110–165	5.4–7.0
Fuel gas cleanup				
Low-Btu gas	185–250	6.7–8.6	170–225	6.1–7.6
Medium-Btu gas	165–230	5.5–7.2	150–200	5.0–6.5
Fluidized bed combustion				
Limestone or dolomite	30–45	1.6–2.1	25–40	1.4–2.0
Particulate control				
ESP-cold	35–45	1.2–1.4	30–40	1.2–1.3
ESP-hot	40–55	1.6–2.0	40–50	1.5–1.8
Fabric filter	60–80	2.2–2.8	55–70	2.0–2.5
Wet scrubber	70–85	2.5–3.0	60–80	2.2–2.6
NO_x control				
Combustion modifications	0.5–10.0	0.05–0.8	0.5–9.0	0.05–0.7
Selective catalytic reduction	35–60	1.7–2.5	35–55	1.6–2.2
Water injection for turbines	4.5–9.0	0.8–0.9	4.5–7.5	0.8–0.9
Thermal discharge control				
Evaporative cooling tower				
Fossil	25–30	1.1–1.2	20–25	0.9–1.1
Nuclear	30–35	1.6–1.7	30–35	1.4–1.6

Source: Based on M. G. Klett, *Typical Costs for Electric Energy Generation and Environmental Controls*, EPA-600/7-79-026, IERL, U.S. Environmental Protection Agency, Research Triangle Park, N.C., January 1979.

*Based on a heat rate of 10,000 Btu/kWh.

A contingency cost is included in the investment cost. If the factory overhead is based chiefly on the investment, a contingency allowance is automatically included.

Interest on borrowed capital is also an indirect cost if borrowing for the plant is necessary. In such instances, the rate is established at the time the money is borrowed. Company records should be a good source of rates for preliminary estimates. At times, prime rates are very high, up to 20 or 21 percent, but such rates are too high for long-term borrowing. A space is provided on the form shown in Table 15.10 in case interest should be included in the estimate.

15.14 Distribution Costs

Loading, packing, and shipping costs are best estimated from other similar company operations, modified for the situations under study. Most companies have cost-control systems and standard costs which provide such information. Published information about distribution costs can be found in Perry [33, sec. 8] and in Aries and Newton [4]. Such costs are sometimes estimated as 1 to 3 percent of sales.

The cost of containers can be obtained from suppliers or from such published sources as Raymus [41] or Peters and Timmerhaus [35]. Table 15.8 gives the sizes and costs of some representative containers as of December 1980. Returnable

Table 15.8 Sizes and approximate costs of representative containers, December 1980

Container size, description	Unit cost	Usable volume, ft^3
55-gal steel drum, new	$17.70	7.35
55-gal steel drum, used, cleaned	9.90	7.35
55-gal aluminum drum	82.00	7.35
55-gal type 304 stainless steel drum	215.60	7.35
55-gal fiber drum, dry products only	7.20	7.35
41-gal fiber drum, dry products only	6.70	5.48
30-gal fiber drum, dry products only	5.70	4.00
15-gal fiber drum, dry products only	2.50	2.00
Multiwall paper bags, polyethylene film	0.32–0.35	1.33
Corrugated cartons, 24 × 16 × 6 in	0.65	1.33
Pallets, expendable, 40 × 48 in to 44 × 50 in	5.40–9.90	
Pallets, warehouse type, 40 × 48 in to 44 × 50 in	11.00–15.20	

Source: Based on G. J. Raymus: "Evaluating the Options for Packaging Chemical Products," *Chemical Engineering,* vol. 80, Oct. 8, 1973, p. 67, and on M. S. Peters and K. D. Timmerhaus, *Plant Design and Economics for Chemical Engineers,* 3d ed., McGraw-Hill Book Company, New York, 1980, p. 573.

containers take a variety of forms, from small drums to tank cars and barges, and their cost includes repair, cleaning, testing, refurbishing, and depreciation or rental.

Transportation costs were discussed briefly in Sec. 15.3. Railroad-car rental rates can be obtained from railroads or leasing companies. When railroad cars are owned or leased by the company, the railroads credit the company with about $0.23 to $0.37 per loaded car-mile traveled by the unit, depending upon the age and the depreciated value of the car.

Distribution charges are segregated from other manufacturing costs to show the effect of various containers and the method of shipment. The cost without distribution costs is the bulk cost, or manufacturing cost before (packaging and) distribution.

15.15 Summary of Operating Cost Estimation

Table 15.9 summarizes the procedures for estimating operating costs when company records are not available. Such methods are useful in the early stages of project evaluation for preparing preliminary estimates. The items in the table follow the order shown in Table 15.10.

Table 15.9 Preliminary operating cost estimates

A. Direct production cost
    1. Materials
        a. Raw materials—estimate from price lists
        b. By-product and scrap credit—estimate from price lists
    2. Utilities—from literature or similar operations
    3. Labor—from manning tables, literature, or similar operations
    4. Supervision—10 to 25 percent of labor
    5. Payroll charges—30 to 45 percent of labor plus supervision
    6. Maintenance—2 to 10 percent of investment per year
    7. Operating supplies—0.5 to 1.0 percent of investment per year
    8. Laboratory—10 to 20 percent of labor per year
    9. Waste disposal—from literature, similar operations, or separate estimate
    10. Royalties—1 to 5 percent of sales
    11. Contingencies—1 to 5 percent of direct costs

B. Indirect costs
    1. Depreciation—5 to 10 percent of investment per year
    2. Real estate taxes—1 to 2 percent of investment per year
    3. Insurance—0.5 to 1.0 percent of investment per year
    4. Interest—10 to 12 percent of investment per year
    5. General plant overhead—50 to 70 percent of labor, supervision, and maintenance

C. Distribution costs
    1. Packaging—estimate from container costs
    2. Shipping—from carriers or 1 to 3 percent of sales

Table 15.10 Solution to Example 15.3

Date 1/23/81   By JHB

PRELIMINARY MANUFACTURING COST ESTIMATE

Location: Not specified

Product: Product A

Process: Oxidation

Cost index: 270   Operating days: 330

Annual production: 2,000,000 lb

Manufacturing capital:

	—
	—
Total	$580,000

Raw materials	Annual quantity	Unit cost	$/year	$/lb
B	1,660,000 lb	$0.60/lb	996,000	
C	1,240,000 lb	$0.45/lb	558,000	
		Gross raw-material cost:	1,554,000	
By-product (Credit)				
D	800,000 lb	$0.20/lb	160,000	
		Total credit:	160,000	
		Net raw-material cost:	1,394,000	

Direct cost	Unit	Quantity	$/Unit	
Steam, hp	thou. lb	none		
Steam, lp	thou. lb	26,000	$2.00	52,000
Water, plant	thou. gal	none		
Water, process	thou. gal	300,000	$0.60	180,000

Electricity	kwhr	4,000,000	$0.035	140,000
Fuel ( )		none		
Inert gas	thou. ft³	none		
Compressed air	thou. ft³	none		
Refrigeration	tons	none		
Labor	work/h	39,600	$8.50	336,600
Supervision	20% of labor			67,320
Payroll charges	40% of labor + supervision			161,570
Maintenance	6% of $580,000			34,800
Operating supplies	0.5% of $580,000			2,900
Laboratory	15% of labor			50,490
Waste disposal	none			
Royalty	none			
Contingencies	3% of direct costs			31,720
		Total direct conversion cost:		1,057,400
Indirect cost				
Depreciation	10% of $580,000			58,000
Real estate taxes	1.5% of $580,000			8,700
Insurance	0.75% of $580,000			4,350
General plant overhead	60% of (labor + super. + maint.)			263,250
		Total indirect conversion cost:		334,300
		Total conversion cost:		1,391,700
		Total manufacturing cost:		2,785,700
Containers ( )	none			
Packaging and shipping expense	none			
		TOTAL COST FOB PLANT:		2,785,700

1.39

Table 15.11 Manufacturing cost checklist*

I. Direct production costs
  A. Raw-material and by-product credits
  B. Utilities
    1. Steam
    2. Water
    3. Electricity
    4. Fuel
    5. Others
  C. Labor
    1. Operating labor
    2. Supervision
    3. Payroll charges
  D. Supplies and miscellaneous
    1. Maintenance
    2. Operating supplies
    3. Laboratory
    4. Waste disposal
    5. Royalty
    6. Contingencies
II. Indirect production costs
  A. Depreciation
  B. Real estate taxes
  C. Insurance
  D. General plant overhead or factory overhead
III. Distribution costs—loading, packing, and shipping

*Matches Table 15.10.

Example 15.3 Estimate the total manufacturing cost to produce 2×10^6 lb/year of product A by the oxidation process. The plant operates 24 h/day for 330 days per year and costs $580,000 to build. The process is batch-operated and consists of a mixing step, a reaction step, a filtration step, and a distillation step. In producing 1 lb of product A, 0.83 lb of product B, costing $0.60/lb, and 0.62 lb of product C, costing $0.45/lb, are consumed. In addition, 0.4 lb of by-product D, worth $0.20/lb, is produced. The balance of the raw material is lost up the stack as carbon dioxide. The plant has average maintenance problems, and the process requires average supervision. There are no interest, royalty, or waste disposal costs. The only utilities requirements in the plant are 150 gal of process water, 2 kWh of electricity, and 13 lb of 100-psig steam per pound of A produced.

SOLUTION The solution to this example is shown in Table 15.10, a prepared form, and Table 15.11, an outline or checklist to match it. Labor is found, from Eq. (15.1), as 39.6 work-h/ton.
    From Table 15.10, the total manufacturing cost is estimated to be $2,785,000/year, or $1.39/lb.

15.16 Shortcut Methods

There is an obvious advantage to shortcut methods of estimating operating costs. One method involves writing an equation which contains all the cost elements discussed in the preceding sections. To make this possible, some sort of average values must be decided upon for the factors to apply to the primary cost elements of materials, utilities, labor, and investment. The use of this method should be discouraged because the results of such an estimate are stereotyped. The method makes no allowances for differences between situations and does not challenge the imagination of the estimator.

There have been attempts to develop shortcut methods, however. Holland, Watson, and Wilkinson [21] describe a method of developing an equation for operating costs. Garcia-Borras [16, 17] reviews various methods of estimating costs, including shortcut capital and operating-cost methods. Sommerville [45] presents a shortcut method for estimating distillation costs. Berry [8] proposed a method of estimating conversion costs based on the number of processing steps and capacity. Care must be exercised in choosing the correct number of processing steps.

Another shortcut method for batch operations has been proposed by Horton [22]. Conversion costs are estimated for ratios of from 5.6 to 1.0 for conversion cost to operating wages. To this conversion cost must be added raw-material costs, royalties, and container costs. The ratio was derived from literature information on work-hours required for various batch processes.

A preliminary or shortcut estimate can be made from a few basic data and estimating other items as a percentage of these. A suggested form and the range of values for a process industry are given in Tables 15.10 and 15.9, respectively.

QUESTIONS

15.1 What does the manufacturing cost include?

15.2 What expenses in addition to the manufacturing cost must be considered in computing the total product cost?

15.3 Define incremental cost.

15.4 Why are incremental costs significant in project appraisal?

15.5 Why is it sometimes necessary to compute the return of the business as a whole after the proposed project is completed if the proposed project itself is profitable on an incremental basis?

15.6 Why are average figures justified for the maintenance cost and the depreciation expense although neither may be constant with time?

15.7 The chapter included some comment on situations in which items are considered to be capital investments rather than operating costs. Why is this distinction important in figuring profitability?

15.8 Your company is already making a product at a raw-material cost of precisely $0.512/lb out of a total operating cost of $1.623. You are responsible for investigating a proposed new project to quadruple the present capacity. Should the price of raw materials be investigated? Why?

15.9 Why is the cost of raw materials or fuels obtained from another company division included as part of the operating cost of the project using them?

15.10 What is the procedure for computing transfer prices between one company division and another?

15.11 Name at least three project characteristics which affect operating labor.

15.12 Why do maintenance cost estimates depend on the size of the project?

15.13 The text mentions that when a unit is being operated at 50 percent capacity, the repair costs will be about 75 percent of the repair cost at 100 percent capacity. Why not 50 percent?

15.14 What project characteristics make high ratios of supervision to labor necessary?

15.15 Why is it desirable to examine the original agreement documents specifying royalty and other patent right payments?

15.16 Define general works expense.

15.17 What is the difference between general works expense and administrative, sales and general expense?

15.18 What is the difference between bulk cost and total manufacturing cost? Why are both computed?

PROBLEMS

15.1 Reconcile Eq. (15.1) with the statement in the text that labor requirements vary to about the 0.2 to 0.25 power of the capacity ratio.

15.2 Calculate the operating cost for the plant in Example 15.3 if the plant is running at one-half capacity.

15.3 Calculate the operating cost for the plant in Example 15.3 if the plant is twice as large.

15.4 Calculate the operating cost for the plant in Example 15.3 if the plant is only one-half as large.

15.5 Repeat Prob. 15.3, assuming that the plant is operating at only one-half of design capacity.

REFERENCES

1. American Association of Cost Engineers: *AACE Bulletin,* Morgantown, W. Va., 1958–1978.
2. American Association of Cost Engineers: *Cost Engineering,* Morgantown, W. Va., 1978 to date.
3. American Association of Cost Engineers: *Cost Engineers' Notebook,* Morgantown, W. Va., 1964 to date.
4. Aries, R. S., and R. D. Newton: *Chemical Engineering Cost Estimation,* McGraw-Hill Book Company, New York, 1955.
5. Baasel, W. D.: *Preliminary Chemical Engineering Plant Design,* American Elsevier Publishing Company, Inc., New York, 1976.
6. Bauman, H. C.: *Fundamentals of Cost Engineering in the Chemical Industry,* Reinhold Publishing Corporation, New York, 1963.
7. Beattie, B. R., and H. S. Foster, Jr.: "Can Prices Tame the Inflationary Tiger?," *American Water Works Association Journal,* vol. 72, August 1980, p. 441.
8. Berry, E. M.: "Estimate Manufacturing Costs," *Chemical Engineering,* vol. 67, June 27, 1960, p. 123.
9. Black, J. H.: "Estimating Operating Costs for New Ventures," *AACE Bulletin,* vol. 4, December 1962, p. 42.
10. Chemical Engineering: *Modern Cost Engineering Methods and Data,* McGraw-Hill Book Company, New York, 1979.
11. Chilton, C. H.: *Cost Engineering in the Process Industries,* McGraw-Hill Book Company, New York, 1960.
12. Clark, R. M., and P. Dorsey: "The Cost of Compliance: An EPA Estimate for Organics Control," *American Water Works Association Journal,* vol. 72, August 1980, p. 450.
13. Deskbook: "Environmental Engineering," *Chemical Engineering,* vol. 84, Oct. 17, 1977, 212 pp.
14. Dybdal, E. C.: "Engineering and Economic Evaluation of Projects," *Chemical Engineering Progress,* vol. 46, February 1950, pp. 57–66.
15. Faith, W. L., D. B. Keyes, and R. L. Clark: *Industrial Chemicals,* 4th ed., John Wiley & Sons, Inc., New York, 1975.
16. Garcia-Borras, T.: "Research-project Evaluations, Part I," *Hydrocarbon Processing,* vol. 55, December 1976, p. 137.

17. ———: "Research-project Evaluations, Part II," *Hydrocarbon Processing,* vol. 56, January 1977, p. 171.
18. Guthrie, K. M.: "Capital and Operating Costs for 54 Chemical Processes," *Chemical Engineering,* vol. 77, June 15, 1970, p. 140.
19. Haines, T. B.: "Direct Operating Requirement for Chemical Processes," *Chemical Engineering Progress,* November 1957, p. 556.
20. Happel, J., and D. G. Jordan: *Chemical Process Economics,* Marcel Dekker, Inc., New York, 1975.
21. Holland, F. A., F. A. Watson, and J. K. Wilkinson: "Manufacturing Costs and How to Estimate Them," *Chemical Engineering,* vol. 81, Apr. 15, 1974, p. 91.
22. Horton, R. L.: "Manufacturing Costs for Batch Produced Chemicals," *Chemical Engineering,* vol. 72, Oct. 11, 1965, p. 236.
23. Isard, W., and E. W. Schooler: *Location Factors in the Petrochemical Industry,* U.S. Department of Commerce, Washington, D.C., 1955.
24. Jenckes, L. C.: "Developing and Evaluating a Manufacturing Cost Estimate," *Chemical Engineering,* vol. 78, Jan. 11, 1971, p. 168.
25. ———: "How to Estimate Operating Costs and Depreciation," *Chemical Engineering,* vol. 77, Dec. 14, 1970, p. 168.
26. ———: "Estimation and Optimization of Operating Costs," *AACE Transactions,* 1970, p. 48.
27. Johnson, P. W., and F. A. Peters: *A Computer Program for Calculating Capital and Operating Costs,* Bureau of Mines Information Circular 8426, U.S. Department of the Interior, Washington, D.C., 1969.
28. Klett, M. G.: *Typical Costs for Electric Energy Generation and Environmental Controls,* EPA-600/7-79-026, IERL, U.S. Environmental Protection Agency, Research Triangle Park, N.C., January 1979.
29. Maxwell, J. D., T. A. Burnett, and H. L. Faucett: *Preliminary Economic Analysis of NO$_x$ Flue Gas Treatment Process,* EPA-600/7-80-021, IERL, U.S. Environmental Protection Agency, Research Triangle Park, N.C., February 1980.
30. McGlamery, C. G., R., L. Torstrick, W. J. Broadfoot, J. P. Simpson, S. V. Tomlinson, and J. F. Young: *Detailed Cost Estimates for Advanced Effluent Desulfurization Processes,* EPA-600/2-75-006, IERL, U.S. Environmental Protection Agency, Research Triangle Park, N.C., January 1975.
31. O'Connell, F. P.: "Chart Gives Operating Labor for Various Plant Capacities," *Chemical Engineering,* vol. 69, Feb. 19, 1962, p. 150.
32. Ostwald, P. F.: *Cost Estimation for Engineering and Management,* Prentice-Hall, Inc., Englewood Cliffs, N.J., 1974.
33. Perry, J. H., ed.: *Chemical Business Handbook,* McGraw-Hill Book Company, New York, 1954.
34. Perry, R. H., and C. H. Chilton, eds.: *Chemical Engineers' Handbook,* 5th ed., McGraw-Hill Book Company, New York, 1973.
35. Peters, M. S., and K. D. Timmerhaus: *Plant Design and Economics for Chemical Engineers,* 3d ed., McGraw-Hill Book Company, New York, 1980.
36. Peterson, J. N., C. C. Chen, and L. B. Evans: "Computer Programs for Chemical Engineers," *Chemical Engineering,* vol. 85, July 31, 1978, p. 79.
37. Popper, H.: *Modern Cost Engineering Techniques,* McGraw-Hill Book Company, New York, 1970.
38. ———: "Process Designers Get More Software," *Chemical and Engineering News,* vol. 48, Mar. 30, 1970, p. 38.
39. ———: *Project Economic Analysis,* Program #198049E, Software Catalog, Texas Instruments, Inc., Lubbock, Tex., September 1980.
40. Radian Corporation: *Sulfur Oxides Control Technology Series: Flue Gas Desulfurization-Dual Alkali Process,* IERL, U.S. Environmental Protection Agency, Research Triangle Park, N.C., October 1980.
41. Raymus, G. J.: "Evaluating the Options for Packaging Chemical Products," *Chemical Engineering,* vol. 80, Oct. 8, 1973, p. 67.
42. Rudd, D. F., and C. C. Watson: *Strategy of Process Engineering,* John Wiley & Sons, Inc., New York, 1968.

43. Scher, C., and D. Netzer: "FGD Good with Coal Conversion," *Hydrocarbon Processing,* vol. 59, December 1980, p. 117.
44. Snyder, T. A., Jr.: "Waste Disposal: What Is Industry's Responsibility?," *Chemical Engineering Progress,* vol. 76, June 1980, p. 24.
45. Sommerville, R. F.: "A New Method Gives Quick, Accurate Estimate of Distillation Costs," *Chemical Engineering,* vol. 79, May 1, 1972, p. 71.
46. Tomlinson, S. V., F. M. Kennedy, F. A. Sudhoff, and R. L. Torstrick: *Definitive SO_x Control Process Evaluations: Limestone, Double Alkali, and Citrate FGD Processes,* EPA-600/7-79-177, IERL, U.S. Environmental Protection Agency, Research Triangle Park, N.C., August 1979.
47. Uhl, V. W.: *Standard Procedure for Cost Analysis of Pollution Control Operations,* vols. I and II, EPA-600/8-79-018 a, b, U.S. Environmental Protection Agency, Research Triangle Park, N.C., June 1979.
48. U.S. Department of Commerce: *Survey of Current Business,* Table S-13, October 1980.
49. Vernon, I. R.: *Realistic Cost Estimating for Manufacturing,* Society of Manufacturing Engineers, Dearborn, Mich., 1968.
50. Wessell, H. E.: "New Graph Correlates Operating Labor Data for Chemical Processes," *Chemical Engineering,* July 1952, p. 209.
51. Winton, J. M.: "Plant Sites: Spreading Out from Houston," *Chemical Week,* vol. 127, Dec. 3, 1980, p. 54.

SIXTEEN

COST CONTROL

D. L. Anderson

CAPITAL-COST CONTROL

16.1 The Nature of Control

Cost Control of capital and operating cost is a major and critical activity of cost engineering. In building a plant, *cost control* means regulating expenditures within the authorized budget while providing a plant of the desired quality within the time limit specified. The plant should be capable of producing a quality product at the specified design capacity.

Cost control cannot end with analysis and prediction of capital costs. Plants and factories must be built, and it is essential that such costs be controlled during their design, procurement, and construction. Plants and factories must be operated, and it is equally important that operating costs be controlled. This chapter is divided into three sections, the first describing the components of capital-cost control; the second the components of operating-cost control; and the third the critical path method.

Most heavy industries and petrochemical construction projects are now built by what is called the *fast track* method. Design, procurement, and construction are done with significant overlap, often appearing to be almost simultaneous. The fast track method is necessary because of today's rapidly developing technology. As the design of certain portions of the plant is completed, procurement can begin on pieces of equipment with long lead time. Similarly, as elements of civil design are completed, such as site work, roads, and drainage, and with a reasonable assurance of continuing production of "approved-for-construction" drawings, the construction team can move to the site and begin. This is typical of how current industrial and petrochemical projects are being managed. If the design were allowed to be completed before the procurement and construction began, the final cost of the plant would be so much greater due to escalation alone that it would not be competitive. Other factors influencing costs are interest cost and obsolescence.

To coordinate the efforts of engineering, procurement, and construction groups requires a group responsible for project general management. This group is responsible for providing the catalyst which blends the engineering, procurement, and construction efforts into a homogeneous team working together, each with knowledge of the other's interface requirements, and including the vendors and subcontractors who will also be contributing to the overall effort.

16.2 Capital Expenditure Control

With the type of project briefly described in Sec. 16.1, capital expenditure control must be brought into effect as soon as the contract is awarded to the design engineer. Thus, the engineer may be made aware that project management will be coordinating his or her efforts with those of the construction group toward the common goal of project completion on time within the budget. Immediately after the award of the contract, the engineer should provide a preliminary budget to project management. The project management cost engineer staff should set up the control system and establish the procedures for an overall project control estimate to be made usually after 6 to 8 months of engineering design and project scope development. Timing depends upon the type of project under consideration, but in no case should the timing in establishing control exceed 6 to 8 months of conceptual development. The engineer's responsibility to provide this estimate should be very clear and made a part of his or her budget. Meanwhile, staff scheduling engineers should establish some major milestones and provide a conceptual schedule so that everyone on the project team is aware of the milestone dates and the magnitude of the project, in terms of construction, equipment, manpower, and the time frame for completion of major components. This tends immediately to set the pace and give direction to the team members. Engineering design and procurement efforts can then be directed toward the long lead time of major equipment. Thus, the elements of capital expenditure control are:

1. Cost control
2. Estimating
3. Planning and scheduling
4. Material management

16.3 Cost Control

The primary requirement of cost control is a trackable budget, which should be developed from a detailed estimate. The estimate must include a detailed defini-

Table 16.1 Major construction activities

Account number	Title
01XX	Site preparation, demolition, salvage, removal for relocation
02XX	Site improvements
03XX	Underground electrical
04XX	Underground piping
07XX	Piling
08XX	Concrete and excavation
12XX	Structural steel
14XX	Architectural (manufacturing and forest products projects)
15XX	Building construction (petroleum and chemical projects)
16XX	Ducts
17XX	Aboveground piping
18XX	Aboveground electrical
19XX	Instrumentation
20XX	Insulation
21XX	Painting
23XX	Paving
4XXX	Major equipment
70XX	Proratables

tion of the project's scope. If the design of the plant has not progressed to the point where certain specific elements are defined, the estimators must complete the definition sufficiently to produce a detailed estimate. This is called *conceptual design* and *conceptual estimating*. The estimate is translated into the budget by the code of accounts. The code of accounts should be standard for project after project and, therefore, well fixed in the minds of project team members. The only development that should occur in the code of accounts is the tailoring needed to permit subsequent capitalization for the owner. These capitalization accounts should be kept separate from the trackable budget used to measure progress. Table 16.1 shows a code of accounts for major activities. Activities are controlled divisions of construction efforts relating to the different types of work involved on the construction project. All direct work performed in the field will be related to a specific activity, such as building concrete forms and welding pipe by size and type.

Figure. 16.1 shows a record of the cash flow of labor and the total cost for a project, but it is not in a form to indicate corrective action at any time.

Figure 16.2 is a typical example of a standard S curve showing cumulative work-hours needed to do a particular part of a project against time. The term *earned value* refers to physical accomplishment and is used in cost engineering to avoid confusion with money or manpower committed. An example of progress made could be the design drawings completed through the reporting period, weighted in terms of estimated work-hours per drawing. At the November vertical

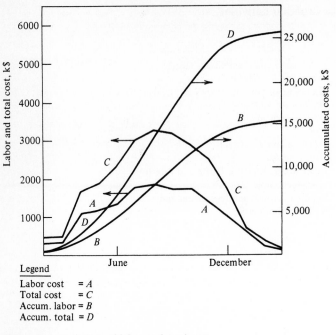

Legend

Labor cost   = A
Total cost   = C
Accum. labor = B
Accum. total = D

Figure 16.1 Cash flow of labor and total cost.

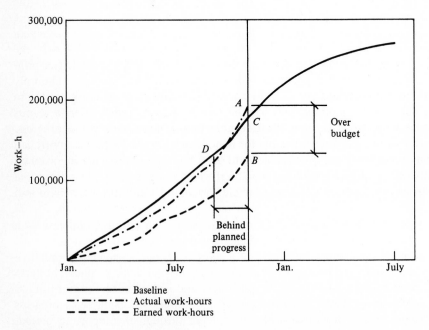

Baseline
Actual work-hours
Earned work-hours

Figure 16.2 Earned value status graph.

line shown in Fig. 16.2, the difference between the actual work-hours and the earned work-hours, points *A* and *B*, is the work-hours over the budget. Similarly, the difference between *C*, the schedule baseline, and *D*, the actual accomplishment (earned value) at the November line discloses how far the project is behind schedule. Obviously, corrective action is necessary.

Variations from the schedule are important and are carefully monitored by charts and tabulations. When a definite trend is noted, a variance is generated. A *variance* is any deviation from the budget. The following are examples of the types of variance which might be experienced on a typical capital project:

Change order (scope change)
Damages (fire, wind, flood)
Productivity (loss or gain)
Quantity overruns

Figure 16.3 is a sample form of a project variance notice.

Computers are now used to print out detailed information periodically, such as weekly or monthly. Each account, such as piping, structural steel, or electrical, is shown, giving information such as the following:

Previous committed cost
Added for period
Total to date
Current estimated cost at completion
Percent expended
Original budget
Over (or under) budget
Percent deviation

Finally, to complete this cost control effort, a record of final costs must be made. This is the history which is to be used in evaluating the estimate for the next similar project. When the code of accounts is standard, this estimate evaluation is then possible. This history becomes a record of the company's experience and is a valuable tool in enhancing its ability to successfully manage complex projects.

At this time, when the costs are known, a capital distribution for the owner should be made. This may consist of several types of reports, as follows:

1. Government (FPC, Federal Power Commission)
2. Insurance underwriters
3. Tax credits
   a. Investment credit
   b. Energy conservation credit
   c. Pollution-abatement devices
4. IRS (Internal Revenue Service) asset write-off (depreciation)

	DATE		P.V.N. NO.	REV. NO.
TO:	B&R JOB NOS.	VARIANCE ORIGINATION ☐ HOME OFFICE ☐ FIELD		
PROJECT		COST/SCHEDULE AREA		

DRAWING, SPECIFICATION. EQUIPMENT NO., OR OTHER REFERENCE

SECTION 1.0 — REASON FOR AND NATURE OF VARIANCE

DESCRIPTION AND/OR SKETCH

2.0 — ORIGINATION

ORIGINATOR | SUPERVISOR

SECTION 3.0

SECTION 3.0 TO BE COMPLETED WHEN INFORMATION IS AVAILABLE

EFFECT ON COST	EFFECT ON SCHEDULE	EFFECT ON CONTRACT

VARIANCE EVALUATION (By Control Team)

HOME OFFICE		MATERIALS $	SUB $	FIELD			TOTAL $
MANHOURS	AMOUNT $			MANHOURS	LABOR $	OTHER $	

SECTION 4.0 — VARIANCE CLASSIFICATION

CLIENT	INTERNAL	ECONOMIC
☐ PROCESS MODIFICATION	☐ ESTIMATE	☐ CHANGE IN MATERIAL COST
☐ DESIGN MODIFICATION	☐ DESIGN	☐ CHANGE IN LABOR COST
☐ SPECIFICATION MODIFICATION	☐ PROCESS	☐ RESULTS OF PROCUREMENT
☐ IMPROVE OPERABILITY		ACTION
☐ SAFETY CONSIDERATIONS		☐ FIELD CONDITIONS
☐ OTHER _____	☐ OTHER _____	☐ OTHER _____
(SPECIFY)	(SPECIFY)	(SPECIFY)

SECTION 5.0

	SCOPE CHANGE ONLY
☐ APPROVED	☐ DISAPPROVED
☐ APPROVED-REQUEST CHANGE ORDER - TO CLIENT	☐ PREPARE CHANGE ORDER
☐ DISAPPROVED	REQUEST ESTIMATE
☐ ADDITIONAL INFORMATION REQUIRED	
☐ REQUEST CONFERENCE	
☐ OTHER (SPECIFY) _____	
RESPONSIBLE MANAGER	CLIENT RESPONSIBLE MANAGER

Figure 16.3 Project variance notice.

The following is an outline of the essential elements of cost control:

1. Project setup
   a. Budget estimate (trackable)
   b. Code of accounts
2. Costing
   a. Labor cost
   b. Material cost
   c. Construction equipment and tools cost
   d. Subcontracts cost
   e. Overhead and indirect cost
3. Reporting
4. Analysis
5. Variancing
6. Finalization
   a. Final cost report—history
   b. Capital distribution—client use

16.4 Estimating

Most capital projects today are divided into two phases. The initial phase provides financing of conceptual design up to a point at which a detailed conceptual estimate can be made—usually 6 to 8 months, depending upon the type and size of the project. During this time, purchase orders may be issued for equipment that requires a long lead time. These purchase orders usually have a cancellation clause in the event that project funds are not subsequently released to execute the detail/design/procurement second phase.

The information provided in the initial phase is enough to use in developing a trackable estimate. Engineering is required to provide scope definition in order to define size, to rate equipment, and to provide foundation sizes, structural steel requirements, instruments by loop, piping, and valves by line. The owners may require this scope development for funding of the second phase of project execution. Design deviation from this conceptual detail forms the basis for variances and affects control of design, engineering, procurement, and construction during the second phase.

The basic elements of a detailed conceptual estimate used to produce a trackable budget are as follows:

1. Estimate scope development
2. Factored allowances
3. Estimate backup or support documentation
4. Estimate review—engineering—construction—owner
5. Qualifications to the estimate

The estimate scope development has been covered somewhat in the above description of information to be provided by the engineering team in the initial phase. However, part of the scope is included in the code of accounts. The standard code of accounts is used, and the detail definition of those accounts determines the needed scope of each account. A listing of requirements from the engineer, with a description, outlines the responsibilities of each design function. Each estimator should have preprinted estimate summary sheets by the code of accounts. If the estimate is computerized, this of course would not be required since the computer would automatically format the estimate in the standard code of accounts. One computer printout is called an *ESLUM spread sheet,* with *ESLUM* an acronym for *EStimated Labor Unit Man-hours,* but it also includes materials. For various items of equipment, it gives the account number and quantity, work-hours, labor work-hours, permanent materials, and construction materials, among others.

Factored allowances can be used in a conceptual estimate but only in determining such costs as temporary facilities, small tools, guard service, fuels, lubricants, and repair parts. Although these factors are part of the company's historical records and are generally accepted, they should never be used to substitute for detail of the facility that is to be built. Other elements must also be applied to a conceptual estimate with the best judgment a project and its company can muster. Some of these factors, which are more difficult to define and sometimes not generally accepted, are essential to complete a detailed conceptual estimate. These are:

Design allowance
Quantity allowance
Escalation
Contingency
Risk

The *design allowance* is an adjustment to the estimated purchased equipment cost to provide for historically normal, routine changes in the design from estimate to delivery. The allowance will vary with the type of equipment and the amount of design information available at the time of the estimate.

The *quantity allowance* is an adjustment applied to the quantities, other than those of process equipment, to provide for the differences which invariably occur between takeoffs at estimate time and installed quantities due to incomplete design and for fabrication scrap or waste. It provides an adjustment to labor, materials, and supplies and is usually based partially on the constructor's work records.

Escalation is an allowance for inflation and is applied to all elements of cost which are affected.

Contingency is a percentage of the escalated base cost to cover errors in human calculation or judgment in estimating the scope of the project. It is a measure of the accuracy of the estimator. Contingencies are anticipated to be spent and form an integral part of the cost of the project.

Risk is an amount of money or a percent of the base cost plus lower-level (above listed) additives required to cover the probability that the base cost plus additives does not adequately define all the expenditures to be covered in seeking funding for the project. It is a judgment of the quality of the estimate made by the person applying the risk.

The estimate backup and support consist of five main elements, all of which contribute to the total capital expenditure:

Construction direct cost
Construction overhead and indirect cost
Engineering cost
Management cost
Owner cost

Each of these elements must be qualified and properly documented. The largest and most time-consuming to develop and qualify is the construction direct cost. Each activity, such as site development, area preparation, underground electrical, concrete, and piping, must be supported in detail. Each foundation sketch, each isometric pipeline, each piece of equipment by number must be supported by price, quantity, and reference number. All these data should be bound and referenced for the estimate backup and support of changes in scope and other variances.

In the estimate review, an estimator may have to support elements of the estimate with these data. This type of support also assures that the estimate cannot be arbitrarily reduced. If the quantities contained in the estimate are correct and supported, and the unit price to extend the quantities is developed from historical data, then logically the only reduction that could be made in the estimate would be in the factors defined above or in the facilities or scope. This is one reason the estimate support is so important. It becomes the definition of the original scope of the project and the basis for all changes in scope and other variances; therefore, it supports control.

The reviews of the estimate are another important step in communicating the scope definition to all members of the project team. The reviews clarify the estimate for the team, and where misunderstandings may occur, they are cleared up in these reviews. Mistakes, omissions, errors, etc. are found and corrected. Weak areas in the estimate are highlighted, and additional work may be required to strengthen a portion or element of detail.

After the estimating team managers are satisfied, through meetings and internal reviews, that the estimate properly reflects the scope of the project, four reviews are usually required:

Engineering review
Construction review
Management review
Owner review

16.5 Planning and Scheduling

Just as the budget is the basic tool in controlling cost, the project plan and schedule budget are the basic tools in controlling time. The *project plan* logically states the sequence for accomplishing the required work. It must consider all work phases of the project, including engineering design, procurement, construction, mechanical checkout, start-up, and special requirements such as tie-ins and early start-up of certain parts of the plant.

Since most capital projects are now being carried out in two phases, conception and execution, there are usually two phases in the planning and scheduling effort. For the first phase, a conceptual schedule or major milestone schedule is developed. This dual-purpose schedule provides control of engineering development, construction planning, and procurement of equipment with long lead time and provides the estimating team with the essential timing of expenditures needed to support those elements of the detail conceptual estimate that are sensitive to timing due to interest, escalation, etc. The second phase is a definitive network schedule and again has a dual purpose. First, it provides control of the following:

1. Engineering, design, sequencing, and progress
2. Procurement
3. Construction timing
   a. 30–60–90 look-ahead schedules
   b. System sequencing for commissioning

Second, it provides for special conditions. Among these are such items as shutdowns, tie-in schedules, and utilities or service scheduling.

A *key item schedule* is a chart against time showing the sequencing of activities, the starting and completion times, and the milestones along the way. Figure 16.4 shows a key item schedule.

16.6 Material Control

The materials management team's responsibilities are divided into three groups:

1. Procurement
2. Material control and tracking
3. Material coordination

The procurement group is responsible for the coordination and consolidation of purchasing and subcontracting. The bulk materials are consolidated in summarized quantity packages for more efficient procurement. These packages are coordinated with engineering and construction requirements, forming inquiries to vendors for bids. The bids are evaluated, and a successful vendor is selected after approval by the owner.

The material control and tracking group divides its responsibilities into three primary functions:

1. Major equipment control
2. Minor equipment and component control
3. Bulk materials or commodities

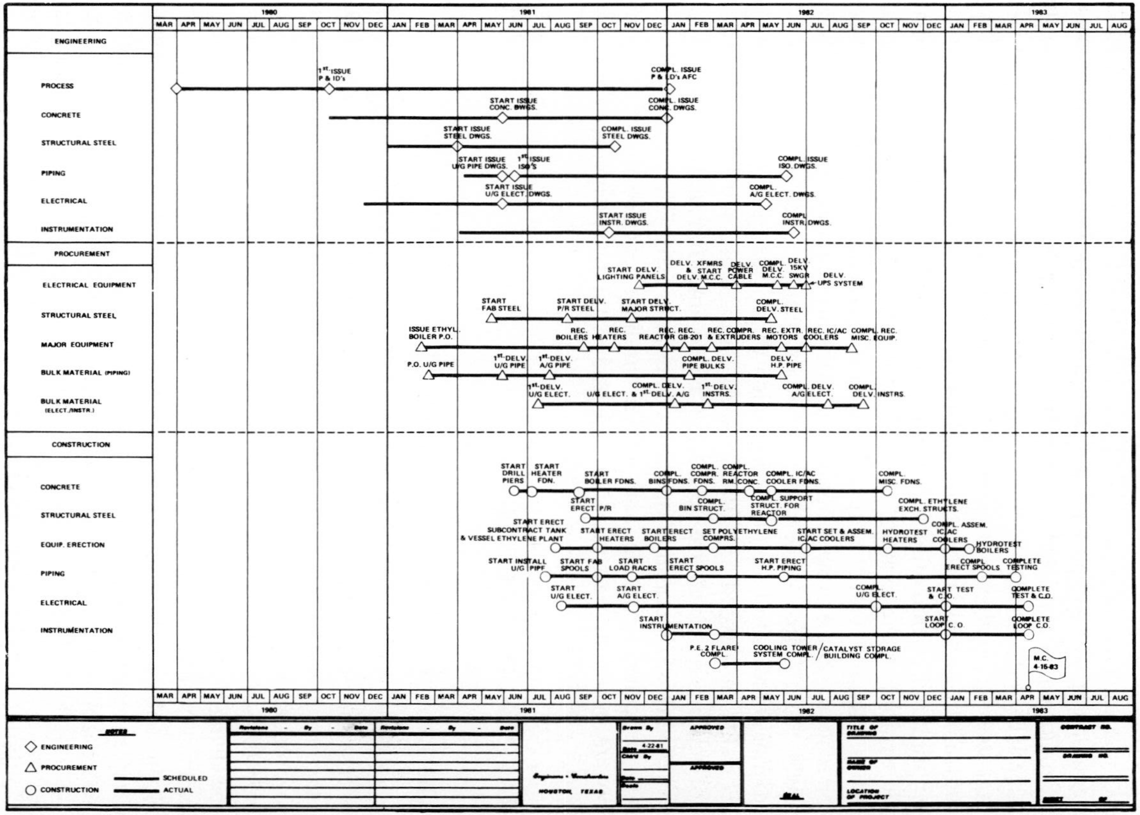

Figure 16.4 Key item schedule.

Major equipment consists of those items of process equipment which perform the essential functions of the manufacturing process. These items are always numbered as shown on the plot plan and equipment arrangement drawings. Examples of major equipment include heaters, exchangers, boilers, motors, pumps, coolers, blowers, fans, tanks, drums, vessels, towers, special machines, and reactors. The minor equipment and components are such items as instruments, traps, components of major equipment items such as filters, and small accessory groups. The bulk materials or commodities include such items as concrete, reinforcing steel, structural steel, pipe, electrical elements such as electrical cable, wire, and conduits, instrument tubing, insulation, and paint.

The primary function of material control and tracking is to assure that materials and equipment are available to the construction forces when required and are used as intended. Good material control substantially aids an efficient construction schedule and assures a minimum of surplus material at the conclusion of the project. Figures 16.5, 16.6, and 16.7 are examples of materials reports.

Material coordination involves inspection to provide quality control and assurance, traffic, and expediting functions. As an example, some projects require pieces of equipment which are very large, bulky, and difficult to handle. When required, the material coordination team comes in to assist. They will provide the necessary inspections, routing, and special handling required to get the equipment on site in the shortest possible time and at the least expense.

OPERATING-COST CONTROL

16.7 The Repetitive Nature of Operating Costs

Operating costs continue repetitively over the life of the plant and will exceed the capital cost manyfold; thus, they are very important. *Operating-cost control* means the regulation of expenditures on materials, labor, utilities, and associated costs to produce a quality product at a competitive price.

Operating-cost control requires a program that is updated periodically to meet the changing conditions for such factors as varying material price and quality, changing product quality, demand and price, increasing maintenance costs, technological obsolescence, and myriad other causes. It must define the objectives, devise a plan, provide a means of comparing performance with the plan, and indicate steps necessary to reduce costs.

16.8 Conjunction with Cost Accounting

Accounting personnel are set up to obtain information on costs. A knowledge of cost accounting, at least the principles developed in Chap. 17, will help the engineer cooperate with the accounting department.

The sections on operating-cost control are limited chiefly to direct material costs, direct labor, and direct production expense. Most of the indirect cost is not

MATERIALS MANAGEMENT

MAJOR EQUIPMENT SUMMARY

C/S AREA: 8788

ENGR. ITEM NO.	DESCRIPTION	ORDER NO.	VENDOR	ORIGINAL PROMISE	LATEST PROMISE	FIELD NEED	RCVD. DATE	SET DATE	COMPLETE DATE
CATEGORY: HEAT EXCHANGERS, CONDENSERS, ETC.									
14 14-5103	WET CRUDE EXCHANGER	P0060	OLD DOMINION FAB.	18APR80	25FEB80V	15APR80			
14 14-5219	HEAVY ENDS COOLER	P0026	BAS-TEX CORP.	08FEB80	26MAY80E	02APR80			
CATEGORY: PUMPS									
23 23-5104	GLYCOL PUMP	P0100	GOULDS PUMP	01MAR80	28MAY80V	15JUN80			

Figure 16.5 Major equipment summary.

MATERIALS MANAGEMENT

MINOR EQUIPMENT SUMMARY

C/S AREA: 8788

ENGR. ITEM NO.	DESCRIPTION	ORDER NO.	VENDOR	ORIGINAL PROMISE	LATEST PROMISE	FIELD NEED	RCVD. DATE	SET DATE	COMPLETE DATE
CATEGORY: INSTRUMENTATION									
IN FE-090-005	ORIFICE PLATE & FLANGE	P61024	DANIEL IND.	01DEC79	15DEC79V		16DEC79		
IN FI-090-005	FLOW INDICATOR	P71000	DUPONT INST.	15DEC79	15DEC79S	15DEC79			
CATEGORY: ELECTRICAL SPECIALTY									
ES 090-13-011-5	600 V SAFETY SWITCH	P62653	NUNN ELECTRIC	14MAR80	14MAR80S	14MARCH80	20MAR80		
ES 090-13-011-6	JUNCTION BOX	P62750	NUNN ELECTRIC	30JUNE80	30JUN80E	01JUL80			

Figure 16.6 Minor equipment summary.

MATERIALS MANAGEMENT

PURCHASE ORDER INVENTORY

STOCK NO.	DESCRIPTION	REQUISITIONED	PURCHASED	RECEIVED	PRICE
50000015	12" Blind Flange, 150 #				
34–12042	P.O.–00020	4	4	4	269.00
34–12160	P.O.–00025	5	5	5	337.00
34–13100	P.O.–00030	4	4	0	269.00

JOB BAL. REQUIRED ISSUED INV. BAL.

3	10	8	1	13	13	9	875.00

Figure 16.7 Purchase order inventory.

subject to control, at least from an engineering point of view. However, the control of indirect expenses is a legitimate subject for the cost engineer regarding items which are controllable, such as the material and labor which go into the indirect cost. The *allocation* of costs can strongly influence apparent costs, as explained in Chap. 17.

Cost control of operating expense, therefore, requires cooperation with the accounting department. Because engineering and technology are major factors, the cost engineer should have the responsibility for operating-cost control.

16.9 Defining Objectives

Table 16.2 gives an outline of operating-cost control. Many of the items in the table are established during the design and construction of the project; hence good operating-cost control begins at the design stage. Good plant layout will minimize operating labor and repair and maintenance costs.

Production schedules require careful review of inventory requirements, sales forecast, input material inventory, quality control, and efficient utilization of labor equipment and material. Similarly, labor schedules should include scheduling of direct operating labor, maintenance labor, and indirect labor for housekeeping, packing, and shipping. The labor schedules should be reviewed frequently for efficient use of both straight time and overtime.

During the early stages of plant operation large expenditures for quality control may be required, but they should diminish with time. However, they should not be reduced below a reasonable frequency that will provide a continuing check on the operating personnel, material, and technical factors.

16.10 Standard Costs

Standards are necessary to judge whether costs are satisfactory. These standards are used for comparison and like all standards can be somewhat arbitrary—

Table 16.2 Outline of operating-cost control

1. Define objectives
   a. Production schedules
   b. Labor-utilization schedules
   c. Input-material quality specifications
   d. Input-material utilization schedules
   e. Inventory requirements
   f. Repair and maintenance requirements
   g. Quality-control requirements
   h. Utility requirements
   i. Product-quality specifications
   j. Waste-disposal requirements
   k. Off-specification material rework

2. Define standards
   a. Standard costs

3. Cost reporting
   a. Periodic reports
   b. Special investigations

4. Corrective action

whether boys are measured by feet or yards does not change a tall boy into a short boy. However, there is some advantage in establishing standard costs in relationship to technological and optimum considerations rather than, say, to some normal historical cost. Standard costs of this nature are sometimes referred to as *industrial engineered standards*. Setting standard costs is a task that requires engineering knowledge of plant operations and is one of the most important functions of an engineer in operating-cost control since it is comparison of actual costs with standard costs which instigates action.

16.11 Cost Reporting

Although the engineers must set standard costs, it is the responsibility of the accounting department to report the comparison of actual against standard costs. Standard forms are used with periodic reporting by the various departments. The object, of course, is to reduce the collection of data to a routine basis.

There are occasions, however, when additional or more detailed information is required. Cost centers can be set up on a temporary basis. Frequently, the cost analysis requires considerable engineering effort, and experimental equipment will have to be installed to establish material and cost data.

16.12 Corrective Action

After a plant has been designed and built, it must be operated so as to produce the desired quality of product at minimum cost with the plant as it exists. Modifications to correct operating inefficiencies, e.g., high maintenance cost, can be effected throughout the life of the plant, but they must be justified on the basis of savings in operating costs.

The plant supervisor or manager is responsible for operating-cost control and for overseeing proper scheduling and use of resources. It is the supervisor's responsibility to review the reports and to see that corrective action is taken where necessary.

The need for corrective action is indicated by comparison of the reported values with the standard costs. The cost engineer can be a valuable advisor in interpreting the comparison. Implementation of corrective action is the responsibility of plant supervisors, who also must follow through to see that corrective action is carried out satisfactorily.

16.13 Classification of Operating-Cost Control Problems

Control problems can vary considerably depending on the type of project and kind of equipment involved and also on material variations, sales-forecast varia-

Table 16.3 Control-problem classification and corrective action required— operating costs

Control problem	Types of corrective action
1. Low yields—material usage too high	a. Check usage directions; see that they are being followed b. Check material-weighing and metering devices c. Check material quality d. Check for leaks, spills, even theft
2. High off-specification production and high waste, reworking, and downgrading	a. Check processing procedures b. Check material quality c. Check equipment cleaning and changeover procedures d. Check control instruments
3. Low production rates	a. Check for excessive rework b. Check maintenance downtime and causes c. Check for excessive changeover and downtime d. Check processing procedures
4. High maintenance costs	a. Review preventive and routine maintenance procedures b. Check that maintenance is performed by qualified personnel c. Emphasize maintenance scheduling against waiting for emergencies. d. Displace equipment requiring excessive maintenance
5. High utility costs	a. Check for proper control and eliminate leaks b. Check that utilities are shut off when not required c. Check utility metering for accuracy
6. High labor costs	a. Check for avoidable overtime b. Check production scheduling for economic labor scheduling c. Check absenteeism and lateness records for chronic offenders d. Check for excessive personnel turnover causing high overtime and training expenditures

tions, inventory requirements, etc. Although the number, magnitude, and character of the problems encountered are seemingly without end, many can be classified. Table 16.3 gives a checklist for some common types of control problems and the control action required.

CRITICAL-PATH METHOD (CPM)

16.14 Network Diagrams

In the late 1950s the Navy, with the help of a consulting firm, developed a program evaluation and review technique (PERT) for scheduling the Navy's Polaris missile program. PERT was originally a method of scheduling to complete a project on time. It did not include costs but did include probabilities for the elapsed time of activities. It has since been modified to include additional factors, such as cost. About the same time, Kelley and Walker developed the critical-path method (CPM) to expedite projects and included costs [4].

Both PERT and CPM are based on network or arrow diagrams and were an immediate success. They have been followed by many modifications but remain the most popular. The remainder of the chapter will be devoted chiefly to CPM, the most common network method used for construction and capital costs.

16.15 A Simple Network

Figure 16.8 shows a network with a normal completion time of 130 days.

Each small circle with a number is an *event* or *node*. The terminal numbers are 1, the start, and 7, the finish. Intermediate numbers correspond to completion of an intermediate task such as finishing the foundations or the third-story plumbing. An arrow represents an *activity*. Thus arrow 2–3 is the activity to go from event 2 to event 3, and number 45 on the arrow is the time in days for activity 2–3

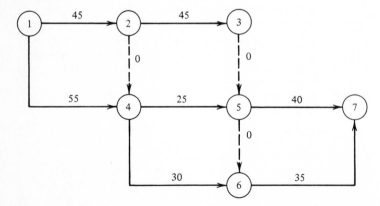

Figure 16.8 CPM network.

to be completed. No activity can be started until all the arrows, or activities, leading to its tail event (from which it starts) have been completed. Thus activity 4–5 cannot be started before activities 1–4 and 2–4 have both been completed. Frequently, it is necessary to use dummy activities or restraints, shown by dotted or solid lines, with zero time requirement. They are necessary to keep the sequence in order. The lengths of the arrows are unrelated to time or cost. The figure shown has seven events and 10 activities, including three dummy activities.

The first step in the analysis is the *forward pass* calculation to find the earliest start time (EST) and the earliest finish time (EFT) for each activity. The result is shown in Fig. 16.9. The calculation is made by starting at event 1 and working forward to the finish event, 7. Thus activity 1–2 has an EST of 0, a 45-day duration, and an EFT of 45. Activity 4–5 has an EST at 55 days because the *highest* EFT into event 4 is 55 days.

The next step is the *backward pass* calculation to find the latest finish time (LFT) and the latest start time (LST). The calculation is made by working backward from the finish to the start. In the backward pass calculation, when there are two or more numbers leading out of an event (LST), the *lowest* number prevails and provides the LST for all activities ending at the event. Figure 16.10 is the result of the backward pass calculation. From the forward pass calculation the network time to finish is 130 days, shown by the end of activity 5–7 event 7. Hence in the backward pass calculation the LFT for both activities 5–7 and 6–7 is at 130 days. Activity 6–7 takes 35 days, so that its LST is at $130 - 35 = 95$ days. At event 5 the *lowest* LST *out* of the event is at 90 days for activity 5–7, and thus 90 days becomes the LFT for both activities, 3–5 and 4–5, ending at event 5.

Figure 16.11 shows the forward and backward pass calculations combined, and Table 16.4 is a tabulation of the results.

Of particular interest is the *total float* for an activity, which represents the time by which an activity may be delayed or extended without affecting the duration of the project. Activity 4–5 has an EFT at 80 but can be allowed to go to an LFT at 90 days without extending the duration of the project beyond 130 days.

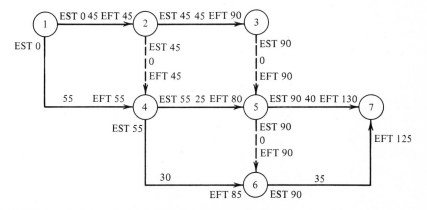

Figure 16.9 Forward-pass calculation for the network.

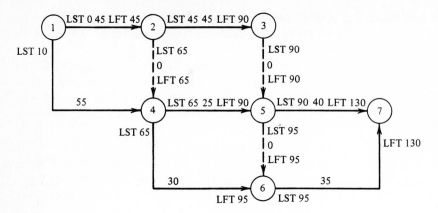

Figure 16.10 Backward-pass calculation for the network.

Thus activity 4–5 has a total float of 10 days, given by the difference between the LFT and the EFT. The total float for all activities is given in column 7 of Table 16.4. Float is sometimes called *slack*.

The *critical path* is the longest duration chain of activities from start to finish. There is no total float for any activity on the critical path. Any increase in duration along the way would delay the finish of the project. Column 7 of Table 16.4 shows that there is no total float along activities 1–2, 2–3, 3–5, and 5–7, and together they constitute the critical path with a duration of 130 days. There can be more than one critical path, but all critical paths have to be equal.

Another term used is *free float,* representing the time an activity can be delayed without affecting the EST of *any* other activity. Free float cannot exceed total float, and may be less. To find the free float for an activity A ending at event E, first find the earliest start time for activities out of event E. Second, note the earliest finish time for activity A. The difference between the two is the free float. There can be no free float along the critical path. The free float for the activities is

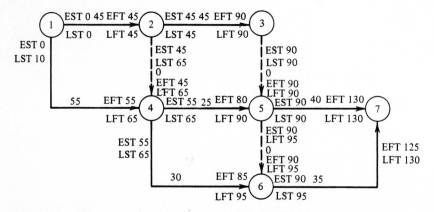

Figure 16.11 Forward and backward pass calculations for the network.

Table 16.4 Results of a CPM network

(1) Activity	(2) Duration, days	(3) Earliest start time, days	(4) Earliest finish time, days	(5) Latest start time, days	(6) Latest finish time, days	(7) Total float, days	(8) Free float, days
1–2	45	0	45	0	45	0	0
1–4	55	0	55	10	65	10	10
2–3	45	45	90	45	90	0	0
3–5	0	90	90	90	90	0	0
2–4	0	45	45	65	65	20	10
4–5	25	55	80	65	90	10	10
5–6	0	90	90	95	95	5	0
4–6	30	55	85	65	95	10	5
5–7	40	90	130	90	130	0	0
6–7	35	90	125	90	130	5	5

given in column 8 of Table 16.4, as determined from Fig. 16.11. For example, the total float for activity 2–4 is 20 days. However, the activity can be lengthened only 10 days without delaying the EST for activities 4–5 and 4–6. The free float is 10 days.

16.16 Advantages of CPM

CPM makes critical activities known at all times, so that those requiring special attention will receive it. On the other hand, activities not on a critical path have some flexibility. CPM is not something that is used merely at the start of a project. The total float and free float along activities change with time on a large project, and CPM is kept up to date continuously. For large projects, government and industry frequently require its use as part of the contract.

CPM is very adaptable to computer programming. In practice, actual dates instead of elapsed dates are printed out, thus allowing for weekends, holidays, and the like. CPM can be combined with costs and is a valuable tool in calculating the most economical way to crash a program at a cost to reduce the time for a project. CPM will keep a project going smoothly with respect to time and expenditure. It cannot correct deficiencies in estimates, but it can disclose them early enough for reassessment and reduce their frequency.

A project can be accomplished in more than one way, and therefore by more than one network diagram. Thus, in setting up a network, the analyst should not use a particular network just because it is feasible. A better network may exist.

16.17 Precedence Diagramming

There is another network method, known as *precedence diagramming,* which uses boxes for the activities. A simple precedence diagram is shown in Fig. 16.12. The

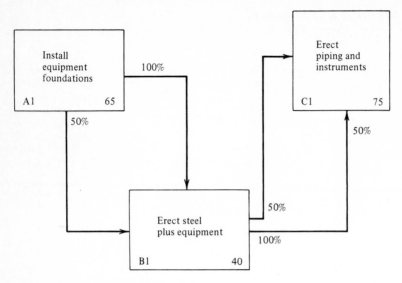

Figure 16.12 A simple precedence diagram.

activity of Box A1 is to install equipment foundations within 65 days. When this activity is 50 percent completed, activity B1 can be started, but B1 cannot be finished until 100 percent of A1 is completed. Activity B1 requires 40 days to complete. Activity B1 can be started at 50 percent of the time required for A1, i.e., at 32.5 days. When activity B1 is 50 percent completed, activity C1 can be started. B1 must be 100 percent completed before 50 percent of C1 can be completed. Activity C1 requires 75 days for completion.

PROBLEMS

16.1 Define the critical path for the following network for the production of an instrument.

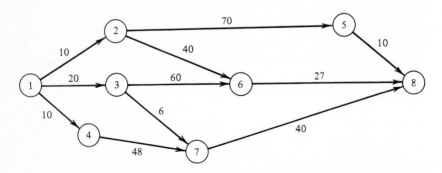

16.2 Define the critical path for the following process (p. 429) to produce an electronic device. Find the free float and total for each operation.

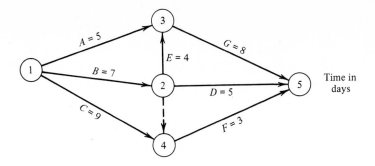

16.3 The following additional information applies to Prob. 16.2:

Operation	Normal		Crash	
	Duration	Operation cost, $	Duration	Operation cost, $
A	5	100	4	150
B	7	200	6	300
C	9	400	8	450
D	5	150	4	200
E	4	100	3	150
F	3	300	2	350
G	8	700	7	850

(*a*) What is the total cost of the project?

(*b*) What would be the minimum additional cost required to finish the project 2 days earlier?

16.4 A network with normal time in days is as follows:

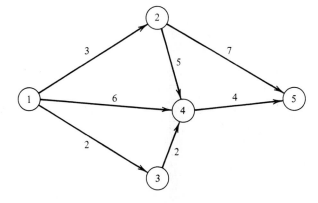

Find the critical path and free float of all activities.

16.5 The following additional information applies to Prob. 16.4. Assume a straight line or linear relationship between cost and reduction in time, as in the last column.

Activity	Normal Days	Normal $	Crash Days	Crash $	Crash $ per day
1–2	3	50	2	100	50
1–3	2	25	1	50	25
1–4	6	140	4	260	60
2–4	5	100	3	180	40
3–4	2	80	2	80	....
2–5	7	115	5	175	30
4–5	4	100	2	240	70
Totals		610		1085	

Find the minimum cost for completing the project in 12, 11, 10, 9, 8, and 7 days.

16.6 For the following network with time in days, find the critical path and the free float of all activities.

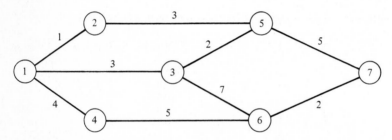

16.7 Return to Prob. 16.6 and assume that labor can be shifted from any activity to any other with an equal change in time, e.g., activity 3–6 can be reduced by 0.1 day by increasing activity 1–2 by 0.1 day. Find the minimum time for the network after replanning.

16.8 Redo Prob. 16.6 with a dummy activity arrow from event 5 to event 6.

16.9 Refer to the diagram for Prob. 16.6. Activity 3–6 can be started when activity 1–3 is one-third completed, and activity 3–5 can be started when activity 1–3 is two-thirds completed. Redraw the network, making sure that event 3 is completed, and find the critical path.

16.10 Three operations are performed in series, with time in hours as shown. B can be started when A is one-third completed, but B cannot be completed until 1 h after A is completed. C can be started when B is one-half completed, but C cannot be more than 50 percent completed until 1 h after B is completed. Show this as a network and find the time for the critical path.

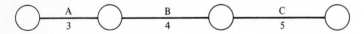

REFERENCES

Cost control

1. Dudick, T. S.: *Cost Controls for Industry,* 2d ed., Prentice-Hall, Inc., Englewood Cliffs, N.J., 1976.
2. Hackney, J. W.: *Control and Management of Capital Projects,* John Wiley & Sons, Inc., New York, 1965.

3. Kavanagh, T. C., F. Muller, and J. J. O'Brien: *Construction Management,* McGraw-Hill Book Company, New York, 1978.

4. McNeill, W. I.: *Effective Cost Control,* Prentice-Hall, Inc., Englewood Cliffs, N.J., 1965.

5. McNesby, E. J.: *Systematic Control of Factory and Management Costs,* Prentice-Hall, Inc., Englewood Cliffs, N.J., 1978.

6. Radke, M.: *Manual of Cost Reduction Techniques,* McGraw-Hill Publishing Company, Maidenhead, England, 1972.

7. Wilson, R. M. S.: *Cost Controls Handbook,* John Wiley & Sons, Inc., New York, 1975.

8. Zeyher, L. R.: *Zeyher's New Guide to Cost Reduction in Plant Operations,* Prentice-Hall, Inc. Englewood Cliffs, N.J., 1977.

CPM and network methods

1. Ahuja, H. N.: *Construction Performance Control by Networks,* John Wiley & Sons, Inc., New York, 1976.

2. Antill, J. M., and R. W. Woodhead: *Critical Path Method in Construction Practice,* 2d ed., John Wiley & Sons, Inc., New York, 1970.

3. Horowitz, J.: *Critical Path Scheduling,* The Ronald Press Company, New York, 1967.

4. Kelley, J. E. Jr., and M. R. Walker, *Critical-path Planning and Scheduling: An Introduction,* Manchley Associates, Ambler, Pa., 1959.

5. Levin, R. I., and C. A. Kirkpatrick: *Quantitative Approaches to Management,* 2d ed., McGraw-Hill Book Company, New York, 1971.

6. O'Brien, J. J.: *CPM in Construction Management,* 2d ed., McGraw-Hill Book Company, New York, 1971.

ASSOCIATED TOPICS

SEVENTEEN

COST ACCOUNTING

R. A. Rogers

BASIC CONCEPTS

17.1 Cost Accounting and the Cost Engineer

The interrelationship between the accounting function and the cost engineering function varies widely between manufacturing and processing industries, between operations and construction, and among individual companies. In some cases, accounting has sole responsibility for recording and reporting cost data; in other instances, the task is performed jointly by accounting and cost engineering; and in still other environments, the cost data are derived solely by cost engineering with only the requirement that the data must be reconcilable with accounting records. The last situation is the most usual in the construction of new facilities and is necessitated by the requirement to project costs based on all known data rather than to rely upon historical data alone. In all these situations, it is essential that there be mutual understanding between the two groups. Cost data frequently must be allocated to different products, processes, or constructed facilities. This requires considerable technical and engineering familiarity with operations as well as knowledge of accounting theory and practice so that the allocations are fair and reasonably equitable.

This chapter will present a brief survey of the concepts and practices of accounting to the extent required to provide cost engineers with the basic, minimum understanding needed for their professional work.

17.2 Accounting Definitions

Accounting is a technique of measuring economic events in terms of money. A general definition is:[1]

> Accounting is the art of recording, classifying and summarizing in a significant manner and in terms of money, transactions and events which are, in part at least, of a financial character and interpreting the results thereof.

Accounting uses what is called a *double-entry system* for recording each transaction in monetary units. The double-entry system can be thought of as a series of

[1] P. Grady, *Inventory of Generally Accepted Principles for Business Enterprises,* AICPA Accounting Research Study 7, American Institute of Certified Public Accountants, New York, 1965, p. 2.

Table 17.1 Pro forma chart of accounts*

Balance sheet accounts	
2000 Assets	3000 Liabilities
2100 Cash	3100 Accounts payable
2200 Accounts receivable	3200 Notes payable
2300 Notes receivable	3300 Taxes payable
2400 Inventory—materials and supplies	3400 Accrued liabilities
2500 Inventory—finished products	3500 Reserve accounts
2600 Work-in-progress	
2700 Equipment	4000 Equity
2800 Buildings and fixtures	4100 Capital stock issued and outstanding
2900 Land	4200 Retained earnings

Profit-and-loss accounts	
5000 Revenues	6300 Heat, light, and power
5100 Sales of finished goods	6400 Communications expense
5200 Other revenues	6500 Reproduction expense
	6600 Insurance
6000 Expenses	6700 Taxes
6100 Cost of goods sold	6800 Depreciation
6200 Salaries and wages	6900 Interest expense

Cost classification accounts (work-in-process)	
7000 Construction work in progress	7600 Electrical systems
7100 Site preparation	7700 Piping systems
7200 Concrete work	
7300 Structural steel	8000 Manufactured goods-in-progress
7400 Heavy equipment	8100 Direct materials
7500 Buildings	8200 Direct labor
	8300 Overhead

*These are summary-level accounts, sometimes called *control accounts*. More detail can be provided by using subaccounts as required.

equations; therefore, the two sides of the equation, the debits and the credits, must be equal for each transaction and cumulatively.

The terms *debit* and *credit* are conventional and have no particular significance as such.

Transactions are classified in accordance with a chart of accounts, which is a list of files by name and code number. Table 17.1 is a skeletal chart of accounts. Each account may be thought of as a record of all the transactions for a particular classification. Each account may have either debit or credit transactions. The difference between the sum of its debit transactions and its credit transactions is called the *balance* and is said to be a *debit balance* or a *credit balance* depending upon which is greater. Certain accounts normally have debit balances, such as the cash account, and others normally have credit balances, such as the accounts payable account.

Another common equation used in accounting is:

$$\text{Assets} - \text{liabilities} = \text{net worth}$$

Accounts are further categorized as balance sheet accounts or profit-and-loss accounts. *Balance sheet accounts* are essentially those identified in the above equation, and their balances are used to prepare a financial statement called a *balance sheet*. Table 17.2 is a pro forma balance sheet. The balance sheet reflects the financial condition of the enterprise as of a given date, usually the end of the month or end of the year. *Profit-and-loss accounts,* or *expense and revenue accounts,* are for the purpose of preparing a financial statement called an *income statement* which reflects the operating results of the enterprise for a given period of time, usually a month or a year. Table 17.3 is a pro forma income statement.

Among the asset accounts is an account called *work-in-process.* This is where construction, manufacturing, or process costs are recorded and carried until the work is complete. This account may be further subdivided into jobs, and the jobs subdivided into elements of cost called *cost codes.* It is these cost accounts with which cost engineers are concerned.

17.3 Types of Accounting

One type of accounting is *financial accounting.* It is essentially a historical record of transactions which have been completed or partially completed. Its function is to report operating results and financial condition of the enterprise to stockholders, Securities and Exchange Commission, Internal Revenue Service, and other regulatory bodies.

Another type of accounting, called *managerial accounting,* is for the purpose of providing financial operating data to the management of the company to assist them in performing the management function. Cost accounting, the principal topic of this chapter, is among the most important parts of managerial accounting. Cost accounting expands the techniques of financial accounting. Its major application is in the planning and control aspects of management, which have in-

Table 17.2
Pro forma Balance Sheet
The Acme Corporation
Dec. 31, 198X
(in thousands of dollars)

Assets	
Current assets:	
Cash	$ 2,131
Accounts receivable	36,292
Notes receivable—due within 1 year	125
Inventories and work-in-process	12,614
Subtotal	51,162
Fixed assets:	
Notes receivable—not due within 1 year	1,250
Equipment—less reserve for depreciation	44,768
Buildings and fixtures—less reserve for depreciation	16,347
Land	5,000
Subtotal	67,365
Total assets	118,527

Liabilities and net worth	
Current liabilities:	
Accounts payable	13,421
Notes payable—due within 1 year	1,200
Taxes payable	2,500
Accrued liabilities	1,248
Subtotal	18,369
Long-term liabilities:	
Notes payable—not due within 1 year	10,000
Total liabilities	28,369
Net worth:	
Capital stock issued and outstanding	72,172
Retained earnings	17,986
Subtotal	90,158
Total liabilities and net worth	118,527

creased in importance during the last few decades since the introduction of scientific techniques to the managerial decision-making process.

17.4 The Mechanics of Accounting

Accounting includes three basic steps: recording, classifying, and summarizing economic data in terms of money.

The chart of accounts must be established by mutual agreement among the

Table 17.3
Pro forma Income Statement
The Acme Corporation
Year ended Dec. 31, 198X
(in thousands of dollars)

Revenues	$88,126
Operating expenses:	
Cost of goods sold	46,218
Salaries and wages	2,729
Heat, light, and power	1,516
Communications expense	127
Reproduction expense	38
Insurance	1,243
Taxes—other than federal income taxes	746
Depreciation	9,982
Subtotal	62,599
Operating income	25,527
Nonoperating expenses:	
Interest expense	1,120
Income before federal income taxes	24,407
Federal income taxes	11,715
Net income	12,692

various groups involved, including particularly accounting, cost engineering, any group charged with responsibility for classifying any of the transactions, and any operating, construction, or management group that will be expected to interpret accounting/cost engineering reports. The chart of accounts as shown in Table 17.1 can be broken down into as much detail as desired by the use of subaccounts. In using the accounts for entry and retrieval of cost information, the cost engineer must be entirely familiar with what each subaccount includes. In practice, a compromise must be made in establishing subaccounts. Too few creates a lack of details in the cost, and too many adds to the accounting cost and can confuse rather than help.

Transactions may originate from several sources, such as (1) payment of a vendor's invoice, which is recorded on a document called a *cash voucher* or an *accounts payable voucher;* (2) payment of salaries and wages based upon employee time sheets; (3) transfer of materials and supplies or capital equipment from inventory accounts on warehouse tickets; (4) equipment use charges processed from equipment time sheets; and (5) calculation of depreciation, insurance, taxes, inventory adjustments, and cost allocations, which are prepared on special forms called *journal vouchers.*

Accounting transaction documents may be prepared in the departments where the transactions originate. Transactions will normally be classified by the personnel closest to the operation and in the best position to know the proper classification. All transaction classifications will be checked by an independent party before entry into the accounts. Errors can be corrected later if discovered, but the best and least expensive method is to prevent as many errors as is economically and practically feasible. Cost data are only as good as the quality of the original data and the quality of the mechanical entry. Data entry errors are kept to a minimum by various checks and balances built into the system.

The summarization step consists of closing the revenue and expense accounts to determine the profit or loss for the period (usually at the end of each month and at the end of each year) and to prepare the income statement. The result is then transferred to the equity account, or retained earnings. The asset and liability account balances are then used to prepare the other main financial statement, the balance sheet. Finally, various cost reports are prepared for both financial and control purposes. The responsibility for the latter may rest partially or wholly with the cost engineer. See Sec. 17.1, which explains the varying relationships which may exist between the cost accountant's work and that of the cost engineer.

Most accounting systems are computerized. This makes it possible to handle great volumes of data on a timely basis and to manipulate the data in various ways to produce meaningful reports and analyses.

17.5 Definitions and Classifications of Cost

The term "cost" has numerous meanings and definitions. The AICPA *Inventory* states:[2]

> Cost is the amount, measured in money, or cash expended or other property transferred, capital stock issued, services performed, or a liability incurred, in consideration of goods or services received. Costs can be classified as unexpired and expired. Unexpired costs (assets) are those which are applicable to the production of future revenues. . . . Expired costs are those which are not applicable to the production of future revenues and for that reason are treated as deductions from current revenues, or are charged against retained earnings.

Costs may be divided into as many classifications as may be useful in preparing meaningful reports for better analysis and understanding. Care should be taken not to use too many classifications; otherwise, reports will be so voluminous as to be useless. It is important to select the appropriate cost type for individual objectives. No general rules are possible because of the wide variety of circumstances.

It was mentioned earlier that the cost estimator needs future-directed, predetermined data rather than purely historical costs. Insufficient attention is often given to this important difference. In an era of great change, as in recent years, the assumption that the past will repeat itself in the future cannot be considered valid even in the short term. The cost estimator, therefore, must attempt to adjust recorded data for estimating purposes after soliciting information from knowledgeable persons within the organization and utilizing available indexes of predicted economic behavior. The more future-directed the cost estimator's figures,

[2]Ibid., p. 228.

the more critical become the predicted activity levels and economic behavior factors.

Direct costs, also called *prime costs,* are directly traceable to the product being manufactured or processed or to the physical asset being constructed, e.g., the fabric in clothing or the concrete in a foundation.

Indirect costs are all the costs of manufacturing, processing, or construction that cannot be classified as direct costs, because it is either impossible or impractical. Each classification of indirect cost is initially accumulated in a separate account, and at the end of the accounting period, the balance is allocated to the individual benefiting activities, such as cost centers, products, or construction activities. There are many bases for allocation, such as direct labor cost, square feet of floor space, or number of employees. An allocation basis for a particular situation must be carefully selected. The end result of any allocation must be reasonable and equitable. The topic is discussed more fully in Secs. 17.10 to 17.14.

17.6 Cost Accounting for Expenditure of Construction Funds

Accounting for the costs of constructing new capital facilities has several purposes. The first is to account for the expenditure of the company's funds. Another is to accumulate costs in a manner which will support depreciation calculations for both financial reporting and income tax computations. Payment of progress fees to contractors also requires cost data for calculating the amount earned under the terms of the contract, such as the percent of work completed by cost classification times appropriate contractual monetary amounts.

Finally, of most interest to the cost engineer is cost data for the construction budget. This information is useful for project cost control purposes. The cost engineer is able to forecast budget overruns by adding to cost accounting's actual cost records all those anticipated costs which can be identified from information such as purchase order commitments, contract commitments, and estimated cost to complete the work.

17.7 Cost Accounting for Manufacturing Operations

The cost of manufacturing operations is accumulated in three separate accounts:

1. Direct material: cost of materials, assemblies, and parts used to complete the project.
2. Direct labor: wages of workers who are helping to complete the product.
3. Overhead (also called *burden*): the cost of all other factors contributing to the completion of the product.

As the product input factors are used in the factory, they are charged to a *work-in-process* account, which may be broken down into the three subaccounts listed above. When the completed product is physically removed from the factory and transferred to the finished-product stores area, its accumulated cost is transferred from the work-in-process account to the *finished-products* account. It follows that the balance in the work-in-process account represents the accumulated cost of the products being manufactured but not yet completed.

The cost of materials is the purchase price and all incidental expenses incurred in transferring them to the purchaser's premises, such as transportation, tariff duties, or insurance. In most cases, purchased materials and parts are charged to a *materials-stores* account, which is credited when the materials and parts are physically transferred to the factory floor and become part of the work-in-process account.

Labor cost is the result of several factors. These include the basic hourly rates and the hours worked, augmented by certain provisions—overtime, social security taxes, etc.—and contractual agreements such as vacation pay, health, pension payments.

The labor cost is the gross pay, although the worker receives a smaller net, or take-home, pay because of deductions for taxes, union dues, and so on.

17.8 Cost Centers and Unit Costs

Costs are accumulated by cost centers, which may or may not coincide with operating departments. *Cost centers* are placed at points in the production process where costs can be measured and recorded as conveniently and accurately as possible. Manufacturing operations are organized by departments headed by a line supervisor who usually has an office with a small staff to care for the clerical work associated with production, such as materials requisitions and labor time control. A department may be broken down into several cost centers to obtain more detailed cost data, or several departments may be combined. The choice of cost centers is important. Cost engineers should maintain liaison with the cost accountants and those who design the cost-accumulation system. The cost engineer's requirements must be known in order to be met within the limitations of cost and practicality.

Cost data must be related to an activity factor such as volume of production. Comparison merely of total costs for different time periods is meaningless. Therefore, all cost reports should present the unit cost as well as the total cost. Selection of a suitable activity factor can be difficult in multiproduct operations when there are substantial differences in the unit costs of the various products.

17.9 Job-order Costing and Process Costing

There are two basic accounting techniques for accumulating unit costs: job-order costing and process costing. *Job-order costing* is used for specialized production jobs, e.g., electric generators manufactured to individual specifications. Costs can be computed accurately, but considerable paperwork and effort are required. For mass production of repetitive units, a job order can cover an accumulation of units. Job orders can be issued for individual customers or for stock.

Process costing is used for continuous-process industries which operate 24 h a day, such as oil, chemical, and steel. Process costing accumulates average costs for a time period, most frequently 1 month. The production cycle continues without interruption, while the cost cycle is cut for each accounting period to determine the result of operations. Process costs use averages and are less accurate than job-order costs but simpler and less expensive to compute. One requirement is the need to estimate inventories at the point of the cost cutoff, where incomplete units are evaluated in terms of completed units.

Process costing is the more widely used technique of cost accumulation. It serves continuous-process industries, which have no alternative, but also mass-production industries, which could use either method. Efficiency is measured by statistical methods which implement the costing technique.

In the job-order costing method, the cost of the basic input factors, direct material and direct labor, are charged to the specific numbered job orders by posting periodically on the job-order sheets, one for each job order. Upon completion of the job order, overhead is applied on the basis of an activity factor, as explained in the next section. Table 17.4 is an example of a job order cost sheet.

Process costs are accumulated by departments (processes) for a time period, usually 1 month. In many instances, materials are introduced to the process at an early stage, and in subsequent processes only the *cost of conversion,* direct labor and overhead, has to be added. Some units may be lost, e.g., for technical reasons

Table 17.4 Job-order cost sheet

Job-Order Cost Sheet

Job order no. _____   Customer _____
Part no. _____   Work started _____
Number of units _____   Work completed _____

| Week of | Material | | Labor | | Applied overhead | Total |
	Requisition no.	Amount	Hours	Amount		
Total						

in chemical industries, and the cost must be absorbed by the salable units. In other instances, an increase in units may occur. Costs are accumulated on process cost sheets, which show input and output for both physical units and costs, and unit costs by individual processes. Table 17.5 is an example of a process cost sheet.

OVERHEAD ALLOCATION

17.10 Techniques of Overhead-Cost Accounting

The *overhead-cost* classification includes all production costs which are not considered prime costs (direct material and direct labor). There is no limit to the number of overhead classifications; however, a few principal groups can be distinguished:

Indirect materials, also known as *supplies,* are materials such as lubricants that do not become a part of the finished product.

Indirect labor, the wages and salaries of employees who are not directly connected with the manufacture of a product, such as supervisors, maintenance workers, and internal transportation. Frequently, fringe benefits are included in this classification.

Facilities costs, both short-term costs of the current year and long-term costs which are depreciated over a number of years. The former includes building maintenance, local real estate taxes, and others; the latter, buildings and equipment.

Service-department costs, for facilities which support production but are not part of production, e.g., accounting, laboratories, stores, cafeteria, and first-aid stations.

It is difficult, if not impossible, to enumerate the many different types of costs which make up the overhead-cost classification. Indeed, overhead accounting is known as one of the most complex problems of cost accounting.

The accumulated overhead costs are allocated in two stages. First, the overhead costs are allocated to the cost centers (departments). These, in turn, allocate the overhead to job-order costs and process costs.

Assuming a large number of cost classifications and a large number of cost centers, the number of computations can become quite extensive. In practice, various shortcut techniques are used to reduce the clerical work.

Cost engineers must recognize that the cost data which finally emerge from this procedure are significantly affected by the measure used for determining the indirect cost and the method used for allocation. Therefore, they should not accept these data unless they are familiar with the techniques used in computing them.

In order to select a *basis for overhead-cost allocation,* a measure must be found that reflects closely the actual benefit provided by a cost classification.

Table 17.5 Process cost sheet

Process Cost Sheet
for the month of _____

	Process 1		Process 2		Process 3	
Quantity schedule (units) **Input:** Transferred from preceding department						
Added in department						
Total						
Output: Transferred to next department						
Work-in-process inventory						
Lost (added) units						
Total						
Cost schedule **Input**	Amount	Units	Amount	Units	Amount	Units
Cost transferred from preceding department						
Cost added in department						
Material						
Labor						
Overhead						
Adjustment for lost (added) units						
Total input						
Output						
Cost transferred to next department						
Work-in-process inventory						
Cost from preceding department						
Cost in department						
Material						
Labor						
Overhead						
Total inventory						
Total output						

Allocation must be made on this particular basis. As an example, consider building maintenance. This can be measured on the basis of square feet and allocated on the basis of floor space occupied by each cost center.

In turn, cost centers must have a measure by which the accumulated overhead is allocated to each job order or process. This may be, e.g., labor hours, or some material quantity for which appropriate data are available. Introduction of computers has greatly simplified the clerical and computational job but has not necessarily improved the quality of the resulting data.

The procedure for allocating overhead, as thus far described, has two disadvantages. It is complicated, and the overhead applicable to units made early in the month cannot be calculated until the end of the month, when the accumulated overhead becomes known. A shortcut method has evolved to overcome this disadvantage. Two estimates are required. First, an estimate is made for each overhead-cost classification covering specific periods, usually a year. Second, the estimate of some measure must be chosen which permits each job order to be charged with the appropriate share of the overhead cost. Direct labor, in the form of dollars or hours, has been the most widely used measure.

A rate can be established by dividing the total estimated overhead cost by the estimated quantity of the selected measure; this is called the *burden rate*. As an illustration, assume that the estimated total overhead cost is $50,000 and that the estimated direct labor cost is $25,000. The burden rate is $50,000/$25,000, or $2, for each direct labor dollar. If the direct labor cost of job order 750 is $2000, the applied overhead is 2000 × $2, or $4000.

It must be kept in mind that the applied burden rate is an estimate, not necessarily the actual overhead. A difference or variance must be expected and reconciled periodically. Other problems arise because overhead-cost classifications react differently to volume fluctuations, as discussed in Sec. 17.11.

There are two principal advantages in using burden rates. The first, the savings in time and cost, has been mentioned. Second, and perhaps more important, is the possibility of obtaining data more satisfactory for determination of operating efficiency. Use of an actual overhead cost subjects the data to fluctuations within the period, such as payment of taxes or the cost of an overhaul. An applied burden rate, on the other hand, equalizes such charges by spreading them over the year. Abrupt overhead costs are eliminated as factors influencing the determination of efficiency.

17.11 Cost Behavior against Activity Volume

The importance of the volume of activity in the accumulation of cost data has already been stressed. Businesspeople have long been aware that profits increase faster than sales when activity is expanding, and vice versa. This is because different cost classifications react differently to changes in production volume and sales volume. Costs are broken down into fixed costs, variable costs, and semivariable costs, as discussed in Chap. 7. One of the principal responsibilities of cost accountants is to study the relationship of the various cost classifications to volume fluctuations.

A number of methods have been developed to accomplish this task. It must be remembered that all are approximations, often with a considerable margin of error. The widely used break-even chart was described in Chap. 7. Charts that are not elaborate incorporate a number of simplifying assumptions which limit their usefulness. Theoretically, in terms of accurate measurement, most cost classifications are semivariable. However, this circumstance can be accommodated by not limiting the charts to the simplest type.

By dividing costs into fixed and variable, one obtains a relationship which is considered applicable only within a limited range of volume fluctuations.

In computing burden rates, the estimated overhead is divided by the estimated quantity of measure, such as estimated direct labor hours for the same period. The overhead will depend upon volume to some extent, and the estimated labor hours much more so and in a complicated fashion. Hence, the burden rate will depend on volume, i.e., a burden figure is correct only for a particular volume.

The distinction between fixed and variable costs refers to short-term considerations. In the long run, all costs can be thought of as variable. Costs are fixed as a result of a previous management decision. The cost estimator should recognize that some fixed costs may be changed by new arrangements.

17.12 Direct Costing and Absorption Costing

The system of applying overhead described so far is known as *absorption costing* because the direct, fixed, and variable costs are absorbed by the work in process and in turn become part of the inventory. If production is greater than sales, inventory of the finished products will be increasing. Suppose volume is high enough so that the overhead applied to the work in process, which is an estimate, is greater than the actual overhead. As a consequence, inventory will be overvalued. Under such conditions, the apparent profit of the organization can be increased by increasing production beyond the normal volume and adding to inventory. This contradicts the basic principle of accounting that profits should be recognized only if they result from sales to outsiders, i.e., arm's-length sales.

Another system of applying overhead known as *direct costing* attempts to overcome this disadvantage but introduces some problems of its own. In direct costing the fixed part of the overhead, and only the fixed part, is not applied to the work in process but is regarded as a period cost and is entirely absorbed as an expense against income in the period. In this way the applied overhead burden includes only variable costs, and no over- or underestimation of the inventory occurs.

There are differences between absorption and direct costing. Inventory will always be held at a lower value under direct costing. Over a period of time, the two methods will give the same cumulative net income because fluctuations in inventory are smoothed out. Over a short period of time, the two can differ depending upon the volume and how much the inventory changes during the period.

There is a difference of opinion among professional accountants about the applicability of the direct-costing technique. It underestimates the cost of the inventory because it does not include the fixed charges. The cost estimator must be alert to the use of these alternative methods, which can have a significant effect on cost and profit data.

17.13 Standard Costs and Budgets

Up to this point, most of the discussion has been concerned with the accumulation of historical costs. A very important aspect is the use of these data for future-directed costs. The true analysis of efficiency depends on the previous determination of measures or yardsticks against which the actual performance is evaluated. These future-directed and estimated costs are particularly important to cost engineers. Since they can depend upon technological factors, cost engineers should cooperate with those who determine such costs.

Standard costs are determined by using scientific techniques and objective quantitative measurement. In contrast, estimated costs are based on individual judgment with limited significance. Standard costs do not necessarily represent expected performance, but rather desired objectives.

Materials standards developed by engineers are valued in accordance with actual or estimated purchase prices to be converted into standard costs. Labor standards, after measurement, are converted into standard costs from appropriate labor rates. Overhead-cost standards are difficult to establish and often are merely estimates based on past performance.

Actual costs are compared to standard costs for control purposes.

Budgets serve for both planning and control purposes. Budgets reflect mainly expected performance and have to be adjusted more frequently than standards. Budget data may be especially valuable to the cost engineer. Comparison of actual costs to budgeted costs indicates how closely actual operations have tracked against the plan.

17.14 Joint Costs

An intricate problem in accounting arises when data are related to more than one activity or transaction and the share of each cannot be measured accurately. When this happens, costs cannot be allocated appropriately; some activities will be overcharged and some undercharged. An interesting case arises in connection with the costing of so-called by-products and joint products.

Several products are often derived from a single source, such as crude oil. Another well-known case is the meat packing industry. The relationship between the products developed is often the result of technological rather than market conditions. Sometimes there is no market for a product that arises in connection with another product which is in great demand.

A product for which there is no demand or only a limited demand is called a *by-product*. The product produced at the same time which is in great demand is the *main product*. Several main products derived from a single source are called *joint products*.

As an illustration, consider the carcass of a slaughtered animal carried through several processes. After it has been divided into several parts, a product remains which cannot be sold but for which there is a market if it is improved. This may be called the by-product. At the next stage, several marketable products become available. The point of separation is called the *split-off point*. The problem arises of how to allocate the accumulated cost at that point to the several products. The most widely used techniques are:

1. In the case of a by-product, the total cost at the split-off point is absorbed by

the main product after any net incidental revenue from the by-product is credited.
2. The allocation for joint products is more difficult. The simplest method is related to some measurable quantity, such as number of units, weights, volume or vitamin content, and so forth, but the resulting gross-margin percentages will fluctuate unreasonably. A method based on the net realizable value is therefore the most common basis for allocation of joint costs.

The important point is that the purpose of joint cost allocation is to determine inventory valuations and to develop an aid in setting prices and estimating the resulting profit from the sales of each joint product. Control is still based upon cost centers, standard costs, budgets, and the applications of cost-control techniques.

The effect of the allocation method for costs is readily shown by an example.

Example 17.1 At a split-off point, $900 must be allocated to product A which amounts to 1000 liters at a sales value of $0.55/liter, and product B, which amounts to 500 liters at a sales value of $0.90/liter. Compare the percentage profits for A and B for an allocation (*a*) based on the volume of products and (*b*) based on the sales value.

SOLUTION
(*a*)

$$\text{Total liters} = 1000 + 500 = 1500$$

$$\text{Cost allocated to A} = 900(1000/1500) = \$600$$

$$\text{Sales value for A} = 1000 \times 0.55 = \$550$$

$$\text{Profit for A} = \$50 \text{ loss}$$

$$\% \text{ profit for A} = 100(50/550) = 9.09\% \text{ loss}$$

$$\text{Cost allocated to B} = 900(500/1500) = \$300$$

$$\text{Sales value for B} = 500 \times 0.90 = \$450$$

$$\text{Profit for B} = \$150$$

$$\% \text{ profit for B} = 100(150/450) = 33.3\%$$

(*b*)

$$\text{Total sales value} = 1000 \times 0.55 + 500 \times 0.90 = \$1000$$

$$\text{Cost allocated to A} = 900(550/1000) = \$495$$

$$\text{Profit for A} = 550 - 495 = \$55$$

$$\% \text{ profit for A} = 100(55/550) = 10\%$$

$$\text{Cost allocated to B} = 900(450/1000) = \$405$$

$$\text{Profit for B} = 450 - 405 = \$45$$

$$\% \text{ profit for B} = 100(45/450) = 10\%$$

An allocation based on volume shows a loss for A and a profit for B. An allocation based on realizable value shows a profit for each, but with a reduced profit for B.

Allocation costs for joint products are usually closely guarded proprietary information, since companies do not wish their competitors to know the apparent profits for individual products and the manner in which they are calculated.

RELATED TOPICS

17.15 Inventory Valuation

This and the following section deal with problems of financial accounting which affect cost figures.

Identical items in materials and supplies inventory may have been acquired at different times and at different prices. In such cases, a rule is required to determine the value at which units shall be transferred to work in process. The rule applied to pricing and has nothing to do with which unit is physically removed from the inventory. Physically, the units are all alike, and the unit removed is generally the one that is most conveniently transferred.

The most common method is called *first-in, first-out,* or *FIFO*. The cost of the oldest unit in inventory, as it appears on the record, is used. When prices are increasing, such as during inflation, the lower acquisition prices are matched with the higher selling prices, resulting in higher accounting profit figures and higher income taxes.

To avoid these unsatisfactory effects, *last-in, first-out* or *LIFO* method was introduced. The cost of the last unit added to inventory, as it appears on the record, is used. Now, the last additions which supposedly reflect current prices are matched with the selling prices of the same period; thus the *paper profits* resulting from the FIFO method are eliminated. The inventory, however, will reflect the prices of the earlier, less expensive additions and, therefore, will be lower than it would be using the FIFO method.

The use of average valuation represents an effort to find a compromise between the two methods. Either a moving average can be used or new averages can be computed periodically.

In order to use material or finished-goods inventory data from the accounting records, one must be aware of the cost differences that may result from different inventory-pricing methods.

17.16 Depreciation

Depreciation was discussed in detail in earlier chapters. It becomes part of the fixed costs of overhead. The importance of depreciation is steadily increasing because of the growing trend toward automation. Thus labor cost is replaced by machine cost, which is reflected in a depreciation charge.

It should be noted here that the manner in which depreciation is taken does affect the current period costs. Moreover, accountants generally prefer to group items together for accounting purposes, and thus a single piece of equipment loses its individuality. Again, in some processes part of the equipment may be fully depreciated, and current costs are not realistic since they make insufficient allowance for wear and tear of equipment.

Cost estimators are not bound by the depreciation expense used by the accounting method. They can set rates for their computations which they consider realistic regardless of whether they are acceptable for tax purposes and therefore accounting procedures.

In concluding this chapter, it can be said that the introduction of more efficient equipment and scientific techniques will cause a new appraisal of present accounting practices. There is no doubt that closer cooperation of cost engineers, who are major users of accounting data, and those who are responsible for designing the systems should be encouraged.

QUESTIONS

17.1 Discuss the reasons why cost engineers need to be familiar with the fundamental concepts of accounting, particularly cost accounting.

17.2 Distinguish between historical and managerial accounting.

17.3 Describe the basic steps for accumulating accounting data.

17.4 Discuss the reasons for the large number of cost definitions and list some of the most important ones.

17.5 Explain the reasons for the use of future-directed cost from the cost engineer's point of view.

17.6 Explain the distinction between direct (prime) costs and indirect (overhead) costs. Give examples.

17.7 Explain the importance of cost centers and their relationship to departments.

17.8 Discuss the importance of presenting unit costs in all cost statements.

17.9 Describe the difference between job-order costs and process costs. Give examples.

17.10 Explain the difficulties in dealing with overhead costs. Give examples of several kinds of overhead costs.

17.11 Discuss the reasons for the need of overhead allocation and the technique used.

17.12 Explain the technique of computing burden rates and the advantages and difficulties arising from their use.

17.13 Explain the importance of the concept of activity volume for cost-accounting purposes. Give examples of variable, semivariable, and fixed costs.

17.14 Explain the benefits and shortcomings of absorption costing compared with direct-costing techniques.

17.15 Explain the techniques and benefits of standard costs and overhead budgets.

17.16 Explain the techniques used for dealing with by-product costs and joint costs. Give examples of each type of cost.

17.17 Discuss the relationship between cost engineers and cost accountants.

REFERENCES

1. Black, H. A., and J. D. Edwards: *Managerial and Cost Accountants' Handbook,* Dow Jones-Irwin, Inc., New York, 1975.
2. Davidson, S.: *Handbook of Modern Accounting,* McGraw-Hill Book Company, New York, 1970.

3. Grady, P.: *Inventory of Generally Accepted Accounting Practices for Business Enterprises,* AICPA Accounting Research Study 7, American Institute of Certified Public Accountants, New York, 1965.
4. Horngren, C. T.: *Cost Accounting, A Managerial Emphasis,* 4th ed., Prentice-Hall, Inc., Englewood Cliffs, N.J., 1977.
5. Louderback, J. B.: *Managerial Accounting,* Wadsworth Publishing Company, Inc., Belmont, Calif., 1975.
6. Moore, C. L.: *Managerial Accounting,* 3d ed., South-Western Publishing Company, Cincinnati, 1972.
7. Nickerson, C. B.: *Managerial Cost Accounting and Analysis,* 2d ed., McGraw-Hill Book Company, New York, 1962.
8. Spiller, E. A., Jr.: *Financial Accounting, Basic Concepts,* 3d ed., Richard D. Irwin, Inc., Homewood, Ill., 1977.
9. Wixom, R., ed.: *Accountants' Handbook,* 5th ed., The Ronald Press Company, New York, 1970.

EIGHTEEN

COST ENGINEERING AND BEYOND

F. C. Jelen

18.1 Elements of Complete Cost

This chapter discusses some topics that generally belong to management, yet must be understood at basic levels in cost engineering. Some of these topics have already been introduced, in part, but are mentioned here for varied emphasis.

The elements of the total cost of manufacturing and selling a product were outlined in Table 15.1. The cost engineer is concerned chiefly with part I of the table, the operating or manufacturing cost, but must be knowledgeable about part II, general expense. The classification in Table 15.1 is subject to variation. Hence the cost engineer must be familiar with the company's classification of costs and the accounting procedure used.

18.2 Start-Up Costs

When a plant or project is supplied with the complete fixed investment and working capital, it is ready for start-up. Start-up does not occur at a precise instant but rather over a period of time; however, one point is usually selected as the start-up time. Some start-up expenses will have occurred earlier, such as training personnel and bringing individual pieces of equipment onstream. Other expenses will occur after the selected start-up time, such as making equipment changes and debugging.

The duration of the start-up can vary widely depending upon the novelty of the process and various technical and personnel considerations. However, it is considered complete when the plant has demonstrated that it is capable of operating at 90 percent of capacity for a predetermined time.

Start-up costs can be divided into two groups for accounting purposes. The first group consists of costs which may be written off immediately as an expense if desired. If a pump is ruined by unexpected corrosion, the cost can be expensed since there is no increase in capitalization. The second group consists of costs which must be capitalized. If an additional processing step is required, a pump needed for that step must be capitalized. Although items in the first group *may* be expensed, they may be capitalized if desired since the government does not care if the immediate tax benefit from the expense is delayed. The trend now is to capitalize most start-up expenses in the belief that the cost of a plant should include all the costs necessary to establish essentially full production capability.

The total start-up expense seldom exceeds 10 percent of the fixed capital cost. Start-up costs will be small with an established technology for a local addition. In contrast, start-up costs can be significant for a plant using a new technology in a new location, perhaps a foreign land. Start-up time can be reduced by more extended technical investigation before the plant is built, such as more pilot plant work. Development work is less expensive in the pilot plant than it is in the actual plant, but once funds are committed to building a plant, delays in a pilot plant can become expensive. Hence the thoroughness with which a process is investigated before a plant is built can be a trade-off between the value of time spent in prior investigation and the start-up cost.

The start-up time may vary from 1 month for well-known processes to as much as 1 year for new ones. It is desirable that skilled crews of engineering supervisors who are production oriented be assigned early in the engineering

Table 18.1 Components of start-up costs

Supervisory and technical personnel hired early
Operating line supervisors hired early
Other overhead personnel, including office personnel, hired early
Sales force hired early
Labor force hired early
Overhead and miscellaneous for all above personnel
Training of personnel
Training, operating, and emergency manuals
Scale models
Outside contracts for operating assistance
Technical assistance from parent company
Consultants and outside work
Contingencies
Travel and living expenses for temporary and transferred personnel
Initial low operating efficiency
Wasted, lost, and spoiled material
Special disposal problems
Engineering cost for design modifications
Installed cost for design modifications
Cost for checking and testing equipment
Cost for supplying temporary resources, such as utilities
Extra cost for initial quality testing and control

phase to consult with the project and design engineers. The purpose is to build into the project features calculated to reduce the start-up time and facilitate the process. Arranging for start-up can be a complicated activity requiring extensive planning and execution. It will be necessary to train not only new operators but also perhaps new trainers.

A list of start-up components is given in Table 18.1. As preparations build to a climax, skilled engineers and operators are stationed at the site during the last stage of construction to start operator-training procedures, make adjustments for convenience in operating, and then supervise final adjustments and the actual start-up program. Pieces of equipment are tested, modified if necessary, and brought onstream individually. Some of this activity may start months ahead of the termination of construction.

Table 18.1 shows that start-up costs can be classified into two types: first, costs arising from technical difficulties requiring engineering and equipment costs; second, costs associated with personnel, involving early employment, training, and provision of temporary habitation.

18.3 Plant Location

The cost of manufacturing, selling, and distributing a product is significantly affected by the location of the plant in which it is made. In an earlier age, manufacturing facilities were located in the market area in which the product was sold. Today, high transportation costs may still limit the location of large bulk-product plants, such as steel mills and fertilizers, to primary market areas, but other factors have increasing weight in site determination for most other products.

The Regional Science Research Institute in Philadelphia indicates that there are about 100 factors involved in the selection of a site for an industrial plant. The most significant are listed in Table 18.2.

The location of a plant is a matter for high-level management and requires attention to economics, politics, personnel, and public relations. If building a new plant requires closing an old plant in another location, the cost of terminating the employment of personnel in the old plant can be a major problem.

For a discussion of factors affecting plant location the reader can find helpful information sources such as local chambers of commerce, utility companies, railroads, labor union organizations for labor rates and working conditions, maps from the U.S. Geodetic Survey (Department of the Interior), Department of Labor for excellent statistics on factory labor rates published in its *Statistical Abstract of the United States,* the Federal Power Commission for electric power rates, the Interstate Commerce Commission, and the Federal Communications Commission. A complete list of the publications of the government agencies can be obtained from the Superintendent of Documents, Government Printing Office, Washington.

Private organizations such as *Fortune Magazine, Engineering News-Record,* The Industrial Conference Board, the American Manufacturers Association, and McGraw-Hill, Inc., publish data useful in industrial site selection.

Table 18.2 Significant factors in selection of a site for a plant

1. Accessibility to market
2. Availability of raw materials
3. Qualified and productive labor force
4. Adequate utilities
   a. Electricity
   b. Water
   c. Steam
   d. Gas
5. Construction costs
6. Fuel availability
   a. Coal
   b. Oil
   c. Gas
7. Effluent disposal
   a. Local sewage plants
   b. Lagoons
   c. Deep wells
   d. Large bodies of diluting waters
   e. Legislation affecting cost of pollution control
8. Transportation facilities
   a. Rail
   b. Truck
   c. Water
   d. Pipeline
   e. Air
9. Legislation affecting business activities
   a. Tax concessions
   b. Restrictive ordinances
   c. Real estate values and taxation
10. Site considerations
    a. Cost of acquisition, preparation, and maintenance
    b. Accessibility
    c. Need for services such as parking, cafeteria, recreational facilities, etc.
11. Taxes
    a. Sales
    b. Business
    c. State and local
    d. Gasoline and oil
    e. Severance and depletion
12. Climate effects
    a. Absenteeism
    b. Heating
    c. Air conditioning

18.4 Contingencies

The term *contingencies* has more than one meaning. In this book, it refers to costs that will probably occur based on past experience, but with some uncertainty regarding the amount. The term is not used as a catchall to cover ignorance. It is poor engineering and poor philosophy to make second-rate estimates and then try to satisfy them by using a large contingency account. The contingency allowance is designed to cover items of cost which are not known exactly at the time of the estimate but which will occur on a statistical basis. For instance, an allowance for water-batch testing of vessels and piping is necessary before process fluids are circulated and must be part of the start-up cost. However, it is known from past experience that some of the vessels will fail structurally during the test, and a contingency account exists to cover that circumstance. When a vessel fails in the test, the cost of the replacement must come out of the job contingency.

In practice, some items are risky because of ignorance or unpredictable circumstances, but the cost is not covered by a contingency. As an example, suppose a construction company is subbidding on the foundations for a project the soil characteristics of which are not known precisely. The proper engineering approach is to perform a soil test from which accurate cost estimates for the foundations can be made. However, the subbidder may not be prepared for this, since he will try to avoid the cost for testing the soil unless he is sure of the contract. In that case, the subbidder may estimate his cost for what he expects the soil to be, increasing his cost to cover the risk. The bid is increased to cover the risk, not to include contingencies. Contingencies will occur independently of the nature of the soil. For example, past experience shows that about 5 percent extra must be allowed to cover emergency conditions, such as delivery of some batches of concrete after 5 P.M. That cost would be covered by a contingency.

18.5 Financing

As time proceeds, a prosperous company will require additional funds for expansion or for new ventures. The funds can be raised from five broad areas:

Profits
Additional common stock
Preferred stock
Loans
Bonds

If profits are used, no financing is required, but profits must be raised on an after-tax basis. Under existing conditions of relatively low profits and high costs for investment, financing new ventures from profits is very limited.

Issuing additional common stock increases the equity of a company and dilutes the position of a stockholder. However, the existing stockholders can be given preference for the new stock to avoid dilution. Adding common stock is the safest way of obtaining new funds since the debt is not increased and dividends can be omitted when necessary. On the other hand, dividends are paid out of profits, i.e., on an after-tax basis.

Issuing preferred stock also increases the equity of a company and does not increase the debt. However, a prescribed dividend is stipulated and must be paid from profits, on an after-tax basis.

Loans are obtained principally from banks and insurance companies and may be short-term or long-term. Interest paid on loans is deductible for tax purposes. However, failure to pay interest or to repay the loan itself on time can lead to bankruptcy. Loans increase debt.

Bonds are similar to loans but are usually sold to the public. They are long-term, generally in $1000 units. The interest paid on bonds is an expense for tax purposes. Bonds increase debt, and failure to pay interest or the face value on maturity can lead to bankruptcy.

The method of financing is a consideration for high-level management, and there are many variations within the types listed above. Cost engineering fundamentally is not concerned with the source of funds. Thus profitability of a venture should be measured independently of the financing. Acceptance or rejection of a venture can be dominated by financing considerations, but the absolute measure of profitability is the same whether the money is borrowed or not.

Consider a 3-year project requiring an investment of 1 M$ at zero time and having a uniform end-of-year income of 0.6438 M$ after expenses but before depreciation and taxes for each of the 3 years. Sum-of-the-years-digits depreciation is to be used for a 3-year tax life, and the tax rate is 46 percent. The discounted cash flow rate of return (DCFFR) is the rate of return that makes the present value after taxes equal to zero, and by Table 3.5, items 1 and 2, is

$$P = 0 = -1(1 - 0.46F_{\text{SDP},r,3}) + 0.6438(1 - 0.46)F_{RP,r,3}$$

and by trial and error, $r = 0.25$. The DCFRR is 25 percent.

If the 1 M$ investment is borrowed at a yearly rate of 15 percent, amounting to a before-tax expense of 0.15 M$/year, and if this is included as an extra expense, the calculation becomes

$$P = 0 = -1(1 - 0.46F_{\text{SDP},r,3}) + 0.6438(1 - 0.46)F_{RP,r,3} - 0.15(1 - 0.46)F_{RP,r,3}$$

$$= -1(1 - 0.46F_{\text{SDP},r,3}) + 0.26665F_{RP,r,3}$$

and by trial and error, $r = 0.134$. The DCFRR now becomes 13.4 percent. The DCFRR of the project itself remains 25 percent but appears to be 13.4 percent only because the finance charge has been considered as an expense. If the interest on the investment is considered, then the source of the investment itself should be considered. In that case, all the investment is borrowed in this venture and there is no initial investment by the company, yet a profit results. The conclusion, then, is that a profit results from no investment, and the DCFRR is infinite!

18.6 Pricing for Profitability

At times, the inverse problem of the preceding section occurs, namely, to find the selling price required to produce a specified profitability. Suppose that in the previous example 1 million items of product are made per year to sell at S dollars per item. Find the value of S that yields a DCFRR of 25 percent. Expenses are $0.5 per item. Then income less expense before taxes will be $(S - 0.5)$ megadollars per year, and

$$P = 0 = -1(1 - 0.46F_{\text{SDP},25\%,3}) + (S - 0.5)(1 - 0.46)F_{RP,25\%,3}$$

$$S = \$1.1438$$

If the company charges \$1.1438 per item, it will have $(1.1438 - 0.5) = \$0.6438$ per item, which agrees with the figure in the previous section of 0.6438 M\$ for 1 million items. The project then pays off at 25 percent per year.

A company is allowed to calculate a required selling price in any way it desires. In some circumstances, as in the next section, the sales price is calculated with the inclusion of the financing cost, or anything else that is desired, such as a dividend to stockholders.

18.7 Levelized Cost

Public utilities, in appearing before regulating commissions, aim principally to establish rates. Hence they are very concerned with what they must charge for their product, kilowatt-hours. Utilities are allowed a return on investment after taxes based on their investment. Interest paid on the investment is an expense for tax purposes and is included in arriving at the price the customer should pay. For internal and economic comparisons, a method based on levelized cost is some-times used. Perhaps it should be called *levelized price* since it is the revenue needed expressed per unit of product produced, e.g., kilowatt-hour, but it is a cost to the consumer. The unit cost to the consumer will vary from year to year, but the *levelized cost* is a cost that has been leveled out for all the years. There are variations in the method, but the following simplified example will illustrate the principle.

Example 18.1 A public utility has an investment of 100 M\$ at the start that will be written off in 3 years using straight-line depreciation. At all times, 50 percent of the investment is borrowed money and 50 percent is equity. The borrowed money is financed at 10 percent per year. The gross expense before taxes and depreciation, and before the financing cost, is 150 M\$/year. The tax rate is 46 percent, and each year 5000×10^6 kWh of electrical energy is sold. The company desires to make 15 percent per year on its equity and in addition 5 percent for dividends to its stockholders, a total of 20 percent on equity. Find the levelized cost.

SOLUTION Let N be the kilowatt-hours in millions sold each year and S the cost to the consumer in dollars per kilowatt-hour. Table 18.3 shows the calculations leading to the revenue required for each year in megadollars and the value of S, the unit cost to the consumer, for each year. The unit cost S will not be constant; however, all the years can be combined on the basis of present value. Then, using megadollars:

Present value of (SN) = present value of (revenue required each year)

If S is to be constant, it becomes the levelized cost S_L and can be taken outside a summation sign. Thus,

$$S_L = \frac{\sum \text{revenue each year}/(1 + r)^n}{\sum \text{kWh each year}/(1 + r)^n}$$

Table 18.3 Costs in megadollars for Example 18.1

	Year 1	Year 2	Year 3
1. Total investment	100	66.67	33.33
2. Equity investment	50	33.33	16.67
3. Financed investment	50	33.33	16.67
4. Gross income	NS	NS	NS
5. Gross expenses	150	150	150
6. Interest expense	5	3.333	1.667
7. Depreciation expense	33.333	33.333	33.333
8. Taxable income (4) − (5) − (6) − (7)	NS−188.33	NS−186.67	NS−185.00
9. Tax at 46%	0.46NS−86.632	0.46NS−85.868	0.46NS−85.1
10. Net profit (8) − (9)	0.54NS−101.698	0.54NS−100.802	0.54NS−99.9
11. 15% of equity for retained earnings	7.5	5.0	2.5
12. 5% of equity dividends	2.5	1.667	0.834
13. Revenue required, NS (10) − (11) + (12)	206.85	199.02	191.17
14. N in kilowatt-hours, millions*	5000	5000	5000
15. Calculated S for year† (13) ÷ (14)	0.0414	0.0398	0.0382

*N = kilowatt-hours sold each year in millions.

†S = cost in dollars per kilowatt-hour as required revenue.

The computation for levelized cost is:

$$S_L = \frac{206.85/(1.15)^1 + 199.02/(1.15)^2 + 191.17/(1.15)^3}{5000/(1.15)^1 + 5000/(1.15)^2 + 5000/(1.15)^3} = 0.0399$$

Thus the company would seek a customer cost of $0.0399/kWh and would have a 15 percent return on its varying equity in the investment and another 5 percent for a dividend.

A practical problem would be spread over many years and expenses, receipts, and energy produced each year would not be constant, but the principle remains the same. There are variations in the method used to compute levelized cost, but the example represents a common use of levelized cost in economic comparisons. In comparing two projects, the one with the lower levelized cost would be preferred.

18.8 Forecasting

In industrial activities, forecasting is normally done by marketing or economics personnel. Cost engineers become involved in forecasting most frequently in connection with venture profitability. Of the many factors involved, those listed in Table 18.4 are the most significant. The table gives the probable variations from forecasts over an assumed 10-year plant life.

The cost of the *fixed capital investment* may be appraised within a range of −10 percent to +25 percent. *Construction time* may vary from −5 percent to as

Table 18.4 Factors affecting profitability forecasts

Factor	Probable variation from forecasts over 10-year plant life, %
1. Cost of fixed capital investment	−10 to +25
2. Construction time	−5 to +50
3. Start-up costs and time	−10 to +100
4. Sales volume	−50 to +150
5. Price of product	−50 to +20
6. Depreciation method	None
7. Plant replacement and maintenance costs	−10 to +100
8. Obsolescence of process or equipment	Indeterminate
9. Income tax rate	−5 to +15
10. Inflation rates	−10 to +100
11. Interest rates	−50 to +50
12. Working capital	−20 to +50
13. Legislation affecting product	Indeterminate
14. Raw-material availability and price	−25 to +50
15. Competition	Indeterminate
16. Salvage value	−100 to +10
17. Intuition or calculated risk	Indeterminate
18. Profit	−100 to +10

high as +50 percent of the estimated time but generally can be predicted more accurately. *Start-up costs* have even a wider variation, ranging from −10 percent to +100 percent, but have a negligible effect on profitability if the gap left by delay in reaching the market with the product is not filled by a competitor.

Forecasts of *sales volume* and *price* for a new product are probably the most challenging factors. Such forecasts are not within the scope of this book, but they constitute the heart of the business process. Predictions of sales volume and prices are made from studies involving market and product research activities. Most companies now employ market research and commercial development specialists to assist production and sales managements in determining the extent of the market, potential sales, and probable price of product over the life of the production assets. Actual sales and price can vary widely from the predicted levels as indicated in the table.

Another factor that is difficult to assess is process or equipment *obsolescence*. Processes have been known to be obsolete within a year after start-up, and in a technological era, specific pieces of equipment become obsolete years before the end of their forecasted lives.

Among the factors affecting profitability, the impact of *legislative changes* must be considered. It is difficult to forecast to what degree this factor will affect profitability. For instance, government interest in health, pollution, quotas, and tariff agreements, among others, are all factors in economic evaluation. In addition, there is the possibility of income tax changes.

Competition from abroad is an increasingly important factor in assessing profitability. Many single products in some industries have already been affected, and the outlook is that increasing competition from abroad will tend to depress prices and profits in the United States. By and large, the individual effect of this factor on profitability is indeterminate at this time.

There is still one other consideration. After all known factors affecting profitability have been carefully weighed, the *go* or *no-go* decision is sometimes reached through a complex mental process which reflects the experience of the individual making the decision. Since it is impossible to characterize this process, it has been referred to as *intuition, calculated risk,* or *educated guess* in the prediction of profitability. Some forecasting can be based on extrapolation of historical data. A good example would be the prediction of steel consumption in the United States per capita for the next 10 years based on the extrapolation of the historical data for the last 20 years. Forecasting frequently requires some subjective inferences about the future, and the value of the forecast is closely related to the correctness of the inferences.

Extrapolation of historical data can be based on a linear relationship of the form

$$y = mx + b$$

but most frequently the relationship is curved, lying above or below a linear relationship. Nonlinear relationships are frequently chosen in exponential forms such as

$$y = ax^b$$
$$y = ab^x$$

which have the advantage of being plotted as straight lines on log-log graph paper.

18.9 Cash Flow for Financing

Cash flow was treated in Chap. 5 as in- and out-of-pocket flow of cash for a project and its relation to profitability. To the controller of a company, cash flow has an importance unrelated to profitability, namely, its significance to funds required by a project from the beginning of construction to the termination of start-up. Expenditures for a large project can extend over years, and funds must be available at all times to meet expenses. Good planning and timing are important, for funds must be available on time. However, if they are available too soon, they must be reinvested outside the project to compensate for the interest expense.

18.10 Profit

Like all terms in common use, *profit* is not easily defined. In a broad sense, profit connotes success and is not a simple concept. It is certainly an important concept to the cost engineer, as well as to the accountant, since costs, as expenses or receipts, are the quantitative basis for defining profit.

In Chap. 5 the difference between profitability and profit was emphasized, profitability being a long-term evaluation of short-term profits. Table 5.15 listed a number of pitfalls in profitability but is applicable to short-term profits as well. Consider the case of depreciation. How depreciation is scheduled over the years influences the apparent profit for the individual years. This is true for other expenses that can be allocated over the years. For example, one power company lost an entire generating station because of a landslide at the river edge. Instead of taking the entire loss in 1 year, the company preferred to allocate the loss over a number of future years. This decision affected the apparent profits for many years ahead.

In Chap. 6 it was shown that profit and profitability are overstated when no allowance for inflation is included. True profits exist only after taxes *and* after an allowance for hidden costs arising from inflation has been included. A problem in the United States is that profits as a percentage have been declining over the years, and hidden costs arising from inflation have compounded the problem. Consequently, internal funds available for expansion and modernization for manufacturing have been disappearing.

Profit has become a disreputable word to the public, but in a capitalist society profit is necessary for survival. In the last decade, the strong desire to show short-term and immediate profits has resulted in a policy of patching. Additional capacity has been obtained by adding to outmoded existing facilities whenever possible. The long-term objectives of modernization and creation of new manufacturing facilities have been slighted. A policy aimed at improving short-term profits and neglecting the long term can only delay the time of disaster and is very costly in the end. Companies exist to prosper and grow, and the measure of this is profit; but apparent profit, particularly over a short time, can be deceiving.

18.11 Planning against Scheduling

Several overlapping stages exist between making the decision to undertake a project and completing it. The first step is planning, which merges into scheduling, but there is a difference between the two. A planner must comprehend the project overall and not get lost in details. A scheduler must know details and enjoy working with them. Although many people have the ability to schedule, only a few are expert planners. The secret of making a good plan is to put the whole plan on a single sheet of paper with major milestones. The master plan will be supplemented by other subplans, perhaps a volume of them, but there must be some master plan that is concise.

Scheduling follows planning and should not be undertaken without good planning. Scheduling will feature major and minor milestones and will encompass considerable detail. Scheduling is usually performed under engineering. It is important to note that in large construction companies procurement is part of engineering, and sometimes the department is called *engineering and procurement*. Scheduling pays close attention to procurement and determines that all equipment and supplies are ordered on time, verifies along the way that deliveries will be on schedule, and arranges for equipment to be inspected at various stages of construction.

18.12 Role of the Computer

The computer has become the single most important tool in cost engineering. It is dominant in cost estimation, scheduling, and control. Computers can do the following:

1. Make complex calculations at fantastic speeds and with great accuracy
2. Rearrange with great speed
3. Compare
4. Make simple decisions
5. Store vast quantities of information
6. Retrieve with great speed
7. Print out details and summations
8. Print out graphics and illustrations
9. Have their stored information updated easily and quickly

On the other hand, despite opinions to the contrary, a computer cannot think. Every step must be programmed. A computer can operate under the guidance of statistics and probability and can even generate randomness internally, yet it only follows orders. A computer can make a very sophisticated analysis, devouring a vast amount of information, but it does so by sequencing myriad single steps. The computer appears brilliant only because of the speed with which it operates.

The computer has two basic capabilities which can be intermixed. First, the computer can be used as a tool for calculation; that was the sole purpose of the original computers. The second basic function of a computer is known as *data processing,* in which a computer serves as a library from which stored information can be retrieved and organized. Some calculation may be involved, but it is usually incidental. Data processing occurs in banking, insurance, credit card accounting, taxation, statistics, information on the spot for sports announcers, and a host of other applications. The spectacular expansion of the storage capacity of computers has enhanced their applicability beyond predictions. Although data processing was not even conceived for the earliest computers, it is estimated that this function now comprises about 95 percent of computer use.

Applications in cost engineering that use computers as calculators are in linear programming, inventory problems, dynamic programming, queuing problems, risk analysis, optimization studies, and profitability analysis.

Applications in cost engineering involving chiefly data processing are in the following fields:

Estimation
Scheduling
Construction and cost accounting
Material control
Cost control

In these fields, computers are a necessity.

In estimation, the computer can store thousands of basic costs, adjust for size and inflation, and print out information in a well-organized manner. The information can easily be kept up to date and expanded, while confusion with deleted material is eliminated.

For scheduling and control, the computer can assimilate a large array of information and reports and permit immediate retrieval in an organized arrangement. Again, the ease with which the information is kept up to date is an advantage.

Many commercial management services are available for estimating and control, and they are all computer oriented. Various programs can be purchased to be run on the user's computer. The computer applications committee of the AACE should be consulted for information about current software programs available.

A current phenomenon is the introduction of the minicomputer. It is suitable for desk-top use but provides an extensive storage capacity and calculating ability at a modest cost.

Programmable pocket calculators can be used advantageously by a cost engineer. The examples and problems in this book have been worked out on such a calculator. All the tables in Appendixes 1 and 2, and a few others, were stored on one side of a 75-mm-long magnetic card. Results are displayed to 10 significant figures, and no interpretation for years, interest rate, or their product is ever necessary.

Computers are still changing rapidly. Minicomputers are becoming full-scale computers. Programmable pocket calculators are becoming small computers and, if coupled to an inexpensive cassette recorder, gain the benefit of substantial storage capacity and program steps.

It is not necessary for cost engineers to be experts in programming, but they should know how to use computers, minicomputers, and programmable pocket calculators in their work. Computers are user oriented, and paradoxically, the more powerful they become internally, the easier they are to use.

18.13 Management Organization

Management can be broken down into three types—functional, projectized, and matrix or hybrid. Consider a construction company in which cost engineering is part of engineering. The functional organization is shown in Fig. 18.1 and is the standard pyramid type, with the organization broken down into functional groups or departments, each reporting to an executive.

The projectized organization is shown in Fig. 18.2, and in this arrangement the organization is by projects.

The matrix organization is shown in Fig. 18.3 and is a combination of the two previous types. In this arrangement, the engineering department reports directly to an executive and to various project coordinators. Engineering has teams working on various projects.

Small projects and projects in the range up to 100 M\$ will generally be handled in the matrix organization, while projects above 100 M\$ will fall in the projectized organization. The functional organization occurs chiefly where there are no projects, such as a manufacturing company.

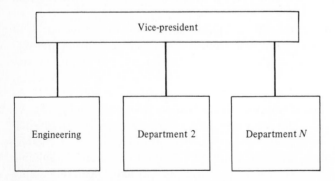

Figure 18.1 A functional organization.

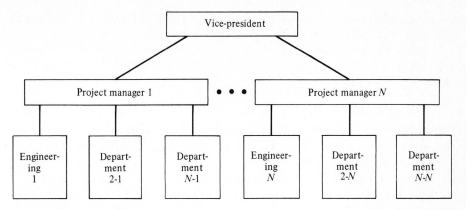

Figure 18.2 A projectized organization.

18.14 Legal Aspects

Although cost engineers are not lawyers, frequently their work becomes associated with legal matters, and some understanding of legal documents becomes necessary. A contract is perhaps the most important of these. One aspect of legal affairs is that whenever a dispute arises, both sides are represented by legal staffs whose only purpose is to win the case. The ultimate decision will be made by a jury composed of average people, not cost engineers. Contracts must therefore be carefully worded. In the end, it is the wording of the contract that will be judged, not what was meant, much less hoped for.

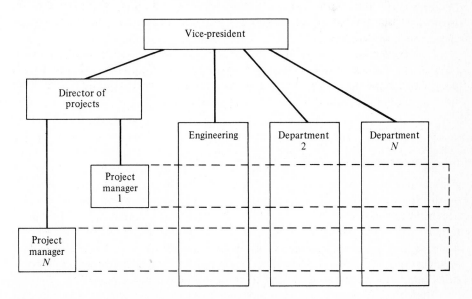

Figure 18.3 A matrix organization.

Construction contracts can be grouped into four classifications, each with variations. The basic types of contracts are:

1. Fixed-fee
2. Cost-plus
3. Shared or target-figure contracts
4. Unit-price contracts

In *fixed-fee* contracts, the work is designated by detailed specifications and drawings which may be supplied by the owner, or by the contractor and accepted by the owner. The cost is fixed and includes the contractor's profit. The risk is taken by the contractor, who has to consider this in bidding. If the bid contains too much for risk, the contractor may not be competitive. If the bid underrates the risk, the contractor may be a heavy loser. The owner assumes little risk. If the job is not within specifications, the contractor can be held liable. The specifications may not be adequate, and if the contractor has performed within the specifications, he or she is not liable. Owners can protect themselves by including performance specifications. Inflation is a serious factor in estimating construction bids, particularly if the construction is spread out over 3 years or more. A fixed-fee contract must include some allowance for inflation and can also include a variation in the fee. Thus there can be a penalty for a late finish and a reward for an early finish. The contract can allow for all kinds of incidental factors, including catastrophic events. Nevertheless, in principle, the fixed-fee contract gives the owner the least risk. The contractor assumes most, if not all, of the risk.

The second type of contract covers a wide variety of agreements known as *cost-plus* or *cost-plus-fee*. The work is described by detailed specifications and drawings, but the ultimate cost to the owner will not be known until the work is completed. It is known as an *open-ended* contract. Limits may be specified on some of the component costs. The contract must state accurately how the costs are to be treated and whether any limitations or exemptions apply. The fee can be a fixed amount or a percentage amount, and many variations exist. There can be a penalty for a late finish and a reward for an early finish. The owner takes most of the risk, and the contractor is well protected against loss. Cost-plus contracts become necessary when the project is unique and when significant changes in specifications are anticipated after the project is started. Many contracts for the military and the government can be handled only on a cost-plus basis.

Between a fixed-fee and a cost-plus-fee contract lies a large gray area known as a *shared* or *target-figure* contract. The fixed-fee contract places most of the risk on the contractor, and the cost-plus-fee contract places most of the risk on the owner. In the shared or target-figure contract, the risk is shared between them to some extent. Here, the owner obtains essentially a fixed fee for the original specifications, but the contractor generally determines the fees for variations in the specifications after the project begins. This type of contract usually contains provisions for sharing savings according to some stipulated agreement.

The unit-price contract is sometimes used in very heavy construction projects in the United States and Canada, but is more frequently used abroad. The contractor's bid is based upon the quantities of work to be done and a unit price for each type of work. A typical example is a contract for constructing a highway. It might be difficult to estimate in advance the exact amounts of fill and excavation required, but a contract based on unit costs, e.g., per cubic meter, would protect both the owner and the contractor. The total cost is fixed by summation of quantities multiplied by a rate.

Technology has become so complicated that it is difficult to prepare drawings and specifications which are completely free from misinterpretation. As a result, *hedge,* or *bludgeoning, clauses* have been rapidly developed. They are used by both owners and vendors. These clauses are general so as to be applicable to any unforeseen condition that may arise. The hedge clause, although general, can be enforced. A cost engineer who is bidding against hedge clauses is required to place an extra cost against every hedge clause in a set of bid documents since these clauses may affect the cost calculation. The use of hedge clauses may be necessary in part, but excessive use can be costly.

Cost is a fundamental factor in all the costs discussed above. Hence cost engineers must be familiar with specifications, drawings, and forensic vernacular, whether they are working for owners or contractors. Cost engineers have other contacts with legal matters. Appraisers frequently appear before regulatory commissions and in court, and these specialists should be masterful expert witnesses.

Every engineer has legal responsibilities to the public as well as to the employer. Thus the design of a bridge must be approved by a professional engineer. Most work done by a cost engineer does not involve public safety, at least directly. However, a cost engineer who shades estimates to favor decisions in which the public has an interest could be violating the law. Inasmuch as cost engineering involves estimating, it becomes difficult to prove guilt since being a poor estimator does not make one a criminal. However, if an estimator willfully distorts quantities and unit costs, he or she becomes subject to the law; and interpretation and enforcement of the law are becoming more stringent.

Legal clauses affecting penalties, brand names, delayed decisions, work by the owner, patents, segregated bids, and so forth are so dispersed throughout the bid documents that the cost engineer must become familiar with all documents to evaluate the risks involved, usually in a comparatively short time. A knowledge of the habits of the owner—asking for bids, requiring prompt approval of shop drawings, permitting substitution of materials, attitude, and so forth—can be of great value. Such information is available from past experience, from competitors, and from suppliers and subcontractors.

Practically every general condition states that the word of the engineer shall be final. Many people contend that the judge's or jury's word is final. Juries, composed of lay people, are unschooled in technical engineering and perhaps are influenced by anything except the facts of a case. There is a ray of hope in arbitration, but a cost engineer might recall that "an ounce of prevention is worth a pound of cure." It is the obligation of the cost engineer to be aware fully of all legal aspects and to advise the employer—even to not bid on the job when the bid documents have many hazards.

Table 18.5 Canons of ethics of engineers

Fundamental Principles of Professional Engineering Ethics

The Engineer, to uphold and advance the honor and dignity of the engineering profession and in keeping with high standards of ethical conduct:

I. Will be honest and impartial, and will serve with devotion his employer, his clients, and the public;

II. Will strive to increase the competence and prestige of the engineering profession;

III. Will use his knowledge and skill for the advancement of human welfare.

Relations with the Public

1.1 The Engineer will have proper regard for the safety, health and welfare of the public in the performance of his professional duties.

1.2 He will endeavor to extend public knowledge and appreciation of engineering and its achievements, and will oppose any untrue, unsupported, or exaggerated statements regarding engineering.

1.3 He will be dignified and modest in explaining his work and merit, will ever uphold the honor and dignity of his profession, and will refrain from self-laudatory advertising.

1.4 He will express an opinion on an engineering subject only when it is founded on adequate knowledge and honest conviction.

1.5 He will preface any ex parte statements, criticisms, or arguments that he may issue by clearly indicating on whose behalf they are made.

Relations with Employers and Clients

2.1 The Engineer will act in professional matters as a faithful agent or trustee for each employer or client.

2.2 He will act fairly and justly toward vendors and contractors, and will not accept from vendors or contractors, any commissions or allowances, directly or indirectly.

2.3 He will inform his employer or client if he is financially interested in any vendor or contractor, or in any inven-

tion, machine, or apparatus, which is involved in a project or work of his employer or client. He will not allow such interest to affect his decisions regarding engineering services which he may be called upon to perform.

2.4 He will indicate to his employer or client the adverse consequences to be expected if his engineering judgment is over-ruled.

2.5 He will undertake only those engineering assignments for which he is qualified. He will engage or advise his employer or client to engage specialists and will cooperate with them whenever his employer's or client's interests are served best by such an arrangement.

2.6 He will not disclose information concerning the business affairs or technical processes of any present or former employer or client without his consent.

2.7 He will not accept compensation—financial or otherwise—from more than one party for the same service, or for other services pertaining to the same work, without the consent of all interested parties.

2.8 The employed engineer will engage in supplementary employment or consulting practice only with the consent of his employer.

Relations with Engineers

3.1 The Engineer will take care that credit for engineering work is given to those to whom credit is properly due.

3.2 He will provide a prospective engineering employee with complete information on working conditions and his proposed status of employment, and after employment will keep him informed of any changes in them.

3.3 He will uphold the principle of appropriate and adequate compensation for those engaged in engineering work, including those in subordinate capacities.

3.4 He will endeavor to provide opportunity for the professional development

Table 18.5 (*continued*)

and advancement of engineers in his employ or under his supervision.	require creative intellectual effort, on a basis that constitutes competition on price alone. Due regard should be given to all professional aspects of the engagement.
3.5 He will not injure maliciously the professional reputation, prospects, or practice of another engineer. However, if he has proof that another engineer has been unethical, illegal, or unfair in his practice, he should so advise the proper authority.	
3.6 He will not compete unfairly with another engineer.	3.8 He will cooperate in advancing the engineering profession by interchanging information and experience with other engineers and students, and by contributing to public communication media, to the efforts of engineering and scientific societies and schools.
3.7 He will not invite or submit price proposals for professional services which	

Approved by Engineers' Council for Professional Development, September 30, 1963

Source: Engineers' Council for Professional Development, *Thirty-third Annual Report,* Sept. 30, 1963, p. 86. (Used with permission).

18.15 Ethics

Most professions have a code of ethics for their members that transcends and supplements legal obligations. The Canons of Ethics of Engineers adopted by the Engineers' Council for Professional Development are given in Table 18.5. They can be accepted without reservations by cost engineers.

One might say that this code of behavior can be reduced to the golden rule—"Do unto others as you would have others do unto you." Inherent in the golden rule, of course, is the corollary, "Honesty is the best policy." These thoughts are the basis of the canons.

The canons are organized into rules governing the individual engineer's relationships with the public, with employer or clients, and with other engineers. Adherence to these rules is important to the individual engineer's personal reputation and to the public image and dignity of the profession.

18.16 Social Aspects

For many years, both the engineering profession and industrial concerns have been giving increasing attention to the role of the engineer and the role of the company in good citizenship. Because engineers do not live in a social vacuum, and because of the impact of the engineer's work on society, engineering curricula have been placing increasing emphasis on the study of the humanities. It is important that engineers, if they are to be more than mere technicians, give consideration to the social implications of their work.

A dramatic and well-known example of this problem is provided by the dangers of large-scale contamination by the products of atomic fission. A feeling of social responsibility prompts prominent engineers and scientists to work toward reduction and control of the contamination and toward peaceful applications of the nuclear energy developed as a wartime weapon.

Industrial leaders are also vitally interested in their role as good citizens and in their social responsibilities. The goodwill and other benefits which eventually derive from this interest probably defy profitability analysis by the methods described in previous chapters. The engineer, as an employee of these industrial concerns, is intimately involved with the activities of these concerns and with the social consequences of their activities. In general, a company's social responsibilities are manifested in community relations, employee relations, and educational activities.

As civilization has progressed from the cave to the modern city, people have become increasingly interdependent. As a result, they have begun to consider the welfare of their neighbors. Thus, a dog-eat-dog society eventually becomes a welfare society.

The effects of continued prosperity, with its accompanying vast increases in population, hasten the change to a welfare society. These changes bring about different sets of values, both economic and social, changes which are obvious when one contrasts our willingness today to tolerate protest groups, opposed to the necessity for nonproductive people to provide for themselves in frontier days. In the days of the battles between the farmer and the Indian, individuals were so busy providing food for themselves that they had no time for writing protest songs and poetry.

The net effect of these influences of technical developments, government, and social changes is a decline in but not elimination of the time value of money. This change, of course, influences and modifies our sense of values.

Cost engineers should, then, be aware of the fact that there are influences which affect their work other than cold mathematics and economics. They should recognize these social effects and nonmathematical aspects on the field of engineering.

NOMENCLATURE

DCFRR	Discounted cash flow rate of return, decimal per year
F_{SDP}	Present value of $1 for sum-of-the-years-digits depreciation
F_{RP}	Factor to convert R to P, years
n	The nth year
N	Units of items or service per year
P	Present value, $
r	Rate of return after taxes, decimal per year
R	Uniform end of year cost, unacost, $/year
S	Unit cost, $ per item
S_L	Levelized cost, $ per item

REFERENCES

1. ASCE: *Ethics, Professionalism and Maintaining Competence,* American Society of Civil Engineers, New York, 1977.
2. ASCE/ICE Joint Conference: *The Engineer in the Community,* The Institution of Civil Engineers, London, 1970.
3. Beauchamp, T. L., and N. E. Bowie: *Ethical Theory and Business,* Prentice-Hall, Inc., Englewood Cliffs, N.J., 1979.
4. Clifton, D. S., Jr., and D. E. Fyffe: *Project Feasibility Analysis,* John Wiley & Sons, Inc., New York, 1977.
5. Cress, K. R.: "Establish a System for Contractor Proposal Evaluation," *Cost Engineering,* vol. 23, June 1981, pp. 153–156.
6. Jeynes, P. H.: *Profitability and Economic Choice,* Iowa State University Press, Ames, 1968.
7. Lock, D.: *Engineer's Handbook of Management Techniques,* Grover Press, Ltd., Epping, England, 1973.
8. O'Brien, J. J.: *Management with Computers,* Van Nostrand Reinhold Company, New York, 1972.
9. Perrucci, R., and J. E. Gerstl: *The Engineer and the Social System,* John Wiley & Sons, Inc., New York, 1969.
10. Sullivan, W. G., and W. W. Claycombe: *Fundamentals of Forecasting,* Reston Publishing Company, Inc., Reston, Va., 1977.
11. Walton, C. C., ed.: *Ethics of Corporate Conduct,* Prentice-Hall, Inc., Englewood Cliffs, N.J., 1977.
12. Weston, J. F., and E. U. Brigham: *Essentials of Managerial Finance,* 5th ed., The Dryden Press, Hinsdale, Ill., 1977.

DISCRETE COMPOUND INTEREST

Table A1.1 Discrete compound-interest table

Discrete compound interest = 2%

	Single payment		Uniform annual series			Uniform gradient series	Depreciation series	
	Compound-interest factor	Present-worth factor	Unacost present-worth factor	Capital-recovery factor	Capitalized-cost factor	Present-worth factor	Sum-of-digits present-worth factor	Straight-line present worth factor
	$(1+i)^n$	$\dfrac{1}{(1+i)^n}$	$\dfrac{(1+i)^n-1}{i(1+i)^n}$	$\dfrac{i(1+i)^n}{(1+i)^n-1}$	$\dfrac{(1+i)^n}{(1+i)^n-1}$	$\dfrac{F_{RP}-nF_{SP}}{i}$	$\dfrac{n-F_{RP}}{0.5n(n+1)i}$	$\dfrac{1}{niF_{PR}}$
	P to S	S to P	R to P	P to R	P to K	G to P	SD to P	SL to P
n	F_{PS}	F_{SP}	F_{RP}	F_{PR}	F_{PK}	F_{GP}	F_{SDP}	F_{SLP}
1	1.0200E 00	9.8039E-01	9.8039E-01	1.0200E 00	5.1000E 01	9.6117E-01	9.8039E-01	9.8039E-01
2	1.0404E 00	9.6117E-01	1.9416E 00	5.1505E-01	2.5752E 01	9.6117E-01	9.7398E-01	9.7074E-01
3	1.0612E 00	9.4232E-01	2.8839E 00	3.4675E-01	1.7338E 01	2.8458E 00	9.6764E-01	9.6129E-01
4	1.0824E 00	9.2385E-01	3.8077E 00	2.6262E-01	1.3131E 01	5.6173E 00	9.6136E-01	9.5193E-01
5	1.1041E 00	9.0573E-01	4.7135E 00	2.1216E-01	1.0608E 01	9.2403E 00	9.5513E-01	9.4269E-01
6	1.1262E 00	8.8797E-01	5.6014E 00	1.7853E-01	8.9263E 00	1.3680E 01	9.4897E-01	9.3357E-01
7	1.1487E 00	8.7056E-01	6.4720E 00	1.5451E-01	7.7256E 00	1.8903E 01	9.4287E-01	9.2457E-01
8	1.1717E 00	8.5349E-01	7.3255E 00	1.3651E-01	6.8255E 00	2.4878E 01	9.3682E-01	9.1569E-01
9	1.1951E 00	8.3676E-01	8.1622E 00	1.2252E-01	6.1258E 00	3.1572E 01	9.3085E-01	9.0692E-01
10	1.2190E 00	8.2035E-01	8.9826E 00	1.1133E-01	5.5663E 00	3.8955E 01	9.2492E-01	8.9826E-01
11	1.2434E 00	8.0426E-01	9.7868E 00	1.0218E-01	5.1089E 00	4.6998E 01	9.1905E-01	8.8971E-01
12	1.2682E 00	7.8849E-01	1.0575E 01	9.4560E-02	4.7280E 00	5.5671E 01	9.1324E-01	8.8128E-01
13	1.2936E 00	7.7303E-01	1.1348E 01	8.8118E-02	4.4059E 00	6.4948E 01	9.0749E-01	8.7295E-01
14	1.3195E 00	7.5788E-01	1.2106E 01	8.2602E-02	4.1301E 00	7.4800E 01	9.0179E-01	8.6473E-01
15	1.3459E 00	7.4301E-01	1.2849E 01	7.7825E-02	3.8913E 00	8.5202E 01	8.9614E-01	8.5662E-01
16	1.3728E 00	7.2845E-01	1.3578E 01	7.3650E-02	3.6825E 00	9.6129E 01	8.9055E-01	8.4861E-01
18	1.4282E 00	7.0016E-01	1.4992E 01	6.6702E-02	3.3351E 00	1.1966E 02	8.7952E-01	8.3289E-01
20	1.4859E 00	6.7297E-01	1.6351E 01	6.1157E-02	3.0578E 00	1.4460E 02	8.6871E-01	8.1757E-01
25	1.6406E 00	6.0953E-01	1.9523E 01	5.1220E-02	2.5610E 00	2.1426E 02	8.4255E-01	7.8094E-01
30	1.8114E 00	5.5207E-01	2.2396E 01	4.4650E-02	2.2325E 00	2.9172E 02	8.1759E-01	7.4655E-01
35	1.9999E 00	5.0003E-01	2.4999E 01	4.0002E-02	2.0001E 00	3.7488E 02	7.9376E-01	7.1425E-01
40	2.2080E 00	4.5289E-01	2.7355E 01	3.6556E-02	1.8278E 00	4.6199E 02	7.7101E-01	6.8389E-01
45	2.4379E 00	4.1020E-01	2.9490E 01	3.3910E-02	1.6955E 00	5.5157E 02	7.4927E-01	6.5534E-01
50	2.6916E 00	3.7153E-01	3.1424E 01	3.1822E-02	1.5912E 00	6.4236E 02	7.2849E-01	6.2847E-01

Table A1.1 (Continued)

Discrete compound interest = 4%

	Single payment		Uniform annual series			Uniform gradient series	Depreciation series	
	Compound-interest factor $(1+i)^n$	Present-worth factor $\dfrac{1}{(1+i)^n}$	Unacost present-worth factor $\dfrac{(1+i)^n - 1}{i(1+i)^n}$	Capital-recovery factor $\dfrac{(1+i)^n}{(1+i)^n - 1}$	Capitalized-cost factor $\dfrac{(1+i)^n}{(1+i)^n - 1}$	Present-worth factor $\dfrac{F_{RP} - nF_{SP}}{i}$	Sum-of-digits present-worth factor $\dfrac{n - F_{RP}}{0.5n(n+1)i}$	Straight-line present-worth factor $\dfrac{1}{niF_{PK}}$
	P to S	S to P	R to P	P to R	P to K	G to P	SD to P	SL to P
n	F_{PS}	F_{SP}	F_{RP}	F_{PR}	F_{PK}	F_{GP}	F_{SDP}	F_{SLP}
1	1.0400E 00	9.6154E-01	9.6154E-01	1.0400E 00	2.6000E 01		9.6154E-01	9.6154E-01
2	1.0816E 00	9.2456E-01	1.8861E 00	5.3020E-01	1.3255E 01	9.2456E-01	9.4921E-01	9.4305E-01
3	1.1249E 00	8.8900E-01	2.7751E 00	3.6035E-01	9.0087E 00	2.7025E 00	9.3712E-01	9.2503E-01
4	1.1699E 00	8.5480E-01	3.6299E 00	2.7549E-01	6.8873E 00	5.2670E 00	9.2526E-01	9.0747E-01
5	1.2167E 00	8.2193E-01	4.4518E 00	2.2463E-01	5.6157E 00	8.5547E 00	9.1363E-01	8.9036E-01
6	1.2653E 00	7.9031E-01	5.2421E 00	1.9076E-01	4.7690E 00	1.2506E 01	9.0222E-01	8.7369E-01
7	1.3159E 00	7.5992E-01	6.0021E 00	1.6661E-01	4.1652E 00	1.7066E 01	8.9102E-01	8.5744E-01
8	1.3686E 00	7.3069E-01	6.7327E 00	1.4853E-01	3.7132E 00	2.2181E 01	8.8004E-01	8.4159E-01
9	1.4233E 00	7.0259E-01	7.4353E 00	1.3449E-01	3.3623E 00	2.7801E 01	8.6926E-01	8.2615E-01
10	1.4802E 00	6.7556E-01	8.1109E 00	1.2329E-01	3.0823E 00	3.3881E 01	8.5868E-01	8.1109E-01
11	1.5395E 00	6.4958E-01	8.7605E 00	1.1415E-01	2.8537E 00	4.0377E 01	8.4830E-01	7.9641E-01
12	1.6010E 00	6.2460E-01	9.3851E 00	1.0655E-01	2.6638E 00	4.7248E 01	8.3812E-01	7.8209E-01
13	1.6651E 00	6.0057E-01	9.9856E 00	1.0014E-01	2.5036E 00	5.4455E 01	8.2812E-01	7.6813E-01
14	1.7317E 00	5.7748E-01	1.0563E 01	9.4669E-02	2.3667E 00	6.1962E 01	8.1830E-01	7.5451E-01
15	1.8009E 00	5.5526E-01	1.1118E 01	8.9941E-02	2.2485E 00	6.9735E 01	8.0867E-01	7.4123E-01
16	1.8730E 00	5.3391E-01	1.1652E 01	8.5820E-02	2.1455E 00	7.7744E 01	7.9921E-01	7.2827E-01
18	2.0258E 00	4.9363E-01	1.2659E 01	7.8993E-02	1.9748E 00	9.4350E 01	7.8080E-01	7.0329E-01
20	2.1911E 00	4.5639E-01	1.3590E 01	7.3582E-02	1.8395E 00	1.1156E 02	7.6306E-01	6.7952E-01
25	2.6658E 00	3.7512E-01	1.5622E 01	6.4012E-02	1.6003E 00	1.5510E 02	7.2138E-01	6.2488E-01
30	3.2434E 00	3.0832E-01	1.7292E 01	5.7830E-02	1.4458E 00	2.0106E 02	6.8322E-01	5.7640E-01
35	3.9461E 00	2.5342E-01	1.8665E 01	5.3577E-02	1.3394E 00	2.4488E 02	6.4823E-01	5.3327E-01
40	4.8010E 00	2.0829E-01	1.9793E 01	5.0523E-02	1.2631E 00	2.8653E 02	6.1607E-01	4.9482E-01
45	5.8412E 00	1.7120E-01	2.0720E 01	4.8262E-02	1.2066E 00	3.2540E 02	5.8647E-01	4.6045E-01
50	7.1067E 00	1.4071E-01	2.1482E 01	4.6550E-02	1.1638E 00	3.6116E 02	5.5917E-01	4.2964E-01

Discrete compound interest = 6%

n	Single payment		Uniform annual series			Uniform gradient series	Depreciation series	
	Compound-interest factor $(1+i)^n$	Present-worth factor $\frac{1}{(1+i)^n}$	Unacost present-worth factor $\frac{(1+i)^n - 1}{i(1+i)^n}$	Capital-recovery factor $\frac{(1+i)^n}{(1+i)^n - 1}$	Capitalized-cost factor $\frac{(1+i)^n}{(1+i)^n - 1}$	Present-worth factor $\frac{F_{RP} - nF_{SP}}{i}$	Sum-of-digits present-worth factor $\frac{n - F_{RP}}{0.5n(n+1)i}$	Straight-line present-worth factor $\frac{1}{niF_{PK}}$
	P to S	S to P	R to P	P to R	P to K	G to P	SD to P	SL to P
	F_{PS}	F_{SP}	F_{RP}	F_{PR}	F_{PK}	F_{GP}	F_{SDP}	F_{SLP}
1	1.0600E 00	9.4340E-01	9.4340E-01	1.0600E 00	1.7667E 01		9.4340E-01	9.4340E-01
2	1.1236E 00	8.9000E-01	1.8334E 00	5.4544E-01	9.0906E 00	8.9000E-01	9.2560E-01	9.1670E-01
3	1.1910E 00	8.3962E-01	2.6730E 00	3.7411E-01	6.2352E 00	2.5692E 00	9.0830E-01	8.9100E-01
4	1.2625E 00	7.9209E-01	3.4651E 00	2.8859E-01	4.8099E 00	4.9455E 00	8.9149E-01	8.6628E-01
5	1.3382E 00	7.4726E-01	4.2124E 00	2.3740E-01	3.9566E 00	7.9345E 00	8.7515E-01	8.4247E-01
6	1.4185E 00	7.0496E-01	4.9173E 00	2.0336E-01	3.3894E 00	1.1459E 01	8.5927E-01	8.1955E-01
7	1.5036E 00	6.6506E-01	5.5824E 00	1.7914E-01	2.9856E 00	1.5450E 01	8.4382E-01	7.9748E-01
8	1.5938E 00	6.2741E-01	6.2098E 00	1.6104E-01	2.6839E 00	1.9842E 01	8.2880E-01	7.7622E-01
9	1.6895E 00	5.9190E-01	6.8017E 00	1.4702E-01	2.4504E 00	2.4577E 01	8.1419E-01	7.5574E-01
10	1.7908E 00	5.5839E-01	7.3601E 00	1.3587E-01	2.2645E 00	2.9602E 01	7.9997E-01	7.3601E-01
11	1.8983E 00	5.2679E-01	7.8869E 00	1.2679E-01	2.1132E 00	3.4870E 01	7.8614E-01	7.1699E-01
12	2.0122E 00	4.9697E-01	8.3838E 00	1.1928E-01	1.9880E 00	4.0337E 01	7.7268E-01	6.9865E-01
13	2.1329E 00	4.6884E-01	8.8527E 00	1.1296E-01	1.8827E 00	4.5963E 01	7.5958E-01	6.8098E-01
14	2.2609E 00	4.4230E-01	9.2950E 00	1.0758E-01	1.7931E 00	5.1713E 01	7.4683E-01	6.6393E-01
15	2.3966E 00	4.1727E-01	9.7122E 00	1.0296E-01	1.7160E 00	5.7555E 01	7.3441E-01	6.4748E-01
16	2.5404E 00	3.9365E-01	1.0106E 01	9.8952E-02	1.6492E 00	6.3459E 01	7.2232E-01	6.3162E-01
18	2.8543E 00	3.5034E-01	1.0828E 01	9.2357E-02	1.5393E 00	7.5357E 01	6.9906E-01	6.0153E-01
20	3.2071E 00	3.1180E-01	1.1470E 01	8.7185E-02	1.4531E 00	8.7230E 01	6.7699E-01	5.7350E-01
25	4.2919E 00	2.3300E-01	1.2783E 01	7.8227E-02	1.3038E 00	1.1597E 02	6.2649E-01	5.1133E-01
30	5.7435E 00	1.7411E-01	1.3765E 01	7.2649E-02	1.2108E 00	1.4236E 02	5.8191E-01	4.5883E-01
35	7.6861E 00	1.3011E-01	1.4498E 01	6.8974E-02	1.1496E 00	1.6574E 02	5.4237E-01	4.1424E-01
40	1.0286E 01	9.7222E-02	1.5046E 01	6.6462E-02	1.1077E 00	1.8596E 02	5.0719E-01	3.7616E-01
45	1.3765E 01	7.2650E-02	1.5456E 01	6.4700E-02	1.0783E 00	2.0311E 02	4.7575E-01	3.4346E-01
50	1.8420E 01	5.4288E-02	1.5762E 01	6.3444E-02	1.0574E 00	2.1746E 02	4.4756E-01	3.1524E-01

Table A1.1 (*Continued*)

Discrete compound interest = 8%

	Single payment		Uniform annual series		Uniform gradient series		Depreciation series	
	Compound-interest factor	Present-worth factor	Unacost present-worth factor	Capital-recovery factor	Capitalized-cost factor	Present-worth factor	Sum-of-digits present-worth factor	Straight-line present-worth factor
	$(1+i)^n$	$\dfrac{1}{(1+i)^n}$	$\dfrac{(1+i)^n-1}{i(1+i)^n}$	$\dfrac{(1+i)^n}{(1+i)^n-1}$	$\dfrac{(1+i)^n}{(1+i)^n-1}$	$\dfrac{F_{RP}-nF_{SP}}{i}$	$\dfrac{n-F_{RP}}{0.5n(n+1)i}$	$\dfrac{1}{niF_{PK}}$
	P to S	S to P	R to P	P to R	P to K	G to P	SD to P	SL to P
n	F_{PS}	F_{SP}	F_{RP}	F_{PR}	F_{PK}	F_{GP}	F_{SDP}	F_{SLP}
1	1.0800E 00	9.2593E-01	9.2593E-01	1.0800E 00	1.3500E 01	·········-01	9.2593E-01	9.2593E-01
2	1.1664E 00	8.5734E-01	1.7833E 00	5.6077E-01	7.0096E 00	8.5734E-01	9.0306E-01	8.9163E-01
3	1.2597E 00	7.9383E-01	2.5771E 00	3.8803E-01	4.8504E 00	2.4450E 00	8.8105E-01	8.5903E-01
4	1.3605E 00	7.3503E-01	3.3121E 00	3.0192E-01	3.7740E 00	4.6501E 00	8.5984E-01	8.2803E-01
5	1.4693E 00	6.8058E-01	3.9927E 00	2.5046E-01	3.1307E 00	7.3724E 00	8.3941E-01	7.9854E-01
6	1.5869E 00	6.3017E-01	4.6229E 00	2.1632E-01	2.7039E 00	1.0523E 01	8.1971E-01	7.7048E-01
7	1.7138E 00	5.8349E-01	5.2064E 00	1.9207E-01	2.4009E 00	1.4024E 01	8.0073E-01	7.4377E-01
8	1.8509E 00	5.4027E-01	5.7466E 00	1.7401E-01	2.1752E 00	1.7806E 01	7.8242E-01	7.1833E-01
9	1.9990E 00	5.0025E-01	6.2469E 00	1.6008E-01	2.0010E 00	2.1808E 01	7.6475E-01	6.9410E-01
10	2.1589E 00	4.6319E-01	6.7101E 00	1.4903E-01	1.8629E 00	2.5977E 01	7.4771E-01	6.7101E-01
11	2.3316E 00	4.2888E-01	7.1390E 00	1.4008E-01	1.7510E 00	3.0266E 01	7.3126E-01	6.4900E-01
12	2.5182E 00	3.9711E-01	7.5361E 00	1.3270E-01	1.6587E 00	3.4634E 01	7.1537E-01	6.2801E-01
13	2.7196E 00	3.6770E-01	7.9038E 00	1.2652E-01	1.5815E 00	3.9046E 01	7.0003E-01	6.0798E-01
14	2.9372E 00	3.4046E-01	8.2442E 00	1.2130E-01	1.5162E 00	4.3472E 01	6.8521E-01	5.8887E-01
15	3.1722E 00	3.1524E-01	8.5595E 00	1.1683E-01	1.4604E 00	4.7886E 01	6.7089E-01	5.7063E-01
16	3.4259E 00	2.9189E-01	8.8514E 00	1.1298E-01	1.4122E 00	5.2264E 01	6.5704E-01	5.5321E-01
18	3.9960E 00	2.5025E-01	9.3719E 00	1.0670E-01	1.3338E 00	6.0843E 01	6.3071E-01	5.2066E-01
20	4.6610E 00	2.1455E-01	9.8181E 00	1.0185E-01	1.2732E 00	6.9090E 01	6.0606E-01	4.9091E-01
25	6.8485E 00	1.4602E-01	1.0675E 01	9.3679E-02	1.1710E 00	8.7804E 01	5.5097E-01	4.2699E-01
30	1.0063E 01	9.9377E-02	1.1258E 01	8.8827E-02	1.1103E 00	1.0346E 02	5.0382E-01	3.7526E-01
35	1.4785E 01	6.7635E-02	1.1655E 01	8.5803E-02	1.0725E 00	1.1609E 02	4.6320E-01	3.3299E-01
40	2.1725E 01	4.6031E-02	1.1925E 01	8.3860E-02	1.0483E 00	1.2604E 02	4.2798E-01	2.9812E-01
45	3.1920E 01	3.1328E-02	1.2108E 01	8.2587E-02	1.0323E 00	1.3373E 02	3.9724E-01	2.6908E-01
50	4.6902E 01	2.1321E-02	1.2233E 01	8.1743E-02	1.0218E 00	1.3959E 02	3.7026E-01	2.4467E-01

Discrete compound interest = 10%

n	Single payment		Uniform annual series			Uniform gradient series	Depreciation series	
	Compound-interest factor $(1+i)^n$ P to S F_{PS}	Present-worth factor $\dfrac{1}{(1+i)^n}$ S to P F_{SP}	Unacost present-worth factor $\dfrac{(1+i)^n - 1}{i(1+i)^n}$ R to P F_{RP}	Capital-recovery factor $\dfrac{(1+i)^n}{(1+i)^n - 1}$ P to R F_{PR}	Capitalized-cost factor $\dfrac{(1+i)^n}{(1+i)^n - 1}$ P to K F_{PK}	Present-worth factor $\dfrac{F_{RP} - nF_{SP}}{i}$ G to P F_{GP}	Sum-of-digits present-worth factor $\dfrac{n - F_{RP}}{0.5n(n+1)i}$ SD to P F_{SDP}	Straight-line present worth factor $\dfrac{1}{niF_{PK}}$ SL to P F_{SLP}
1	1.1000E 00	9.0909E-01	9.0909E-01	1.1000E 00	1.1000E 01	....	9.0909E-01	9.0909E-01
2	1.2100E 00	8.2645E-01	1.7355E 00	5.7619E-01	5.7619E 00	8.2645E-01	8.8154E-01	8.6777E-01
3	1.3310E 00	7.5131E-01	2.4869E 00	4.0211E-01	4.0211E 00	2.3291E 00	8.5525E-01	8.2895E-01
4	1.4641E 00	6.8301E-01	3.1699E 00	3.1547E-01	3.1547E 00	4.3781E 00	8.3013E-01	7.9247E-01
5	1.6105E 00	6.2092E-01	3.7908E 00	2.6380E-01	2.6380E 00	6.8618E 00	8.0614E-01	7.5816E-01
6	1.7716E 00	5.6447E-01	4.3553E 00	2.2961E-01	2.2961E 00	9.6842E 00	7.8421E-01	7.2588E-01
7	1.9487E 00	5.1316E-01	4.8684E 00	2.0541E-01	2.0541E 00	1.2763E 01	7.6128E-01	6.9549E-01
8	2.1436E 00	4.6651E-01	5.3349E 00	1.8744E-01	1.8744E 00	1.6029E 01	7.4030E-01	6.6687E-01
9	2.3579E 00	4.2410E-01	5.7590E 00	1.7364E-01	1.7364E 00	1.9421E 01	7.2022E-01	6.3989E-01
10	2.5937E 00	3.8554E-01	6.1446E 00	1.6275E-01	1.6275E 00	2.2891E 01	7.0099E-01	6.1446E-01
11	2.8531E 00	3.5049E-01	6.4951E 00	1.5396E-01	1.5396E 00	2.6396E 01	6.8257E-01	5.9046E-01
12	3.1384E 00	3.1863E-01	6.8137E 00	1.4676E-01	1.4676E 00	2.9901E 01	6.6491E-01	5.6781E-01
13	3.4523E 00	2.8966E-01	7.1034E 00	1.4078E-01	1.4078E 00	3.3377E 01	6.4798E-01	5.4641E-01
14	3.7975E 00	2.6333E-01	7.3667E 00	1.3575E-01	1.3575E 00	3.6800E 01	6.3174E-01	5.2619E-01
15	4.1772E 00	2.3939E-01	7.6061E 00	1.3147E-01	1.3147E 00	4.0152E 01	6.1616E-01	5.0707E-01
16	4.5950E 00	2.1763E-01	7.8237E 00	1.2782E-01	1.2782E 00	4.3416E 01	6.0120E-01	4.8898E-01
18	5.5599E 00	1.7986E-01	8.2014E 00	1.2193E-01	1.2193E 00	4.9640E 01	5.7302E-01	4.5563E-01
20	6.7275E 00	1.4864E-01	8.5136E 00	1.1746E-01	1.1746E 00	5.5407E 01	5.4697E-01	4.2568E-01
25	1.0835E 01	9.2296E-02	9.0770E 00	1.1017E-01	1.1017E 00	6.7696E 01	4.8994E-01	3.6308E-01
30	1.7449E 01	5.7309E-02	9.4269E 00	1.0608E-01	1.0608E 00	7.7077E 01	4.4243E-01	3.1423E-01
35	2.8102E 01	3.5584E-02	9.6442E 00	1.0369E-01	1.0369E 00	8.3987E 01	4.0247E-01	2.7555E-01
40	4.5259E 01	2.2095E-02	9.7791E 00	1.0226E-01	1.0226E 00	8.8953E 01	3.6855E-01	2.4444E-01
45	7.2890E 01	1.3719E-02	9.8628E 00	1.0139E-01	1.0139E 00	9.2456E 01	3.3949E-01	2.1917E-01
50	1.1739E 02	8.5186E-03	9.9148E 00	1.0086E-01	1.0086E 00	9.4889E 01	3.1439E-01	1.9830E-01

Table A1.1 (*Continued*)

Discrete compound interest = 12%

	Single payment		Uniform annual series			Uniform gradient series	Depreciation series	
	Compound-interest factor	Present-worth factor	Unacost present-worth factor	Capital-recovery factor	Capitalized-cost factor	Present-worth factor	Sum-of-digits present-worth factor	Straight-line present worth factor
	$(1+i)^n$	$\dfrac{1}{(1+i)^n}$	$\dfrac{(1+i)^n-1}{i(1+i)^n}$	$\dfrac{(1+i)^n}{(1+i)^n-1}$	$\dfrac{(1+i)^n}{(1+i)^n-1}$	$\dfrac{F_{RP}-nF_{SP}}{i}$	$\dfrac{n-F_{RP}}{0.5n(n+1)i}$	$\dfrac{1}{niF_{PR}}$
	P to S	S to P	R to P	P to R	P to K	G to P	SD to P	SL to P
n	F_{PS}	F_{SP}	F_{RP}	F_{PR}	F_{PK}	F_{GP}	F_{SDP}	F_{SLP}
1	1.1200E 00	8.9286E-01	8.9286E-01	1.1200E 00	9.3333E 00		8.9286E-01	8.9286E-01
2	1.2544E 00	7.9719E-01	1.6901E 00	5.9170E-01	4.9308E 00	7.9719E-01	8.6097E-01	8.4503E-01
3	1.4049E 00	7.1178E-01	2.4018E 00	4.1635E-01	3.4696E 00	2.2208E 00	8.3079E-01	8.0061E-01
4	1.5735E 00	6.3552E-01	3.0373E 00	3.2923E-01	2.7436E 00	4.1273E 00	8.0221E-01	7.5934E-01
5	1.7623E 00	5.6743E-01	3.6048E 00	2.7741E-01	2.3117E 00	6.3970E 00	7.7512E-01	7.2096E-01
6	1.9738E 00	5.0663E-01	4.1114E 00	2.4323E-01	2.0269E 00	8.9302E 00	7.4944E-01	6.8523E-01
7	2.2107E 00	4.5235E-01	4.5638E 00	2.1912E-01	1.8260E 00	1.1644E 01	7.2507E-01	6.5197E-01
8	2.4760E 00	4.0388E-01	4.9676E 00	2.0130E-01	1.6775E 00	1.4471E 01	7.0194E-01	6.2095E-01
9	2.7731E 00	3.6061E-01	5.3282E 00	1.8768E-01	1.5640E 00	1.7356E 01	6.7995E-01	5.9203E-01
10	3.1058E 00	3.2197E-01	5.6502E 00	1.7698E-01	1.4749E 00	2.0254E 01	6.5906E-01	5.6502E-01
11	3.4785E 00	2.8748E-01	5.9377E 00	1.6842E-01	1.4035E 00	2.3129E 01	6.3918E-01	5.3979E-01
12	3.8960E 00	2.5668E-01	6.1944E 00	1.6144E-01	1.3453E 00	2.5952E 01	6.2026E-01	5.1620E-01
13	4.3635E 00	2.2917E-01	6.4235E 00	1.5568E-01	1.2973E 00	2.8702E 01	6.0224E-01	4.9412E-01
14	4.8871E 00	2.0462E-01	6.6282E 00	1.5087E-01	1.2573E 00	3.1362E 01	5.8507E-01	4.7344E-01
15	5.4736E 00	1.8270E-01	6.8109E 00	1.4682E-01	1.2235E 00	3.3920E 01	5.6869E-01	4.5406E-01
16	6.1304E 00	1.6312E-01	6.9740E 00	1.4339E-01	1.1949E 00	3.6367E 01	5.5306E-01	4.3587E-01
18	7.6900E 00	1.3004E-01	7.2497E 00	1.3794E-01	1.1495E 00	4.0908E 01	5.2390E-01	4.0276E-01
20	9.6463E 00	1.0367E-01	7.4694E 00	1.3388E-01	1.1157E 00	4.4968E 01	4.9724E-01	3.7347E-01
25	1.7000E 01	5.8823E-02	7.8431E 00	1.2750E-01	1.0625E 00	5.3105E 01	4.3992E-01	3.1373E-01
30	2.9960E 01	3.3384E-02	8.0552E 00	1.2414E-01	1.0345E 00	5.8782E 01	3.9286E-01	2.6851E-01
35	5.2800E 01	1.8940E-02	8.1755E 00	1.2232E-01	1.0193E 00	6.2605E 01	3.5482E-01	2.3359E-01
40	9.3051E 01	1.0747E-02	8.2438E 00	1.2130E-01	1.0109E 00	6.5116E 01	3.2273E-01	2.0609E-01
45	1.6399E 02	6.0980E-03	8.2825E 00	1.2074E-01	1.0061E 00	6.6734E 01	2.9563E-01	1.8406E-01
50	2.8900E 02	3.4602E-03	8.3045E 00	1.2042E-01	1.0035E 00	6.7762E 01	2.7252E-01	1.6609E-01

Discrete compound interest = 14%

	Single payment		Uniform annual series			Uniform gradient series	Depreciation series	
	Compound-interest factor $(1+i)^n$	Present-worth factor $\dfrac{1}{(1+i)^n}$	Unacost present-worth factor $\dfrac{(1+i)^n - 1}{i(1+i)^n}$	Capital-recovery factor $\dfrac{(1+i)^n}{(1+i)^n - 1}$	Capitalized-cost factor $\dfrac{(1+i)^n}{(1+i)^n - 1}$	Present-worth factor $\dfrac{F_{RP} - nF_{SP}}{i}$	Sum-of-digits present-worth factor $\dfrac{n - F_{RP}}{0.5n(n+1)i}$	Straight-line present worth factor $\dfrac{1}{ni}F_{PK}$
	P to S	S to P	R to P	P to R	P to K	G to P	SD to P	SL to P
n	F_{PS}	F_{SP}	F_{RP}	F_{PR}	F_{PK}	F_{GP}	F_{SDP}	F_{SLP}
1	1.1400E 00	8.7719E-01	8.7719E-01	1.1400E 00	8.1429E 00	. . . .	8.7719E-01	8.7719E-01
2	1.2996E 00	7.6947E-01	1.6467E 00	6.0729E-01	4.3378E 00	7.6947E-01	8.4128E-01	8.2333E-01
3	1.4815E 00	6.7497E-01	2.3216E 00	4.3073E-01	3.0767E 00	2.1194E 00	8.0758E-01	7.7388E-01
4	1.6890E 00	5.9208E-01	2.9137E 00	3.4320E-01	2.4515E 00	3.8957E 00	7.7592E-01	7.2843E-01
5	1.9254E 00	5.1937E-01	3.4331E 00	2.9128E-01	2.0806E 00	5.9731E 00	7.4615E-01	6.8662E-01
6	2.1950E 00	4.5559E-01	3.8887E 00	2.5716E-01	1.8368E 00	8.2511E 00	7.1814E-01	6.4811E-01
7	2.5023E 00	3.9964E-01	4.2883E 00	2.3319E-01	1.6657E 00	1.0649E 01	6.9176E-01	6.1261E-01
8	2.8526E 00	3.5056E-01	4.6389E 00	2.1557E-01	1.5398E 00	1.3103E 01	6.6689E-01	5.7986E-01
9	3.2519E 00	3.0751E-01	4.9464E 00	2.0217E-01	1.4441E 00	1.5563E 01	6.4343E-01	5.4960E-01
10	3.7072E 00	2.6974E-01	5.2161E 00	1.9171E-01	1.3694E 00	1.7991E 01	6.2128E-01	5.2161E-01
11	4.2262E 00	2.3662E-01	5.4527E 00	1.8339E-01	1.3100E 00	2.0357E 01	6.0035E-01	4.9570E-01
12	4.8179E 00	2.0756E-01	5.6603E 00	1.7667E-01	1.2619E 00	2.2640E 01	5.8056E-01	4.7169E-01
13	5.4924E 00	1.8207E-01	5.8424E 00	1.7116E-01	1.2226E 00	2.4825E 01	5.6182E-01	4.4941E-01
14	6.2613E 00	1.5971E-01	6.0021E 00	1.6661E-01	1.1901E 00	2.6901E 01	5.4408E-01	4.2872E-01
15	7.1379E 00	1.4010E-01	6.1422E 00	1.6281E-01	1.1629E 00	2.8862E 01	5.2725E-01	4.0948E-01
16	8.1372E 00	1.2289E-01	6.2651E 00	1.5962E-01	1.1401E 00	3.0706E 01	5.1129E-01	3.9157E-01
18	1.0575E 01	9.4561E-02	6.4674E 00	1.5462E-01	1.1044E 00	3.4036E 01	4.8173E-01	3.5930E-01
20	1.3743E 01	7.2762E-02	6.6231E 00	1.5099E-01	1.0785E 00	3.6914E 01	4.5500E-01	3.3116E-01
25	2.6462E 01	3.7790E-02	6.8729E 00	1.4550E-01	1.0393E 00	4.2344E 01	3.9840E-01	2.7492E-01
30	5.0950E 01	1.9627E-02	7.0027E 00	1.4280E-01	1.0200E 00	4.5813E 01	3.5326E-01	2.3342E-01
35	9.8100E 01	1.0194E-02	7.0700E 00	1.4144E-01	1.0103E 00	4.7932E 01	3.1667E-01	2.0200E-01
40	1.8888E 02	5.2943E-03	7.1050E 00	1.4075E-01	1.0053E 00	4.9238E 01	2.8654E-01	1.7763E-01
45	3.6368E 02	2.7497E-03	7.1232E 00	1.4039E-01	1.0028E 00	4.9996E 01	2.6140E-01	1.5829E-01
50	7.0023E 02	1.4281E-03	7.1327E 00	1.4020E-01	1.0014E 00	5.0438E 01	2.4015E-01	1.4265E-01

Table A1.1 (*Continued*)

Discrete compound interest = 16%

	Single payment		Uniform annual series		Uniform gradient series		Depreciation series	
	Compound-interest factor	Present-worth factor	Uncost present-worth factor	Capital-recovery factor	Capitalized-cost factor	Present-worth factor	Sum-of-digits present-worth factor	Straight-line present worth factor
	$(1+i)^n$	$\dfrac{1}{(1+i)^n}$	$\dfrac{(1+i)^n-1}{i(1+i)^n}$	$\dfrac{(1+i)^n}{(1+i)^n-1}$	$\dfrac{(1+i)^n}{(1+i)^n-1}$	$\dfrac{F_{RP}-nF_{SP}}{i}$	$\dfrac{n-F_{RP}}{0.5n(n+1)i}$	$\dfrac{1}{niF_{PK}}$
	P to S	S to P	R to P	P to R	P to K	G to P	SD to P	SL to P
n	F_{PS}	F_{SP}	F_{RP}	F_{PR}	F_{PK}	F_{GP}	F_{SDP}	F_{SLP}
1	1.1600E 00	8.6207E-01	8.6207E-01	1.1600E 00	7.2500E 00		8.6207E-01	8.6207E-01
2	1.3456E 00	7.4316E-01	1.6052E 00	6.2296E-01	3.8935E 00	7.4316E-01	8.2243E-01	8.0262E-01
3	1.5609E 00	6.4066E-01	2.2459E 00	4.4526E-01	2.7829E 00	2.0245F 00	7.8553E-01	7.4863E-01
4	1.8106E 00	5.5229E-01	2.7982E 00	3.5738E-01	2.2336E 00	3.6814E 00	7.5114E-01	6.9955E-01
5	2.1003E 00	4.7611E-01	3.2743E 00	3.0541E-01	1.9088E 00	5.5858E 00	7.1904E-01	6.5486E-01
6	2.4364E 00	4.1044E-01	3.6847E 00	2.7139E-01	1.6962E 00	7.6380E 00	6.8907E-01	6.1412E-01
7	2.8262E 00	3.5383E-01	4.0386E 00	2.4761E-01	1.5476E 00	9.7610E 00	6.6103E-01	5.7694E-01
8	3.2784E 00	3.0503E-01	4.3436E 00	2.3022E-01	1.4389E 00	1.1896E 01	6.3476E-01	5.4295E-01
9	3.8030E 00	2.6295E-01	4.6065E 00	2.1708E-01	1.3568E 00	1.4000E 01	6.1026E-01	5.1184E-01
10	4.4114E 00	2.2668E-01	4.8332E 00	2.0690E-01	1.2931E 00	1.6040E 01	5.8713E-01	4.8332E-01
11	5.1173E 00	1.9542E-01	5.0286E 00	1.9886E-01	1.2429E 00	1.7994E 01	5.6547E-01	4.5715E-01
12	5.9360E 00	1.6846E-01	5.1971E 00	1.9241E-01	1.2026E 00	1.9847E 01	5.4510E-01	4.3309E-01
13	6.8858E 00	1.4523E-01	5.3423E 00	1.8718E-01	1.1699E 00	2.1590E 01	5.2594E-01	4.1095E-01
14	7.9875E 00	1.2520E-01	5.4675E 00	1.8290E-01	1.1431E 00	2.3217E 01	5.0789E-01	3.9054E-01
15	9.2655E 00	1.0793E-01	5.5755E 00	1.7936E-01	1.1210E 00	2.4728E 01	4.9086E-01	3.7170E-01
16	1.0748E 01	9.3041E-02	5.6685E 00	1.7641E-01	1.1026E 00	2.6124E 01	4.7479E-01	3.5428E-01
18	1.4462E 01	6.9144E-02	5.8178E 00	1.7188E-01	1.0743E 00	2.8583E 01	4.4525E-01	3.2321E-01
20	1.9461E 01	5.1385E-02	5.9288E 00	1.6867E-01	1.0542E 00	3.0632E 01	4.1878E-01	2.9644E-01
25	4.0874E 01	2.4465E-02	6.0971E 00	1.6401E-01	1.0251E 00	3.4284E 01	3.6352E-01	2.4388E-01
30	8.5850E 01	1.1648E-02	6.1772E 00	1.6189E-01	1.0118E 00	3.6423E 01	3.2020E-01	2.0591E-01
35	1.8031E 02	5.5459E-03	6.2153E 00	1.6089E-01	1.0056E 00	3.7633E 01	2.8556E-01	1.7758E-01
40	3.7872E 02	2.6405E-03	6.2335E 00	1.6042E-01	1.0026E 00	3.8299E 01	2.5737E-01	1.5564E-01
45	7.9544E 02	1.2572E-03	6.2421E 00	1.6020E-01	1.0013E 00	3.8660E 01	2.3405E-01	1.3871E-01
50	1.6707E 03	5.9855E-04	6.2463E 00	1.6010E-01	1.0006E 00	3.8852E 01	2.1448E-01	1.2493E-01

Discrete compound interest = 20%

	Single payment		Uniform annual series		Uniform gradient series		Depreciation series	
	Compound-interest factor	Present-worth factor	Unacost present-worth factor	Capital-recovery factor	Capitalized-cost factor	Present-worth factor	Sum-of-digits present-worth factor	Straight-line present worth factor
	$(1+i)^n$	$\dfrac{1}{(1+i)^n}$	$\dfrac{(1+i)^n - 1}{i(1+i)^n}$	$\dfrac{(1+i)^n}{(1+i)^n - 1}$	$\dfrac{(1+i)^n}{(1+i)^n - 1}$	$\dfrac{F_{RP} - nF_{SP}}{i}$	$\dfrac{n - F_{RP}}{0.5n(n+1)i}$	$\dfrac{1}{niF_{PK}}$
	P to S	S to P	R to P	P to R	P to K	G to P	SD to P	SL to P
n	F_{PS}	F_{SP}	F_{RP}	F_{PR}	F_{PK}	F_{GP}	F_{SDP}	F_{SLP}
1	1.2000E 00	8.3333E-01	8.3333E-01	1.2000E 00	6.0000E 00	0	8.3333E-01	8.3333E-01
2	1.4400E 00	6.9444E-01	1.5278E 00	6.5455E-01	3.2727E 00	6.9444E-01	7.8704E-01	7.6389E-01
3	1.7280E 00	5.7870E-01	2.1065E 00	4.7473E-01	2.3736E 00	1.8519E 00	7.4460E-01	7.0216E-01
4	2.0736E 00	4.8225E-01	2.5887E 00	3.8629E-01	1.9314E 00	3.2986E 00	7.0563E-01	6.4718E-01
5	2.4883E 00	4.0188E-01	2.9906E 00	3.3438E-01	1.6719E 00	4.9061E 00	6.6980E-01	5.9812E-01
6	2.9860E 00	3.3490E-01	3.3255E 00	3.0071E-01	1.5035E 00	6.5806E 00	6.3678E-01	5.5425E-01
7	3.5832E 00	2.7908E-01	3.6046E 00	2.7742E-01	1.3871E 00	8.2551E 00	6.0632E-01	5.1494E-01
8	4.2998E 00	2.3257E-01	3.8372E 00	2.6061E-01	1.3030E 00	9.8831E 00	5.7817E-01	4.7964E-01
9	5.1598E 00	1.9381E-01	4.0310E 00	2.4808E-01	1.2404E 00	1.1434E 01	5.5211E-01	4.4789E-01
10	6.1917E 00	1.6151E-01	4.1925E 00	2.3852E-01	1.1926E 00	1.2887E 01	5.2796E-01	4.1925E-01
11	7.4301E 00	1.3459E-01	4.3271E 00	2.3110E-01	1.1555E 00	1.4233E 01	5.0553E-01	3.9337E-01
12	8.9161E 00	1.1216E-01	4.4392E 00	2.2526E-01	1.1263E 00	1.5467E 01	4.8467E-01	3.6993E-01
13	1.0699E 01	9.3464E-02	4.5327E 00	2.2062E-01	1.1031E 00	1.6588E 01	4.6524E-01	3.4867E-01
14	1.2839E 01	7.7887E-02	4.6106E 00	2.1689E-01	1.0845E 00	1.7601E 01	4.4712E-01	3.2933E-01
15	1.5407E 01	6.4905E-02	4.6755E 00	2.1388E-01	1.0694E 00	1.8509E 01	4.3019E-01	3.1170E-01
16	1.8488E 01	5.4088E-02	4.7296E 00	2.1144E-01	1.0572E 00	1.9321E 01	4.1435E-01	2.9560E-01
18	2.6623E 01	3.7561E-02	4.8122E 00	2.0781E-01	1.0390E 00	2.0680E 01	3.8561E-01	2.6734E-01
20	3.8338E 01	2.6084E-02	4.8696E 00	2.0536E-01	1.0268E 00	2.1739E 01	3.6025E-01	2.4348E-01
25	9.5396E 01	1.0483E-02	4.9476E 00	2.0212E-01	1.0106E 00	2.3428E 01	3.0850E-01	1.9790E-01
30	2.3738E 02	4.2127E-03	4.9789E 00	2.0085E-01	1.0042E 00	2.4263E 01	2.6904E-01	1.6596E-01
35	5.9067E 02	1.6930E-03	4.9915E 00	2.0034E-01	1.0017E 00	2.4661E 01	2.3816E-01	1.4262E-01
40	1.4698E 03	6.8038E-04	4.9966E 00	2.0014E-01	1.0007E 00	2.4847E 01	2.1344E-01	1.2491E-01
45	3.6573E 03	2.7343E-04	4.9986E 00	2.0005E-01	1.0003E 00	2.4932E 01	1.9323E-01	1.1108E-01
50	9.1004E 03	1.0988E-04	4.9995E 00	2.0002E-01	1.0001E 00	2.4970E 01	1.7647E-01	9.9989E-02

Table A1.1 (*Continued*)

Discrete compound interest $= 25\%$

	Single payment		Uniform annual series		Uniform gradient series		Depreciation series	
	Compound-interest factor $(1+i)^n$	Present-worth factor $\dfrac{1}{(1+i)^n}$	Unacost present-worth factor $\dfrac{(1+i)^n-1}{n(1+i)^n}$	Capital-recovery factor $\dfrac{(1+i)^n}{(1+i)^n-1}$	Capitalized-cost factor $\dfrac{(1+i)^n}{(1+i)^n-1}$	Present-worth factor $\dfrac{F_{RP}-nF_{SP}}{i}$	Sum-of-digits present-worth factor $\dfrac{n-F_{RP}}{0.5n(n+1)i}$	Straight-line present worth factor $\dfrac{1}{niF_{PR}}$
	P to S	S to P	R to P	P to R	P to K	G to P	SD to P	SL to P
n	F_{PS}	F_{SP}	F_{RP}	F_{PR}	F_{PK}	F_{GP}	F_{SDP}	F_{SLP}
1	1.2500E 00	8.0000E-01	8.0000E-01	1.2500E 00	5.0000E 00		8.0000E-01	8.0000E-01
2	1.5625E 00	6.4000E-01	1.4400E 00	6.9444E-01	2.7778E 00	6.4000E-01	7.4667E-01	7.2000E-01
3	1.9531E 00	5.1200E-01	1.9520E 00	5.1230E-01	2.0492E 00	1.6640E 00	6.9867E-01	6.5067E-01
4	2.4414E 00	4.0960E-01	2.3616E 00	4.2344E-01	1.6938E 00	2.8928E 00	6.5536E-01	5.9040E-01
5	3.0518E 00	3.2768E-01	2.6893E 00	3.7185E-01	1.4874E 00	4.2035E 00	6.1619E-01	5.3786E-01
6	3.8147E 00	2.6214E-01	2.9514E 00	3.3882E-01	1.3553E 00	5.5142E 00	5.8088E-01	4.9190E-01
7	4.7684E 00	2.0972E-01	3.1611E 00	3.1634E-01	1.2654E 00	6.7725E 00	5.4861E-01	4.5159E-01
8	5.9605E 00	1.6777E-01	3.3289E 00	3.0040E-01	1.2016E 00	7.9469E 00	5.1901E-01	4.1611E-01
9	7.4506E 00	1.3422E-01	3.4631E 00	2.8876E-01	1.1550E 00	9.0207E 00	4.9217E-01	3.8479E-01
10	9.3132E 00	1.0737E-01	3.5705E 00	2.8007E-01	1.1203E 00	9.9870E 00	4.6760E-01	3.5705E-01
11	1.1642E 01	8.5899E-02	3.6564E 00	2.7349E-01	1.0940E 00	1.0846E 01	4.4507E-01	3.3240E-01
12	1.4552E 01	6.8719E-02	3.7251E 00	2.6845E-01	1.0738E 00	1.1602E 01	4.2435E-01	3.1043E-01
13	1.8190E 01	5.4976E-02	3.7801E 00	2.6454E-01	1.0582E 00	1.2262E 01	4.0527E-01	2.9078E-01
14	2.2737E 01	4.3980E-02	3.8241E 00	2.6150E-01	1.0460E 00	1.2833E 01	3.8765E-01	2.7315E-01
15	2.8422E 01	3.5184E-02	3.8593E 00	2.5912E-01	1.0365E 00	1.3326E 01	3.7136E-01	2.5724E-01
16	3.5527E 01	2.8147E-02	3.8874E 00	2.5724E-01	1.0290E 00	1.3748E 01	3.5625E-01	2.4296E-01
18	5.5511E 01	1.8014E-02	3.9279E 00	2.5459E-01	1.0183E 00	1.4415E 01	3.2917E-01	2.1822E-01
20	8.6736E 01	1.1529E-02	3.9539E 00	2.5292E-01	1.0117E 00	1.4893E 01	3.0564E-01	1.9769E-01
25	2.6470E 02	3.7779E-03	3.9849E 00	2.5095E-01	1.0038E 00	1.5562E 01	2.5865E-01	1.5940E-01
30	8.0779E 02	1.2379E-03	3.9950E 00	2.5031E-01	1.0012E 00	1.5832E 01	2.2370E-01	1.3317E-01
35	2.4652E 03	4.0565E-04	3.9984E 00	2.5010E-01	1.0004E 00	1.5937E 01	1.9684E-01	1.1424E-01
40	7.5232E 03	1.3292E-04	3.9995E 00	2.5003E-01	1.0001E 00	1.5977E 01	1.7561E-01	9.9987E-02
45	2.2959E 04	4.3565E-05	3.9998E 00	2.5001E-01	1.0000E 00	1.5991E 01	1.5861E-01	8.8885E-02
50	7.0065E 04	1.4272E-05	3.9999E 00	2.5000E-01	1.0000E 00	1.5997E 01	1.4431E-01	7.9999E-02

Discrete compound interest = 30%

	Single payment		Uniform annual series		Uniform gradient series		Depreciation series	
	Compound-interest factor $(1+i)^n$	Present-worth factor $\frac{1}{(1+i)^n}$	Unacost present-worth factor $\frac{(1+i)^n-1}{i(1+i)^n}$	Capital-recovery factor $\frac{(1+i)^n}{(1+i)^n-1}$	Capitalized-cost factor $\frac{(1+i)^n}{(1+i)^n-1}$	Present-worth factor $\frac{F_{RP}-nF_{SP}}{i}$	Sum-of-digits present-worth factor $\frac{n-F_{RP}}{0.5n(n+1)i}$	Straight-line present worth factor $\frac{1}{niF_{PK}}$
	P to S	S to P	R to P	P to R	P to K	G to P	SD to P	SL to P
n	F_{PS}	F_{SP}	F_{RP}	F_{PR}	F_{PK}	F_{GP}	F_{SDP}	F_{SLP}
1	1.3000E 00	7.6923E-01	7.6923E-01	1.3000E 00	4.3333E 00	......	7.6923E-01	7.6923E-01
2	1.6900E 00	5.9172E-01	1.3609E 00	7.3478E-01	2.4493E 00	5.9172E-01	7.1006E-01	6.8047E-01
3	2.1970E 00	4.5517E-01	1.8161E 00	5.5063E-01	1.8354E 00	1.5020E 00	6.5772E-01	6.0537E-01
4	2.8561E 00	3.5013E-01	2.1662E 00	4.6163E-01	1.5388E 00	2.5524E 00	6.1125E-01	5.4156E-01
5	3.7129E 00	2.6933E-01	2.4356E 00	4.1058E-01	1.3686E 00	3.6297E 00	5.6987E-01	4.8711E-01
6	4.8268E 00	2.0718E-01	2.6427E 00	3.7839E-01	1.2613E 00	4.6656E 00	5.3290E-01	4.4046E-01
7	6.2749E 00	1.5937E-01	2.8021E 00	3.5687E-01	1.1896E 00	5.6218E 00	4.9975E-01	4.0030E-01
8	8.1573E 00	1.2259E-01	2.9247E 00	3.4192E-01	1.1397E 00	6.4800E 00	4.6994E-01	3.6559E-01
9	1.0604E 01	9.4300E-02	3.0190E 00	3.3124E-01	1.1041E 00	7.2343E 00	4.4304E-01	3.3544E-01
10	1.3786E 01	7.2534E-02	3.0915E 00	3.2346E-01	1.0782E 00	7.8872E 00	4.1869E-01	3.0915E-01
11	1.7922E 01	5.5795E-02	3.1473E 00	3.1773E-01	1.0591E 00	8.4452E 00	3.9660E-01	2.8612E-01
12	2.3298E 01	4.2922E-02	3.1903E 00	3.1345E-01	1.0448E 00	8.9173E 00	3.7648E-01	2.6586E-01
13	3.0288E 01	3.3017E-02	3.2233E 00	3.1024E-01	1.0341E 00	9.3135E 00	3.5812E-01	2.4794E-01
14	3.9374E 01	2.5398E-02	3.2487E 00	3.0782E-01	1.0261E 00	9.6437E 00	3.4131E-01	2.3205E-01
15	5.1186E 01	1.9537E-02	3.2682E 00	3.0598E-01	1.0199E 00	9.9172E 00	3.2588E-01	2.1788E-01
16	6.6542E 01	1.5028E-02	3.2832E 00	3.0458E-01	1.0153E 00	1.0143E 01	3.1169E-01	2.0520E-01
18	1.1246E 02	8.8924E-03	3.3037E 00	3.0269E-01	1.0090E 00	1.0479E 01	2.8648E-01	1.8354E-01
20	1.9005E 02	5.2618E-03	3.3158E 00	3.0159E-01	1.0053E 00	1.0702E 01	2.6483E-01	1.6579E-01
25	7.0564E 02	1.4172E-03	3.3286E 00	3.0043E-01	1.0014E 00	1.0977E 01	2.2227E-01	1.3314E-01
30	2.6200E 03	3.8168E-04	3.3321E 00	3.0011E-01	1.0004E 00	1.1069E 01	1.9117E-01	1.1107E-01
35	9.7279E 03	1.0280E-04	3.3330E 00	3.0003E-01	1.0001E 00	1.1098E 01	1.6755E-01	9.5228E-02
40	3.6119E 04	2.7686E-05	3.3333E 00	3.0001E-01	1.0000E 00	1.1107E 01	1.4905E-01	8.3331E-02
45	1.3411E 05	7.4567E-06	3.3333E 00	3.0000E-01	1.0000E 00	1.1110E 01	1.3419E-01	7.4074E-02
50	4.9793E 05	2.0083E-06	3.3333E 00	3.0000E-01	1.0000E 00	1.1111E 01	1.2200E-01	6.6667E-02

Table A1.1 (*Continued*)

Discrete compound interest = 40%

	Single payment		Uniform annual series			Uniform gradient series	Depreciation series	
	Compound-interest factor	Present-worth factor	Uncost present-worth factor	Capital-recovery factor	Capitalized-cost factor	Present-worth factor	Sum-of-digits present-worth factor	Straight-line present-worth factor
	$(1+i)^n$	$\dfrac{1}{(1+i)^n}$	$\dfrac{(1+i)^n-1}{i(1+i)^n}$	$\dfrac{(1+i)^n}{(1+i)^n-1}$	$\dfrac{(1+i)^n}{(1+i)^n-1}$	$\dfrac{F_{RP}-nF_{SP}}{i}$	$\dfrac{n-F_{RP}}{0.5n(n+1)i}$	$\dfrac{1}{niF_{PR}}$
	P to S	S to P	R to P	P to R	P to K	G to P	SD to P	SL to P
n	F_{PS}	F_{SP}	F_{RP}	F_{PR}	F_{PK}	F_{GP}	F_{SDP}	F_{SLP}
1	1.4000E 00	7.1429E-01	7.1429E-01	1.4000E 00	3.5000E 00		7.1429E-01	7.1429E-01
2	1.9600E 00	5.1020E-01	1.2245E 00	8.1667E-01	2.0417E 00	5.1020E-01	6.4620E-01	6.1224E-01
3	2.7440E 00	3.6443E-01	1.5889E 00	6.2936E-01	1.5734E 00	1.2391E 00	5.8795E-01	5.2964E-01
4	3.8416E 00	2.6031E-01	1.8492E 00	5.4077E-01	1.3519E 00	2.0200E 00	5.3769E-01	4.6231E-01
5	5.3782E 00	1.8593E-01	2.0352E 00	4.9136E-01	1.2284E 00	2.7637E 00	4.9414E-01	4.0703E-01
6	7.5295E 00	1.3281E-01	2.1680E 00	4.6126E-01	1.1532E 00	3.4278E 00	4.5619E-01	3.6133E-01
7	1.0541E 01	9.4865E-02	2.2628E 00	4.4192E-01	1.1048E 00	3.9970E 00	4.2296E-01	3.2326E-01
8	1.4758E 01	6.7760E-02	2.3306E 00	4.2907E-01	1.0727E 00	4.4713E 00	3.9371E-01	2.9132E-01
9	2.0661E 01	4.8400E-02	2.3790E 00	4.2034E-01	1.0509E 00	4.8585E 00	3.6783E-01	2.6433E-01
10	2.8925E 01	3.4572E-02	2.4136E 00	4.1432E-01	1.0358E 00	5.1696E 00	3.4484E-01	2.4136E-01
11	4.0496E 01	2.4694E-02	2.4383E 00	4.1013E-01	1.0253E 00	5.4166E 00	3.2431E-01	2.2166E-01
12	5.6694E 01	1.7639E-02	2.4559E 00	4.0718E-01	1.0180E 00	5.6106E 00	3.0590E-01	2.0466E-01
13	7.9371E 01	1.2599E-02	2.4685E 00	4.0510E-01	1.0128E 00	5.7618E 00	2.8933E-01	1.8988E-01
14	1.1112E 02	8.9993E-03	2.4775E 00	4.0363E-01	1.0091E 00	5.8788E 00	2.7435E-01	1.7696E-01
15	1.5557E 02	6.4281E-03	2.4839E 00	4.0259E-01	1.0065E 00	5.9688E 00	2.6075E-01	1.6560E-01
16	2.1780E 02	4.5915E-03	2.4885E 00	4.0185E-01	1.0046E 00	6.0376E 00	2.4837E-01	1.5553E-01
18	4.2688E 02	2.3426E-03	2.4941E 00	4.0094E-01	1.0023E 00	6.1299E 00	2.2669E-01	1.3856E-01
20	8.3668E 02	1.1952E-03	2.4970E 00	4.0048E-01	1.0012E 00	6.1828E 00	2.0837E-01	1.2485E-01
25	4.4999E 03	2.2223E-04	2.4999E 00	4.0009E-01	1.0002E 00	6.2347E 00	1.7308E-01	9.9978E-02
30	2.4201E 04	4.1320E-05	2.4999E 00	4.0002E-01	1.0000E 00	6.2486E 00	1.4765E-01	8.3330E-02
35	1.3010E 05	7.6824E-06	2.5000E 00	4.0000E-01	1.0000E 00	6.2493E 00	1.2897E-01	7.1428E-02
40	7.0004E 05	1.4285E-06	2.5000E 00	4.0000E-01	1.0000E 00	6.2498E 00	1.1433E-01	6.2500E-02
45	3.7650E 06	2.6561E-07	2.5000E 00	4.0000E-01	1.0000E 00	6.2500E 00	1.0266E-01	5.5558E-02
50	2.0249E 07	4.9385E-08	2.5000E 00	4.0000E-01	1.0000E 00	6.2500E 00	9.3117E-02	5.0000E-02

Discrete compound interest = 50%

	Single payment		Uniform annual series		Uniform gradient series		Depreciation series	
	Compound-interest factor	Present-worth factor	Unacost present-worth factor	Capital-recovery factor	Capitalized-cost factor	Present-worth factor	Sum-of-digits present-worth factor	Straight-line present worth factor
	$(1+i)^n$	$\dfrac{1}{(1+i)^n}$	$\dfrac{(1+i)^n-1}{i(1+i)^n}$	$\dfrac{(1+i)^n}{(1+i)^n-1}$	$\dfrac{(1+i)^n}{(1+i)^n-1}$	$\dfrac{F_{RP}-nF_{SP}}{i}$	$\dfrac{n-F_{RP}}{0.5n(n+1)i}$	$\dfrac{1}{niF_{PK}}$
	P to S	S to P	R to P	P to R	P to K	G to P	SD to P	SL to P
n	F_{PS}	F_{SP}	F_{RP}	F_{PR}	F_{PK}	F_{GP}	F_{SDP}	F_{SLP}
1	1.5000E 00	6.6667E-01	6.6667E-01	1.5000E 00	3.0000E 00	......4E-01	6.6667E-01	6.6667E-01
2	2.2500E 00	4.4444E-01	1.1111E 00	9.0000E-01	1.8000E 00	4.4444E-01	5.9259E-01	5.5556E-01
3	3.3750E 00	2.9630E-01	1.4074E 00	7.1053E-01	1.4211E 00	1.0370E 00	5.3086E-01	4.6914E-01
4	5.0625E 00	1.9753E-01	1.6049E 00	6.2308E-01	1.2462E 00	1.6296E 00	4.7901E-01	4.0123E-01
5	7.5938E 00	1.3169E-01	1.7366E 00	5.7583E-01	1.1517E 00	2.1564E 00	4.3512E-01	3.4733E-01
6	1.1391E 01	8.7791E-02	1.8244E 00	5.4812E-01	1.0962E 00	2.5953E 00	3.9767E-01	3.0407E-01
7	1.7086E 01	5.8528E-02	1.8829E 00	5.3108E-01	1.0622E 00	2.9465E 00	3.6550E-01	2.6899E-01
8	2.5629E 01	3.9018E-02	1.9220E 00	5.2030E-01	1.0406E 00	3.2196E 00	3.3767E-01	2.4025E-01
9	3.8443E 01	2.6012E-02	1.9480E 00	5.1335E-01	1.0267E 00	3.4277E 00	3.1342E-01	2.1644E-01
10	5.7665E 01	1.7342E-02	1.9653E 00	5.0882E-01	1.0176E 00	3.5838E 00	2.9217E-01	1.9653E-01
11	8.6498E 01	1.1561E-02	1.9769E 00	5.0585E-01	1.0117E 00	3.6994E 00	2.7343E-01	1.7972E-01
12	1.2975E 02	7.7073E-03	1.9846E 00	5.0388E-01	1.0078E 00	3.7842E 00	2.5681E-01	1.6538E-01
13	1.9462E 02	5.1382E-03	1.9897E 00	5.0258E-01	1.0052E 00	3.8459E 00	2.4198E-01	1.5306E-01
14	2.9193E 02	3.4255E-03	1.9931E 00	5.0172E-01	1.0034E 00	3.8904E 00	2.2870E-01	1.4237E-01
15	4.3789E 02	2.2837E-03	1.9954E 00	5.0114E-01	1.0023E 00	3.9224E 00	2.1674E-01	1.3303E-01
16	6.5684E 02	1.5224E-03	1.9970E 00	5.0076E-01	1.0015E 00	3.9452E 00	2.0593E-01	1.2481E-01
18	1.4779E 03	6.7664E-04	1.9986E 00	5.0034E-01	1.0007E 00	3.9729E 00	1.8715E-01	1.1104E-01
20	3.3253E 03	3.0073E-04	1.9994E 00	5.0015E-01	1.0003E 00	3.9868E 00	1.7143E-01	9.9970E-02
25	2.5251E 04	3.9602E-05	1.9999E 00	5.0002E-01	1.0000E 00	3.9979E 00	1.4154E-01	7.9997E-02
30	1.9175E 05	5.2151E-06	2.0000E 00	5.0000E-01	1.0000E 00	3.9997E 00	1.2043E-01	6.6666E-02
35	1.4561E 06	6.8676E-07	2.0000E 00	5.0000E-01	1.0000E 00	3.9999E 00	1.0476E-01	5.7143E-02
40	1.1057E 07	9.0438E-08	2.0000E 00	5.0000E-01	1.0000E 00	4.0000E 00	9.2683E-02	5.0000E-02
45	8.3967E 07	1.1909E-08	2.0000E 00	5.0000E-01	1.0000E 00	4.0000E 00	8.3092E-02	4.4444E-02
50	6.3762E 08	1.5683E-09	2.0000E 00	5.0000E-01	1.0000E 00	4.0000E 00	7.5294E-02	4.0000E-02

TWO

CONTINUOUS COMPOUND INTEREST

Table A2.1 Continuous compound interest: Value of

$$F_{PS,\overline{i},n} = e^{in}$$

$in \to$	0.00	0.01	0.02	0.03	0.04	0.05	0.06	0.07	0.08	0.09
0	1.0000E 00	1.0101E 00	1.0202E 00	1.0305E 00	1.0408E 00	1.0513E 00	1.0618E 00	1.0725E 00	1.0833E 00	1.0942E 00
0.1	1.1052E 00	1.1163E 00	1.1275E 00	1.1388E 00	1.1503E 00	1.1618E 00	1.1735E 00	1.1853E 00	1.1972E 00	1.2092E 00
0.2	1.2214E 00	1.2337E 00	1.2461E 00	1.2586E 00	1.2712E 00	1.2840E 00	1.2969E 00	1.3100E 00	1.3231E 00	1.3364E 00
0.3	1.3499E 00	1.3634E 00	1.3771E 00	1.3910E 00	1.4049E 00	1.4191E 00	1.4333E 00	1.4477E 00	1.4623E 00	1.4770E 00
0.4	1.4918E 00	1.5068E 00	1.5220E 00	1.5373E 00	1.5527E 00	1.5683E 00	1.5841E 00	1.6000E 00	1.6161E 00	1.6323E 00
0.5	1.6487E 00	1.6653E 00	1.6820E 00	1.6989E 00	1.7160E 00	1.7333E 00	1.7507E 00	1.7683E 00	1.7860E 00	1.8040E 00
0.6	1.8221E 00	1.8404E 00	1.8589E 00	1.8776E 00	1.8965E 00	1.9155E 00	1.9348E 00	1.9542E 00	1.9739E 00	1.9937E 00
0.7	2.0138E 00	2.0340E 00	2.0544E 00	2.0751E 00	2.0959E 00	2.1170E 00	2.1383E 00	2.1598E 00	2.1815E 00	2.2034E 00
0.8	2.2255E 00	2.2479E 00	2.2705E 00	2.2933E 00	2.3164E 00	2.3396E 00	2.3632E 00	2.3869E 00	2.4109E 00	2.4351E 00
0.9	2.4596E 00	2.4843E 00	2.5093E 00	2.5345E 00	2.5600E 00	2.5857E 00	2.6117E 00	2.6379E 00	2.6645E 00	2.6912E 00
1.0	2.7183E 00	2.7456E 00	2.7732E 00	2.8011E 00	2.8292E 00	2.8577E 00	2.8864E 00	2.9154E 00	2.9447E 00	2.9743E 00
1.1	3.0042E 00	3.0344E 00	3.0649E 00	3.0957E 00	3.1268E 00	3.1582E 00	3.1899E 00	3.2220E 00	3.2544E 00	3.2871E 00
1.2	3.3201E 00	3.3535E 00	3.3872E 00	3.4212E 00	3.4556E 00	3.4903E 00	3.5254E 00	3.5609E 00	3.5966E 00	3.6328E 00
1.3	3.6693E 00	3.7062E 00	3.7434E 00	3.7810E 00	3.8190E 00	3.8574E 00	3.8962E 00	3.9354E 00	3.9749E 00	4.0149E 00
1.4	4.0552E 00	4.0960E 00	4.1371E 00	4.1787E 00	4.2207E 00	4.2631E 00	4.3060E 00	4.3492E 00	4.3929E 00	4.4371E 00
1.5	4.4817E 00	4.5267E 00	4.5722E 00	4.6182E 00	4.6646E 00	4.7115E 00	4.7588E 00	4.8066E 00	4.8550E 00	4.9037E 00
1.6	4.9530E 00	5.0028E 00	5.0531E 00	5.1039E 00	5.1552E 00	5.2070E 00	5.2593E 00	5.3122E 00	5.3656E 00	5.4195E 00
1.7	5.4739E 00	5.5290E 00	5.5845E 00	5.6407E 00	5.6973E 00	5.7546E 00	5.8124E 00	5.8709E 00	5.9299E 00	5.9895E 00
1.8	6.0496E 00	6.1104E 00	6.1719E 00	6.2339E 00	6.2965E 00	6.3598E 00	6.4237E 00	6.4883E 00	6.5535E 00	6.6194E 00
1.9	6.6859E 00	6.7531E 00	6.8210E 00	6.8895E 00	6.9588E 00	7.0287E 00	7.0993E 00	7.1707E 00	7.2427E 00	7.3155E 00
2.0	7.3891E 00	7.4633E 00	7.5383E 00	7.6141E 00	7.6906E 00	7.7679E 00	7.8460E 00	7.9248E 00	8.0045E 00	8.0849E 00

Table A2.1 (*Continued*)

$F_{PS,\bar{i},n} = e^{in}$

$in \rightarrow$ $\downarrow$	0.00	0.01	0.02	0.03	0.04	0.05	0.06	0.07	0.08	0.09
2.1	8.1662E 00	8.2482E 00	8.3311E 00	8.4149E 00	8.4994E 00	8.5849E 00	8.6711E 00	8.7583E 00	8.8463E 00	8.9352E 00
2.2	9.0250E 00	9.1157E 00	9.2073E 00	9.2999E 00	9.3933E 00	9.4877E 00	9.5831E 00	9.6794E 00	9.7767E 00	9.8749E 00
2.3	9.9742E 00	1.0074E 01	1.0176E 01	1.0278E 01	1.0381E 01	1.0486E 01	1.0591E 01	1.0697E 01	1.0805E 01	1.0913E 01
2.4	1.1023E 01	1.1134E 01	1.1246E 01	1.1359E 01	1.1473E 01	1.1588E 01	1.1705E 01	1.1822E 01	1.1941E 01	1.2061E 01
2.5	1.2182E 01	1.2305E 01	1.2429E 01	1.2554E 01	1.2680E 01	1.2807E 01	1.2936E 01	1.3066E 01	1.3197E 01	1.3330E 01
2.6	1.3464E 01	1.3599E 01	1.3736E 01	1.3874E 01	1.4013E 01	1.4154E 01	1.4296E 01	1.4440E 01	1.4585E 01	1.4732E 01
2.7	1.4880E 01	1.5029E 01	1.5180E 01	1.5333E 01	1.5487E 01	1.5643E 01	1.5800E 01	1.5959E 01	1.6119E 01	1.6281E 01
2.8	1.6445E 01	1.6610E 01	1.6777E 01	1.6945E 01	1.7116E 01	1.7288E 01	1.7462E 01	1.7637E 01	1.7814E 01	1.7993E 01
2.9	1.8174E 01	1.8357E 01	1.8541E 01	1.8728E 01	1.8916E 01	1.9106E 01	1.9298E 01	1.9492E 01	1.9688E 01	1.9886E 01
3.0	2.0086E 01	2.0287E 01	2.0491E 01	2.0697E 01	2.0905E 01	2.1115E 01	2.1328E 01	2.1542E 01	2.1758E 01	2.1977E 01
3.1	2.2198E 01	2.2421E 01	2.2646E 01	2.2874E 01	2.3104E 01	2.3336E 01	2.3571E 01	2.3807E 01	2.4047E 01	2.4288E 01
3.2	2.4533E 01	2.4779E 01	2.5028E 01	2.5280E 01	2.5534E 01	2.5790E 01	2.6050E 01	2.6311E 01	2.6576E 01	2.6843E 01
3.3	2.7113E 01	2.7385E 01	2.7660E 01	2.7938E 01	2.8219E 01	2.8503E 01	2.8789E 01	2.9079E 01	2.9371E 01	2.9666E 01
3.4	2.9964E 01	3.0265E 01	3.0569E 01	3.0877E 01	3.1187E 01	3.1500E 01	3.1817E 01	3.2137E 01	3.2460E 01	3.2786E 01
3.5	3.3115E 01	3.3448E 01	3.3784E 01	3.4124E 01	3.4467E 01	3.4813E 01	3.5163E 01	3.5517E 01	3.5874E 01	3.6234E 01
3.6	3.6598E 01	3.6966E 01	3.7338E 01	3.7713E 01	3.8092E 01	3.8475E 01	3.8861E 01	3.9252E 01	3.9646E 01	4.0045E 01
3.7	4.0447E 01	4.0854E 01	4.1264E 01	4.1679E 01	4.2098E 01	4.2521E 01	4.2948E 01	4.3380E 01	4.3816E 01	4.4256E 01
3.8	4.4701E 01	4.5150E 01	4.5604E 01	4.6063E 01	4.6525E 01	4.6993E 01	4.7465E 01	4.7942E 01	4.8424E 01	4.8911E 01
3.9	4.9402E 01	4.9899E 01	5.0400E 01	5.0907E 01	5.1419E 01	5.1935E 01	5.2457E 01	5.2985E 01	5.3517E 01	5.4055E 01
4.0	5.4598E 01	5.5147E 01	5.5701E 01	5.6261E 01	5.6826E 01	5.7397E 01	5.7974E 01	5.8557E 01	5.9145E 01	5.9740E 01
4.1	6.0340E 01	6.0947E 01	6.1559E 01	6.2178E 01	6.2803E 01	6.3434E 01	6.4072E 01	6.4715E 01	6.5366E 01	6.6023E 01

$$F_{PS,\overline{i,n}} = e^{in}$$

in → ↓	0.00	0.01	0.02	0.03	0.04	0.05	0.06	0.07	0.08	0.09
4.2	6.6686E 01	6.7357E 01	6.8033E 01	6.8717E 01	6.9408E 01	7.0105E 01	7.0810E 01	7.1522E 01	7.2240E 01	7.2966E 01
4.3	7.3700E 01	7.4440E 01	7.5189E 01	7.5944E 01	7.6708E 01	7.7478E 01	7.8257E 01	7.9044E 01	7.9838E 01	8.0640E 01
4.4	8.1451E 01	8.2269E 01	8.3096E 01	8.3931E 01	8.4775E 01	8.5627E 01	8.6488E 01	8.7357E 01	8.8235E 01	8.9121E 01
4.5	9.0017E 01	9.0922E 01	9.1836E 01	9.2759E 01	9.3691E 01	9.4632E 01	9.5583E 01	9.6544E 01	9.7514E 01	9.8494E 01
4.6	9.9484E 01	1.0048E 02	1.0149E 02	1.0251E 02	1.0354E 02	1.0458E 02	1.0564E 02	1.0670E 02	1.0777E 02	1.0885E 02
4.7	1.0995E 02	1.1105E 02	1.1217E 02	1.1330E 02	1.1443E 02	1.1558E 02	1.1675E 02	1.1792E 02	1.1910E 02	1.2030E 02
4.8	1.2151E 02	1.2273E 02	1.2397E 02	1.2521E 02	1.2647E 02	1.2774E 02	1.2902E 02	1.3032E 02	1.3163E 02	1.3295E 02
4.9	1.3429E 02	1.3564E 02	1.3700E 02	1.3838E 02	1.3977E 02	1.4117E 02	1.4259E 02	1.4403E 02	1.4547E 02	1.4694E 02
	.0	.1	.2	.3	.4	.5	.6	.7	.8	.9
5.0	1.4841E 02	1.6402E 02	1.8127E 02	2.0034E 02	2.2141E 02	2.4469E 02	2.7043E 02	2.9887E 02	3.3030E 02	3.6504E 02
6.0	4.0343E 02	4.4586E 02	4.9275E 02	5.4457E 02	6.0185E 02	6.6514E 02	7.3510E 02	8.1241E 02	8.9785E 02	9.9227E 02
7.0	1.0966E 03	1.2120E 03	1.3394E 03	1.4803E 03	1.6360E 03	1.8080E 03	1.9982E 03	2.2083E 03	2.4406E 03	2.6973E 03
8.0	2.9810E 03	3.2945E 03	3.6410E 03	4.0239E 03	4.4471E 03	4.9148E 03	5.4317E 03	6.0029E 03	6.6342E 03	7.3320E 03
9.0	8.1031E 03	8.9553E 03	9.8971E 03	1.0938E 04	1.2088E 04	1.3360E 04	1.4765E 04	1.6318E 04	1.8034E 04	1.9930E 04
10.0	2.2026E 04	2.4343E 04	2.6903E 04	2.9733E 04	3.2860E 04	3.6316E 04	4.0135E 04	4.4356E 04	4.9021E 04	5.4176E 04

Table A2.2 Continuous compound interest: Value of

$$F_{SP,\overline{i},n} = e^{-in}$$

in → ↓	0.00	0.01	0.02	0.03	0.04	0.05	0.06	0.07	0.08	0.09
0	1.0000E 00	9.9005E-01	9.8020E-01	9.7045E-01	9.6079E-01	9.5123E-01	9.4176E-01	9.3239E-01	9.2312E-01	9.1393E-01
0.1	9.0484E-01	8.9583E-01	8.8692E-01	8.7810E-01	8.6936E-01	8.6071E-01	8.5214E-01	8.4366E-01	8.3527E-01	8.2696E-01
0.2	8.1873E-01	8.1058E-01	8.0252E-01	7.9453E-01	7.8663E-01	7.7880E-01	7.7105E-01	7.6338E-01	7.5578E-01	7.4826E-01
0.3	7.4082E-01	7.3345E-01	7.2615E-01	7.1892E-01	7.1177E-01	7.0469E-01	6.9768E-01	6.9073E-01	6.8386E-01	6.7706E-01
0.4	6.7032E-01	6.6365E-01	6.5705E-01	6.5051E-01	6.4404E-01	6.3763E-01	6.3128E-01	6.2500E-01	6.1878E-01	6.1263E-01
0.5	6.0653E-01	6.0050E-01	5.9452E-01	5.8860E-01	5.8275E-01	5.7695E-01	5.7121E-01	5.6553E-01	5.5990E-01	5.5433E-01
0.6	5.4881E-01	5.4335E-01	5.3794E-01	5.3259E-01	5.2729E-01	5.2205E-01	5.1685E-01	5.1171E-01	5.0662E-01	5.0158E-01
0.7	4.9659E-01	4.9164E-01	4.8675E-01	4.8191E-01	4.7711E-01	4.7237E-01	4.6767E-01	4.6301E-01	4.5841E-01	4.5384E-01
0.8	4.4933E-01	4.4486E-01	4.4043E-01	4.3605E-01	4.3171E-01	4.2741E-01	4.2316E-01	4.1895E-01	4.1478E-01	4.1066E-01
0.9	4.0657E-01	4.0252E-01	3.9852E-01	3.9455E-01	3.9063E-01	3.8674E-01	3.8289E-01	3.7908E-01	3.7531E-01	3.7158E-01
1.0	3.6788E-01	3.6422E-01	3.6059E-01	3.5701E-01	3.5345E-01	3.4994E-01	3.4646E-01	3.4301E-01	3.3960E-01	3.3622E-01
1.1	3.3287E-01	3.2956E-01	3.2628E-01	3.2303E-01	3.1982E-01	3.1664E-01	3.1349E-01	3.1037E-01	3.0728E-01	3.0422E-01
1.2	3.0119E-01	2.9820E-01	2.9523E-01	2.9229E-01	2.8938E-01	2.8650E-01	2.8365E-01	2.8083E-01	2.7804E-01	2.7527E-01
1.3	2.7253E-01	2.6982E-01	2.6714E-01	2.6448E-01	2.6185E-01	2.5924E-01	2.5666E-01	2.5411E-01	2.5158E-01	2.4908E-01
1.4	2.4660E-01	2.4414E-01	2.4171E-01	2.3931E-01	2.3693E-01	2.3457E-01	2.3224E-01	2.2993E-01	2.2764E-01	2.2537E-01
1.5	2.2313E-01	2.2091E-01	2.1871E-01	2.1654E-01	2.1438E-01	2.1225E-01	2.1014E-01	2.0805E-01	2.0598E-01	2.0393E-01
1.6	2.0190E-01	1.9989E-01	1.9790E-01	1.9593E-01	1.9398E-01	1.9205E-01	1.9014E-01	1.8825E-01	1.8637E-01	1.8452E-01
1.7	1.8268E-01	1.8087E-01	1.7907E-01	1.7728E-01	1.7552E-01	1.7377E-01	1.7204E-01	1.7033E-01	1.6864E-01	1.6696E-01
1.8	1.6530E-01	1.6365E-01	1.6203E-01	1.6041E-01	1.5882E-01	1.5724E-01	1.5567E-01	1.5412E-01	1.5259E-01	1.5107E-01
1.9	1.4957E-01	1.4808E-01	1.4661E-01	1.4515E-01	1.4370E-01	1.4227E-01	1.4086E-01	1.3946E-01	1.3807E-01	1.3670E-01
2.0	1.3534E-01	1.3399E-01	1.3266E-01	1.3134E-01	1.3003E-01	1.2873E-01	1.2745E-01	1.2619E-01	1.2493E-01	1.2369E-01

$$F_{SP,\bar{i},n} = e^{-in}$$

in → ↓	0.00	0.01	0.02	0.03	0.04	0.05	0.06	0.07	0.08	0.09
2.1	1.2246E-01	1.2124E-01	1.2003E-01	1.1884E-01	1.1765E-01	1.1648E-01	1.1533E-01	1.1418E-01	1.1304E-01	1.1192E-01
2.2	1.1080E-01	1.0970E-01	1.0861E-01	1.0753E-01	1.0646E-01	1.0540E-01	1.0435E-01	1.0331E-01	1.0228E-01	1.0127E-01
2.3	1.0026E-01	9.9261E-02	9.8274E-02	9.7296E-02	9.6328E-02	9.5369E-02	9.4420E-02	9.3481E-02	9.2551E-02	9.1630E-02
2.4	9.0718E-02	8.9815E-02	8.8922E-02	8.8037E-02	8.7161E-02	8.6294E-02	8.5435E-02	8.4585E-02	8.3743E-02	8.2910E-02
2.5	8.2085E-02	8.1268E-02	8.0460E-02	7.9659E-02	7.8866E-02	7.8082E-02	7.7305E-02	7.6536E-02	7.5774E-02	7.5020E-02
2.6	7.4274E-02	7.3535E-02	7.2803E-02	7.2078E-02	7.1361E-02	7.0651E-02	6.9948E-02	6.9252E-02	6.8563E-02	6.7881E-02
2.7	6.7206E-02	6.6537E-02	6.5875E-02	6.5219E-02	6.4570E-02	6.3928E-02	6.3292E-02	6.2662E-02	6.2039E-02	6.1421E-02
2.8	6.0810E-02	6.0205E-02	5.9606E-02	5.9013E-02	5.8426E-02	5.7844E-02	5.7269E-02	5.6699E-02	5.6135E-02	5.5576E-02
2.9	5.5023E-02	5.4476E-02	5.3934E-02	5.3397E-02	5.2866E-02	5.2340E-02	5.1819E-02	5.1303E-02	5.0793E-02	5.0287E-02
3.0	4.9787E-02	4.9292E-02	4.8801E-02	4.8316E-02	4.7835E-02	4.7359E-02	4.6888E-02	4.6421E-02	4.5959E-02	4.5502E-02
3.1	4.5049E-02	4.4601E-02	4.4157E-02	4.3718E-02	4.3283E-02	4.2852E-02	4.2426E-02	4.2004E-02	4.1586E-02	4.1172E-02
3.2	4.0762E-02	4.0357E-02	3.9955E-02	3.9557E-02	3.9164E-02	3.8774E-02	3.8388E-02	3.8006E-02	3.7628E-02	3.7254E-02
3.3	3.6883E-02	3.6516E-02	3.6153E-02	3.5793E-02	3.5437E-02	3.5084E-02	3.4735E-02	3.4390E-02	3.4047E-02	3.3709E-02
3.4	3.3373E-02	3.3041E-02	3.2712E-02	3.2387E-02	3.2065E-02	3.1746E-02	3.1430E-02	3.1117E-02	3.0807E-02	3.0501E-02
3.5	3.0197E-02	2.9897E-02	2.9599E-02	2.9305E-02	2.9013E-02	2.8725E-02	2.8439E-02	2.8156E-02	2.7876E-02	2.7598E-02
3.6	2.7324E-02	2.7052E-02	2.6783E-02	2.6516E-02	2.6252E-02	2.5991E-02	2.5733E-02	2.5476E-02	2.5223E-02	2.4972E-02
3.7	2.4724E-02	2.4478E-02	2.4234E-02	2.3993E-02	2.3754E-02	2.3518E-02	2.3284E-02	2.3052E-02	2.2823E-02	2.2596E-02
3.8	2.2371E-02	2.2148E-02	2.1928E-02	2.1710E-02	2.1494E-02	2.1280E-02	2.1068E-02	2.0858E-02	2.0651E-02	2.0445E-02
3.9	2.0242E-02	2.0041E-02	1.9841E-02	1.9644E-02	1.9448E-02	1.9255E-02	1.9063E-02	1.8873E-02	1.8686E-02	1.8500E-02
4.0	1.8316E-02	1.8133E-02	1.7953E-02	1.7774E-02	1.7597E-02	1.7422E-02	1.7249E-02	1.7077E-02	1.6907E-02	1.6739E-02
4.1	1.6573E-02	1.6408E-02	1.6245E-02	1.6083E-02	1.5923E-02	1.5764E-02	1.5608E-02	1.5452E-02	1.5299E-02	1.5146E-02

Table A2.2 (*Continued*)

$F_{SP,\overline{i},n} = e^{-in}$

in ↓ n →	0.00	0.01	0.02	0.03	0.04	0.05	0.06	0.07	0.08	0.09
4.2	1.4996E-02	1.4846E-02	1.4699E-02	1.4552E-02	1.4408E-02	1.4264E-02	1.4122E-02	1.3982E-02	1.3843E-02	1.3705E-02
4.3	1.3569E-02	1.3434E-02	1.3300E-02	1.3168E-02	1.3037E-02	1.2907E-02	1.2778E-02	1.2651E-02	1.2525E-02	1.2401E-02
4.4	1.2277E-02	1.2155E-02	1.2034E-02	1.1914E-02	1.1796E-02	1.1679E-02	1.1562E-02	1.1447E-02	1.1333E-02	1.1221E-02
4.5	1.1109E-02	1.0998E-02	1.0889E-02	1.0781E-02	1.0673E-02	1.0567E-02	1.0462E-02	1.0358E-02	1.0255E-02	1.0153E-02
4.6	1.0052E-02	9.9518E-03	9.8528E-03	9.7548E-03	9.6577E-03	9.5616E-03	9.4665E-03	9.3723E-03	9.2790E-03	9.1867E-03
4.7	9.0953E-03	9.0048E-03	8.9152E-03	8.8265E-03	8.7386E-03	8.6517E-03	8.5656E-03	8.4804E-03	8.3960E-03	8.3125E-03
4.8	8.2297E-03	8.1479E-03	8.0668E-03	7.9865E-03	7.9071E-03	7.8284E-03	7.7505E-03	7.6734E-03	7.5970E-03	7.5214E-03
4.9	7.4466E-03	7.3725E-03	7.2991E-03	7.2265E-03	7.1546E-03	7.0834E-03	7.0129E-03	6.9431E-03	6.8741E-03	6.8057E-03

in ↓ n →	.0	.1	.2	.3	.4	.5	.6	.7	.8	.9
5.0	6.7379E-03	6.0967E-03	5.5166E-03	4.9916E-03	4.5166E-03	4.0868E-03	3.6979E-03	3.3460E-03	3.0276E-03	2.7394E-03
6.0	2.4788E-03	2.2429E-03	2.0294E-03	1.8363E-03	1.6616E-03	1.5034E-03	1.3604E-03	1.2309E-03	1.1138E-03	1.0078E-03
7.0	9.1188E-04	8.2510E-04	7.4659E-04	6.7554E-04	6.1125E-04	5.5308E-04	5.0045E-04	4.5283E-04	4.0973E-04	3.7074E-04
8.0	3.3546E-04	3.0354E-04	2.7465E-04	2.4852E-04	2.2487E-04	2.0347E-04	1.8411E-04	1.6659E-04	1.5073E-04	1.3639E-04
9.0	1.2341E-04	1.1167E-04	1.0104E-04	9.1424E-05	8.2724E-05	7.4852E-05	6.7729E-05	6.1283E-05	5.5452E-05	5.0175E-05
10.0	4.5400E-05	4.1080E-05	3.7170E-05	3.3633E-05	3.0432E-05	2.7536E-05	2.4916E-05	2.2545E-05	2.0400E-05	1.8458E-05

Table A2.3 Continuous compound interest: Value of

$$F_{\overline{RP},\overline{i},n} = \frac{1 - e^{-in}}{in}$$

in→ i↓	0.00	0.01	0.02	0.03	0.04	0.05	0.06	0.07	0.08	0.09
0.0	1.0000E 00	9.9502E-01	9.9007E-01	9.8515E-01	9.8026E-01	9.7541E-01	9.7059E-01	9.6580E-01	9.6105E-01	9.5632E-01
0.1	9.5163E-01	9.4696E-01	9.4233E-01	9.3773E-01	9.3316E-01	9.2861E-01	9.2410E-01	9.1962E-01	9.1517E-01	9.1074E-01
0.2	9.0635E-01	9.0198E-01	8.9764E-01	8.9333E-01	8.8905E-01	8.8480E-01	8.8057E-01	8.7637E-01	8.7220E-01	8.6806E-01
0.3	8.6394E-01	8.5985E-01	8.5578E-01	8.5175E-01	8.4773E-01	8.4375E-01	8.3979E-01	8.3585E-01	8.3194E-01	8.2806E-01
0.4	8.2420E-01	8.2037E-01	8.1656E-01	8.1277E-01	8.0901E-01	8.0527E-01	8.0156E-01	7.9787E-01	7.9420E-01	7.9056E-01
0.5	7.8694E-01	7.8334E-01	7.7977E-01	7.7622E-01	7.7269E-01	7.6918E-01	7.6570E-01	7.6224E-01	7.5880E-01	7.5538E-01
0.6	7.5198E-01	7.4861E-01	7.4525E-01	7.4192E-01	7.3861E-01	7.3531E-01	7.3204E-01	7.2879E-01	7.2556E-01	7.2235E-01
0.7	7.1916E-01	7.1599E-01	7.1284E-01	7.0971E-01	7.0660E-01	7.0351E-01	7.0044E-01	6.9739E-01	6.9435E-01	6.9134E-01
0.8	6.8834E-01	6.8536E-01	6.8240E-01	6.7946E-01	6.7654E-01	6.7363E-01	6.7074E-01	6.6787E-01	6.6502E-01	6.6218E-01
0.9	6.5937E-01	6.5657E-01	6.5378E-01	6.5102E-01	6.4827E-01	6.4554E-01	6.4282E-01	6.4012E-01	6.3744E-01	6.3477E-01
1.0	6.3212E-01	6.2949E-01	6.2687E-01	6.2427E-01	6.2168E-01	6.1911E-01	6.1655E-01	6.1401E-01	6.1149E-01	6.0898E-01
1.1	6.0648E-01	6.0400E-01	6.0154E-01	5.9909E-01	5.9665E-01	5.9423E-01	5.9182E-01	5.8943E-01	5.8705E-01	5.8469E-01
1.2	5.8234E-01	5.8000E-01	5.7768E-01	5.7537E-01	5.7308E-01	5.7080E-01	5.6853E-01	5.6627E-01	5.6403E-01	5.6181E-01
1.3	5.5959E-01	5.5739E-01	5.5520E-01	5.5302E-01	5.5086E-01	5.4871E-01	5.4657E-01	5.4445E-01	5.4233E-01	5.4023E-01
1.4	5.3815E-01	5.3607E-01	5.3400E-01	5.3195E-01	5.2991E-01	5.2788E-01	5.2587E-01	5.2386E-01	5.2187E-01	5.1988E-01
1.5	5.1791E-01	5.1595E-01	5.1401E-01	5.1207E-01	5.1014E-01	5.0823E-01	5.0632E-01	5.0443E-01	5.0255E-01	5.0068E-01
1.6	4.9881E-01	4.9696E-01	4.9512E-01	4.9329E-01	4.9148E-01	4.8967E-01	4.8787E-01	4.8608E-01	4.8430E-01	4.8253E-01
1.7	4.8077E-01	4.7903E-01	4.7729E-01	4.7556E-01	4.7384E-01	4.7213E-01	4.7043E-01	4.6874E-01	4.6706E-01	4.6539E-01
1.8	4.6372E-01	4.6207E-01	4.6043E-01	4.5879E-01	4.5716E-01	4.5555E-01	4.5394E-01	4.5234E-01	4.5075E-01	4.4917E-01
1.9	4.4760E-01	4.4603E-01	4.4448E-01	4.4293E-01	4.4139E-01	4.3986E-01	4.3834E-01	4.3682E-01	4.3532E-01	4.3382E-01
2.0	4.3233E-01	4.3085E-01	4.2938E-01	4.2791E-01	4.2646E-01	4.2501E-01	4.2357E-01	4.2213E-01	4.2071E-01	4.1929E-01

Table A2.3 (Continued)

$$F_{\overline{RP},i,n} = \frac{1 - e^{-in}}{in}$$

$in \rightarrow$	0.00	0.01	0.02	0.03	0.04	0.05	0.06	0.07	0.08	0.09
2.1	4.1788E-01	4.1647E-01	4.1508E-01	4.1369E-01	4.1231E-01	4.1094E-01	4.0957E-01	4.0821E-01	4.0686E-01	4.0552E-01
2.2	4.0418E-01	4.0285E-01	4.0153E-01	4.0021E-01	3.9890E-01	3.9760E-01	3.9631E-01	3.9502E-01	3.9373E-01	3.9245E-01
2.3	3.9119E-01	3.8993E-01	3.8868E-01	3.8743E-01	3.8618E-01	3.8495E-01	3.8372E-01	3.8250E-01	3.8128E-01	3.8007E-01
2.4	3.7887E-01	3.7767E-01	3.7648E-01	3.7529E-01	3.7411E-01	3.7294E-01	3.7177E-01	3.7061E-01	3.6946E-01	3.6831E-01
2.5	3.6717E-01	3.6603E-01	3.6490E-01	3.6377E-01	3.6265E-01	3.6154E-01	3.6043E-01	3.5932E-01	3.5823E-01	3.5714E-01
2.6	3.5605E-01	3.5497E-01	3.5389E-01	3.5282E-01	3.5176E-01	3.5070E-01	3.4964E-01	3.4859E-01	3.4755E-01	3.4651E-01
2.7	3.4548E-01	3.4445E-01	3.4343E-01	3.4241E-01	3.4140E-01	3.4039E-01	3.3939E-01	3.3839E-01	3.3740E-01	3.3641E-01
2.8	3.3542E-01	3.3445E-01	3.3347E-01	3.3250E-01	3.3154E-01	3.3058E-01	3.2963E-01	3.2868E-01	3.2773E-01	3.2679E-01
2.9	3.2585E-01	3.2492E-01	3.2400E-01	3.2307E-01	3.2215E-01	3.2124E-01	3.2033E-01	3.1943E-01	3.1853E-01	3.1763E-01
3.0	3.1674E-01	3.1585E-01	3.1497E-01	3.1409E-01	3.1321E-01	3.1234E-01	3.1147E-01	3.1061E-01	3.0975E-01	3.0890E-01
3.1	3.0805E-01	3.0720E-01	3.0636E-01	3.0552E-01	3.0469E-01	3.0386E-01	3.0303E-01	3.0221E-01	3.0139E-01	3.0057E-01
3.2	2.9976E-01	2.9895E-01	2.9815E-01	2.9735E-01	2.9655E-01	2.9576E-01	2.9497E-01	2.9419E-01	2.9341E-01	2.9263E-01
3.3	2.9185E-01	2.9108E-01	2.9032E-01	2.8955E-01	2.8879E-01	2.8803E-01	2.8728E-01	2.8653E-01	2.8578E-01	2.8504E-01
3.4	2.8430E-01	2.8357E-01	2.8283E-01	2.8210E-01	2.8138E-01	2.8065E-01	2.7993E-01	2.7922E-01	2.7850E-01	2.7779E-01
3.5	2.7709E-01	2.7638E-01	2.7568E-01	2.7498E-01	2.7429E-01	2.7360E-01	2.7291E-01	2.7223E-01	2.7154E-01	2.7086E-01
3.6	2.7019E-01	2.6951E-01	2.6884E-01	2.6818E-01	2.6751E-01	2.6685E-01	2.6619E-01	2.6554E-01	2.6489E-01	2.6424E-01
3.7	2.6359E-01	2.6294E-01	2.6230E-01	2.6166E-01	2.6103E-01	2.6040E-01	2.5976E-01	2.5914E-01	2.5851E-01	2.5789E-01
3.8	2.5727E-01	2.5665E-01	2.5604E-01	2.5543E-01	2.5482E-01	2.5421E-01	2.5361E-01	2.5301E-01	2.5241E-01	2.5181E-01
3.9	2.5122E-01	2.5063E-01	2.5004E-01	2.4945E-01	2.4887E-01	2.4829E-01	2.4771E-01	2.4714E-01	2.4656E-01	2.4599E-01
4.0	2.4542E-01	2.4485E-01	2.4429E-01	2.4373E-01	2.4317E-01	2.4261E-01	2.4206E-01	2.4150E-01	2.4095E-01	2.4041E-01
4.1	2.3986E-01	2.3932E-01	2.3878E-01	2.3824E-01	2.3770E-01	2.3717E-01	2.3663E-01	2.3610E-01	2.3557E-01	2.3505E-01

$$F_{\overline{RP},\,\overline{i}n} = \frac{1 - e^{-in}}{in}$$

$\overline{i}n \downarrow$ / $\overline{i}n \rightarrow$	0.00	0.01	0.02	0.03	0.04	0.05	0.06	0.07	0.08	0.09
4.2	2.3452E-01	2.3400E-01	2.3348E-01	2.3297E-01	2.3245E-01	2.3194E-01	2.3143E-01	2.3092E-01	2.3041E-01	2.2991E-01
4.3	2.2940E-01	2.2890E-01	2.2840E-01	2.2791E-01	2.2741E-01	2.2692E-01	2.2643E-01	2.2594E-01	2.2545E-01	2.2497E-01
4.4	2.2448E-01	2.2400E-01	2.2352E-01	2.2304E-01	2.2257E-01	2.2209E-01	2.2162E-01	2.2115E-01	2.2068E-01	2.2022E-01
4.5	2.1975E-01	2.1929E-01	2.1883E-01	2.1837E-01	2.1791E-01	2.1746E-01	2.1700E-01	2.1655E-01	2.1610E-01	2.1565E-01
4.6	2.1521E-01	2.1476E-01	2.1432E-01	2.1388E-01	2.1344E-01	2.1300E-01	2.1256E-01	2.1213E-01	2.1169E-01	2.1126E-01
4.7	2.1083E-01	2.1040E-01	2.0998E-01	2.0955E-01	2.0913E-01	2.0870E-01	2.0828E-01	2.0787E-01	2.0745E-01	2.0703E-01
4.8	2.0662E-01	2.0621E-01	2.0580E-01	2.0539E-01	2.0498E-01	2.0457E-01	2.0417E-01	2.0376E-01	2.0336E-01	2.0296E-01
4.9	2.0256E-01	2.0216E-01	2.0177E-01	2.0137E-01	2.0098E-01	2.0059E-01	2.0020E-01	1.9981E-01	1.9942E-01	1.9904E-01
	.0	.1	.2	.3	.4	.5	.6	.7	.8	.9
5.0	1.9865E-01	1.9488E-01	1.9125E-01	1.8774E-01	1.8435E-01	1.8108E-01	1.7791E-01	1.7485E-01	1.7189E-01	1.6903E-01
6.0	1.6625E-01	1.6357E-01	1.6096E-01	1.5844E-01	1.5599E-01	1.5361E-01	1.5131E-01	1.4907E-01	1.4690E-01	1.4478E-01
7.0	1.4273E-01	1.4073E-01	1.3879E-01	1.3689E-01	1.3505E-01	1.3326E-01	1.3151E-01	1.2981E-01	1.2815E-01	1.2654E-01
8.0	1.2496E-01	1.2342E-01	1.2192E-01	1.2045E-01	1.1902E-01	1.1762E-01	1.1626E-01	1.1492E-01	1.1362E-01	1.1234E-01
9.0	1.1110E-01	1.0988E-01	1.0868E-01	1.0752E-01	1.0637E-01	1.0526E-01	1.0416E-01	1.0309E-01	1.0204E-01	1.0101E-01
10.0	9.9995E-02	9.9006E-02	9.8036E-02	9.7084E-02	9.6151E-02	9.5235E-02	9.4337E-02	9.3456E-02	9.2591E-02	9.1741E-02
11.0	9.0908E-02	9.0089E-02	8.9284E-02	8.8494E-02	8.7718E-02	8.6956E-02	8.6206E-02	8.5469E-02	8.4745E-02	8.4033E-02
12.0	8.3333E-02	8.2644E-02	8.1967E-02	8.1300E-02	8.0645E-02	8.0000E-02	7.9365E-02	7.8740E-02	7.8125E-02	7.7519E-02
13.0	7.6923E-02	7.6336E-02	7.5757E-02	7.5188E-02	7.4627E-02	7.4074E-02	7.3529E-02	7.2993E-02	7.2464E-02	7.1942E-02
14.0	7.1429E-02	7.0922E-02	7.0422E-02	6.9930E-02	6.9444E-02	6.8965E-02	6.8493E-02	6.8027E-02	6.7568E-02	6.7114E-02
15.0	6.6667E-02	6.6225E-02	6.5789E-02	6.5359E-02	6.4935E-02	6.4516E-02	6.4103E-02	6.3694E-02	6.3291E-02	6.2893E-02
16.0	6.2500E-02	6.2112E-02	6.1728E-02	6.1350E-02	6.0976E-02	6.0606E-02	6.0241E-02	5.9880E-02	5.9524E-02	5.9172E-02
17.0	5.8824E-02	5.8430E-02	5.8140E-02	5.7803E-02	5.7471E-02	5.7143E-02	5.6818E-02	5.6497E-02	5.6180E-02	5.5866E-02
18.0	5.5556E-02	5.5249E-02	5.4945E-02	5.4645E-02	5.4348E-02	5.4054E-02	5.3763E-02	5.3476E-02	5.3191E-02	5.2910E-02
19.0	5.2632E-02	5.2356E-02	5.2083E-02	5.1813E-02	5.1546E-02	5.1282E-02	5.1020E-02	5.0761E-02	5.0505E-02	5.0251E-02
20.0	5.0000E-02	4.9751E-02	4.9505E-02	4.9261E-02	4.9020E-02	4.8780E-02	4.8544E-02	4.8309E-02	4.8077E-02	4.7847E-02

Table A2.4 Continuous compound interest: Value of

$$F_{SDP,\bar{i},n} = \frac{2}{in}\left(1 - \frac{1 - e^{-in}}{in}\right)$$

in↓ / in→	0.00	0.01	0.02	0.03	0.04	0.05	0.06	0.07	0.08	0.09
0	1.0000E 00	9.9668E-01	9.9337E-01	9.9007E-01	9.8680E-01	9.8354E-01	9.8030E-01	9.7707E-01	9.7386E-01	9.7066E-01
0.1	9.6748E-01	9.6432E-01	9.6117E-01	9.5804E-01	9.5492E-01	9.5182E-01	9.4873E-01	9.4566E-01	9.4261E-01	9.3956E-01
0.2	9.3654E-01	9.3353E-01	9.3053E-01	9.2755E-01	9.2458E-01	9.2163E-01	9.1869E-01	9.1576E-01	9.1285E-01	9.0995E-01
0.3	9.0707E-01	9.0420E-01	9.0135E-01	8.9851E-01	8.9568E-01	8.9287E-01	8.9007E-01	8.8728E-01	8.8451E-01	8.8175E-01
0.4	8.7900E-01	8.7627E-01	8.7355E-01	8.7084E-01	8.6814E-01	8.6546E-01	8.6279E-01	8.6014E-01	8.5749E-01	8.5486E-01
0.5	8.5225E-01	8.4964E-01	8.4705E-01	8.4446E-01	8.4189E-01	8.3934E-01	8.3679E-01	8.3426E-01	8.3174E-01	8.2923E-01
0.6	8.2673E-01	8.2425E-01	8.2177E-01	8.1931E-01	8.1686E-01	8.1442E-01	8.1199E-01	8.0957E-01	8.0717E-01	8.0477E-01
0.7	8.0239E-01	8.0002E-01	7.9766E-01	7.9530E-01	7.9297E-01	7.9064E-01	7.8832E-01	7.8601E-01	7.8371E-01	7.8143E-01
0.8	7.7915E-01	7.7689E-01	7.7463E-01	7.7239E-01	7.7015E-01	7.6793E-01	7.6572E-01	7.6351E-01	7.6132E-01	7.5914E-01
0.9	7.5696E-01	7.5480E-01	7.5264E-01	7.5050E-01	7.4837E-01	7.4624E-01	7.4413E-01	7.4202E-01	7.3992E-01	7.3784E-01
1.0	7.3576E-01	7.3369E-01	7.3163E-01	7.2958E-01	7.2754E-01	7.2551E-01	7.2349E-01	7.2148E-01	7.1947E-01	7.1748E-01
1.1	7.1549E-01	7.1351E-01	7.1154E-01	7.0958E-01	7.0763E-01	7.0569E-01	7.0375E-01	7.0183E-01	6.9991E-01	6.9800E-01
1.2	6.9610E-01	6.9421E-01	6.9233E-01	6.9045E-01	6.8859E-01	6.8673E-01	6.8488E-01	6.8303E-01	6.8120E-01	6.7937E-01
1.3	6.7755E-01	6.7574E-01	6.7394E-01	6.7214E-01	6.7036E-01	6.6858E-01	6.6680E-01	6.6504E-01	6.6328E-01	6.6153E-01
1.4	6.5979E-01	6.5806E-01	6.5633E-01	6.5461E-01	6.5290E-01	6.5120E-01	6.4950E-01	6.4781E-01	6.4613E-01	6.4445E-01
1.5	6.4278E-01	6.4112E-01	6.3947E-01	6.3782E-01	6.3618E-01	6.3455E-01	6.3292E-01	6.3130E-01	6.2969E-01	6.2808E-01
1.6	6.2648E-01	6.2489E-01	6.2330E-01	6.2172E-01	6.2015E-01	6.1859E-01	6.1703E-01	6.1547E-01	6.1393E-01	6.1239E-01
1.7	6.1085E-01	6.0933E-01	6.0781E-01	6.0629E-01	6.0478E-01	6.0328E-01	6.0179E-01	6.0030E-01	5.9881E-01	5.9733E-01
1.8	5.9586E-01	5.9440E-01	5.9294E-01	5.9149E-01	5.9004E-01	5.8860E-01	5.8716E-01	5.8573E-01	5.8431E-01	5.8289E-01
1.9	5.8148E-01	5.8007E-01	5.7867E-01	5.7728E-01	5.7589E-01	5.7450E-01	5.7312E-01	5.7175E-01	5.7039E-01	5.6902E-01
2.0	5.6767E-01	5.6632E-01	5.6497E-01	5.6363E-01	5.6230E-01	5.6097E-01	5.5964E-01	5.5833E-01	5.5701E-01	5.5570E-01

$$F_{SDP,\bar{i},n} = \frac{2}{in}\left(1 - \frac{1 - e^{-in}}{in}\right)$$

$\stackrel{in \to}{\downarrow}$	0.00	0.01	0.02	0.03	0.04	0.05	0.06	0.07	0.08	0.09
2.1	5.5440E-01	5.5310E-01	5.5181E-01	5.5052E-01	5.4924E-01	5.4797E-01	5.4669E-01	5.4543E-01	5.4416E-01	5.4291E-01
2.2	5.4165E-01	5.4041E-01	5.3916E-01	5.3793E-01	5.3669E-01	5.3547E-01	5.3424E-01	5.3302E-01	5.3181E-01	5.3060E-01
2.3	5.2940E-01	5.2820E-01	5.2700E-01	5.2581E-01	5.2463E-01	5.2345E-01	5.2227E-01	5.2110E-01	5.1993E-01	5.1877E-01
2.4	5.1761E-01	5.1646E-01	5.1531E-01	5.1416E-01	5.1302E-01	5.1188E-01	5.1075E-01	5.0962E-01	5.0850E-01	5.0738E-01
2.5	5.0627E-01	5.0516E-01	5.0405E-01	5.0295E-01	5.0185E-01	5.0076E-01	4.9967E-01	4.9858E-01	4.9750E-01	4.9642E-01
2.6	4.9535E-01	4.9428E-01	4.9321E-01	4.9215E-01	4.9109E-01	4.9004E-01	4.8899E-01	4.8794E-01	4.8690E-01	4.8586E-01
2.7	4.8483E-01	4.8380E-01	4.8277E-01	4.8175E-01	4.8073E-01	4.7972E-01	4.7871E-01	4.7770E-01	4.7669E-01	4.7569E-01
2.8	4.7470E-01	4.7370E-01	4.7271E-01	4.7173E-01	4.7075E-01	4.6977E-01	4.6879E-01	4.6782E-01	4.6685E-01	4.6589E-01
2.9	4.6493E-01	4.6397E-01	4.6302E-01	4.6207E-01	4.6112E-01	4.6018E-01	4.5924E-01	4.5830E-01	4.5737E-01	4.5644E-01
3.0	4.5551E-01	4.5458E-01	4.5366E-01	4.5275E-01	4.5183E-01	4.5092E-01	4.5002E-01	4.4911E-01	4.4821E-01	4.4731E-01
3.1	4.4642E-01	4.4553E-01	4.4464E-01	4.4376E-01	4.4287E-01	4.4200E-01	4.4112E-01	4.4025E-01	4.3938E-01	4.3851E-01
3.2	4.3765E-01	4.3679E-01	4.3593E-01	4.3508E-01	4.3423E-01	4.3338E-01	4.3253E-01	4.3169E-01	4.3085E-01	4.3001E-01
3.3	4.2918E-01	4.2835E-01	4.2752E-01	4.2670E-01	4.2587E-01	4.2505E-01	4.2424E-01	4.2342E-01	4.2261E-01	4.2180E-01
3.4	4.2100E-01	4.2020E-01	4.1940E-01	4.1860E-01	4.1780E-01	4.1701E-01	4.1622E-01	4.1544E-01	4.1465E-01	4.1387E-01
3.5	4.1309E-01	4.1232E-01	4.1154E-01	4.1077E-01	4.1001E-01	4.0924E-01	4.0848E-01	4.0772E-01	4.0696E-01	4.0620E-01
3.6	4.0545E-01	4.0470E-01	4.0395E-01	4.0321E-01	4.0247E-01	4.0173E-01	4.0099E-01	4.0025E-01	3.9952E-01	3.9879E-01
3.7	3.9806E-01	3.9733E-01	3.9661E-01	3.9589E-01	3.9517E-01	3.9446E-01	3.9374E-01	3.9303E-01	3.9232E-01	3.9161E-01
3.8	3.9091E-01	3.9021E-01	3.8951E-01	3.8881E-01	3.8811E-01	3.8742E-01	3.8673E-01	3.8604E-01	3.8536E-01	3.8467E-01
3.9	3.8399E-01	3.8331E-01	3.8263E-01	3.8196E-01	3.8128E-01	3.8061E-01	3.7994E-01	3.7928E-01	3.7861E-01	3.7795E-01
4.0	3.7729E-01	3.7663E-01	3.7597E-01	3.7532E-01	3.7467E-01	3.7402E-01	3.7337E-01	3.7273E-01	3.7208E-01	3.7144E-01
4.1	3.7080E-01	3.7016E-01	3.6953E-01	3.6889E-01	3.6826E-01	3.6763E-01	3.6700E-01	3.6638E-01	3.6575E-01	3.6513E-01

Table A2.4 (*Continued*)

$$F_{SDP,\bar{i},n} = \frac{2}{in}\left(1 - \frac{1 - e^{-in}}{in}\right)$$

in → / ↓	0.00	0.01	0.02	0.03	0.04	0.05	0.06	0.07	0.08	0.09
4.2	3.6451E-01	3.6389E-01	3.6328E-01	3.6266E-01	3.6205E-01	3.6144E-01	3.6083E-01	3.6023E-01	3.5962E-01	3.5902E-01
4.3	3.5842E-01	3.5782E-01	3.5722E-01	3.5663E-01	3.5603E-01	3.5544E-01	3.5485E-01	3.5426E-01	3.5368E-01	3.5309E-01
4.4	3.5251E-01	3.5193E-01	3.5135E-01	3.5077E-01	3.5019E-01	3.4962E-01	3.4905E-01	3.4848E-01	3.4791E-01	3.4734E-01
4.5	3.4678E-01	3.4621E-01	3.4565E-01	3.4509E-01	3.4453E-01	3.4397E-01	3.4342E-01	3.4287E-01	3.4231E-01	3.4176E-01
4.6	3.4121E-01	3.4067E-01	3.4012E-01	3.3958E-01	3.3904E-01	3.3850E-01	3.3796E-01	3.3742E-01	3.3688E-01	3.3635E-01
4.7	3.3582E-01	3.3529E-01	3.3476E-01	3.3423E-01	3.3370E-01	3.3318E-01	3.3265E-01	3.3213E-01	3.3161E-01	3.3109E-01
4.8	3.3058E-01	3.3006E-01	3.2955E-01	3.2903E-01	3.2852E-01	3.2801E-01	3.2750E-01	3.2700E-01	3.2649E-01	3.2599E-01
4.9	3.2548E-01	3.2498E-01	3.2448E-01	3.2399E-01	3.2349E-01	3.2299E-01	3.2250E-01	3.2201E-01	3.2152E-01	3.2103E-01
	.0	.1	.2	.3	.4	.5	.6	.7	.8	.9
5.0	3.2054E-01	3.1573E-01	3.1106E-01	3.0651E-01	3.0209E-01	2.9779E-01	2.9360E-01	2.8953E-01	2.8555E-01	2.8169E-01
6.0	2.7792E-01	2.7424E-01	2.7066E-01	2.6716E-01	2.6375E-01	2.6043E-01	2.5718E-01	2.5401E-01	2.5091E-01	2.4789E-01
7.0	2.4494E-01	2.4205E-01	2.3923E-01	2.3647E-01	2.3377E-01	2.3113E-01	2.2855E-01	2.2602E-01	2.2355E-01	2.2113E-01
8.0	2.1876E-01	2.1644E-01	2.1417E-01	2.1194E-01	2.0976E-01	2.0762E-01	2.0552E-01	2.0347E-01	2.0145E-01	1.9947E-01
9.0	1.9753E-01	1.9563E-01	1.9376E-01	1.9193E-01	1.9013E-01	1.8837E-01	1.8663E-01	1.8493E-01	1.8326E-01	1.8162E-01
10.0	1.8000E-01	1.7841E-01	1.7686E-01	1.7532E-01	1.7382E-01	1.7234E-01	1.7088E-01	1.6945E-01	1.6804E-01	1.6665E-01
11.0	1.6529E-01	1.6395E-01	1.6263E-01	1.6133E-01	1.6005E-01	1.5879E-01	1.5755E-01	1.5633E-01	1.5513E-01	1.5394E-01
12.0	1.5278E-01	1.5163E-01	1.5050E-01	1.4938E-01	1.4828E-01	1.4720E-01	1.4613E-01	1.4508E-01	1.4404E-01	1.4302E-01
13.0	1.4201E-01	1.4102E-01	1.4004E-01	1.3907E-01	1.3812E-01	1.3717E-01	1.3625E-01	1.3533E-01	1.3443E-01	1.3353E-01
14.0	1.3265E-01	1.3178E-01	1.3093E-01	1.3008E-01	1.2924E-01	1.2842E-01	1.2760E-01	1.2680E-01	1.2600E-01	1.2522E-01
15.0	1.2444E-01	1.2368E-01	1.2292E-01	1.2218E-01	1.2144E-01	1.2071E-01	1.1999E-01	1.1927E-01	1.1857E-01	1.1788E-01
16.0	1.1719E-01	1.1651E-01	1.1584E-01	1.1517E-01	1.1452E-01	1.1387E-01	1.1322E-01	1.1259E-01	1.1196E-01	1.1134E-01
17.0	1.1073E-01	1.1012E-01	1.0952E-01	1.0892E-01	1.0834E-01	1.0776E-01	1.0718E-01	1.0661E-01	1.0605E-01	1.0549E-01
18.0	1.0494E-01	1.0439E-01	1.0385E-01	1.0332E-01	1.0279E-01	1.0226E-01	1.0175E-01	1.0123E-01	1.0072E-01	1.0022E-01
19.0	9.9723E-02	9.9230E-02	9.8741E-02	9.8258E-02	9.7779E-02	9.7304E-02	9.6835E-02	9.6369E-02	9.5909E-02	9.5452E-02
20.0	9.5000E-02	9.4552E-02	9.4108E-02	9.3669E-02	9.3233E-02	9.2802E-02	9.2374E-02	9.1951E-02	9.1531E-02	9.1115E-02

THREE

PROBABILITY

Table A3.1 Cumulative probability for Z or less for a standard normal distribution

Z	0	1	2	3	4	5	6	7	8	9
−3.9	.0000	.0000	.0000	.0000	.0000	.0000	.0000	.0000	.0000	.0000
−3.8	.0001	.0001	.0001	.0001	.0001	.0001	.0001	.0001	.0001	.0001
−3.7	.0001	.0001	.0001	.0001	.0001	.0001	.0001	.0001	.0001	.0001
−3.6	.0002	.0002	.0001	.0001	.0001	.0001	.0001	.0001	.0001	.0001
−3.5	.0002	.0002	.0002	.0002	.0002	.0002	.0002	.0002	.0002	.0002
−3.4	.0003	.0003	.0003	.0003	.0003	.0003	.0003	.0003	.0003	.0002
−3.3	.0005	.0005	.0005	.0004	.0004	.0004	.0004	.0004	.0004	.0003
−3.2	.0007	.0007	.0006	.0006	.0006	.0006	.0006	.0005	.0005	.0005
−3.1	.0010	.0009	.0009	.0009	.0008	.0008	.0008	.0008	.0007	.0007
−3.0	.0013	.0013	.0013	.0012	.0012	.0011	.0011	.0011	.0010	.0010
−2.9	.0019	.0018	.0017	.0017	.0016	.0016	.0015	.0015	.0014	.0014
−2.8	.0026	.0025	.0024	.0023	.0023	.0022	.0021	.0021	.0020	.0019
−2.7	.0035	.0034	.0033	.0032	.0031	.0030	.0029	.0028	.0027	.0026
−2.6	.0047	.0045	.0044	.0043	.0041	.0040	.0039	.0038	.0037	.0036
−2.5	.0062	.0060	.0059	.0057	.0055	.0054	.0052	.0051	.0049	.0048
−2.4	.0082	.0080	.0078	.0075	.0073	.0071	.0069	.0068	.0066	.0064
−2.3	.0107	.0104	.0102	.0099	.0096	.0094	.0091	.0089	.0087	.0084
−2.2	.0139	.0136	.0132	.0129	.0126	.0122	.0119	.0116	.0113	.0110
−2.1	.0179	.0174	.0170	.0166	.0162	.0158	.0154	.0150	.0146	.0143
−2.0	.0228	.0222	.0217	.0212	.0207	.0202	.0197	.0192	.0188	.0183

Z	0	1	2	3	4	5	6	7	8	9
−1.9	.0287	.0281	.0274	.0268	.0262	.0256	.0250	.0244	.0238	.0233
−1.8	.0359	.0352	.0344	.0336	.0329	.0322	.0314	.0307	.0300	.0294
−1.7	.0446	.0436	.0427	.0418	.0409	.0401	.0392	.0384	.0375	.0367
−1.6	.0548	.0537	.0526	.0516	.0505	.0495	.0485	.0475	.0465	.0455
−1.5	.0668	.0655	.0643	.0630	.0618	.0606	.0594	.0582	.0570	.0559
−1.4	.0808	.0793	.0778	.0764	.0749	.0735	.0722	.0708	.0694	.0681
−1.3	.0968	.0951	.0934	.0918	.0901	.0885	.0869	.0853	.0838	.0823
−1.2	.1151	.1131	.1112	.1093	.1075	.1056	.1038	.1020	.1003	.0985
−1.1	.1357	.1335	.1314	.1292	.1271	.1251	.1230	.1210	.1190	.1170
−1.0	.1587	.1562	.1539	.1515	.1492	.1469	.1446	.1423	.1401	.1379
−0.9	.1841	.1814	.1788	.1762	.1736	.1711	.1685	.1660	.1635	.1611
−0.8	.2119	.2090	.2061	.2033	.2005	.1977	.1949	.1922	.1894	.1867
−0.7	.2420	.2389	.2358	.2327	.2297	.2266	.2236	.2206	.2177	.2148
−0.6	.2743	.2709	.2676	.2643	.2611	.2578	.2546	.2514	.2483	.2451
−0.5	.3085	.3050	.3015	.2981	.2946	.2912	.2877	.2843	.2810	.2776
−0.4	.3446	.3409	.3372	.3336	.3300	.3264	.3228	.3192	.3156	.3121
−0.3	.3821	.3783	.3745	.3707	.3669	.3632	.3594	.3557	.3520	.3483
−0.2	.4207	.4168	.4129	.4090	.4052	.4013	.3974	.3936	.3897	.3859
−0.1	.4602	.4562	.4522	.4483	.4443	.4404	.4364	.4325	.4286	.4247
−0.0	.5000	.4960	.4920	.4880	.4840	.4801	.4761	.4721	.4681	.4641

Table A3.1 (*Continued*)

Z	0	1	2	3	4	5	6	7	8	9
0.0	.5000	.5040	.5080	.5120	.5160	.5199	.5239	.5279	.5319	.5359
0.1	.5398	.5438	.5478	.5517	.5557	.5596	.5636	.5675	.5714	.5753
0.2	.5793	.5832	.5871	.5910	.5948	.5987	.6026	.6064	.6103	.6141
0.3	.6179	.6217	.6255	.6293	.6331	.6368	.6406	.6443	.6480	.6517
0.4	.6554	.6591	.6628	.6664	.6700	.6736	.6772	.6808	.6844	.6879
0.5	.6915	.6950	.6985	.7019	.7054	.7088	.7123	.7157	.7190	.7224
0.6	.7257	.7291	.7324	.7357	.7389	.7422	.7454	.7486	.7517	.7549
0.7	.7580	.7611	.7642	.7673	.7703	.7734	.7764	.7794	.7823	.7852
0.8	.7881	.7910	.7939	.7967	.7995	.8023	.8051	.8078	.8106	.8133
0.9	.8159	.8186	.8212	.8238	.8264	.8289	.8315	.8340	.8365	.8389
1.0	.8413	.8438	.8461	.8485	.8508	.8531	.8554	.8577	.8599	.8621
1.1	.8643	.8665	.8686	.8708	.8729	.8749	.8770	.8790	.8810	.8830
1.2	.8849	.8869	.8888	.8907	.8925	.8944	.8962	.8980	.8997	.9015
1.3	.9032	.9049	.9066	.9082	.9099	.9115	.9131	.9147	.9162	.9177
1.4	.9192	.9207	.9222	.9236	.9251	.9265	.9278	.9292	.9306	.9319
1.5	.9332	.9345	.9357	.9370	.9382	.9394	.9406	.9418	.9430	.9441
1.6	.9452	.9463	.9474	.9484	.9495	.9505	.9515	.9525	.9535	.9545
1.7	.9554	.9564	.9573	.9582	.9591	.9599	.9608	.9616	.9625	.9633
1.8	.9641	.9648	.9656	.9664	.9671	.9678	.9686	.9693	.9700	.9706
1.9	.9713	.9719	.9726	.9732	.9738	.9744	.9750	.9756	.9762	.9767

Z	0	1	2	3	4	5	6	7	8	9
2.0	.9772	.9778	.9783	.9788	.9793	.9798	.9803	.9808	.9812	.9817
2.1	.9821	.9826	.9830	.9834	.9838	.9842	.9846	.9850	.9854	.9857
2.2	.9861	.9864	.9868	.9871	.9874	.9878	.9881	.9884	.9887	.9890
2.3	.9893	.9896	.9898	.9901	.9904	.9906	.9909	.9911	.9913	.9916
2.4	.9918	.9920	.9922	.9925	.9927	.9929	.9931	.9932	.9934	.9936
2.5	.9938	.9940	.9941	.9943	.9945	.9946	.9948	.9949	.9951	.9952
2.6	.9953	.9955	.9956	.9957	.9959	.9960	.9961	.9962	.9963	.9964
2.7	.9965	.9966	.9967	.9968	.9969	.9970	.9971	.9972	.9973	.9974
2.8	.9974	.9975	.9976	.9977	.9977	.9978	.9979	.9979	.9980	.9981
2.9	.9981	.9982	.9982	.9983	.9984	.9984	.9985	.9985	.9986	.9986
3.0	.9987	.9987	.9987	.9988	.9988	.9989	.9989	.9989	.9990	.9990
3.1	.9990	.9991	.9991	.9991	.9992	.9992	.9992	.9992	.9993	.9993
3.2	.9993	.9993	.9994	.9994	.9994	.9994	.9994	.9995	.9995	.9995
3.3	.9995	.9995	.9995	.9996	.9996	.9996	.9996	.9996	.9996	.9997
3.4	.9997	.9997	.9997	.9997	.9997	.9997	.9997	.9997	.9997	.9998
3.5	.9998	.9998	.9998	.9998	.9998	.9998	.9998	.9998	.9998	.9998
3.6	.9998	.9998	.9999	.9999	.9999	.9999	.9999	.9999	.9999	.9999
3.7	.9999	.9999	.9999	.9999	.9999	.9999	.9999	.9999	.9999	.9999
3.8	.9999	.9999	.9999	.9999	.9999	.9999	.9999	.9999	.9999	.9999
3.9	1.000	1.000	1.000	1.000	1.000	1.000	1.000	1.000	1.000	1.000

Table A3.2 Random numbers: The first 1000 decimal places of π

14159	26535	89793	23846	26433	83279	50288	41971	69399	37510
58209	74944	59230	78164	06286	20899	86280	34825	34211	70679
82148	08651	32823	06647	09384	46095	50582	23172	53594	08128
48111	74502	84102	70193	85211	05559	64462	29489	54930	38196
44288	10975	66593	34461	28475	64823	37867	83165	27120	19091
45648	56692	34603	48610	45432	66482	13393	60726	02491	41273
72458	70066	06315	58817	48815	20920	96282	92540	91715	36436
78925	90360	01133	05305	48820	46652	13841	46951	94151	16094
33057	27036	57595	91953	09218	61173	81932	61179	31051	18548
07446	23799	62749	56735	18857	52724	89122	79381	83011	94912
98336	73362	44065	66430	86021	39494	63952	24737	19070	21798
60943	70277	05392	17176	29317	67523	84674	81846	76694	05132
00056	81271	45263	56082	77857	71342	75778	96091	73637	17872
14684	40901	22495	34301	46549	58537	10507	92279	68925	89235
42019	95611	21290	21960	86403	44181	59813	62977	47713	09960
51870	72113	49999	99837	29780	49951	05973	17328	16096	31859
50244	59455	34690	83026	42522	30825	33446	85035	26193	11881
71010	00313	78387	52886	58753	32083	81420	61717	76691	47303
59825	34904	28755	46873	11595	62863	88235	37875	93751	95778
18577	80532	17122	68066	13001	92787	66111	95909	21642	01989

FOUR

LEARNING CURVE

Table A4.1 Values for the exponential learning-curve function

n	90% learning ratio			80% learning ratio			70% learning ratio		
	$E(N)$	$\Sigma E(N)$	$\Sigma E(N)/N$	$E(N)$	$\Sigma E(N)$	$\Sigma E(N)/N$	$E(N)$	$\Sigma E(N)$	$\Sigma E(N)/N$
1	1.0000E 00	1.0000E 00	1.0000E 00	1.0000E 00	1.0000E 00	1.0000E 00	1.0000E 00	1.0000E 00	1.0000E 00
2	9.0000E-01	1.9000E 00	9.5000E-01	8.0000E-01	1.8000E 00	9.0000E-01	7.0000E-01	1.7000E 00	8.5000E-01
3	8.4621E-01	2.7462E 00	9.1540E-01	7.0210E-01	2.5021E 00	8.3403E-01	5.6818E-01	2.2682E 00	7.5606E-01
4	8.1000E-01	3.5562E 00	8.8905E-01	6.4000E-01	3.1421E 00	7.8553E-01	4.9000E-01	2.7582E 00	6.8955E-01
5	7.8299E-01	4.3392E 00	8.6784E-01	5.9564E-01	3.7377E 00	7.4755E-01	4.3685E-01	3.1950E 00	6.3901E-01
6	7.6159E-01	5.1008E 00	8.5013E-01	5.6168E-01	4.2994E 00	7.1657E-01	3.9773E-01	3.5928E 00	5.9879E-01
7	7.4395E-01	5.8447E 00	8.3496E-01	5.3449E-01	4.8339E 00	6.9056E-01	3.6740E-01	3.9601E 00	5.6574E-01
8	7.2900E-01	6.5737E 00	8.2172E-01	5.1200E-01	5.3459E 00	6.6824E-01	3.4300E-01	4.3031E 00	5.3789E-01
9	7.1600E-01	7.2898E 00	8.0998E-01	4.9295E-01	5.8389E 00	6.4876E-01	3.2283E-01	4.6260E 00	5.1400E-01
10	7.0469E-01	7.9945E 00	7.9945E-01	4.7651E-01	6.3154E 00	6.3154E-01	3.0579E-01	4.9318E 00	4.9318E-01
11	6.9455E-01	8.6890E 00	7.8991E-01	4.6211E-01	6.7775E 00	6.1613E-01	2.9116E-01	5.2229E 00	4.7481E-01
12	6.8543E-01	9.3745E 00	7.8120E-01	4.4935E-01	7.2268E 00	6.0224E-01	2.7841E-01	5.5013E 00	4.5844E-01
13	6.7714E-01	1.0052E 01	7.7320E-01	4.3792E-01	7.6647E 00	5.8960E-01	2.6717E-01	5.7685E 00	4.4373E-01
14	6.6955E-01	1.0721E 01	7.6580E-01	4.2759E-01	8.0923E 00	5.7802E-01	2.5718E-01	6.0257E 00	4.3041E-01
15	6.6257E-01	1.1384E 01	7.5891E-01	4.1820E-01	8.5105E 00	5.6737E-01	2.4821E-01	6.2739E 00	4.1826E-01
16	6.5610E-01	1.2040E 01	7.5249E-01	4.0960E-01	8.9201E 00	5.5751E-01	2.4010E-01	6.5149E 00	4.0712E-01
17	6.5008E-01	1.2690E 01	7.4646E-01	4.0168E-01	9.3218E 00	5.4834E-01	2.3273E-01	6.7467E 00	3.9687E-01
18	6.4446E-01	1.3334E 01	7.4080E-01	3.9436E-01	9.7162E 00	5.3979E-01	2.2598E-01	6.9727E 00	3.8737E-01
19	6.3918E-01	1.3974E 01	7.3545E-01	3.8755E-01	1.0104E 01	5.3178E-01	2.1978E-01	7.1925E 00	3.7855E-01
20	6.3422E-01	1.4608E 01	7.3039E-01	3.8121E-01	1.0485E 01	5.2425E-01	2.1405E-01	7.4065E 00	3.7033E-01
25	6.1307E-01	1.7713E 01	7.0853E-01	3.5478E-01	1.2309E 01	4.9234E-01	1.9083E-01	8.4040E 00	3.3616E-01
30	5.9631E-01	2.0727E 01	6.9090E-01	3.3556E-01	1.4020E 01	4.6733E-01	1.7375E-01	9.3050E 00	3.1017E-01
35	5.8250E-01	2.3666E 01	6.7617E-01	3.1836E-01	1.5643E 01	4.4694E-01	1.6050E-01	1.0133E 01	2.8951E-01
40	5.7080E-01	2.6543E 01	6.6357E-01	3.0497E-01	1.7193E 01	4.2984E-01	1.4984E-01	1.0902E 01	2.7256E-01

n	90% learning ratio			80% learning ratio			70% learning ratio		
	$E(N)$	$\Sigma E(N)$	$\Sigma E(N)/N$	$E(N)$	$\Sigma E(N)$	$\Sigma E(N)/N$	$E(N)$	$\Sigma E(N)$	$\Sigma E(N)/N$
45	5.6067E-01	2.9366E 01	6.5257E-01	2.9362E-01	1.8684E 01	4.1519E-01	1.4103E-01	1.1625E 01	2.5832E-01
50	5.5176E-01	3.2142E 01	6.4284E-01	2.8383E-01	2.0122E 01	4.0243E-01	1.3358E-01	1.2307E 01	2.4614E-01
60	5.3668E-01	3.7574E 01	6.2623E-01	2.6765E-01	2.2868E 01	3.8113E-01	1.2162E-01	1.3574E 01	2.2624E-01
70	5.2425E-01	4.2871E 01	6.1244E-01	2.5469E-01	2.5471E 01	3.6387E-01	1.1235E-01	1.4738E 01	2.1054E-01
80	5.1372E-01	4.8054E 01	6.0067E-01	2.4397E-01	2.7957E 01	3.4947E-01	1.0489E-01	1.5819E 01	1.9774E-01
90	5.0460E-01	5.3140E 01	5.9044E-01	2.3490E-01	3.0346E 01	3.3718E-01	9.8719E-02	1.6833E 01	1.8703E-01
100	4.9659E-01	5.8141E 01	5.8141E-01	2.2706E-01	3.2651E 01	3.2651E-01	9.3509E-02	1.7791E 01	1.7791E-01
150	4.6690E-01	8.2156E 01	5.4771E-01	1.9928E-01	4.3234E 01	2.8822E-01	7.5900E-02	2.1972E 01	1.4648E-01
200	4.4693E-01	1.0496E 02	5.2482E-01	1.8165E-01	5.2720E 01	2.6360E-01	6.5456E-02	2.5482E 01	1.2741E-01
300	4.2021E-01	1.4820E 02	4.9401E-01	1.5942E-01	6.9663E 01	2.3221E-01	5.3130E-02	3.1342E 01	1.0447E-01
400	4.0223E-01	1.8927E 02	4.7317E-01	1.4532E-01	8.4849E 01	2.1212E-01	4.5819E-02	3.6260E 01	9.0649E-02
500	3.8882E-01	2.2879E 02	4.5757E-01	1.3525E-01	9.8847E 01	1.9769E-01	4.0849E-02	4.0577E 01	8.1153E-02
600	3.7819E-01	2.6711E 02	4.4519E-01	1.2754E-01	1.1197E 02	1.8661E-01	3.7191E-02	4.4468E 01	7.4114E-02
800	3.6201E-01	3.4103E 02	4.2629E-01	1.1626E-01	1.3627E 02	1.7034E-01	3.2074E-02	5.1355E 01	6.4194E-02
1000	3.4994E-01	4.1217E 02	4.1217E-01	1.0820E-01	1.5867E 02	1.5867E-01	2.8594E-02	5.7401E 01	5.7401E-02

FIVE

ANSWERS TO PROBLEMS

2.1 $500 interest payment first month; $217.35 principal payment first month; $497.83 interest payment second month; $219.52 principal payment second month

2.2 $952.38

2.3 $23.81

2.4 $55

2.5 $1006.29

2.6 $1494.47

2.7 (a) 8.2432 percent; (b) 8.300 percent; (c) 8.3278 percent; (d) 8.3287 percent; (e) 8.3287 percent

2.8 $22,104

2.9 (a) $20,407; (b) $20,592

2.10 (a) $1382.40; (b) $1449.50

2.11 29,180 M$

2.12 (a) 8 percent; (b) 7.08 percent

2.13 $16,531 net receipt

2.14 $32,214

2.15 $5334.90

2.16 $7014.20

2.17 $5868.40

2.18 (a) $1610.40; (b) $1519.20

2.19 $5995.50

2.20 $13,276

2.21 9.4 percent per year

2.22 8.8 years

2.23 Choose $4000 cash now plus $7000 in 4 years. The present-value advantage is $1219

2.24 2.6 percent per year

2.25 9.38 percent per year

2.26 12.63 years

2.27 79.7 h

2.28 (a) $128,780; (b) 11.9 years from a selling price of $50,000

2.29 $59,381

2.30 $38,231

2.31 9.18 percent per year

2.32 $25,667

2.33 $28,234

2.34 $11,462

2.35 $469.32

2.36 $1845.60

2.37 It can be reduced to F_{RP}

2.38 $9298.30

2.39 $6540.20

2.40 7.64 years. Could be seven payments of $1000 plus one last payment of $363

2.41 0.875 percent monthly rate; 10.5 percent nominal rate; 11.02 percent effective rate

2.42 $20,275

2.43 $1250 at the end of the tenth year

2.44 14.1 percent per year

2.45 (a) $925.65; (b) $1081.63

2.46 4.4 percent per year nominal rate

2.47 8 percent per year

2.48 $860.99

2.49 5 years

2.50 (a) $838.35; (b) $7391.10; (c) $1419.60

2.51 $2266

2.52 73.2 percent per year

2.53 (a) 12.7 percent per year; (b) 23 percent per year

2.54 A royalty is more economical. The unacost is $80,000 against $85,396

2.55 12.5 years for machine B; machine C never pays off

2.56 (a) $P_{A,10} = \$57,695$ $P_{B,10} = \$53,029$
   (b) $P_{A,4} = \$28,479$ $P_{B,4} = \$26,176$
   (c) $R_A = \$8598$ $R_B = \$7903$
   (d) $K_A = \$107,480$ $K_B = \$98,790$
   (e) 4.6 years; (f) approximately 15.5 percent per year

2.59 Patch twice for 2 years' added life

2.60 The new catalyst must last 3 years

2.61 $125,330 as a first cost now for a 3-year life

2.62 Plan B is better. The present value for 5 years is 4.659 M$ against 2.839 M$

2.63 Machine B is more economical. The unacost is $6059 against $7825

2.64 4 years

2.65 $315,400

2.66 −1.5 percent per year

2.67 $4.56 per part
2.68 Invest 300 k$ and save 43.77 k$/year
2.69 $4652
2.70 $2141/year
2.71 4.27 M$
2.72 Use the machine for 3 years
2.73 (a) −3.4 percent per year; (b) 3935 k$

3.1 (a) 10 years; (b) 0.73728 M$
3.2

Year	Twice straight-line declining-balance depreciation	Units of production depreciation
1	$22,500	$2,500
2	11,250	10,000
3	5,625	20,000
4	625	7,500

3.3 Present values are $8223 for straight-line depreciation, $9413 for twice straight-line declining balance with switch, and $8993 for sum-of-the-years-digits depreciation
3.4 $F_{DB} = 0.1059$
3.5 $F_{DB} = 0.27522$, but the maximum permissible value is 0.25
3.6 Depreciation by years, in order, is $3333.30, $2667.70, $2000.00, $1333.30, and $666.70, which checks the ordinary calculation for sum-of-the-years-digits depreciation.
3.7

Year	Depreciation
1	$2500.0
2	2833.3
3	2166.7
4	1500.0
5	833.3
6	166.7

3.8 $15,000
3.9 Cash flow in dollars, by years, is:

	1	2	3	4
SL depreciation	25,565	25,565	25,565	25,565
SD depreciation	30,050	27,060	24,070	21,080

3.10 Additions to the reserve for future taxes, in dollars by years, is:

1	2	3	4	5
+613.33	+306.67	—0—	−306.67	−613.33

3.11 Decimal depreciation by years, is:

1	0.45455
2	0.29091
3	0.16364
4	0.07273
5	0.01818

Not permitted since it exceeds twice the straight-line declining-balance method

3.12 (*a*) 0.308 M$; (*b*) 0.445 M$

3.13 Cash flows in megadollars: 1.856 for percentage depletion against 1.580 for cost depletion

3.14 2.768 M$

3.15 The minimum shift is 140 k$, but it can be as high as 160 k$

3.16 $10.53/ton

3.17 9.5 percent annual rate compounded semiannually

3.18 $941.69

3.19 $921.42

3.20 $3771

3.21 $550.50/year before taxes

3.22 Must last about 20 years

3.23 3.6 percent per year after taxes

3.24 Machine A is more economical. The after-tax unacost is $4620.70 against $8056.10

3.25 (*a*) 7.4 percent; (*b*) 13.6 percent

3.26 Pay the royalty. The after-tax unacost is $43,200 against $54,000

3.27 12.5 years for machine B; machine C never pays off

3.28 Patch four times to add a 4-year life

3.29 The new catalyst should last 3.5 years

3.30 $111,950 is the first cost

3.31 Plan B is better. The present value after taxes for 5 years is 1.7177 M$ against 0.6811 M$

3.32 Machine B is more economical. The after-tax unacost is $3942 against $4834

3.33 The lowest after-tax unacost is 33.238 k$/year for 4 years

3.34 (*a*) 4.9 percent per year; (*b*) 4.4 percent per year

3.35 (*a*) $2901.10 after-tax loss; (*b*) $2930.80 after-tax loss

3.36 $6864.8/year before taxes

3.37 3.32 M$ before taxes

3.38 Minimum for 4 years

3.39 1781 k$/year

3.40 The advantage for an immediate write-off as a present value after taxes is $30,788 - 27,467 = 3321

4.1
$$P = \frac{2Q}{(in)^2}[1 - e^{-in}(in + 1)]$$

4.2
$$P = \frac{\bar{R}_2 - \bar{R}_1}{i^2 n}[1 - e^{-in}(in + 1)] + n\bar{R}_1\left[\frac{1 - e^{-in}}{in}\right]$$

4.3
$$P = Y\frac{1 - e^{-ikn}}{e^{in} - 1}$$

4.4 $59,292
4.5 i continuous = 13.2 percent per year
4.6 i continuous = 13.8 percent per year, which is better than the value for the preceding problem
4.7 $32.95/month
4.8 $4018 before taxes
4.9 Savings for machine A as a uniform flow after taxes:
7714.8 − 4376 = $3338.40/year or $6182/year before taxes
4.10 14.1 percent
4.12

	ψ	$1 - 0.46\,\psi$
Simulated depreciation	0.64278	0.70432
End-of-year depreciation	0.58417	0.73128

4.14 (*a*) $1100; (*b*) $1097
4.15 (*a*) $997,620; (*b*) $1,454,800; (*c*) $1,147,700

5.1 2.2 years
5.2 (*a*) 2.6 years; (*b*) 2.8 years
5.3 NPV = 0.587 M$; DCFRR = 36 percent
5.4

	A		B		C	
		Rank		Rank		Rank
Discounted cash flow	28%	1	26%	2	23%	3
Present worth at 10%	371	3	394	1	389	2

5.5 $4.15 per part
5.6 NPV A = −2855 k$; NPV B = −2742 k$. Choose plan B
5.7 P = $4055
5.8 Plan A NPV = −72.64 k$; plan B NPV = 613.60 k$. Build a full-size plant now.
5.9 20 percent
5.10 11.3 percent
5.11 6.9 percent
5.12 NPV = $111,000 at 0 percent, $24,000 at 4 percent, −$2000 at 6 percent. DCFRR = 5.8 percent
5.13 (*a*) 13.7 percent; (*b*) 4.4 years
5.14 42.7 percent
5.15 13.5 percent
5.16 NPV = 20 percent = +376 k$. Accept project B
5.17 24.7 percent
5.18 For keeping the old machine, the NPV = −$475. Install a new machine, but the decision is close
5.20 DCFRR = 6.7 percent and 49.1 percent
5.21 NPV = 0 only for DCFRR = ∞, meaningless
5.22 (*a*) B/C = 1.86; (*b*) B/C = 2.01; (*c*) 1934 k$ present value

5.23 1.385
5.24 Limestone \$70/kW sodium carbonate \$80/kW

6.1 \$50,225
6.2 \$231,430
6.3 4.8 years
6.4 \$2110.40/year before taxes at the end of year 1, increasing by a factor of 1.10 each year
6.13 8 percent per year
6.14 \$42,286 against \$50,225 given for Prob. 6.1
6.15 \$45,889
6.16 \$231,430
6.17 The productivity rate gain p may vary within a system

7.1 60,000 m^3/year
7.2 $NV = 5.31$ M\$ at $N = 2.5$. The incremental variable cost $= 1.875$ M\$ per unit ΔN at $N = 2.5$
7.3 \$1.98/t
7.4 55 percent
7.5 13.5, say, 14
7.6 (a) $N = 32.1$ percent; (b) $N = 26.5$ percent; (c) $N = 40.9$ percent; (d) \$33,325 extra profit
7.7 0.0429 fractional increase
7.8 $Z_g = \$133,400$; $Z_t = \$72,036$; capitalized value \$400,200. The depreciation charge affects the capitalized value
7.9 $Z_g = \$400,000$
7.10 (a) At 75 percent capacity, \$0.629/kg; at 100 percent capacity, \$0.532/kg; (b) \$0.24/day/Δkg; (c) \$19.4 $\times$ 10^{-6}/kg/Δkg
7.11 (a) \$0.3/kg; (b) \$0.2625/kg
7.12 (a) \$40,000; (b) $NS = \$538,460$; (c) $NS = \$661,540$
7.13 (a) Do not shut down; $Z_g = -\$180,000$ against $-\$225,000$; (b) \$77.50
7.14 1 − 5000: buy; 5001 − 25,000: make; more than 25,000: buy
7.15 The maximum profit occurs at a selling price of \$15
7.16 The maximum profit is \$3125 at a sales price of \$15, with \$3125 for advertising, and selling 1250 units
7.17 Break-even at 20 percent production; \$17,280/year after taxes at capacity production; \$25,380/year after taxes at 50 percent above capacity production
7.18 160 percent above rated production
7.19 (a) $N = 2037$ break-even point; $N = 12,681$ profit limit point; (b) maximum profit $Z_g = \$130,940$ at $N = 7663$; (c) the average profit is \$13.589/ton; the incremental profit is -13.1 Δ\$ per Δton
7.20 (a) $AN - (B + D)N^3 - CN^2 - C_F = 0$; (b) $A - 3(B + D)N^2 - 4CN^3 = 0$; (c) $-2(B + D)N - 3CN^2 + (C_F/N^2) = 0$; (d) $3CN^2 + 2DN - (C_F/N^2) = 0$; (e) $4CN^3 + 3DN^2$
7.21 (a) 101 or 9703 parts per period; (b) $N = 4902$ parts per period with $Z_g = \$235,100$ per period; (c) $N = 7071$ parts per period with a minimum average cost of \$2.828
7.22 (a) $100 - 0.02Q$; (b) 50; (c) $Q = 2500$ for maximum profit with $Z_g = \$32,500$
7.23 (a) Assign the next unit to A; Δ\$/Δ unit $= 2.70$ against 2.95; (b) 6417 units on

A and 5083 units on B; (c) 2.783 $\Delta\$/\Delta$ unit on each machine; (d) 3750 units on each machine

7.24 The optimum mix is $N_A = 500$ and $N_B = 900$, with a maximum profit of \$5150 for 8 h. The choice is to make a single A or 3B because A takes three times as long to make, and

$$\frac{\partial C_T}{\partial N_A} = 6 = 3\frac{\partial D_T}{\partial N_B} = 6$$

7.25 (a) \$50,000 per period for A, labor, and \$50,000 per period for B, materials; (b) same

8.1 Mean 2.187; variance 0.273
8.2 Mean 2.252; variance 0.3082
8.3 Mean 121.5; variance 928.3
8.4 (a) 0.23; (b) 34 percent
8.5 (a) 0.194; (b) 0.349; (c) 0.93; (d) 0.264
8.6 (a) 0.083; (b) 0.5; (c) 0.417
8.7 0.433
8.9 (a) 0.0724; (b) 0.2496; (c) 0.7374
8.10 (a) Mean 4, variance 16; (b) $P_c(X) = 1 - e^{-X/4}$
8.11

X	$P_d(X)$	X	$P_d(X)$
-5	0.0022	2	0.1760
-3	0.0270	3	0.1210
-1	0.1210	5	0.0270
0	0.1760	7	0.0022
1	0.1944		

8.12 20.2 percent
8.13 92 percent
8.14 (a) \$136,200; (b) \$151,200
8.15 14.4 percent
8.16 \$5661
8.17 \$87,380
8.18

Total min	Random numbers	Total min	Random numbers
52.5	01–03	127.5	48–67
67.5	04–08	142.5	68–86
82.5	09–17	157.5	87–94
97.5	18–31	172.5	95–00
112.5	32–47		

8.19

X	$P_c(X)$	X	$P_c(X)$
0	0.0228	2	0.9772
0.5	0.1587	2.25	0.9938
1	0.5000	2.5	0.9987
1.5	0.8413		

8.20

X	$P_c(X)$	X	$P_c(X)$
0	0	2.5	0.71
0.1	0.05	3.0	0.77
0.4	0.18	4.0	0.86
1.0	0.39	5.0	0.92
1.5	0.52	7.0	0.97
2.0	0.63		

8.21 7.0

8.22 0.40

8.23 0.40

8.24 The expected cost is 18.48 M$. The probability of a cost between 18.48 M$ and 20 percent over that is 9.1 percent

8.25 (a) 25.24 k$; (b) 1 percent error in B leads to a 25.2 percent change in the NPV, 1 percent error in C leads to an 11.1 percent change in the NPV, 1 percent error in D leads to a 4.3 percent change in the NPV; (c) 14 percent

8.26 3.62 M$ approximately

8.27

X	$P(X)$ Poisson	$P(X)$ Binomial
0	0.3678	0.3672
1	0.3678	0.3686
2	0.1839	0.1843
3	0.0613	0.0612
4	0.0153	0.0152
5	0.0031	0.0030
6	0.00051	0.00049

8.28 $Y = 0.5X + 4.0$

8.29 3.62 M$

9.1 R existent $= \$11,400$; R displacer $= \$9623$

9.2 20 percent per year approximately

9.3 R existent $= \$6800$; R displacer $= \$5868$; after taxes

9.4 R existent $= \$928.80$; R displacer $= \$2756$; after taxes

9.5 R existent for 1 year $= \$1922$; R displacer $= \$2756$; after taxes

9.6 5 years

9.7 $R = \$2204/\text{year}$; $K = \$18,368$

9.8 One decreases and one increases by a fixed amount each year

9.9 13 years

9.10 Minimum at 11 years with $K/C_i = 1.3826$

9.11 284 bulbs per week

9.12 The minimum cost is $217 per period by displacing the group every third period

9.13 132.1 h

9.14 93.5 h

9.15 (a) 149.2 h; (b) 100.4 h

9.16 (a) 165.1 h; (b) 180.7 h

9.17

Lot	Rule of thumb	Geometric mean	True
1	4.67	6.05	5.53
2	27	24.55	26.19
3	60	55.77	59.07
4	90	88.30	90.27

9.18 From the plot, $E_M = 1800N^{-0.28}$ approximately; by least squares, $E_N = 1605N^{-0.242}$

9.19 The learning ratio is 84.6 percent

9.20 The new learning curve probably will have a smaller slope and will have to be reestablished

10.1 63.0 m

10.2 16.8 MW

10.3 0.068 m

10.4 Invest $126,000. The additional investment returns less than 17 percent

10.5 15 stories

10.6 Build a 10-Mkg plant. This is the only acceptable option

10.7 14,180 parts per batch

10.8 11,000 parts per batch, approximately

10.9 $T = 0.016$ h per stamping; $533,000/year

10.10 5

10.11 2.25 m; $42,719

10.12 $0.24/kg of Z

10.13 3.7 years

10.14 $N = 2.13$, say, two effects

10.15 (a) 5.5 h, including cleaning time; (b) 5139 kg of cake per cycle; (c) 6 h, including cleaning time

10.16 (a) $A_{in} = 900$ kg/day; $A_{out} = 450$ kg/day; $B_{out} = 450$ kg/day; efficiency $= 0.50$

(b) Cumulative amounts for the first 10 days:
$A_{in} = 9500$ kg; $A_{out} = 2333$ kg; $B_{out} = 7167$ kg
Averages for the first 10 days:
$A_{in} = 950$ kg/day; $A_{out} = 233.3$ kg/day; $B_{out} = 716.7$ kg/day; average efficiency 0.754

(c) Amounts for the tenth day only:
$A_{in} = 905$ kg; $A_{out} = 429.5$ kg; $B_{out} = 475.5$ kg; efficiency 0.525

10.17 53.5 h operating time plus 10 h cleaning

10.18 (a) $1886/h with $T = 1.47$ h; (b) 354 jewels/hour

10.19 $112,000/year for a cycle of 3.06 days

10.20 $R = 1.92$

10.21 The optimum cycle is 182 days operating plus 3 days of cleaning

10.22 0.8039 m³; $431/1000 m³

10.23 (a) 3088 kg/day; (b) $24,150/day

10.24 31 h running plus 5 h rejuvenating

10.25 5.9 min

10.26 25,000 kg/batch

10.27 Operate 5.1 days, clean 1 day
10.28 510 days
10.29 About 0.66 radians or 37.8 degrees
10.30 The area divided by the base is the average height or average efficiency. It does matter if input to the equipment is not constant

10.31
$$\frac{C}{L} = \frac{0.7}{0.6} \frac{a}{b}$$

10.32 (a) 434.3 h; (b) 112.4 m²
10.33 $F = 4.64 \times 10^5$ kg/h; $\Delta T = 49.9°C$
10.34 Maximum area $= AB/4$ by either method
10.35 (a) or (b) $X = 3.54$, $Y = 3.54$ maximum; $X = -3.54$, $Y = -3.54$ maximum; $X = -3.54$, $Y = +3.54$ minimum; $X = +3.54$, $Y = -3.54$ minimum
10.36 $X = 0.25$, $Y = 0.25$, $Z = 0.50$
10.37 Tabulate Y against X. The minimum value for Y is 0.2698 at $X = 1.54$. At $X = 1.54$ and for $A = B = C = 2$, for equal small changes in A, B, or C, raising B will have the greatest effect in lowering Y

11.1 (a) 100 items; (b) \$16/year; (c) five items
11.2 (a) 100 items; (b) \$16.64/year; (c) nine items

11.3
$$Q_{\text{opt}} = \sqrt{\frac{C_o U}{C_h}} \qquad C_{t,\text{opt}} = 2\sqrt{C_h C_o U}$$

11.4 100 items
11.5 $Q_{\text{opt}} = 24,361$ items for a 61-day supply

11.6
$$Q_{\text{opt}} = \sqrt{\frac{C_o UD}{C_h(D - U)}} \qquad C_{t,\text{opt}} = 2\sqrt{\frac{C_h C_o U(D - U)}{D}} + C_h S$$

11.7 $N = 4500$ approximately
11.8 Stock five
11.9 10 days
11.10 74.1
11.11 Five machinists
11.12 0.428
11.13 400 kg
11.14 8.9 Mkg
11.15 27 items
11.16 Approximately 43 percent
11.17 The minimum cost is \$21.54/week for a stock of three
11.18 (d); $T_s = 4$ min per automobile
11.19 Two pumps; C_t minimum = \$0.1131/min; Three pumps; C_t minimum = \$0.1386/min
11.20 \$115.26/h
11.21 0.142
11.22 0.14 approximately
11.23 127 tests per hour capability
11.24 \$5670/week with three crews
11.25 14 or 15

11.26 0.55 approximately
11.27 The minimum cost is for the sequence ABDC

12.3
$$X_1 + 3X_2 \geq 8$$
$$3X_1 + 4X_2 \geq 19$$
$$3X_1 + X_2 \geq 7$$
$$Z = 50X_1 + 25X_2$$

12.4 Plot the first three equations of Prob. 12.3 as equalities. The feasible region is above and to the right
12.5 The minimum cost is 150 using one unit of hog liver and four units of castor oil (as well as five units of slack variable X_3)
12.6 The optimal solution is: $X_1 = 1$, $X_2 = 4$, $X_3 = 5$; $Z = 150$
12.7 Same as Prob. 12.6
12.8 15 of the most difficult and 75 of the least difficult for maximum grade of 390
12.9 The minimum cost is 800 k$ for 40 radio advertisements
12.10 17.25 of A and 30.8 of B; $Z = \$155/\text{day}$
12.11 $X_1 = 400$, $X_2 = 500$, $X_3 = 100$
12.12 14,000 bbl/day crude A to process D; 12,000 bbl/day crude B to process C; 6000 bbl/day crude B to process D
12.13 The optimum is 5.05 M$ with A = 0, B = 2, C = 1
12.14 (a) The optimum is 4 M$ with A = 0, B = 2, C = 0; (b) The optimum is 2 M$ with A = 0, B = 1, C = 0
12.15 The optimal solution is BJIDA
12.17 The optimum is either zero item 1, three item 2, and zero item 3; or two item 1, zero item 2, and zero item 3
12.18 Overall, the best operation is at 80 percent efficiency; type A, unit 1, with unit 2 at a temperature of 430°C for a total 5-year income of $560,000
12.19 The minimum cost is $100,000

| | Out of #1 | Out of #2 | |
Into #1	Into #2	Into #3	Out of #3
20%	50%	60%	99%

12.20

	Operating cost, %	Fixed cost, %
(a)	23.8	76.2
(b)	23.8	76.2
(c)	22.4	77.6
(d)	22.4	77.6

12.21
$$v = 2.889A^{0.4444}B^{0.2222}C^{0.3333}$$
$$X_1 = 1.284A^{-0.5556}B^{0.2222}C^{0.3333}$$
$$X_2 = 0.9813A^{0.2222}B^{0.1111}c^{-0.3333}$$

13.1 1.70 to 1.735

13.2 The minimum is between 1.70 and 1.75 with 199 experiments

13.3 1.7188 to 1.7578

13.4

	Range for X	F_r
(a)	1.8571 to 2.4286	0.2857
(b)	2 to 2.5	0.25
(c)	2 to 2.25	0.125
(d)	2.077 to 2.2308	0.0769

13.5 $X = Y = 0$

13.6 $X = Y = 0$

13.7 Place $U = 5Y$; thus, $Z = X^2 + U^2$

13.8 $X = 3$, $Y = 2$, $Z = 1$, for which $f = -6$

13.9 Same as Prob. 13.8

13.10 Same as Prob. 13.8

13.11 (a)
$$-2A_m + 1A_{m+1} + 1A_{m+2} = 0$$

(b)
$$\frac{-1}{(1 + 2X)(1 - X)}$$

(c)
$$-\tfrac{1}{3}[(-1)^n 2^{n+1} + 1]$$

(d)
$$\Sigma = \frac{-1 + (-2A_{n-1} + A_n)X^{n+1} - 2A_n X^{n+2}}{1 - X - 2X^2}$$

with A_n and A_{n-1} given by part (c)

13.12 The scale of relation is:
$$-A_m + 3A_{m+1} - 3A_{m+2} + A_{m+3} = 0$$

The next term of the series is $36U^7$

13.13
$$\frac{1}{1 - 3U + 3U^2 - U^3}$$

13.16 (a)
$$\frac{1}{1 - U - U^2}$$

(b)
$$\frac{1}{\sqrt{5}}\left[\left(\frac{1 + \sqrt{5}}{2}\right)^{N+1} - \left(\frac{1 - \sqrt{5}}{2}\right)^{N+1}\right]$$

13.17 $(1 + \sqrt{5})/2$

13.18 $\Sigma = F_{N+2} - 1$ where F_{N+2} can be obtained from Prob. 13.16, part (b)

14.1 878 k$

14.2 220 k$

14.3

1970 = 100	1976 = 152
1971 = 105	1977 = 162
1972 = 109	1978 = 174
1973 = 114	1979 = 190
1974 = 131	1980 = 207
1975 = 144	

14.4 1.507 M$
14.5 1443 k$
14.6 $300,000
14.7 0.69
14.8 $5540
14.9 7.26 M$
14.10 11.3 M$
14.11 18.6 M$
14.12 32.3 M$
14.13 23.2 M$
14.14 Low, 19.4 M$; average, 23.1 M$; high, 26.6 M$
14.15 66 M$
14.16 29.9 M$
14.17 45.6 M$

15.2 $1,560,000/year or $1.57/lb
15.3 $4,740,000/year or $1.19/lb
15.4 $1,730,000/year or $1.73/lb
15.5 $2,600,000/year or $1.30/lb

16.1 The critical path is 1-3-6-8 at 107 periods
16.2 The critical path is 1-2-3-5 at 19 periods

Activity	Total float	Free float
1-2	0	0
1-3	6	6
1-4	7	0
2-3	0	0
2-4	9	2
2-5	7	7
3-5	0	0
4-5	7	7

16.3 (*a*) $1950; (*b*) $150
16.4 Critical path is 1-2-4-5 at 12 days total

Activity	Total float	Free float
1-2	0	0
1-3	4	0
1-4	2	2
2-4	0	0
3-4	4	4
2-5	2	2
4-5	0	0

16.5 12 days, $610; 11 days, $650; 10 days, $690; 9 days, $770; 8 days, $870; 7 days, $1000

16.6 The critical path is 1–3–6–7 at 12 days

Activity	Total float	Free float
1–2	3	0
1–3	0	0
1–4	1	0
2–5	3	1
3–5	2	0
3–6	0	0
4–6	1	1
5–7	2	2
6–7	0	0

16.7 10.5 days

16.8 The critical path is 1–3–6–5–7 at 15 days

Activity	Total float	Free float
1–2	6	0
1–3	0	0
1–4	1	0
2–5	6	6
3–5	5	5
3–6	0	0
4–6	1	1
5–7	0	0
6–5	0	0
6–7	3	3

16.9 Replace 1–3 by 1–3_1–3_2–3_3, all 1 day apart. Add the dummy activity from 3_3 to 7. The critical path is 1–4–6–7 at 11 days

16.10 The critical path is 8.5 h

Name Index

Subject Index